Lecture Notes in Physics

Volume 976

The Lecture Notes in Physics

The series Lecture Notes in Physics (LNP), founded in 1969, reports new developments in physics research and teaching-quickly and informally, but with a high quality and the explicit aim to summarize and communicate current knowledge in an accessible way. Books published in this series are conceived as bridging material between advanced graduate textbooks and the forefront of research and to serve three purposes:

- to be a compact and modern up-to-date source of reference on a well-defined topic;
- to serve as an accessible introduction to the field to postgraduate students and nonspecialist researchers from related areas;
- to be a source of advanced teaching material for specialized seminars, courses and schools.

Both monographs and multi-author volumes will be considered for publication. Edited volumes should however consist of a very limited number of contributions only. Proceedings will not be considered for LNP.

Volumes published in LNP are disseminated both in print and in electronic formats, the electronic archive being available at springerlink.com. The series content is indexed, abstracted and referenced by many abstracting and information services, bibliographic networks, subscription agencies, library networks, and consortia.

Proposals should be sent to a member of the Editorial Board, or directly to the responsible editor at Springer:

Dr Lisa Scalone
Springer Nature
Physics
Tiergartenstrasse 17
69121 Heidelberg, Germany
lisa.scalone@springernature.com

More information about this series at http://www.springer.com/series/5304

Sintayehu Tesfa

Quantum Optical Processes

From Basics to Applications

 Springer

Sintayehu Tesfa
Physics Department
Jazan University
Jazan, Saudi Arabia

ISSN 0075-8450   ISSN 1616-6361   (electronic)
Lecture Notes in Physics
ISBN 978-3-030-62347-0   ISBN 978-3-030-62348-7   (eBook)
https://doi.org/10.1007/978-3-030-62348-7

This Springer imprint is published by the registered company Springer Nature Switzerland AG.
The registered company address is: Gewerbestrasse 11, 6330 Cham, Switzerland

Preface

This book is intended to establish a link between basic theories in cavity quantum electrodynamics, quantum optics and quantum information, and corresponding applications. It is mainly written to harmonize the discussion scattered across these fields with quantum measurement theory to update the ideas, techniques, and approaches in conventional quantum optics; formulate the working principles of continuous variable quantum information processing; highlight and project the corresponding application with the aim of designing and characterizing simple, realistic, and feasible quantum optical schemes, mechanisms, processes, and operations; and bridge the gap between the discrete and continuous variable approaches. It is particularly aimed at elucidating the emerging interest of exploiting curious properties of the quantized light in processing quantum information, specifically focusing on the discussion of the appraisal of quantized light, atom–radiation interaction, and quantum features of light along with attainable manipulations, implementations, and realizations.

Hoping to make it accessible to multidisciplinary readership, adequate mathematical derivations, simplified examples, illustrations and demonstrations, and pertinent interpretations and foresight are provided as concisely as possible. While trying to establish the intended goal, the original ideas, proposals, and conjectures as presented by respective proponent researchers and subsequent developments are carefully crafted and selected citations are provided at the end of each chapter. In doing so, an attempt has been made to make this book as self-contained as possible. In line with this, topics ranging from the general quantization of electromagnetic field to the corresponding optical processes and operations that lead to applications are considered. This book generally comprises an Introduction and eleven chapters divided into four parts (for more details: see Chap. 1). The Introduction is designed to summarize the overall approach, the rationale and goal, the subject matter or the content, and the overview of the topics included in each part.

The first part, for example, deals with description of quantized light in terms of certain mathematical representations such as coherent state, quasi-statistical distribution, and Gaussian continuous variable. It also includes the approaches required to study quantum features of radiation and photon measurement techniques. This part is mainly designed to outline the mathematical approaches and procedures, and formulations in quantum optics that facilitate the understanding of the nature and application of the quantized light (intended for readers who are not expert in quantum optics). Whereas the second part is devised to expose the interaction of quantized light with atom in different settings and configurations in such a way that the learnt idea could serve as the launching pad for foreseeing and analyzing the manipulations or implementations during application. Based on the properties of the quantized light and its interaction with atom as established in preceding parts, the third part is devoted to the quantification of various quantum features of light, the characterization of physical systems that generate quantized light, and provision of some practical examples. Hoping to pave the way for the application of the quantized light, the fourth part that meant to introduce quantum optical processes, operations, general overview, and foresight of quantum communication and computation is included.

Even though this book is devised to contain materials usually absent from standard quantum optics textbooks such as photon measurement and cavity quantum electrodynamics, it has more than sufficient material to set up a postgraduate course for quantum optics. Incidentally, I would like to stress that the selected topics are arranged keeping in mind that the scope of the conversation needs to transcend conventional quantum optics in a way it can render the interest of wider scientific communities. In light of this and hoping to connect the resource in quantum optics with theories in quantum information, the content and presentation are chosen in view of the readership with good knowledge of quantum theory and quantum optics or quantum information, and planning to direct their research interest to the field of continuous variable quantum information processing.

I am grateful for the support, encouragement, and valuable comments I have been lavished with throughout my professional career from my friends, colleagues, and students at Addis Ababa, Dilla, Adama, and Jazan Universities, all of whose names I am unable to list here. I am especially indebted to Fesseha Kassahun, Mulugeta Bekele, P. Singh, Daniel Bekele, Tilahun Tesfaye, Eyob Alebachew, Chernet Amente, Misrak Getahun, Ali Al Kamli, and Jabir Hakami who have influenced my worldview in science and research in one way or the other. I would like also to extend my appreciation to the colleagues and students who showed enthusiasm to collaborate on projects conceived from what is discussed here and helped me to improve the way of presentation and selection of contents, particularly to Tewodros, Deribe, Mekonnen, Shunke, Menwuyelet, and Sitotaw. The writing of this book would have never been realized without the love, care, and encouragement of my beloved wife, Mekdes Gebre, and my children, Abel, Bezawit, Dagim, and Betsegah Sintayehu.

Addis Ababa, Ethiopia

Sintayehu Tesfa

Contents

About the Author

Sintayehu Tesfa is assistant professor at Jazan University where he teaches general physics courses and he is also adjunct professor at Adama and Addis Ababa Universities where he advises students and teaches quantum optics. Earlier, he taught quantum optics at Dilla University, and he has been a guest scientist at Max Planck Institute of Complex Systems, Dresden. For the past 15 years, he has been working on quantum optics and quantum information with emphasis on exploring nonclassical properties of light. Particularly, he is one of the pioneers in the study of the quantification of the degree of quantum entanglement and discord of the light generated by nonlinear and linear optical sources such as parametric up-conversion and correlated emission laser.

Introduction

It might be alluring to proclaim at the outset that quantization of electromagnetic field, and its interaction with matter, led to a prolific research activities and a birth of quantum electrodynamics and quantum optics. Particularly, with the dawning of quantum optics, a significant progress in the understanding of quantum theory and manipulation of atomic and optical properties has been successfully made. Even though such attention has began with the advent of quantum theory, the research on nonclassical interaction of atoms with light is getting finer just recently, and then the capability to witness the quantum behavior of light and to think of its application have grown tremendously over the years unabated. Even after so many achievements, the research in the field of quantum optics is still in an enthralling era since the deepest fundamental traits of quantum physics is currently tested with the help of quantum optical table top experiments in which the interconnectedness of the parts of the optical system have found a central role in the development of a versatile technical possibilities (for historical account on the advance of quantum optics; see for instance [1]). The outcomes of the research in quantum optics have also inspired the onset of the aspiration to harness the nonclassical features of light to improve the way information should be processed. One may thus imagine optical systems as obvious choice specially for quantum communication since photons can be taken as an excellent carrier of quantum information due to their comparatively low decoherence rate. This may lead to inquire whether one can construct integrated systems for communication and computation in which all the processing tasks utilize optical structures.

With the insight that optical mechanisms are one of the most promising schemes to implement quantum information processing tasks, the main purpose of this book is directed towards outlining the way how the appraised nonclassical properties of light can modify the outcomes of the application; and is also intended to expound the background and pertinent formulations of the emerging interest of exploiting the curious properties of the quantized light. Specifically, basic descriptions, characterizations, measurement techniques, possible sources, nonclassical features,

S. Tesfa, *Quantum Optical Processes*, Lecture Notes in Physics 976,
https://doi.org/10.1007/978-3-030-62348-7_1

practical implications, corresponding processes and applications of the quantized light, and its interaction with matter would be explored. With the aid of the quantum traits such as coherent superposition, entanglement, nonlocality, decoherence and no-cloning, quantum optical processes such as continuous variable entanglement swapping, teleportation and telecloning from which follow practical aspects such as quantum gate operations, cryptography and error correction would also be discussed. As illustration, the advantages and associated challenges including the foresight of implementing continuous variable quantum communication and computation protocols would be highlighted. In the process, a great care has been taken to balance the presentation in such a way that the book would be accessible to communities of quantum optics and atomic physics in one side, and quantum information in the other by borrowing ideas from each. Hence aiming at connecting the physical resources in quantum optics with theories in quantum information, the topics of discussion are chosen and organized upon taking in mind the readership with a good knowledge of quantum theory and quantum optics or quantum information, and planning to direct their interest and career to the field of continuous variable quantum information.

1.1   Rationale and Goal

To realize the goal of abating the setback imposed by existing ways of availing computing facilities, recent vast research activities indicate that quantum aspects of light can provide a paradigm shift in communication and computation [2–4]. Such observation may necessitate the quantization of electromagnetic field that can be designated in terms of a coherent state representation [5, 6]. Once the electromagnetic field is quantized, it is required to look into the interaction of the quantum of radiation with an atom to be able to comprehend about quantum features of radiation, and subsequent applications [7]. For example, communication and utilization of most devices required for daily use are made available via the atom-radiation interaction whose understanding depends on the way the atom and radiation are treated. To obtain the most interesting result of the corresponding process, one usually assumes both the atom and radiation to be treated as quantum systems mainly as trapped systems [8]. Carefully designed interaction in such a setup can lead to the generation of light with nonclassical properties such as sub-Poissonian photon statistics, squeezing, nonlocality, entanglement, quantum discord and steerability [9]. So the notion such as the quantification of the nonclassical features of radiation is one of the active fields of research in quantum optics. To apprehend various quantum information processing tasks, one may thus need to have a good background in quantum optics and cavity quantum electrodynamics.

One can also proclaim that if the hardwork put into theoretical quantum optics does not amount to application, it may not be as exciting as it is currently. One may then recognize the need for harmonizing the discussion on nonclassical optical processes with application as attempted to achieve in this book. The aspects of the application emanate mainly from the way one describes, represents or measures a quantized light [10]. The quantized electromagnetic radiation can also be modeled as a collection of noninteracting quantum harmonic oscillators that can be taken as a system of Gaussian continuous variables [2, 11]. One can cite for example Gaussian operations that are effectively epitomized by interaction Hamiltonian at most quadratic in the optical mode's annihilation and creation operators. This hints that various light induced quantum processes can be conceived from the characteristics of the basic wave mixing approaches and measurement schemes [3]. These view points can be put together as Gaussian operations that map Gaussian states onto Gaussian states should include beam splitting, squeezing transformation, homodyne detection and phase space displacement [11, 12]. So digging into the fundamental principles and theories of photon measurement is believed to be vital to get an insight of the study of continuous variable optical processing.

Even though a quantized light can generally exhibit a number of nonclassical features, the blueprint of the application is mainly drawn from the unique properties related to squeezing, nonlocality and entanglement, where the idea of using squeezing for application is tantamount to encoding the information onto the quadrature with minimum noise fluctuation, and then minimize the inherent loss of quantum information in the process of encoding and decoding. The other important attribute of a quantized light is quantum nonlocality: the quantized light fails to grasp a local description of events, which leads to Bell analysis that can be taken as a basis for realizing quantum optical processes. Entanglement as a resource can in the same context facilitate the manipulation of quantum system at a very far location without directly interacting with it as in quantum teleportation. The assured technical possibility of entangling beams of light via a nonlinear interaction can be taken as a prime justification to consider light as a viable candidate to test the theoretical proposals in quantum information processing; the ambition sought to be galvanized with the approaches in quantum optics.

Pertaining to the possibility of inducing coherent superposition between atomic and optical quantum states, quantum information processing is expected to differ in a major way from the conventional classical counterpart on which the aspiration and commitment to delve into a continuous variable quantum information processing task hang on. The inspiration to examine the fundamental principles of quantum information in view of optical process can also arise from the observation that the intricacy in the foundation of quantum theory can be exploited to perform important tasks more reliably and efficiently. Quantum correlation for instance can be deployed as a resource while sending information over a relatively distant locations, transferring between two nodes and sharing among various receivers. Incidentally, there is an abundant evidence which shows that quantum correlations can be utilized to establish cryptographic codes difficult to break, and aid in computing certain functions in a fewer steps than any classical computer can do [13].

Quantum information can also be perceived as interpretation of quantum process from thermodynamic point of view; and expected to lead to a wider conceptual analysis of quantum phenomena such as quantum decoherence, discord and entanglement steering. The assertion that communication and information are intrinsically physical envisages that a quantum optical element can be embedded into the building block of theoretical foundation of quantum information processing with the hope that the utilization of a nonclassical feature of light can enhance the inherent performance. The research in quantum information can also be taken as an effort to map a classical information theory onto the corresponding quantum world. It is hence contended that one of the main themes of this book is to address quantum theory from information theoretic point of view; the perception on which it is organized and written.

It might be required at some point to extend the exposition on quantum information processing to the regime of continuous variable; in the process, widen the perspective from discrete to continuous variable and broaden the scope from finite to infinite dimensions. The main motivation to proceed with a continuous variable stems from the realization that implementation procedures in quantum communication and computation; namely, preparation, manipulation and measurement of quantum states can be achieved with the help of continuous quadrature amplitude of the quantized electromagnetic field [3]. The emerging entanglement and the resulting entanglement based quantum processes are unfortunately imperfect [12]. It is thus found necessary to formulate various ways of quantifying the fidelity of quantum optical processes. As a foresight, one may emphasis that optical processes enjoy a significant fidelity that makes them dependable resource in the actual quantum operations; a similar kind of assertion is sought to be illustrated in much part of this book.

1.2   Subject Matter

Hoping to make this book as self contained as possible, topics ranging from the quantization of electromagnetic field to application are considered. The first part that comprises Chaps. 2, 3, and 4 is devoted to outline the general description of quantized light in terms of mathematical representation and relate it with measurement techniques. This part is designed to formulate the theoretical background required to understand the nature and application of the quantized light. The second part that comprises Chaps. 5 and 6 deals with the interaction of radiation with atom in different configurations and perspectives in the way that it serves as the springboard for analyzing the manipulations in application. Besides, the third part that comprises Chaps. 7 and 8 is devoted to the discussion of the quantification of various quantum features of radiation and the viable physical systems that can generate quantized light. This much of the discussion might suffice to appreciate the quantum features of light as in conventional quantum optics. Even then, aiming at paving the way for application of quantized light as in quantum information processing, the fourth part that comprises Chaps. 9, 10, 11, and 12, and meant to introduce quantum optical processes and operations, and highlight the implications of going into the quantum regime and general foresight of the pertinent application is included.

1.2.1 Appraisal of Quantized Light

Since quantization of electromagnetic field subtly validates the particle nature of light it might be imperative looking into how light can be represented in quantum realm [5, 14]. So after the electromagnetic field is quantized in such a way that the term related to the vacuum fluctuation is evident from the obtained result, various ways by which the quantum state of light can be designated are presented (see Chap. 2). Specifically, based on coherent state representation, the defining characteristics of light that can be epitomized by number state, chaotic state, squeezed state and variants of these including the displaced squeezed state are introduced. Besides, in relation to the observation that a precise knowledge of the number of available photons is not technically possible, some phase space distribution functions that portray the density operator in a certain operator ordering are introduced [6]. With the hope of making the way of going from one procedure to the other easy, the relation among various distribution functions is also formulated. In addition, owing to the fact that the technique of finding the required expectation values or correlation functions is one of the essential approaches applicable in quantum optics and quantum information, different examples expected to illustrate the intended procedures are provided. As part of the characterization of quantized light, the description of properties of light in terms of continuous variable Gaussian state associated with the quadrature operators is also outlined [11].

The quantum system one seeks to study is often enclosed in the cavity where the reflection from the walls is exploited while establishing mathematically manageable approaches and generating experimentally measurable data. One should note that there are different ways of studying the dynamics of the quantum system enclosed in a cavity (see Chap. 3). For instance, for a system exposed to damping mechanism, quantum Langevin equation and/or master equation can be employed. As a result, various approaches by which the master equation can be derived are outlined; after that, the pertinent quantum Langevin equation is derived; and then, the general input-output relation including the causality effect is addressed [15]. Various approaches such as propagator formulation, Feynman path integral, characteristic function and stochastic differential equation are also discussed. Besides, special techniques such as linearization procedure would be looked at.

On the basis of the assertion that a complete description of quantum system requires a deeper inspection than just measuring one or specific observable, and noting that the process of photon measurement relies on photoelectric effect, the fundamental theory of photon measurement, technique of wave mixing, photon addition and detection along with the corresponding conundrum and contention would be discussed (see Chap. 4 and also [10, 16]). Markedly, for quantized light, since the intensity and phase are related by uncertainty relation (do not commute), it is not admissible to precisely measure them at the same time [17]. One may thus require to know the correlation between various beams of light to better understand properties of the radiation generated by quantum system. The ways of evaluating the correlations among beams of photon would hence be discussed with the emphasis given to distinguish between classical and quantum correlations [18]. Moreover, the approaches and physical schemes that can be used to separate the light based on its different aspects; and then, recombine after certain delay in time or length are highlighted.

1.2.2   Atom-Radiation Interactions

What one reads in a measurement process and subsequent interpretation of the outcome should reflect what one knows about the interaction of light with matter. In line with this, despite the fact that the two-level approximation ignores the real feature of the atomic system, and so fails to provide a complete picture of atom-radiation interaction phenomenon, relentless research over the years shows that there is enormous richness in the dynamical process that can be encompassed within such approximation [19]. When a two-level atom is coupled to external radiation, in the context of quantum optics, the interest is often geared towards what happens to the radiation rather than the atom or the coupled system. The focus in Chap. 5 is, however, to describe a free two-level atom in terms of the probability amplitudes, density operator, atomic operators, occupations of each energy levels and atomic coherence. Emphasis would also be given to the properties of atomic operators and the accompanying transition mechanisms along with the physical phenomena that led to different atomic damping roots in the realm of semiclassical approach. To overcome the limitation imposed by a semiclassical treatment, the rational thing to do is to extend the discussion to the quantum regime.

Once the atom is placed in the cavity and the radiation is treated as a quantum system, the information about the interaction of the radiation with atom can be inferred from the bare atomic states or state of the emitted radiation or atomic dressed state. One should note that the atomic dressed state can fairly describe the available emission-absorption events and probabilities. The main theme of Chap. 6 is thus designed to comprehend the basics of cavity quantum electrodynamics, where unique properties of both the radiation and the atom are sought for when the walls of the cavity are taken to be a very good as well as a lossy reflector. One may note that the enclosure ensures the preservation of the quantum coherence of atom-radiation interaction over a considerable time scale [8]. In a complex atomic scheme, it is the rate of decay of each atomic energy level and the emerging cross term that determine the correlation between various emission roots—the idea used to envision the origin of atomic coherence that leads to quantum features of the associated light. The knowledge of the population statistics can then be taken as a viable indicator to generate radiation with specific coherence properties via externally pumping the energy levels participating in the interaction. The interaction of radiation with the enclosing system specially leads to damping whose effect on the population dynamics and the properties of the generated radiation is shown to be quite significant. With this in mind, the origin and consequences of atomic coherence in the multi-level scenario and nonlinearity associated with interaction of several atoms would also be discussed.

The other interesting attribute of confining an atom in the cavity is the possibility of transferring coherence from driving laser field to atomic energy levels whose significance and implication would be discussed. In case the atom is placed in the cavity for quite sometime, the emission-absorption events can take place repeatedly, and so one may lose the knowledge of the energy level the atom occupies—the idea that exemplifies the conception of atomic coherent superposition [20]. In case many atoms are placed in the cavity, not all of them absorb the radiation and excited to the upper energy level at the same time. The absorption-emission process in this case would be more complicated since there is a real chance for the radiation emitted by one of the atoms to be absorbed by the other. The exchange of radiation among atoms in the long run may induce correlation among the accompanying energy levels that depends not only on the interaction of the atom with radiation but also on the energy level the neighboring atoms occupy; the conception that leads to the discussion on a dipole blockade [21].

1.2.3   Quantum Aspects of Light

Despite the abundant theoretical formulations and implementations based on single and two-mode squeezing, the possibility of experimenting with a more complex nonclassical behavior such as higher-order and multi-mode squeezing is thought to be still interesting and worthy of attention [22]. In light of this, the existing criteria that meant to characterize the lowest order nonclassicality are extended to a higher-order, and more complex and demanding scenarios (see Chap. 7). On account of the observation that the research on the quantum features of light is attracting a considerable interest due to the relative simplicity and high efficiency in the generation, manipulation and detection of a continuous variable optical entangled state, various criteria that can be applicable to quantify the degree of quantumness or nonclassicality such as high order squeezing, nonlocality, entanglement, quantum discord and entanglement steering are forwarded [23].

In view of the proposal that with the dawning of quantum information processing of continuous variable the need for generating a bright light with a robust quantum feature is growing, various mechanisms such as parametric oscillator, multi-wave mixer and correlated emission laser that can generate a light with nonclassical properties are explored (see Chap. 8). The main source of light with nonclassical properties can in general be linked to a nonlinear optical process such as pumping a crystal with an external coherent radiation and a linear optical process generated by correlated emission of multi-level atoms [24]. A critical scrutiny may reveal that the nonclassicality could be related to the optical coherence induced while coupling the atomic energy levels of the source. With the intention of developing the procedure one can use to analyze the nonclassicality, the techniques of quantifying the nonclassical features of radiation with the help of the master equation, quantum Langevin equation, stochastic differential equation and the linearization procedure are also provided as demonstration along with the corresponding examples.

1.2.4   Exploring Quantum Optical Processes

It is clear from existing works that while attempting to carry out quantum operations, there are operations such as perfectly copying quantum state that one is unable to do [25]. This situation precludes a complete transfer of quantum state to other location; the assertion often taken as the basis for the field of quantum information processing. This claim may envisage the need for devising a means of mitigating the challenges imposed by quantum no-cloning, no-broadcasting or no-deleting by designing an alternative quantum procedure, and also by invoking some processes that can be done in the realm of quantum theory and expected to enhance the fidelity of communication and computation. Similar implication of adopting quantum features of radiation to access the information imprinted onto the quantum system such as quantum decoherence would be discussed (see Chap. 10).

Contrary to the ramification of quantum no-cloning, copying the quantum state happens to be one of the operations required to carry out a meaningful implementation [26]. The procedure of copying quantum state is also expedient whenever there is a need for distributing quantum information among several parties and an interest to enhance the fidelity of transmission over a lossy quantum channel [27]. Whatever one seeks, quantum principles do not preclude a partial or an imperfect copying of the quantum state. Pertaining to this, the prevailing physical situation near perfect (optimal) copying of quantum state would be explored (see Chap. 11). Besides, the deceptively simple observation that the change made by the experimenter on one of the entangled parties can alter the properties of the other without necessarily requiring to directly communicate can be taken as a basis to explain much of the intended quantum optical processes and applications. For example, when each particle in the composition is made to entangle with its partner, the appropriate measurement collapses the state of the remaining particles onto an entangled state; the idea that leads to the notion of entanglement swapping [28].

With respect to application, since two systems that do not directly interact can be entangled with the help of entanglement swapping, it should be possible to ascertain that the content of quantum information can be transferred from the sending to receiving site with the aid of entanglement between the particles as a resource—a process that constitutes the notion of quantum teleportation and telecloning if used along with quantum cloning [29]. In case the eavesdropper manages to intercept the sender's qubit on the way to the receiver, she can access only part of the entangled state which can be taken as a basis for the conception of quantum secrecy. Much of these quantum operations can be practically implemented by making use of polarization based photonic qubits and the corresponding continuous variable version. The main theme of Chap. 11 would hence be to introduce quantum optical processes that can be implemented by using coherent light.

The perception that superposition, entanglement and interference can be deployed as a resource to carry out certain algorithm may also encourage to foresee that the computation that relies on classical principles may not able to account for every possibility that a physically realizable computing device can offer. It so becomes evident that the interaction among quantum systems can be utilized in implementing the unitary evolution required to perform a universal quantum logic gate operation (see Chap. 9 and also [30]). When it comes to quantum optical systems, existing works clearly show the possibility of performing arbitrary single-qubit operations by using sets of half and quarter wave plates, beam splitters and reflectors [31]. Taking this as motivation, the potential of a continuous variable quantum optical properties in enhancing ceratin information processing tasks such as error correction and cryptography would be highlighted (see Chap. 12). Based on the fact that the resulting state is in the superposition of the clean and logically erred states, the mechanism by which the fidelity of the computed or communicated information would be enhanced by making use of the quantum aspects of the involved states would also be explored along with a general discussion on continuous variable quantum gate operation, computation, secured communication, error correction and network [32].

1.3 Foresight

Large part of this book is put aside to illustrate continuous variable quantum features as potential resource for implementing information processing tasks. Despite the availability of a considerable size of work, the way and the mechanism by which the information can be manipulated or processed with the help of quantum properties of radiation are still fresh and enduring. By and large, identifying and characterizing the suitable optical candidate for practical utilization are deemed to be the integral part of the field of modern quantum optics. Quantum information processing tasks specially rely on the sharing and manipulation of a maximally entangled qubit pairs between the sender and receiver, which requires a coherent manipulation of a large number of coupled quantum systems that can be realized by optically pumping the atomic ensemble into states with strong atom-atom and atom-radiation interactions. If combined with the emerging dipole blockade, such a mechanism is expected to allow to generate a superposition of a collective spin states in the ensemble of the atoms, to coherently convert these states into the states of photon wave packets of prescribed direction, and to perform quantum gate operations between distant qubits (see Chap. 9).

In the context of quantum optics, constructing a series of horizontally and vertically polarized photons can faithfully represent a computational bit series since photonic qubits are stable, have long coherence time, high mobility and experience low interaction with the environment. Such attributes and the possibility that it can be manipulated by using beam splitters, reflectors and squeezers would arguably make light an ideal carrier of the content of quantum information. Nonetheless, even though one can utilize light to create secured communication channel via the technique of quantum key distribution, there are still practical challenges to realize a long distance continuous variable quantum communication and a workable quantum computation structure since the required entangling operations between photonic qubits can be implemented only probabilistically. Another challenge also emerges while attempting to use photonic qubits as a storage device due to the speed at which the light travels since many algorithms require a longer storage time. But in the regime of continuous variable, the circuit to be constructed would be deterministic contrary to the single-photon case. Markedly, the required measurements are straightforward in the single-photon approach but non-Gaussian measurements pose serious challenge in the continuous variable approach. It is not really obvious to assert which of these problems represent the biggest hurdle in building a large scale operational system.

Despite the actual and imagined challenges, there are fortunately a number of interesting avenues for research (for some example; see Chap. 12). For one, incorporating a continuous variable scheme in the setup is expected to expand the scope and focus of the research. For the other, it should be noted that various physical platforms or systems, other than what is discussed in this book, can be imagined with the view of constructing quantum devices that can be utilized in realizing operational quantum optical devices [33]. So one of the main purposes of this book is to lay the foundation and incite inspiration so that the research in a continuous variable quantum information processing can be advanced by exploiting the physical systems, mathematical approaches and practical techniques available in quantum optics.

References

1. D.F. Walls, G.J. Milburn, *Quantum Optics*, 2nd edn. (Springer, Berlin, 2008); D. Browne, Prog. Quant. Elec. **54**, 2 (2017); A. Zeilinger, Phys. Scr. **92**, 072501 (2017)
2. P. Kok, B.W. Covett, *Introduction to Optical Quantum Information Processing* (Cambridge University, New York, 2010)
3. S.L. Braunstein, P. van Loock, Rev. Mod. Phys. **77**, 513 (2005)
4. H.A. Bochor, T.C. Ralph, *A Guide to Experiments in Quantum Optics*, 2nd revised and Elarged edn. (Wiley-VCH Verlag GmbH and Co. KGaA, Weinheim, 2004); S. Slussarenkoa, G.J. Prydeb, Appl. Phys. Rev. **6**, 041303 (2019)
5. A. Einstein, Ann. D. Phys. **17**, 132 (1905) (English translation: A.B. Arons, M.B. Peppard, Am. J. Phys. **33**, 367 (1965)); R.J. Glauber, Phys. Rev. **131**, 2766 (1963)
6. H. Weyl, Z. Phys. **46**, 1 (1927); E.P. Winger, Phys. Rev. **40**, 749 (1932); E.C.G. Sudarshan, Phys. Rev. Lett. **10**, 277 (1963); K.E. Cahill, R.J. Glauber, Phys. Rev. **177**, 1882 (1969).
7. H.A. Kramers, W. Heisenberg, Z. Phys. **31**, 681 (1925); E.T. Jaynes, F.W. Cummings, Proc. IEEE **51**, 89 (1963); B.R. Mollow, Phys. Rev. **1969**, 188 (1969)
8. J.I. Cirac, P. Zoller, Phys. Rev. Lett. **74**, 4091 (1995); M.D. Lukin et al., ibid. **87**, 037901 (2001); L.H. Pedersen, K. Molmer, Phys. Rev. A **79**, 012320 (2009); J.P. Home et al., Science **325**, 1227 (2009); L. Isenhower et al., Phys. Rev. Lett. **104**, 010503 (2010); T. Wilk et al., ibid., 010502 (2010); A.V. Gorshkov et al., ibid. **107**, 133602 (2011); K. Molmer, L. Isenhower, M. Saffman, J. Phys. B **44**, 184016 (2011); D. Kielpinski et al., Quant. Inf. Proc. **15**, 5315 (2016)
9. S. Tesfa, Phys. Rev. A **74**, 043816 (2006); J. Phys. B **42**, 215506 (2009): Opt. Commun. **285**, 830 (2012)
10. M. Schlosshauer, Rev. Mod. Phys. **76**, 1267 (2005)
11. G. Adesso, F. Illuminati, J. Phys. A **40**, 7821 (2007); C. Weedbrook et al., Rev. Mod. Phys. **84**, 621 (2012)
12. J.W. Pan et al., Rev. Mod. Phys. **84**, 777 (2012)

13. P.W. Shor, *Algorithms for Quantum Computation: discrete logarithm and factoring, in Proceeding of 35th Annual Symposium on Foundations of Computer Science, Santa Fe* (1994); SIAM J. Comput. **26**, 1484 (1997); P. Kok et al., Rev. Mod. Phys. **79**, 135 (2007)

14. H.P. Yuen, Phys. Rev. A **13**, 2226 (1976); C.M. Cave, Phys. Rev. D **23**, 1693 (1981); G.S. Agarwal, K. Tara, Phys. Rev. A **43**, 492 (1991); R.L. de Matos Filho, W. Vogel, ibid. **54**, 4560 (1996); M. Dakna et al., ibid. **55**, 3184 (1997); V.V. Dodonov et al., ibid. **58**, 4087 (1998); M. Bellini, A. Zavatta, Progress in Optics **55**, 41 (2010)

15. W.P. Schleich, *Quantum Optics in Phase Space* (Wiley-VCH Verlag, Berlin, 2001); S. Tesfa, *A Nondegenerate Three-Level Cascade Laser Coupled to a Two-Mode Squeezed Vacuum Reservoir* (PhD Thesis, Addis Ababa University, 2008)

16. R.J. Glauber, Phys. Rev. **130**, 2529 (1963); W.H. Zurek, Phys. Rev. D **24**, 1516 (1981); ibid. **26**, 1862 (1982); D.J. Starling et al., Phys. Rev. A **80**, 041803 (2009); A.M. Burke et al., Phys. Rev. Lett. **104**, 176801 (2010); X.Y. Xu et al., ibid. **111**, 033604 (2013); Y. Aharonov, E. Cohen, A.C. Elitzur, Phys. Rev. A **89**, 052105 (2014)

17. N. Imoto, H.A. Haus, Y. Yamamoto, Phys. Rev. A **32**, 2287 (1985)

18. K.H. Kagalwala et al., Nat. Photonics **7**, 72 (2013); F. De Zela, Phys. Rev. A **89**, 013845 (2014); X.F. Qian et al., Phys. Rev. Lett. **117**, 153901 (2016)

19. M. Lewenstein, T.W. Mossberg, R.J. Glauber, Phys. Rev. Lett. **59**, 775 (1987); H.J. Carmichael, A.S. Lan, D.F. Walls, J. Mod. Opt. **34**, 821 (1987); O. Kocharovskaya et al., Phys. Rev. A **49**, 4928 (1994); A.K. Kudlaszyk, R. Tanas, J. Mod. Opt. **48**, 347 (2001); S. Tesfa, ibid. **54**, 1759 (2007); ibid. **56**, 105 (2009)

20. X. Maitre et al., Phys. Rev. Lett. **79**, 769 (1997); M. Hennrich et al., ibid. **85**, 4872 (2000); A. Kuhn, M. Hennrich, G. Rempe, ibid. **89**, 067901 (2002)

21. R.H. Dicke, Phys. Rev. **93**, 99 (1954); K.A. Safinya et al., Phys. Rev. Lett. **47**, 405 (1981); M. Gross, S. Haroche, Phys. Rep. **93**, 301 (1982); M. Saffman, T.G. Walker, Phys. Rev. A **72**, 042302 (2005); L.H. Pedersen, K. Molmer, ibid. **79**, 012320 (2009); M. Saffman, T.G. Walker, K. Molmer, Rev. Mod. Phys. **82**, 2313 (2010); Y. Miroshnychenko et al., Phys. Rev. A **82**, 013405 (2010)

22. C.K. Hong, L. Mandel, Phys. Rev. Lett. **54**, 323 (1985); M. Hillery, Opt. Commun. **62**, 135 (1987); Phys. Rev. A **36**, 3796 (1987); H. Fan, G. Yu, Phys. Rev. A **65**, 033829 (2002)

23. J.S. Bell, Physics Physique Fizika **1**, 195 (1965); J.F. Clauser et al., Phys. Rev. Lett. **23**, 880 (1969); M.D. Reid, Phys. Rev. A **40**, 913 (1989); Z.Y. Ou et al., Phys. Rev. Lett. **68**, 3663 (1992); R. Simon, ibid. **84**, 2726 (2000); L.M. Duan et al., ibid., 2722 (2000); G. Vidal, R.F. Wener, Phys. Rev. A **65**, 032314 (2002); G. Adesso, A. Serafini, F. Illuminati, ibid. **70**, 022318 (2004); K. Dechoum et al., ibid. **70**, 053807 (2004); J. Fiurasek, N.J. Cerf, Phys. Rev. Lett. **93**, 063601 (2004); A.A. Klyachko, B. Öztop, A.S. Shumovsky, Phys. Rev. A **75**, 032315 (2007); S. Lou, ibid. **77**, 042303 (2008)

24. N.M. Kroll, Phys. Rev. **127**, 1207 (1962); J.K. Oshman, S. Harris, IEEE J. Quantum Electron. **4**, 491 (1968); M.T. Raiford, Phys. Rev. A **2**, 1541 (1970); D. Stoler, Phys. Rev. Lett. **33**, 1397 (1974); L. Gilles et al., Phys. Rev. A **55**, 2245 (1997); S. Chaturvedi, K. Dechoum, P.D. Drummond, ibid. **65**, 033805 (2002); S. Tesfa, Eur. Phys. J. D **46**, 351 (2008); J. Phys. B **41**, 065506 (2008)

25. W.K. Wootters, W. H. Zurek, Nature **299**, 802 (1982); H. Barnum et al., Phys. Rev. Lett. **76**, 2818 (1996); S. Bandyopadhyay; G. Kar, Phys. Rev. A **60**, 3296 (1999); H. Barnum et al., Phys. Rev. Lett. **99**, 240501 (2007); S. Luo, N. Li, X. Cao, Phys. Rev. A **79**, 054305 (2009)

26. B.C. Sanders, Phys. Rev. A **45**, 6811 (1992); V. Buzek, M. Hillery, ibid. **54**, 1844 (1996); N. Gisin, Phys. Lett. A **242**, 1 (1998); B.C. Sanders, D.A. Rice, Phys. Rev. A **61**, 013805 (1999); N.J. Cerf, A. Ipe, X. Rottenberg, Phys. Rev. Lett. **85**, 1754 (2000); J.S.N. Nielsen et al., ibid. **97**, 083604 (2006); K. Huang et al., ibid. **115**, 023602 (2015)

27. C. Simon, G. Weihs, A. Zeilinger, Phys. Rev. Lett. **84**, 2993 (2000); J. Mod. Opt. **47**, 233 (2000)

28. X. Jia et al., Phys. Rev. Lett. **93**, 250503 (2004); N. Takei et al., ibid. **94**, 220502 (2005); J.H. Obermaier, P. van Loock, Phys. Rev. A **83**, 012319 (2011)

29. S.L. Braunstein, H.J. Kimble, Phys. Rev. Lett. **80**, 869 (1998); M. Murao et al., Phys. Rev. A **59**, 156 (1999); J. Zhang, C. Xie, K. Peng, ibid. **73**, 042315 (2006)
30. A. Barenco et al., Phys. Rev. A **52**, 3457 (1995); S. Lloyd, Phys. Rev. Lett. **75**, 346 (1995); H.F. Hofmann, ibid. **94**, 160504 (2005)
31. D. Gottesman, I.L. Chuang, Nature **402**, 390 (1999); E. Knill, R. Laflamme, G.J. Milburn, ibid. **409**, 46 (2001)
32. D. Gottesman, Phys. Rev. A **57**, 127 (1998); A.P. Lund, T.C. Ralph, H.L. Haselgrove, Phys. Rev. Lett. **100**, 030503 (2008); N.C. Menicucci, ibid. **112**, 120504 (2014)
33. L.M. Duan et al., Nature **414**, 413 (2001); J.F. Sherson et al., ibid. **443**, 557 (2006); S. Takeda et al., ibid. **500**, 315 (2013); U.L. Andersen et al., Nat. Phys. **11**, 713 (2015)

Part I

Appraisal of Quantized Light

Representations of Quantized Light

Light can be epitomized in classical realm by ordinary mathematical function that can be expressed in terms of frequency, wavelength and amplitude of the wave train. Such description can explain much of the optical phenomena related to reflection, refraction, diffraction, dispersion and interference. Even then, in connection to the observation that the classical approach may not able to capture every aspect of light, there has been enduring effort to understand some of the controversial traits of light, which evinces that light behaves like a continuous variable electromagnetic wave while traveling but exhibits discreteness of energy when it interacts with matter: the idea that gives birth to the notion of the quantum of light or photon [1]; the sources, nonclassical features and applications of which we seek to explore. To achieve such a vast goal, one may begin with the characterization of the quantum of light hoping that in the process the emerging distinct peculiar properties of the quantumness would be evident. This consideration can be taken as justification for expediting the quantization of the electromagnetic field. The electromagnetic field would thus be quantized following the procedure that inherently leads to the appearance of terms that can be linked to vacuum fluctuations (see Sect. 2.1).

One may note that even after the advent of quantum electrodynamics, there still lacks a precise explanation to simple question such as what constitutes a quantum of light [2]. In spite of such shortcomings, we opt to stick to the technical description of the quantized electromagnetic field rather than dwelling on the underlying philosophical conundrum (see Chap. 4). In doing so, the quantized picture of light would be built on the foundation that the energy of the radiation takes discrete values when emitted and absorbed. The notion of the quantization of light should also be understood in this context when one attempts to explain the available experimental results such as black body radiation, photoelectric effect, spontaneous and stimulated emissions, Compton effect and Lamb shifts [3]. It might also be necessary and appropriate to look into various ways of representing the state of the quantum of light since quantization of the electromagnetic field envisages the conception of the particle nature of light.

© The Author(s), under exclusive license to Springer Nature Switzerland AG 2020   15
S. Tesfa, *Quantum Optical Processes*, Lecture Notes in Physics 976,
https://doi.org/10.1007/978-3-030-62348-7_2

One of the key ways of denoting the quantized light is based on the coherent state representation introduced by Schrödinger, and later adopted by Glauber to illustrate the coherence of light [4]. For example, the well celebrated Jaynes-Cummings model—which exemplifies the interaction of a two-level atom with a single-mode coherent light, and leads to the collapses and revivals of Rabi oscillations—is one of the interesting outcomes of the coherent representation of light (see Chaps. 5 and 6). The other representation that worths mentioning is the squeezed state for which the quantum noise is presumed to reduce below the classical limit in one of the quadratures at the expense of the increased noise in the conjugate quadrature, and found to enable for instance precision measurement in gravitational wave detection and noiseless communications (for more potential application; see Chaps. 11 and 12). With this in mind, the overall characterization of light in terms of number state, chaotic state, coherent state, squeezed state and variants of these including the displaced squeezed state is presented (see Sect. 2.2 and also [5]).

There is also a great need for a mathematical procedure that can be helpful in interpreting the outcome of an experiment without necessarily knowing the actual number of photons. In connection to the operator ordering problem, the pertinent c-number distribution function that can be used to study the property of the radiation and designate the density operator in a prescribed operator ordering would be discussed (see Sect. 2.3). Owing to the fact that the technique of finding the required expectation values by applying the pertinent distribution function is one of the vital approaches applicable throughout, various examples that thought to illustrate the intended procedures are also provided. With the hope of facilitating going from one procedure to the other with ease, relation among different distribution functions is also established. Moreover, as part of the representation of quantized light, the way of describing properties of light in terms of a continuous variable Gaussian state linked to the quadrature operators is outlined in Sect. 2.4.

2.1   Quantization of Electromagnetic Field

To be familiar with the exciting fields of quantum optics and quantum information processing to which much part of this book is devoted, it might be prerequisite having an acclimatization with the quantization of electromagnetic field and some of its peculiar features such as squeezing and entanglement. To study the quantum features of light and the interaction of light with atom, it is found essential to quantize the electromagnetic field [5]. To do so, one may begin with free space Maxwell's equations:

$$\nabla.\mathbf{E} = 0, \qquad \nabla X \mathbf{E} = -\frac{\partial \mathbf{B}}{\partial t}, \qquad \nabla.\mathbf{B} = 0, \qquad \nabla X \mathbf{B} = \varepsilon_0 \mu_0 \frac{\partial \mathbf{E}}{\partial t}, \qquad (2.1)$$

where $\mathbf{E}$ is the electric field, $\mathbf{B}$ is the magnetic field, ε_0 and μ_0 are the permittivity and permeability of the free space. It is also customary obtaining the Lagrangian function by defining the vector $\mathbf{A}$ and scalar potentials[1] Φ in terms of the electric and magnetic fields as

$$\mathbf{E} = -\frac{\partial \mathbf{A}}{\partial t} - \nabla\Phi, \qquad \mathbf{B} = \nabla X \mathbf{A}. \qquad (2.2)$$

It is common knowledge that the electromagnetic field can be quantized with the aid of either the electric or magnetic field, although we opt to use the electric field where the preference is mainly historic. So to be consistent with the existing physical reality, the Coulomb's or radiative gauge in which $\nabla.\mathbf{A} = 0$ is imposed implying that $\Phi = 0$.

With the help of this gauge selection, one can see that the wave equation takes the form

$$\nabla^2 \mathbf{A} - \frac{1}{c^2}\frac{\partial^2 \mathbf{A}}{\partial t^2} = 0 \qquad (2.3)$$

whose solution can be proposed by expanding the vector potential in plane modes as

$$\mathbf{A}(\mathbf{r}, t) = \sum_{k,s} \left[\alpha_k(t)\mathbf{u}_{k,s}(\mathbf{r}) + \alpha_k^*(t)\mathbf{u}_{k,s}^*(\mathbf{r})\right] \qquad (2.4)$$

with the mode function defined by

$$\mathbf{u}_{k,s}(\mathbf{r}) = \frac{1}{\sqrt{V}}e^{i\mathbf{k}.\mathbf{r}}\hat{e}_{k,s} \qquad (2.5)$$

and $\alpha_k(t) = \alpha(0)e^{-i\omega_k t}$, where $s = 1, 2$ stands for the direction of the field, V for the normalization or quantization volume and $\mathbf{k}$ for the wave vector (note that $\hbar$ is set to 1 throughout).

In view of Coulomb's gauge and Maxwell's equation, one can verify that the unit vectors $\hat{k}$, $\hat{e}_{k,1}$ and $\hat{e}_{k,2}$ form orthonormal sets with

$$\hat{e}_{k,1} X \hat{e}_{k,2} = \hat{k}, \qquad \hat{k} X \hat{e}_{k,1} = \hat{e}_{k,2}, \qquad \hat{e}_{k,2} X \hat{k} = \hat{e}_{k,1} \qquad (2.6)$$

and upon insisting that such conditions to be consistent with the underlying physical reality, the modes are taken to be orthonormal:

$$\sum_{k,s} u_{k,s}(\mathbf{r})_l \, u_{k,s}^*(\mathbf{r}')_m = \delta_{l,m}(\mathbf{r} - \mathbf{r}'). \qquad (2.7)$$

---

[1]Note that the form of Maxwell's equations remains the same if one applies these potentials as well and it is possible to employ different approaches to quantize the radiation field.

With these considerations, the equal time commutation relations[2]

$$[A_m(\mathbf{r}, t), A_l(\mathbf{r}', t)] = [\Pi_m(\mathbf{r}, t), \Pi_l(\mathbf{r}', t)] = 0, \tag{2.8}$$

$$[A_m(\mathbf{r}, t), E_l(\mathbf{r}', t)] = -i\hat{I}\delta_{m,l}^{(3)}(\mathbf{r} - \mathbf{r}') \tag{2.9}$$

are expected (see Eq. (2.17)).

One can thus define the pertinent annihilation operator as

$$\hat{a}_{k,s}(t) = \int d^3r \; \mathbf{u}_{k,s}^*(\mathbf{r}) \left[\sqrt{\frac{\varepsilon_0 \omega_k}{2}} \mathbf{A}(\mathbf{r}, t) - i \sqrt{\frac{\varepsilon_0}{2\omega_k}} \mathbf{E}(\mathbf{r}, t) \right], \tag{2.10}$$

where $\omega_k = |\mathbf{k}|c$ is the angular frequency. Then applying equal time commutation relations for the field operators, it may not be difficult to verify that

$$[\hat{a}_{k,s}, \hat{a}_{k',s'}] = 0, \qquad [\hat{a}_{k,s}, \hat{a}_{k',s'}^\dagger] = \hat{I}\delta_{k,k'}\delta_{s,s'} \tag{2.11}$$

in which $\hat{a}_{k,s}^\dagger$ and $\hat{a}_{k,s}$ can be interpreted as the creation and annihilation operators for the radiation field, which are often dubbed as boson operators and $\hat{I}$ is identity operator.

With this in mind, the operators of the vector potential and electric field can be put as

$$\hat{\mathbf{A}}(\mathbf{r}, t) = \sum_{k,s} \frac{1}{\sqrt{2\varepsilon_0 \omega_k V}} \left[\hat{a}_{k,s} e^{i\mathbf{k}.\mathbf{r}} + \hat{a}_{k,s}^\dagger e^{-i\mathbf{k}.\mathbf{r}} \right] \hat{e}_{k,s}, \tag{2.12}$$

$$\hat{\mathbf{E}}(\mathbf{r}, t) = \sum_{k,s} \sqrt{\frac{\omega_k}{2\varepsilon_0 V}} \left[\hat{a}_{k,s} e^{i\mathbf{k}.\mathbf{r}} - \hat{a}_{k,s}^\dagger e^{-i\mathbf{k}.\mathbf{r}} \right] \hat{e}_{k,s}. \tag{2.13}$$

One may note that the magnetic field operator can also be obtained from Eq. (2.12) and definition of the magnetic field in terms of the vector potential when required.

Once the electromagnetic field is quantized, quantizing the corresponding energy is expected to be straightforward. It so happens that the quantization of the energy can be associated with the variational principle that can be closely stated as $\delta S = 0$, where S is the classical action defined by $S = \int_{t_1}^{t_2} L dt$ in which L is the Lagrangian and $t_2 - t_1$ is the time required by the field to cover the intended path between the initial and final points. The Lagrangian in turn can be expressed as a function of Φ

---

[2]One can also start with the assumption that the electromagnetic field can be designated by harmonic oscillator, and then justify the accompanying commutation relations by invoking variable transformation $\alpha_k \longrightarrow \sqrt{\frac{2\pi}{\omega_k}} \hat{a}_k$.

and **A** while the Lagrangian density $\mathcal{L}$ is given by equation of the form

$$\partial_t \left(\frac{\partial \mathcal{L}}{\partial(\partial_t A_\mu)} \right) + \sum_{j=1}^{3} \partial_j \left(\frac{\partial \mathcal{L}}{\partial(\partial_j A_\mu)} \right) = \frac{\partial \mathcal{L}}{\partial A_\mu} = 0 \tag{2.14}$$

where $\mu = 0, 1, 2, 3$ with $A_0 = \Phi$, $A_1 = A_x$, $A_2 = A_y$ and $A_3 = A_z$. With correct choice of the Lagrangian density, it should be possible to regain Maxwell's equations. Without going into details, the Lagrangian density that can be used to regain Maxwell's equations is proposed to be

$$\mathcal{L} = \frac{\varepsilon_0 E^2}{2} - \frac{B^2}{2\mu_0}. \tag{2.15}$$

Once the proper Lagrangian density is known, the next step is obtaining the Hamiltonian density that can be expressed as

$$\mathcal{H} = \sum_{j=1}^{3} \partial_t A_j \Pi_j - \mathcal{L}, \tag{2.16}$$

where Π_j is the conjugate canonical momenta that can be defined as

$$\Pi_\mu = \frac{\partial \mathcal{L}}{\partial(\partial_t A_\mu)}. \tag{2.17}$$

Upon varying the index μ, it may not be difficult to verify that $\Pi_j = -\varepsilon_0 E_j$ and $\Pi_0 = 0$, and write

$$\mathcal{H} = \frac{\varepsilon_0 E^2}{2} + \frac{B^2}{2\mu_0}. \tag{2.18}$$

After that, application of Eqs. (2.12) and (2.13) leads to the Hamiltonian

$$\hat{H} = \frac{1}{2} \sum_{k,s} \omega_k \left[\hat{a}_{k,s}^\dagger \hat{a}_{k,s} + \hat{a}_{k,s} \hat{a}_{k,s}^\dagger \right], \tag{2.19}$$

which can be rewritten upon using Eq. (2.11) as

$$\hat{H} = \sum_{k,s} \omega_k \left[\hat{a}_{k,s}^\dagger \hat{a}_{k,s} + \frac{1}{2} \right], \tag{2.20}$$

which typifies the quantized Hamiltonian of the radiation field having different polarizations and wavelengths. One may note that the last term is linked to the vacuum fluctuations and can be dropped when one seeks to deal with the dynamics of the radiation.

2.2   Quantum States of Light

It is customary in quantum theory embodying the dynamics of a system by an infinite dimensional complex state vector. To understand the fundamental traits of the quantized light, and to be able to foresee its application, it becomes imperative having a clear picture of the underlying peculiar properties of the quantum states of light such as vacuum, coherent, thermal, squeezed and number states [6]. These states are categorized as pure implying that they have some unique properties and also obtaining the pertinent distribution functions is much easier when compared to the case of mixed states.

2.2.1   Single-Mode Pure States

2.2.1.1 Number State
A number state is taken as the eigenstate of the harmonic oscillator and portrayed in terms of the number operator: $\hat{N} = \hat{a}^\dagger \hat{a}$ which designates the number of available photons in a given state. Markedly, a number state is the eigenstate of a number operator: $\hat{N}|n\rangle = n|n\rangle$, where n is the number of photons that can be generated from the vacuum state by successive application of the raising operator

$$|n\rangle = \frac{(\hat{a}^\dagger)^n}{\sqrt{n}}|0\rangle. \tag{2.21}$$

The number state also satisfies the eigenvalue equations:

$$\hat{a}|n\rangle = \sqrt{n}|n-1\rangle, \qquad \hat{a}^\dagger|n\rangle = \sqrt{n+1}|n+1\rangle. \tag{2.22}$$

Based on the fact that the number states form a complete orthonormal bases, one can also propose that $\langle n|m\rangle = \delta_{nm}$ and write the completeness relation as

$$\hat{I} = \sum_{n=0}^{\infty} |n\rangle\langle n|. \tag{2.23}$$

Even though the number state can be a useful representation for a system whose number of available photons is very small, it is not the most suitable description for optical fields in which the number of photons is large. Since the number of photons in the field is often considerably large, the involved experimental challenges might have hampered the generation of photon number states with more than quite a few number of photons. However, owing to the fact that practically realizable optical fields are either superposition or mixture of number states, the number state may still be required in quantum information manipulations (see Chaps. 9, 11 and 12).

2.2.1.2 Coherent State

A coherent state is presumed to possess an indefinite number of photons but has precisely defined phase in contrast to the number state that has a specific number of photons and a random phase.[3] Thus one can imagine a coherent state to be generated by the action of the displacement operator,

$$\hat{D}(\alpha) = e^{\alpha \hat{a}^\dagger - \alpha^* \hat{a}}, \tag{2.24}$$

on the vacuum state as

$$|\alpha\rangle = \hat{D}(\alpha)|0\rangle, \tag{2.25}$$

where α is a complex number (c-number) that corresponds to the operator $\hat{a}$.

This definition evinces that a coherent state can be interpreted as the displaced form of the harmonic oscillator;

$$\hat{D}^\dagger(\alpha)\hat{D}(\alpha) = \hat{I}, \qquad \hat{D}^\dagger(\alpha) = \hat{D}^{-1}(\alpha) = \hat{D}(-\alpha), \tag{2.26}$$

$$\hat{D}^\dagger(\alpha)\hat{a}^\dagger\hat{D}(\alpha) = \hat{a}^\dagger + \alpha^*, \qquad \hat{D}^\dagger(\alpha)\hat{a}\hat{D}(\alpha) = \hat{a} + \alpha. \tag{2.27}$$

By making use of the Baker-Hausdroff's operator identity (see Appendix 1 of Chap. 3), it is also possible to see that

$$\hat{D}(\alpha) = e^{-\frac{\alpha^*\alpha}{2}} e^{\alpha \hat{a}^\dagger} e^{-\alpha^* \hat{a}}. \tag{2.28}$$

A coherent state can also be taken as the eigenvector of the harmonic oscillator annihilation operator:

$$\hat{a}|\alpha\rangle = \alpha|\alpha\rangle. \tag{2.29}$$

So on the basis of a power series expansion of the form

$$e^{\alpha \hat{a}^\dagger} = \sum_{n=0} \frac{(\alpha \hat{a}^\dagger)^n}{n!}, \tag{2.30}$$

---

[3] A coherent state is a minimum uncertainty state in which the variances of the quadrature operators satisfy minimum uncertainty condition and taken as the benchmark for identifying quantum properties since it is very close to the pertinent classical state (see Sect. 4.2.1 and also [4]).

one can express a coherent state in terms of a number state as

$$\langle n|\alpha\rangle = \frac{\alpha^n}{\sqrt{n!}} \exp\left[-\frac{\alpha^*\alpha}{2}\right]. \tag{2.31}$$

With the aid of the completeness relation of a number state, one can see that

$$|\alpha\rangle = e^{-\frac{\alpha^*\alpha}{2}} \sum_{n=0} \frac{\alpha^n}{\sqrt{n!}}|n\rangle. \tag{2.32}$$

In light of earlier discussion, it may not be difficult to show that

$$\langle\alpha|\beta\rangle = \exp\left[\alpha^*\beta - \frac{\alpha^*\alpha}{2} - \frac{\beta^*\beta}{2}\right]. \tag{2.33}$$

One can see from Eq. (2.33) that the coherent state is over complete, which can be related to the overlap of a zero energy fluctuations. It is also possible to propose a completeness relation for a coherent state as

$$\hat{I} = \int d^2\alpha \, |\alpha\rangle\langle\alpha|, \tag{2.34}$$

where the integration is taken over real and imaginary parts,

$$d^2\alpha = d(\mathrm{Re}(\alpha)) \, d(\mathrm{Im}(\alpha)), \tag{2.35}$$

since α is a complex parameter.

2.2.1.3 Chaotic State

Most radiations available in nature that can be generated from conventional sources and light produced by such as a laser machine operating below threshold constitute thermal light that can be perceived as the best example of chaotic light. Contrary to the observation that a thermal light in a cavity maintained at equilibrium with the walls of the cavity at temperature T to be depicted in classical physics by the energy density distribution, in the context of quantum optics, the state of the thermal light is embodied by density operator. So one may need to obtain the density operator for a chaotic light upon utilizing the method of Lagrange multipliers and maximization of the entropy.

One may then note that the von Neumann's entropy can be defined in terms of the density operator as

$$\varXi = -\kappa \, Tr(\hat{\rho}\ln\hat{\rho}), \tag{2.36}$$

where κ is Boltzmann constant, Tr stands for the trace operation and $\hat{\rho}$ is the density operator one looks for.

One can infer from thermodynamics that at equilibrium entropy should be maximum: $\delta \Xi = 0$, which implies that $Tr[(\ln \hat{\rho} + 1)\delta \hat{\rho}] = 0$. Owing to the fact that $Tr(\hat{\rho}) = 1$, one can see that $Tr(\lambda \hat{\rho}) = \lambda$, and for a constant λ; $\delta Tr(\lambda \hat{\rho}) = 0$. In the same token, the expectation value of the Hamiltonian is calculated with the help of $\langle \hat{H} \rangle = Tr(\hat{\rho}\hat{H})$; and for a system in equilibrium, the mean of the Hamiltonian would be constant. These considerations hence lead to

$$\delta \langle \hat{H} \rangle = Tr(\hat{H} \delta \hat{\rho}) = 0, \tag{2.37}$$

which corresponds to the realization that the operators in the Schrödinger picture are constants. It is also straightforward to note that $Tr(\beta \hat{H} \delta \hat{\rho}) = 0$, where β is another constant.

Then combination of these ideas shows that

$$Tr\left[(1 + \lambda + \beta \hat{H} + \ln \hat{\rho})\delta \hat{\rho}\right] = 0. \tag{2.38}$$

For arbitrary $\delta \hat{\rho}$, it happens that $1 + \lambda + \beta \hat{H} + \ln \hat{\rho} = 0$; implying that

$$\hat{\rho} = e^{-(1+\lambda)-\beta \hat{H}}. \tag{2.39}$$

On account of the fact that $Tr(\hat{\rho}) = 1$, one can see that

$$\hat{\rho} = \frac{e^{-\beta \hat{H}}}{Tr(e^{-\beta \hat{H}})}. \tag{2.40}$$

To be consistent with thermodynamical phenomena, one may note that it is admissible to set $\beta = (\kappa T)^{-1}$.

In relation to the Hamiltonian of the radiation (2.20), the density operator in the number state can be expressed as

$$\hat{\rho} = \frac{\sum_{n=0}^{\infty} e^{-\beta \omega \hat{a}^\dagger \hat{a}} |n\rangle \langle n|}{\sum_{m=0}^{\infty} \langle m| e^{-\beta \omega \hat{a}^\dagger \hat{a}} |m\rangle} = \frac{\sum_{n=0}^{\infty} e^{-\beta \omega n} |n\rangle \langle n|}{\sum_{m=0}^{\infty} e^{-\beta \omega m}}, \tag{2.41}$$

where the zero point energy of the radiation is assumed to be shifted by $\omega/2$, and can be rewritten upon defining $x = e^{-\beta \omega}$ and using the fact that $\sum_{k=0}^{\infty} x^k = (1 - x)^{-1}$ for $x < 1$ as

$$\hat{\rho} = \left(1 - e^{-\beta \omega}\right) \sum_{n=0}^{\infty} e^{-n\beta \omega} |n\rangle \langle n|. \tag{2.42}$$

Equation (2.42) is thus designated as the density operator for the chaotic light.

The same density operator can also be expressed in terms of the mean photon number

$$\bar{n} = \langle \hat{a}^\dagger \hat{a} \rangle \tag{2.43}$$

and number state as

$$\hat{\rho} = \sum_{n=0}^{\infty} \frac{\bar{n}^n}{(1+\bar{n})^{1+n}} |n\rangle \langle n|. \tag{2.44}$$

Since the number state can be related to coherent light as depicted by Eq. (2.32), it might be insightful noting that the density operator for a thermal light can also be expressed in terms of a complex continuous variable when required.

2.2.1.4 Squeezed State

The electromagnetic field can be designated by a complex amplitude that describes the magnitude and phase of the field in which the fluctuations in real and imaginary parts associated with the position and momentum quadrature components should obey the well celebrated uncertainty relation. In case the uncertainties in the real and imaginary parts are equal, the pertinent minimum uncertainty states are referred to as coherent states. But when the requirement of equal uncertainties is not physically attainable, there would be a whole new class of minimum uncertainty states christened as squeezed state. So a given state is said to be in squeezed state when the uncertainty in one of the field quadrature components is less than what it should be in the corresponding coherent state.

A squeezed vacuum state can be generated under general condition from an ordinary vacuum state by using the squeeze operator defined by

$$\hat{S}(\zeta) = \exp\left[\frac{1}{2}\left(\zeta^* \hat{a}^{\dagger 2} - \zeta \hat{a}^2\right)\right], \tag{2.45}$$

where $\zeta = re^{i\theta}$ is the complex squeeze parameter with property that

$$|\zeta\rangle = \hat{S}(\zeta)|0\rangle. \tag{2.46}$$

Squeezing can also be perceived as the quantum fluctuation or noise in one of the quadratures is below the vacuum level at the expense of enhanced fluctuation in the conjugate quadrature while the product of the variances in the two quadratures satisfies the uncertainty relation (see also Sect. 7.1). In line with this, the quadrature operators can be taken as two coupled Hermitian operators defined by

$$\hat{a}_+ = \hat{a} + \hat{a}^\dagger, \qquad \hat{a}_- = i(\hat{a}^\dagger - \hat{a}). \tag{2.47}$$

One can verify that these operators satisfy the commutation relation $[\hat{a}_+, \hat{a}_-] = 2i\hat{I}$.

With this information, a single-mode light is said to be in squeezed state if either[4] of

$$\Delta a_{\pm}^2 < 1 \qquad (2.40)$$

while $\Delta a_-^2 \Delta a_+^2 \geq 1$, where the variance of any operator can be expressed as

$$\Delta O^2 = \langle \hat{O}^2 \rangle - \langle \hat{O} \rangle^2. \qquad (2.49)$$

2.2.2 Generation of Combined States

By successive application of the displacement and squeeze operators in different orders, it should be possible to generate various generic quantum states of light [7]. In the following, we list some of which appear to be more realistic and expected to display potential for application without offering in-depth discussion. For starter, consider the action of the squeeze operator on coherent state:

$$|\zeta, \alpha\rangle = \hat{S}(\zeta)\hat{D}(\alpha)|0\rangle. \qquad (2.50)$$

This state is referred to as two-photon coherent or Yuen state.

In the same way, it is possible to define a displaced squeezed state as

$$|\zeta, \alpha\rangle = \hat{D}(\alpha)\hat{S}(\zeta)|0\rangle. \qquad (2.51)$$

Notice that the order of the action of the displacement and squeeze operators is reversed. Likewise, a photon-added coherent state can be obtained by successive utilization of the photon creation operator on coherent state:

$$|\alpha, m\rangle = \hat{a}^{\dagger^m}\hat{D}(\alpha)|0\rangle. \qquad (2.52)$$

Generalization for a two-mode or multi-mode case can also be done in the same manner. It might hence worthwhile summarizing some possible states.

---

[4]It may also suffice to use the variances to envisage whether there is squeezing. But in case one seeks to be more precisely consistent with the uncertainty principle, $\Delta a_{\pm} < 1$ and $\Delta a_- \Delta a_+ \geq 1$ with $\Delta O = \sqrt{\langle \hat{O}^2 \rangle - \langle \hat{O} \rangle^2}$ need to be used instead. The quadrature operators can also be provided often with additional constant where the commutation relation and the criterion for witnessing the squeezing should be accordingly modified.

Name	Single mode	Two-mode
Vacuum states	$|0\rangle$	$|0_a, 0_b\rangle$
Coherent state (Glauber)	$\hat{D}(\alpha)|0\rangle$	$\hat{D}_{ab}(\alpha, \beta)|0_a, 0_b\rangle$
Squeezed states	$\hat{S}(\zeta)|0\rangle$	$\hat{S}_{ab}|0_a, 0_b\rangle$
Yuen states	$\hat{S}(\zeta)\hat{D}(\alpha)|0\rangle$	$\hat{S}_{ab}\hat{D}_{ab}(\alpha, \beta)|0_a, 0_b\rangle$
Displaced squeezed states	$\hat{D}(\alpha)\hat{S}(\zeta)|0\rangle$	$\hat{D}_{ab}(\alpha, \beta)\hat{S}_{ab}|0_a, 0_b\rangle$
Photon-added coherent states	$\hat{a}^{\dagger^m}\hat{D}(\alpha)|0\rangle$	$\hat{a}^{\dagger^m}\hat{D}_{ab}(\alpha, \beta)|0_a, 0_b\rangle$
Photon-added squeezed states	$\hat{a}^{\dagger^m}\hat{S}(\zeta)|0\rangle$	$\hat{a}^{\dagger^m}\hat{S}_{ab}|0_a, 0_b\rangle$
Photon-added Yuen states	$\hat{a}^{\dagger^m}\hat{S}(\zeta)\hat{D}(\alpha)|0\rangle$	$\hat{a}^{\dagger^m}\hat{S}_{ab}\hat{D}_{ab}(\alpha, \beta)|0_a, 0_b\rangle$
Photon-added displaced squeezed states	$\hat{a}^{\dagger^m}\hat{D}(\alpha)\hat{S}(\zeta)|0\rangle$	$\hat{a}^{\dagger^m}\hat{D}_{ab}(\alpha, \beta)\hat{S}_{ab}|0_a, 0_b\rangle$

Lists of some of the quantum states of the quantized light. Note that it is also possible to use $\hat{b}^{\dagger}$ instead of $\hat{a}^{\dagger}$ in defining various two-mode states

2.2.3 Two-Mode Squeezed State

Up to now, some vital generic states of the quantized light are presented. Even then, it might be conceivable to think of other ways of exploring the properties of light since in some cases looking for a more complicated states of light may be required. For instance, owing to the current urge for improved quantum information processing procedures, the need for two-mode and three-mode squeezed radiations is quite enormous (for possible source; see Chap. 8). One of the well studied quantum features of the superposed radiation is a two-mode squeezing (for higher-order squeezing; see Sect. 7.1). It might be thus imperative exploring the means of quantifying the degree of two-mode squeezing in terms of a relevant quadrature [8]. One may hence note that a two-mode squeezing is similar to its single-mode counterpart except that, for a multi-mode case, it is the noise level of one of the combined quadratures that should be below the classical limit.

A two-mode squeezed vacuum state can be comfortably portrayed by a state vector of the form

$$|a, b\rangle = \hat{S}_{ab}(r)|0, 0\rangle, \tag{2.53}$$

where r is the two-mode squeeze parameter in which the squeeze operator $\hat{S}_{ab}(r)$ can be defined as

$$\hat{S}_{ab}(r) = e^{ir\left(\hat{a}\hat{b} - \hat{a}^{\dagger}\hat{b}^{\dagger}\right)}. \tag{2.54}$$

So making use of the pertinent density operator may evince that

$$\langle \hat{a} \rangle = \langle 0, 0|\hat{a}(r)|0, 0\rangle, \qquad \langle \hat{b} \rangle = \langle 0, 0|\hat{b}(r)|0, 0\rangle, \tag{2.55}$$

where

$$\hat{a}(r) = \hat{S}_{ab}^\dagger(r)\hat{a}\hat{S}_{ab}(r) = \hat{a}\cosh r - \hat{b}^\dagger \sinh r, \tag{2.56}$$

$$\hat{b}(r) = \hat{S}_{ab}^\dagger(r)\hat{b}\hat{S}_{ab}(r) = \hat{b}\cosh r - \hat{a}^\dagger \sinh r, \tag{2.57}$$

in which $\hat{a} = \hat{a}(0)$ and $\hat{b} = \hat{b}(0)$ (see also Sect. 2.3).

In the same way, the squeezing properties of a two-mode light can be exemplified by two quadrature operators:

$$\hat{c}_+ = \frac{1}{\sqrt{2}}(\hat{a}^\dagger + \hat{a} + \hat{b}^\dagger + \hat{b}), \qquad \hat{c}_- = \frac{i}{\sqrt{2}}(\hat{a}^\dagger - \hat{a} + \hat{b}^\dagger - \hat{b}), \tag{2.58}$$

where $\hat{a}$ and $\hat{b}$ are the boson operators that represent the radiation in different modes, and satisfy the commutation relation $[\hat{c}_+, \hat{c}_-] = 2i\hat{I}$. With this characterization, a two-mode light is said to be in squeezed state provided that either of

$$\Delta c_\pm < 1, \tag{2.59}$$

and $\Delta c_+ \Delta c_- \geq 1$.

By applying the definition of the position and momentum operators in terms of the boson operator,

$$\hat{Q}_i = \frac{1}{\sqrt{2}}(\hat{a}_i^\dagger + \hat{a}_i), \qquad \hat{P}_i = \frac{i}{\sqrt{2}}(\hat{a}_i^\dagger - \hat{a}_i), \tag{2.60}$$

it should also be possible to recast the squeeze operator as

$$\hat{S}_{12}(r) = e^{ir\left(\hat{Q}_1\hat{P}_2 + \hat{Q}_2\hat{P}_1\right)}, \tag{2.61}$$

which can be taken as the generalized form of the squeeze operator for a two-mode radiation. It might also be appealing to contemplate to define the squeeze operator for a multi-mode radiation in the same manner (for extension to multi-mode and higher-order squeezing; see Sect. 7.1).

For now, the two-mode quadrature operators can be expressed in terms of the position and momentum operators as

$$\hat{c}_+ = \frac{1}{2}[\hat{Q}_1 + \hat{Q}_2], \qquad \hat{c}_- = \frac{1}{2}[\hat{P}_1 - \hat{P}_2]. \tag{2.62}$$

Since it is admissible to cast the squeeze operator into different forms as desired, the quantum and statistical properties of a two-mode squeezed radiation can be studied with the aid of similar states—which can be utilized in extending the discussion on a squeezed light to a higher-order scenario.

2.3   Phase Space Representation

It should be evident by now that a quantized light can be described in terms of quantum operators known to have a fundamental ordering problem related to the commutation relation. The operator ordering more often poses a significant mathematical challenge but can be overcome with the aid of c-number representation.[5] One of the approaches that can be applied is the phase space distribution function in which the density operator is expressed in terms of classical phase space parameters as a c-number function. When compared to operator calculations, the results following from employing the distribution function are presumed to be more directly related to the relevant experimental outcomes. In addition to the usual classical interpretation such as photon number distribution, one should note that the quasi-statistical distribution functions are recently found application in the study of quantum entanglement and related issues of continuous variable systems.

To expedite the description of various quasi-statistical distribution functions, it turns out to be advantageous to begin with the characteristic function

$$\Phi(c, t) = \left\langle e^{ic\hat{O}(t)} \right\rangle = Tr\left(\hat{\rho}e^{ic\hat{O}(t)}\right), \tag{2.63}$$

where c is a real constant and $\hat{O}(t)$ is the operator that designates the system to be studied. So the characteristic function would be the Fourier transform of the probability distribution:

$$\Phi(c, t) = \int_{-\infty}^{\infty} e^{ic\alpha} \wp(\alpha, t) d\alpha. \tag{2.64}$$

Equation (2.64) may hint that the quasi-statistical distribution function can be obtained once the characteristic function related to the prescribed ordering is known. Besides, it might not be difficult to infer that the characteristic function can be utilized to calculate various orders of moments directly. For instance, the nth order moment of a given operator can be obtained by using

$$\langle \hat{O}^n(t) \rangle = \frac{\partial^n \Phi(c, t)}{\partial(ic)^n} \Big|_{c=0} \tag{2.65}$$

from which one can see the characteristic function as the moment generating function.

---

[5]The phase space formulation of quantum optics can be represented by position and momentum parameters on equal footing in contrast to the Schrödinger or Heisenberg pictures; and hence, the name. The distribution function is also believed to fully characterize the quantum state, and so enables one to express the quantum expectation values as average of classical observable [9].

2.3.1   Distribution Functions

Different distribution functions might be conceived at some point based on the manner in which the operators are ordered. Even then, the result of the analysis should turn out to be the same irrespective of the quasi-statistical distribution one opted to apply.

2.3.1.1   P-function

One of the widely used quasi-statistical distributions is the P-function taken to be the c-number equivalent of the anti-normally ordered density operator. Imagine that the density operator can be expanded in a power series as

$$\hat{\rho} = \sum_{l,m} C_{l,m} \hat{a}^l \hat{a}^{\dagger m},$$
(2.66)

where $C_{l,m}$ are coefficients, and the corresponding P-function can be defined as

$$P(\alpha) = \frac{\langle \alpha | \hat{\rho} | \alpha \rangle}{\pi}.$$
(2.67)

$\hat{\rho}$ is presumed to be arranged in anti-normal order and $1/\pi$ is inserted to ensure normalization so that its interpretation is consistent with the notion of classical distribution function.

Employing the completeness relation for the coherent state given by Eq. (2.34) leads straightaway to

$$\hat{\rho} = \int \frac{d^2\alpha}{\pi} \sum_{l,m} C_{l,m} \alpha^l \alpha^{*m} | \alpha \rangle \langle \alpha |,$$
(2.68)

from which follows

$$P(\alpha^*, \alpha) = \frac{1}{\pi} \sum_{l,m} C_{l,m} \alpha^l \alpha^{*m};$$
(2.69)

and hence,

$$\hat{\rho} = \int d^2\alpha \, P(\alpha, \alpha^*) | \alpha \rangle \langle \alpha |.$$
(2.70)

So based on the fact that the trace of the density operator is unity, it would not be hard to verify that

$$\int d^2\alpha \, P(\alpha, \alpha^*) = 1 :$$
(2.71)

the P-function is normalized in the same way as any other classical distribution function [10].

It is also proven to be useful to express the P-function in terms of the normally ordered characteristic function. We thus opt to begin with the characteristic function

$$\Phi_n(z, z^*) = Tr\left(\hat{\rho} e^{z\hat{a}^\dagger} e^{-z^*\hat{a}}\right). \tag{2.72}$$

Upon recalling that the density operator is anti-normally ordered, and inserting the completeness relation for the coherent state, one can see that

$$\Phi_n(z, z^*) = \int d^2\alpha \ P(\alpha, \alpha^*) \exp\left[z\alpha^* - z^*\alpha\right], \tag{2.73}$$

where the P-function can be expressed as the Fourier transform of the characteristic function:

$$P(\alpha, \alpha^*) = \frac{1}{\pi^2} \int d^2z \ \Phi_n(z, z^*) \exp\left[z\alpha^* - z^*\alpha\right]. \tag{2.74}$$

Then following the intended approach, the P-function is found to be

$$P(\alpha, \alpha^*) = \delta^2(\alpha), \qquad \text{for coherent light,} \tag{2.75}$$

$$P(\alpha, \alpha^*) = \frac{1}{\pi\bar{n}} \exp\left[-\frac{\alpha^*\alpha}{\bar{n}}\right], \qquad \text{for chaotic light,} \tag{2.76}$$

$$P(\alpha, \alpha^*) = \frac{2}{\pi \sinh r} \int d^2z \exp(z^*\alpha - z\alpha^*), \qquad \text{for squeezed light.} \tag{2.77}$$

It may not be difficult to see from Fig. 2.1 that the P-function for a chaotic light exhibits Gaussian behavior. Critical scrutiny may also reveal that the width of the Gaussian profile would widened but its height diminished with the increasing mean photon number of the chaotic light. On the other hand, one may note that Eq. (2.77) is not a well behaved distribution function: the P-function for the squeezed light is not well defined, which heralds the pertinent nonclassical features.

2.3.1.2 R-function

R-function can be taken as a c-number equivalent of the density operator of arbitrary ordering and defined as

$$R(\alpha^*, \beta) = \frac{\zeta(\alpha^*, \beta)}{\pi} \exp\left[-\frac{\alpha^*\alpha}{2} - \frac{\beta^*\beta}{2} + \beta\alpha^*\right], \tag{2.78}$$

where $\zeta(\alpha^*, \beta)$ is the c-number equivalent of the normally ordered function of $\hat{a}^\dagger$ and $\hat{b}$, and the ordering process is carried out by replacing $\hat{b}$ by $\beta + \frac{\partial}{\partial\beta^*}$ and $\hat{a}^\dagger$ by α^*.

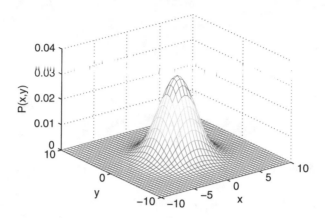

Fig. 2.1 Plot of the P-function of a chaotic light for $\bar{n} = 10$. Note that a variable transformation $\alpha = x + iy$ is utilized

One of the main importance of the R-representation is to express an operator function of arbitrary ordering into c-number equivalent of the normal ordering:

$$\zeta(\hat{a}^\dagger, \hat{b}) = \frac{1}{\pi^2} \int d^2\alpha d^2\beta \, R(\alpha^*, \beta)|\alpha\rangle\langle\beta| \exp\left[-\frac{\alpha^*\alpha}{2} - \frac{\beta^*\beta}{2}\right]. \qquad (2.79)$$

It might be helpful noting that the R-function can behave well when the P-function does not, that is, the R-function can be useful in transforming equations of quantum operators including the density operator into an ordinary quasi-statistical distribution function in case the system under consideration exhibits quantum features that cannot be captured by the P-function.[6]

2.3.1.3 Q-function

Q-function on the other hand can be taken as a c-number equivalent of the density operator for the normal ordering and defined as

$$Q(\alpha) = \frac{\langle\alpha|\hat{\rho}|\alpha\rangle}{\pi}, \qquad (2.80)$$

where $\hat{\rho}$ can be expanded in terms of the boson operators in the normal order and the Q-function can be obtained by applying the anti-normally ordered characteristic

---

[6]Starting with the normalization of the density operator, one can verify that $\int R(\alpha^*, \alpha)e^{-\alpha^*\alpha} \, d^2\alpha = \pi$. One may note that the normalization condition for R-function has additional Gaussian weight factor. Due to this weight factor, R-function is not usually given the status of the quasi-statistical distribution function.

function:

$$Q(\alpha, t) = \frac{1}{\pi^2} \int d^2z \, \phi_a(z, t) \exp(z^*\alpha - z\alpha^*), \qquad (2.81)$$

where the relevant characteristic function is expressible as

$$\phi_a(z, t) = Tr\left(\hat{\rho}(0)e^{-z^*\hat{a}(t)}e^{z\hat{a}^\dagger(t)}\right). \qquad (2.82)$$

With operator identity, $e^{\hat{A}}e^{\hat{B}} = e^{\hat{A}+\hat{B}+\frac{1}{2}[\hat{A},\,\hat{B}]}$, it is possible to see that

$$\phi_a(z, t) = e^{-\frac{1}{2}zz^*} Tr\left(\hat{\rho}(0)\exp[z\hat{a}^\dagger(t) - z^*\hat{a}(t)]\right). \qquad (2.83)$$

As illustration, imagine the system to be initially in the vacuum state; and later, placed in a lossless cavity that interacts with squeezed light. It thus turns out to be appealing relating the squeeze operator with a degenerate parametric oscillator that can be described by the interaction Hamiltonian of the form

$$\hat{H} = \frac{i\varepsilon}{2}\left(\hat{a}^2 - \hat{a}^{\dagger 2}\right), \qquad (2.84)$$

where ε is the measure of the strength of the coupling of the classical radiation with a nonlinear crystal (see Fig. 8.1). The time development of the operator that designates the squeezed radiation can be obtained from the Hamiltonian (2.84) by applying the Heisenberg equation:

$$\frac{d\hat{a}}{dt} = -\varepsilon\hat{a}^\dagger. \qquad (2.85)$$

To solve this differential equation, it might be helpful differentiating once again with respect to time;

$$\frac{d^2\hat{a}}{dt^2} = \varepsilon^2\hat{a}. \qquad (2.86)$$

The solution of Eq. (2.86) can be written as

$$\hat{a}(t) = A\cosh\varepsilon t + B\sinh\varepsilon t, \qquad (2.87)$$

where A and B are constants to be determined from initial condition: $\hat{a}(t = 0) = \hat{a}$, $A = \hat{a}$ and $B = \hat{a}^\dagger$. Taking this into consideration, one can write

$$\hat{a}(t) = \hat{a}\cosh\varepsilon t - \hat{a}^\dagger\sinh\varepsilon t. \qquad (2.88)$$

Once the time evolution of the pertinent operator is known, the characteristic function (2.83) can be put in the form

$$\psi_a(z, t) = e^{-\frac{1}{2}zz} \ Tr\Big(\hat{\rho}(0) \exp\big[z(a^\dagger \cosh \varepsilon t - a \sinh \varepsilon t)$$
$$- z^*(\hat{a} \cosh \varepsilon t - \hat{a}^\dagger \sinh \varepsilon t)\big]\Big). \tag{2.89}$$

It should be clear from Eq. (2.89) that the characteristic function can be determined provided that the state of the radiation at $t = 0$ is known.

To pave the way for incorporating the effects of the radiation in the cavity prior to interaction, different possible initial conditions need to be considered. For example, for a system initially in the vacuum state, the characteristic function takes the form

$$\phi_a(z, t) = \exp\left[-z^*z \cosh^2 \varepsilon t - \left(\frac{z^2 + z^{*2}}{2}\right) \sinh \varepsilon t \cosh \varepsilon t\right]. \tag{2.90}$$

One can hence check that the pertinent Q-function has the form

$$Q(\alpha, r) = \frac{\mathrm{sech}\, r}{\pi} \exp\left[-\alpha^*\alpha - \left(\frac{\alpha^2 + \alpha^{*2}}{2}\right) \tanh r\right], \tag{2.91}$$

where εt is equated to r based on the resemblance of the Hamiltonians.[7] Note that Eq. (2.91) designates the Q-function of squeezed vacuum. It is straightforward to see from Fig. 2.2 that the Q-function for a squeezed light also exhibits a Gaussian behavior. Contrary to the result for a chaotic light, the current Gaussian function has a different width along different axes that can be related to the squeezing phenomenon (relate with Fig. 2.1). One may note that the height of the Gaussian distribution function would be enhanced while the width narrowed with decreasing squeezed parameter.

The Q-function for the light generated from a degenerate parametric oscillator when the signal mode is initially in a coherent state is also found following a similar procedure to be

$$Q(\alpha, r) = \frac{\mathrm{sech}\, r}{\pi} \exp\Big[-\alpha^*\alpha - \gamma^*\gamma$$
$$- \tanh r \left(\frac{\alpha^2 + \alpha^{*2} - \gamma^2 - \gamma^{*2}}{2}\right) + \mathrm{sech}\, r(\gamma\alpha^* + \gamma^*\alpha)\Big]. \tag{2.92}$$

---

[7]This Q-function can also be obtained following other straightforward approaches such as comparing the definition of the Q-function in terms of the density operator for the squeezed light; $Q(\alpha) = \langle\alpha|r\rangle\langle r|\alpha\rangle/\pi$.

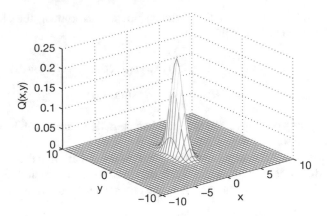

Fig. 2.2 Plot of the Q-function of a squeezed light for $r = 0.8$ when a variable transformation $\alpha = x + iy$ is employed

One can also see in the same way

$$Q(\alpha, r) = \frac{\operatorname{sech} r}{\pi} \exp\left[-\alpha^*\alpha - \lambda^*\lambda - \frac{\tanh r}{2}\left(\alpha^2 + \alpha^{*2} + \lambda^2 + \lambda^{*2}\right)\right.$$
$$\left. + \tanh r (\lambda^*\alpha^* + \lambda\alpha) + \lambda\alpha^* + \lambda^*\alpha\right], \tag{2.93}$$

$$Q(\alpha, r) = \frac{1}{\pi\sqrt{(1+\bar{n})^2\cosh^2 r - \bar{n}^2\sinh^2 r}}$$
$$\times \exp\left[-\alpha^*\alpha\,\frac{(1+\bar{n})\cosh^2 r + \bar{n}\sinh^2 r}{(1+\bar{n})^2\cosh^2 r - \bar{n}^2\sinh^2 r}\right.$$
$$\left. -(\alpha^{*2} + \alpha^2)\frac{(1/2+\bar{n})\cosh r \sinh r}{(1+\bar{n})^2\cosh^2 r - \bar{n}^2\sinh^2 r}\right], \tag{2.94}$$

for displaced squeezed state and thermally seeded signal mode of a degenerate parametric oscillator where the cavity is taken to be initially in a chaotic state; and later, allowed to interact with the squeezed light. This approach perhaps embodies the mixing of the squeezed and chaotic light beams where the Q-function for coherent and chaotic lights would be

$$Q(\alpha) = \frac{1}{\pi} \exp\left[-\alpha^*\alpha\right], \tag{2.95}$$

$$Q(\alpha) = \frac{1}{\pi(1+\bar{n})} \exp\left[-\frac{\alpha^*\alpha}{1+\bar{n}}\right]. \tag{2.96}$$

2.3.1.4 Wigner Function

The Wigner function which is a c-number equivalent of the symmetrically ordered density operator can be defined as

$$W(\alpha) = \frac{1}{\pi^2} \int d^2 z \phi_s(z) e^{z^*\alpha - z\alpha^*}, \tag{2.97}$$

where $\phi_s(z) = Tr(\hat{\rho} e^{z\hat{a}^\dagger - z^*\hat{a}})$ is the symmetrically order characteristic function that can be expressed with the aid of the operator identity $e^{\hat{A}+\hat{B}} = e^{-\frac{1}{2}[\hat{A},\hat{B}]}e^{\hat{A}}e^{\hat{B}}$ as

$$\phi_s(z) = e^{\frac{-z^*z}{2}} \phi_n(z) = e^{\frac{z^*z}{2}} \phi_a(z). \tag{2.98}$$

The Wigner function can hence be put in either

$$W(\alpha) = \frac{1}{\pi^2} \int d^2 z \exp\left[-\frac{z^*z}{2} + z^*\alpha - z\alpha^* \right] \phi_a(z) \tag{2.99}$$

or

$$W(\alpha) = \frac{1}{\pi^2} \int d^2 z \exp\left[\frac{z^*z}{2} + z^*\alpha - z\alpha^* \right] \phi_n(z) \tag{2.100}$$

form. What remains to determine basically is the characteristic function associated with the normal or anti-normal ordering; and then, carry out the resulting integrations. Since the associated characteristic functions that describe different options have been already obtained, it should be straightforward to obtain the pertinent Wigner function.

2.3.2   Relation Among Distribution Functions

It may be evident from earlier discussion that there is a lucid relationship among various distribution functions. It should also be clear that the rigor required to obtain different quasi-statistical distribution functions can be different and one of the distribution functions can be more appropriate to study a given situation than the others. Calculating the required distribution function starting from scratch every time may not be advisable and economical. It hence appears imperative devising a means of finding each distribution function once one of them is fairly known.

To establish the intended relation among various distribution functions, one can thus begin with the definition of the density operator in terms of the P-function (2.70). Right away, taking the trace of the density operator over the coherent state leads to

$$Q(\beta^*, \beta) = \int \frac{d^2\alpha}{\pi} P(\alpha^*, \alpha) \exp\left[-\alpha^*\alpha - \beta^*\beta + \alpha^*\beta + \beta^*\alpha \right]. \tag{2.101}$$

Equation (2.101) indicates that the Q-function can be obtained once the corresponding P-function is known.

But to express the P-function in terms of the Q-function, it is advisable to begin with

$$\langle -\beta | \hat{\rho} | \beta \rangle = \int d^2\alpha \, P(\alpha^*, \alpha)\langle -\beta | \alpha \rangle \langle \alpha | \beta \rangle, \tag{2.102}$$

in which

$$\langle -\beta | \alpha \rangle = \exp\left[-\frac{\alpha^*\alpha}{2} - \frac{\beta^*\beta}{2} - \beta^*\alpha \right]. \tag{2.103}$$

Upon making use of Eq. (2.103) and the definition of the Q-function,

$$\langle -\beta | \hat{\rho} | \beta \rangle = \pi \, Q(-\beta^*, \beta) e^{-2\beta^*\beta}, \tag{2.104}$$

one can verify that

$$Q(-\beta^*, \beta) = \int \frac{d^2\alpha}{\pi} \, P(\alpha^*, \alpha) \exp\left[-\alpha^*\alpha + \beta^*\beta + \alpha^*\beta - \beta^*\alpha \right]. \tag{2.105}$$

Taking the Fourier transform of Eq. (2.105) after that leads to

$$P(\alpha^*, \alpha) = e^{\alpha^*\alpha} \int \frac{d^2\beta}{\pi^2} Q(-\beta^*, \beta) \exp\left[-\beta^*\beta + \beta^*\alpha - \beta\alpha^* \right], \tag{2.106}$$

which insinuates the possibility of obtaining the P-function once the Q-function is known. In the same way, making use of the definition of the anti-normally ordered characteristic function can allow to represent the Wigner-function in terms of the P-function:

$$W(\alpha^*, \alpha) = 2 \int \frac{d^2\beta}{\pi} \, P(\beta^*, \beta) \exp\left[-2|\alpha - \beta|^2 \right]. \tag{2.107}$$

Following the same approach, it might be possible to establish a complete relation among various distribution functions. In light of this proposal, parameterized integral expansion of the density operator can be introduced in which the weight function is identified with the P, Q and Wigner functions upon varying the order parameter s: $s = +1, 0, -1$ [11]—the idea utilized in studying photon mixing at the detector (see Sect. 4.5). To illustrate the intended approach, one can introduce the displacement operator in terms of the s-ordered products as

$$\hat{D}(\alpha, s) = \exp\left[\frac{s}{2}\alpha^*\alpha + \alpha\hat{a}^\dagger - \alpha^*\hat{a} \right], \tag{2.108}$$

where comparison with the displacement operator reveals that $s = -1$ stands for anti-normal ordering, $s = 0$ symmetric ordering and $s = 1$ normal ordering. Consistent with this, the Fourier transform of the generalized displacement operator can be expressed as

$$\hat{T}(\alpha, s) = \frac{1}{\pi} \int \hat{D}(z, s) \exp\left[\alpha z^* - \alpha^* z\right] d^2 z. \tag{2.109}$$

Let us also introduce the function $C(\alpha, s)$ as expectation value of this operator which can be equated with the characteristic function while the density operator can be expressed as

$$\hat{\rho} = \frac{1}{\pi} \int C(\alpha, s)\hat{T}(\alpha, -s) d^2 \alpha. \tag{2.110}$$

It may not be difficult to infer that the dependence of the function $C(\alpha, s)$ on a parameter s can be epitomized by the convolution integral of the form

$$C(\alpha, s) = \frac{2}{\pi(s' - s)} \int \exp\left[-\frac{2|\alpha - \beta|^2}{s' - s}\right] C(\beta, s') d^2 \beta \tag{2.111}$$

for $\mathrm{Re}(s) < \mathrm{Re}(s')$. Such restriction suggests that one can make transformation among various quasi-statistical distribution functions as long as the appropriate condition is satisfied.

2.3.3 Applications of Distribution Functions

One can insist that a quasi-statistical distribution function would have a potential to determine relevant orders of moments without directly facing the challenges posed by operator ordering. We hence shall employ the Q-function to demonstrate some of its applications.

2.3.3.1 Photon Number Distribution

Existing experience indicates that reconstruction of the photon number distribution of optical states can provide a significant amount of information on the nature of the beam of light.[8] Despite the available vast applications, photon detectors that operate as a perfect photon counter are rare to come by due to the limitation imposed by quantum efficiency mainly linked to the fact that, in a photon detection process, only smaller fraction of the incoming photons can be captured; and hence, photon

---

[8]Quantum tomography can also provide an alternative way of measuring photon number distribution by implementing homodyne detection that requires appropriate mode matching of the signal with a suitable local oscillator at beam splitter without disregarding the involved challenge specially in case of pulsed optical fields.

distribution can provide only a probabilistic figure of merit. Even then, a photon number distribution can generally be defined as

$$P(n) = \langle n|\hat{\rho}|n\rangle, \qquad (2.112)$$

which stands for the probability of finding n photons of the radiation that can be described by the density operator $\hat{\rho}$. For a single-mode radiation, by introducing the completeness relation for the coherent state twice in Eq. (2.112), one can see that

$$P(n) = \int \frac{d^2\alpha}{\pi} \frac{d^2\beta}{\pi} \langle n|\alpha\rangle\langle\alpha|\hat{\rho}|\beta\rangle\langle\beta|n\rangle. \qquad (2.113)$$

It may not be difficult to see from Eq. (2.113) that the photon number distribution can be expressed in terms of already introduced distribution functions. In case the density operator is assumed to be expandable in a power series of $\hat{a}$ and $\hat{a}^\dagger$ in the normal order, Eq. (2.113) can be put in the form

$$P(n) = \int \frac{d^2\alpha}{\pi} d^2\beta \langle n|\alpha\rangle Q(\alpha^*, \beta, t)\langle\alpha|\beta\rangle\langle\beta|n\rangle. \qquad (2.114)$$

It might be straightforward to see that expanding the Q-function in a power series leads to

$$P(n) = \frac{1}{n!} \int \frac{d^2\alpha}{\pi} d^2\beta \sum_{l,m}^{\infty} C_{lm}\alpha^{*l}\beta^m (\alpha\beta^*)^n \exp[\alpha^*\beta - \alpha^*\alpha - \beta^*\beta], \qquad (2.115)$$

from which follows

$$P(n) = \frac{\pi}{n!} \frac{\partial^{2n}}{\partial\alpha^{*n}\partial\alpha^n}[Q(\alpha, \alpha^*)e^{\alpha^*\alpha}]|_{\alpha^*=\alpha=0}. \qquad (2.116)$$

For a light generated by a degenerate parametric amplifier (see Fig. 8.1) in which the signal mode is taken to be initially in a chaotic state, the photon number distribution can be rewritten with the aid of Eq. (2.94) as

$$P(n) = \frac{1}{n!\sqrt{a}} \frac{\partial^{2n}}{\partial\alpha^{*n}\partial\alpha^n} \left(\sum_{i,j,k=0}^{\infty} \frac{\alpha^{i+2j}\alpha^{*i+2k}b^i c^{j+k}}{i!j!k!a^{i+j+k}}\right)|_{\alpha^*=\alpha=0}, \qquad (2.117)$$

where

$$a = (1+\bar{n})^2 \cosh^2(r) - \bar{n}^2 \sinh^2 r, \qquad (2.118)$$

$$b = (1+\bar{n})\bar{n}, \qquad (2.119)$$

$$c = -\left(\frac{1}{2}+\bar{n}\right)\cosh r \sinh r. \qquad (2.120)$$

Then with the help of the mathematical relation $\frac{\partial^n x^m}{\partial x^n} = \frac{m! x^{m-n}}{(m-n)!}$, it is possible to see that

$$P(n) \quad \frac{1}{n!\sqrt{a}}$$

$$\times \sum_{i,j,k=0}^{\infty} \frac{b^i c^{j+k}}{i! j! k! a^{i+j+k}} \frac{(i+2j)!(i+2k)!}{(i+2j-n)!(i+2k-n)!} \alpha^{i+2j-n} \alpha^{*i+2k-n} \Big|_{\alpha^*=\alpha=0}.$$

$$(2.121)$$

Equation (2.121) would be different from zero for $\alpha^* = \alpha = 0$ provided that $i + 2j = n$ and $i + 2k = n$, which implies that $j = k$, $j = (n-i)/2$ and $i + j + k = n$, that is,

$$P(n) = \frac{n!}{a^{\frac{2n+1}{2}}} \sum_{i=0}^{n} \frac{b^i c^{n-i}}{i! \left[\left(\frac{n-i}{2} \right)! \right]^2}, \qquad (2.122)$$

where the summation is extended up to n since the factorial is meaningful only for $n \geq i$.

One may note that Eq. (2.122) reduces for $\bar{n} = 0$ (squeezed light) to

$$P(n) = \frac{(-1)^n n! \tanh^n r}{2^n \left[\frac{n}{2}! \right]^2 \cosh r}, \qquad (2.123)$$

and for $r = 0$ (thermal light) to

$$P(n) = \frac{\bar{n}^n}{(1+\bar{n})^{n+1}}. \qquad (2.124)$$

In the same way, one may show the photon number distribution of a coherent light to be

$$P(n) = \frac{(\alpha^* \alpha)^n}{n!} e^{-\alpha^* \alpha}. \qquad (2.125)$$

One may see from Figs. 2.3 and 2.4 that the probability of getting a large number of photons together is much less than getting smaller numbers, that is, the probability of getting a larger number of photon is almost zero irrespective of the mean photon number. Deeper scrutiny may also reveal that for a light with a smaller squeezed parameter, it is even unlikely to get a larger number of photons that might be related to the nature of the mean photon number of the squeezed light: $\bar{n} = \sinh^2(r)$. It might also be possible to deduce from the result depicted in Fig. 2.5 that the photon number distribution for a coherent light would be large for very small mean photon number and the case when one photon is available.

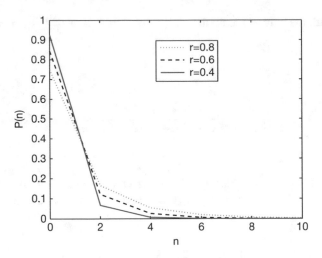

Fig. 2.3 Plots of the photon number distribution of a squeezed light for different values of r

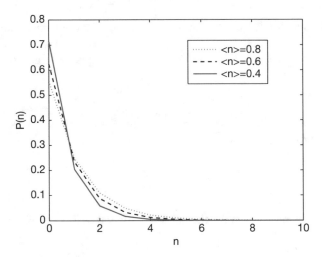

Fig. 2.4 Plots of the photon number distribution of a chaotic light for different values of $\bar{n} = \langle n \rangle$

2.3.3.2 Joint Probability of Finding n and m Photons

To determine the joint probability for finding n and m photons in the cavity modes a and b simultaneously, one may begin with

$$
P(n, m, t) = \frac{\pi^2}{n!\,m!} \frac{\partial^{2n}}{\partial\alpha^{*n}\,\partial\alpha^n} \frac{\partial^{2m}}{\partial\beta^{*m}\,\partial\beta^m} \left[Q(\alpha, \beta, t) \exp\left(\alpha^*\alpha + \beta^*\beta\right) \right]_{\alpha^*=\alpha=\beta^*=\beta=0},
$$

$$
(2.126)
$$

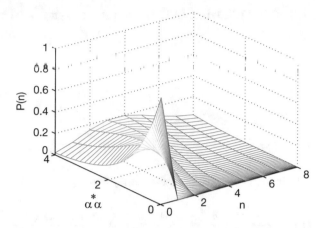

Fig. 2.5 Plot of the photon number distribution of a coherent light

where the pertinent Q-function for a two-mode radiation can be defined in terms of the characteristic function as [12]

$$Q(\alpha, \beta, t) = \frac{1}{\pi^4} \int d^2z d^2\eta \, \zeta(z, \eta, t) \exp\left[z^*\alpha + \eta^*\beta - z\alpha^* - \eta\beta^*\right] \quad (2.127)$$

in which the corresponding anti-normally ordered characteristic function can be put in the Schrödinger picture in the form

$$\zeta(z, \eta, t) = Tr\left\{\hat{\rho}(0)e^{-z^*\hat{a}(t)}e^{-\eta^*\hat{b}(t)}e^{z\hat{a}^\dagger(t)}e^{\eta\hat{b}^\dagger(t)}\right\}. \quad (2.128)$$

Upon citing the operator identity, $e^{\hat{A}}e^{\hat{B}} = e^{\hat{B}}e^{\hat{A}}e^{[\hat{A},\hat{B}]}$, it is possible to express Eq. (2.128) in terms of c-number variables associated with the normal ordering when the cavity is initially maintained at vacuum state as

$$\zeta(z, \eta, t) = \exp\left[-az^*z - b\eta^*\eta + c(z\eta + z^*\eta^*)\right], \quad (2.129)$$

where

$$a = \langle\alpha^*(t)\alpha(t)\rangle + 1, \qquad b = \langle\beta^*(t)\beta(t)\rangle + 1, \qquad c = \langle\alpha(t)\beta(t)\rangle. \quad (2.130)$$

Then substitution of Eq. (2.129) into (2.127); and then, carrying out the integration yield

$$Q(\alpha, \beta) = \frac{uv - w^2}{\pi^2} \exp\left[-u\alpha^*\alpha - v\beta^*\beta + w(\alpha\beta + \alpha^*\beta^*)\right], \qquad (2.131)$$

in which

$$u = \frac{b}{ab - c^2}, \qquad v = \frac{a}{ab - c^2}, \qquad w = \frac{c}{ab - c^2}. \qquad (2.132)$$

Once the Q-function is obtained, one can verify that

$$P(n, m, t) = \frac{uv - w^2}{n!m!} \frac{\partial^{2n}}{\partial\alpha^{*n}\partial\alpha^n} \frac{\partial^{2m}}{\partial\beta^{*m}\partial\beta^m} \sum_{i,j,k,l=0}^{\infty} \frac{(1 - u)^i (1 - v)^j w^{k+l}}{i!j!k!l!}$$

$$\times \alpha^{i+k}\alpha^{*i+l}\beta^{j+k}\beta^{*j+l}|_{\alpha=\alpha^*=\beta=\beta^*=0}, \qquad (2.133)$$

where performing the differentiation, and then employing the condition $\alpha = \alpha^* = \beta = \beta^* = 0$ result

$$P(n, m, t) = n!m!(uv - w^2) \sum_{i=n-m}^{n} \frac{(1 - u)^i (1 - v)^{i+m-n} w^{2(n-i)}}{i!(i + m - n)![(n - i)!]^2}. \qquad (2.134)$$

This is the sought for joint probability of finding n and m photons that can be explicitly written once the parameters u, v and w are known for the system under consideration.

2.3.3.3 Photon Count Distribution

Out of the entire photons produced by an optical mechanism, it is only the amount less than the produced photons that would be available for detection while the remaining photons are just scattered away undetected. Photon count distribution hence can be perceived as a function that represents the probability for counting or detecting m photons of the radiation out of the available n photons. It may be appropriate for clarity taking the light propagating in an open space and contains n photons. If out of these photons only m of them are available for counting, definitely $n - m$ of them evade detection, and so $m \leq n$. For any counting in which only m of them showup out of n similar particles, there are $\frac{n!}{m!(n-m)!}$ different ways of detection. Hence the probability of detecting a single photon in prescribed scenario is $\eta^m (1 - \eta)^{n-m}$, where η represents the probability for detecting a single photon and often equated to quantum efficiency.

One can then propose that if $P(n)$ is the probability that a light mode contains n photons, the photon count distribution to be expressed as

$$\varrho(m) = \sum_{n=m}^{\infty} \frac{n!}{m!(n-m)!} \eta^m (1-\eta)^{n-m} P(n), \qquad (2.135)$$

which can be expressed in terms of the phase space distribution functions. This assertion may insinuate that one can use the definition of the expectation value for normally ordered operators to express the photon count distribution:

$$\varrho(m) = \frac{\eta^m}{m!} \langle : (\hat{a}^\dagger \hat{a})^m \exp(-\eta \hat{a}^\dagger \hat{a}) : \rangle, \qquad (2.136)$$

where :: stands for normal ordering of the operators placed in between. One may note that Eq. (2.136) can be rewritten in terms of the normally ordered density operator as

$$\varrho(m) = \frac{\eta^m}{m!} Tr[\hat{\rho} : (\hat{a}^\dagger \hat{a})^m \exp(-\eta \hat{a}^\dagger \hat{a}) :]. \qquad (2.137)$$

Since the density operator is assumed to be in the normal order, and so every thing is arranged in the normal order.

Employing a power series expansion, and then upon applying the completeness relation for a coherent state twice, one gets

$$\varrho(m) = \frac{\eta^m}{m!} \int \frac{d^2\alpha d^2\beta}{\pi^2} \langle \alpha | \hat{\rho} | \beta \rangle \exp(-\alpha^*\alpha - \beta^*\beta + \beta^*\alpha - \eta\beta^*\alpha)\alpha^m \beta^{*m}, \qquad (2.138)$$

where application of the definition of the Q-function results

$$\varrho(m) = \frac{\eta^m}{m!} \frac{d^m}{da^m} \int d^2\alpha \ Q(\alpha^*, a\alpha) \exp[(a-1)\alpha^*\alpha]|_{a=1-\eta}. \qquad (2.139)$$

Despite the elegance of the pertinent mathematical representation, obtaining the photon count distribution in practice can be formidable task at times.

2.3.3.4 Mean Photon Number

Owing to the inherent statistical behavior of quantum theory, the measurement outcomes in a given experimental setup may not take the same value if the experiment happens to be repeated several times. In case one seeks to reproduce the result of the experiment, it is the statistical mean of the measured values that perhaps can be repeated. With such underlying subtlety in quantum measurement theory (for conceptual conundrum of quantum measurement theory and suggested remedies; see Chap. 4), the expected value from the experiment is interpreted as statistical mean: usually dubbed as expectation value.

The mean photon number of a beam of light for example can be defined as

$$\bar{n} = \langle \hat{n} \rangle = Tr(\hat{\rho}\hat{n}) \tag{2.140}$$

(relate with the discussion in Sect. 2.2). Upon inserting the completeness relation for a number state into Eq. (2.140), and then exploiting the properties of the trace operation, one can see that

$$\bar{n} = \sum_{n=0}^{\infty} n P(n). \tag{2.141}$$

Since it closely resembles the classical average, one can take it as a natural definition of the mean photon number.

With the aid of Eq. (2.70), one also gets

$$\langle \hat{A}(\hat{a}^{\dagger}, \hat{a}) \rangle = \int P(\alpha^*, \alpha) Tr(|\alpha\rangle\langle\alpha|\hat{A}(\hat{a}^{\dagger}, \hat{a})) d^2\alpha, \tag{2.142}$$

where $\hat{A}(\hat{a}^{\dagger}, \hat{a})$ is taken as arbitrary normally ordered function of $\hat{a}^{\dagger}$ and $\hat{a}$ that can be expanded in a power series. Then evoking the trace operation implies that

$$\langle\alpha|\hat{A}(\hat{a}^{\dagger}, \hat{a})|\alpha\rangle = \sum_{l,m} C_{l,m} \alpha^{*l} \alpha^m = A_n(\alpha^*, \alpha). \tag{2.143}$$

One may note that $A_n(\alpha^*, \alpha)$ is the c-number function corresponding to the normal ordering. The mean photon number can thus be defined in terms of the P-function as

$$\bar{n} = \int \alpha^* \alpha \, P(\alpha^*, \alpha) d^2\alpha. \tag{2.144}$$

One can also begin with the general definition

$$\langle \hat{A}(\hat{a}^{\dagger}, \hat{a}) \rangle = Tr(\hat{\rho}\hat{A}(\hat{a}^{\dagger}, \hat{a})). \tag{2.145}$$

Upon taking the density operator to be expandable in the normal order, and $\hat{A}(\hat{a}^{\dagger}, \hat{a})$ in anti-normal order combination of $\hat{a}^{\dagger}$ and $\hat{a}$, the mean of an operator $\hat{A}(\hat{a}^{\dagger}, \hat{a})$ can be rewritten as

$$\langle \hat{A}(\hat{a}^{\dagger}, \hat{a}) \rangle = \sum_{j,k} \sum_{l,m} C_{jk} C_{lm} Tr(\hat{a}^{\dagger j} \hat{a}^k \hat{a}^l \hat{a}^{\dagger m}) \tag{2.146}$$

in which applying the completeness relation for a coherent state, and then carrying out the trace operation result

$$\langle \hat{A}(\hat{a}^\dagger, \hat{a}) \rangle = \int Q(\alpha^*, \alpha) A_a(\alpha^*, \alpha) d^2\alpha, \tag{2.147}$$

where

$$A_a(\alpha^*, \alpha) = \sum_{l,m} C_{lm} \alpha^l \alpha^{*m} \tag{2.148}$$

is the c-number function associated with anti-normal ordering.

Since the photon number $\hat{a}^\dagger \hat{a}$ can be expressed using the boson commutation relation in the anti-normal ordering as $\hat{a}\hat{a}^\dagger - 1$, the mean photon number can be evaluated by using

$$\bar{n} = \int Q(\alpha^*, \alpha)(\alpha\alpha^* - 1) d^2\alpha. \tag{2.149}$$

In the same token, the expectation value of a given operator $\hat{A}$ can be expressed in case the operator can be put in the normal order as

$$\langle \hat{A}(\hat{a}^\dagger, \hat{a}) \rangle = \sum_{j,k} \sum_{l,m} C_{jk} C_{lm} Tr\left(\hat{a}^{\dagger j} \hat{a}^k \hat{a}^{\dagger l} \hat{a}^m\right).$$

One can see that the introduction of the completeness relation for a coherent state leads to

$$\langle \hat{A}(\hat{a}^\dagger, \hat{a}) \rangle = \int \frac{d^2\alpha}{\pi} \sum_{l,m} C_{lm} \alpha^{*l} \left(\alpha + \frac{\partial}{\partial \alpha^*}\right)^m A_n(\alpha^*, \alpha), \tag{2.150}$$

where $A_n(\alpha^*, \alpha)$ is the c-number variable pertinent to the normal ordering of the operator (see Appendix 1 of Chap. 3). Then employing the definition of the Q-function indicates

$$\langle \hat{A}(\hat{a}^\dagger, \hat{a}) \rangle = \int Q\left(\alpha^*, \alpha + \frac{\partial}{\partial \alpha^*}, t\right) A_n(\alpha^*, \alpha) d^2\alpha. \tag{2.151}$$

Note that with the aid of the Q-functions provided earlier, the mean photon number for coherent light, number state and squeezed light turn out to be

$$\bar{n} = \alpha^*\alpha, \qquad \bar{n} = n, \qquad \bar{n} = \sinh^2 r. \tag{2.152}$$

2.3.3.5 Quantification of the Degree of Squeezing

Quantification of the degree of squeezing is one of the ways by which the strength of the quantum features of the radiation can be theoretically revealed or predicted (see Chap. 7). Quantification of the degree of squeezing in practical framework encompasses detailed calculation of moments such as correlations and mean photon number. To do so, one can begin with the fact that the degree of squeezing of a single-mode radiation can be quantified with the help of two Hermitian operators (quadrature operators) that can be denoted and defined as $\hat{a}_+ = \hat{a} + \hat{a}^\dagger$ and $\hat{a}_- = i(\hat{a}^\dagger - \hat{a})$. One may note that the quadrature variances associated with the anti-normal ordering can be put in the form

$$\Delta \hat{a}_\pm^2 = 2\langle \hat{a}\hat{a}^\dagger \rangle \pm \left[\langle \hat{a}^2 \rangle + \langle \hat{a}^{\dagger 2} \rangle - \langle \hat{a}^\dagger \rangle^2 - \langle \hat{a} \rangle^2 \right] - 2\langle \hat{a} \rangle \langle \hat{a}^\dagger \rangle - 1. \tag{2.153}$$

As illustration, the quadrature variances for a chaotic light can be obtained by using[9]

$$Q(\alpha^*, \alpha) = \frac{1}{\pi(1 + \bar{n})} \exp \left[-\frac{\alpha^* \alpha}{1 + \bar{n}} \right], \tag{2.154}$$

as a result

$$\langle \hat{a}\hat{a}^\dagger \rangle = 1 + \bar{n}, \qquad \langle \hat{a}^2 \rangle = 0, \qquad \langle \hat{a} \rangle = 0. \tag{2.155}$$

The quadrature fluctuations of a chaotic light hence can be captured by

$$\Delta a_\pm^2 = 1 + 2\bar{n}. \tag{2.156}$$

Since the mean of the photon number cannot be negative, $\Delta a_\pm^2 \geq 1$. This outcome entails that a chaotic light does not exhibit squeezing feature.

The same calculation for the light that can be described by Eq. (2.91) yields

$$\Delta \hat{a}_\pm^2 = e^{\mp 2r}. \tag{2.157}$$

Since r is taken to be a positive parameter, it is straightforward to realize that

$$\Delta a_+^2 \leq 1, \qquad \Delta a_-^2 \geq 1. \tag{2.158}$$

This result indicates that the radiation generated by a degenerate parametric oscillator exhibits squeezing feature in the plus quadrature. Owing to the observation that the degree of squeezing would be better when $\Delta a_+^2 \to 0$, one can observe that

---

[9]The Q-function for a chaotic light can be obtained either by deriving following the outlined procedure or setting $r = 0$ in Eq. (2.94).

the degree of squeezing for this system increases exponentially with the squeeze parameter.

2.4 Gaussian Continuous Variable

Except for the number state, the number of photons involved in the discussion of the properties of radiation is presumed to be unknown. One may thus need to designate an ordinary light by a variable whose relevant degrees of freedom correspond to an operator having a continuous spectrum. This proposal entails that the eigenstates of the boson operator of the electromagnetic radiation can span an infinite dimensional Hilbert space, which can also be modeled as a collection of noninteracting harmonic oscillators with different frequencies in which each oscillator is alluded to as a mode of the system; and hence, can be taken as a system of continuous variable.

In the regime of multi-dimension, N canonical bosonic modes can be described by Hamiltonian $\mathcal{H} = \bigotimes_{k=1}^{N} \mathcal{H}_k$ in Hilbert space that can be constructed from the tensor product of an infinite dimensional Fock space each having a single mode. The quantized electromagnetic field can be designated at some point as a system of arbitrary number of harmonic oscillators of different frequencies:

$$\hat{H} = \sum_{k=1}^{N} \omega_k \left(\hat{a}_k^\dagger \hat{a}_k + \frac{1}{2} \right), \tag{2.159}$$

where $\hat{a}_k$ is the annihilation operator of the photon in mode k, and satisfies the boson commutation relations: $[\hat{a}_k^\dagger, \hat{a}_{k'}] = \hat{I}\delta_{kk'}$ and $[\hat{a}_k, \hat{a}_{k'}] = 0$ (see also Sect. 2.1).

The notion of continuous variable in quantum optics hence heralds the description of the quantized electromagnetic field variables such as amplitude and phase that can be denoted by the position and momentum quadratures [13]. The eigenstates of the position and momentum operators constructed from the boson operators however may not be readily conceived, and so one needs to craft a suitable approximations of these states. In view of the presumption that the assembly of the device useable in application should take such consideration into account while optimizing the deviation of the actual states from the ideal position and momentum eigenstates, it might be advisable utilizing the position and momentum quadratures of the electromagnetic field modes. One may note that the quadrature operators obey the same commutation relations as the canonical positions and momenta, but should not be taken as it is synonymous with the physical position and momentum of field excitations.

The quadrature operators for each mode thus can be expressed in practice as

$$\hat{q}_k = \frac{1}{\sqrt{2}}(\hat{a}_k^\dagger + \hat{a}_k), \qquad \hat{p}_k = \frac{i}{\sqrt{2}}(\hat{a}_k^\dagger - \hat{a}_k), \tag{2.160}$$

and satisfy the eigenvalue equations $\hat{q} = q|q\rangle$ and $\hat{p} = p|p\rangle$ with continuous eigenvalues q and p (see Sect. 2.2). Besides, the two bases that correspond to the quadratures can be connected by Fourier transform which can be constructed for instance form squeezed coherent states as

$$|q\rangle = \frac{1}{2\sqrt{2\pi}} \int e^{-\frac{ipq}{2}} |p\rangle \, dp, \qquad |p\rangle = \frac{1}{2\sqrt{2\pi}} \int e^{\frac{ipq}{2}} |q\rangle \, dq. \qquad (2.161)$$

The phase space for quadrature operators can thus be defined in the same way as classical phase space for position and momentum, that is, the probability distributions in the classical phase space becomes the quasi-statistical distributions over the quadrature phase space.

Once the quadrature operators are defined, one may wish to look into some vital transformations such as translation, rotation and squeezing. To begin with, the translation operation can be implemented by applying the associated displacement operator as

$$\hat{D}^{\dagger}(\alpha)\hat{q}\hat{D}(\alpha) = \hat{q} + \sqrt{2}\mathrm{Re}(\alpha), \qquad \hat{D}^{\dagger}(\alpha)\hat{p}\hat{D}(\alpha) = \hat{p} + \sqrt{2}\mathrm{Im}(\alpha), \qquad (2.162)$$

where $\hat{D}_k(\alpha) = e^{\alpha\hat{a}_k^{\dagger} - \alpha^*\hat{a}_k}$ is displacement operator for each mode and

$$\alpha = \frac{1}{\sqrt{2}}(q + ip) \qquad (2.163)$$

is complex amplitude corresponding to the operator.

Equation (2.162) indicates that the expectation values of $\hat{q}$ and $\hat{p}$ can be displaced by a simple translation as much as the amount that depends on the argument of the displacement operator, and pertinent variances of $\hat{q}$ and $\hat{p}$ would not be affected by such translation. One can hence make use of the displacement of the quadrature,

$$\hat{D}(\alpha)|q\rangle = e^{i2\pi\mathrm{Im}(\alpha)}|q + \mathrm{Re}(\alpha)\rangle, \qquad (2.164)$$

when required.

The other possible operation is rotation or phase shift by arbitrary angle θ that can be expressed as

$$\hat{\rho}_{\theta} = \hat{R}(\theta)\hat{\rho}\hat{R}^{\dagger}(\theta), \qquad (2.165)$$

where $\hat{R}(\theta) = e^{-i\theta\hat{n}}$ is the rotation operator. One may see that the number operator is the generator of rotation, and so the rotation-transformation of the quadrature operators can be written as

$$\hat{R}^{\dagger}(\theta)\hat{q}\hat{R}(\theta) = \hat{q}\cos\theta + \hat{p}\sin\theta, \qquad \hat{R}^{\dagger}(\theta)\hat{p}\hat{R}(\theta) = \hat{p}\cos\theta - \hat{q}\sin\theta \qquad (2.166)$$

(see the Appendix).

It is also possible to transform $\hat{q}$ and $\hat{p}$ with the aid of squeeze operator;

$$\hat{S}(r)\hat{q}\hat{S}^\dagger(r) = \hat{q}e^r, \qquad \hat{S}(r)\hat{p}\hat{S}^\dagger(r) = \hat{p}e^{-r}. \tag{2.167}$$

Equation (2.167) may indicate that, as long as $r \neq 0$, one of the quadrature operators would be reduced below the vacuum limit at the expense of the other as expected. It is also possible to express[10] that

$$\hat{S}(r)|q\rangle = e^{-\frac{r}{2}}|qe^{-r}\rangle. \tag{2.168}$$

To put the mathematical framework of the notion of continuous variable on a strong footing, one may need to adhere to the fact that for N-mode radiation, the quadrature operators can be grouped as

$$\hat{R} = \left(\hat{q}_1, \hat{p}_1, \ldots, \hat{q}_N, \hat{p}_N\right)^T, \tag{2.169}$$

which enables writing the commutation relations between the quadrature operators in a more compact form as

$$\left[\hat{R}_k, \hat{R}_l\right] = i\Omega_{kl}, \tag{2.170}$$

where Ω_{kl} is a constant that can be expressed in a symplectic form as

$$\Omega = \bigoplus_{k=1}^{N} \begin{pmatrix} 0 & 1 \\ -1 & 0 \end{pmatrix}. \tag{2.171}$$

It might be worthy emphasizing that the states of a continuous variable system can be taken as a set of a positive trace operators that can be described by the density operator on the Hilbert space. Even then, a complete description of a quantum state of an infinite dimensional system can be expressed in terms of one of its s-ordered characteristic functions

$$\Phi_s(z) = Tr\left[\hat{\rho}\hat{D}_k(z)\right]e^{\frac{1}{2}sz^*z}, \tag{2.172}$$

---

[10]The operations of translation, rotation and squeezing in the phase space correspond to the state evolution generated by linear and quadratic Hamiltonian of a single-mode radiation that can be realized in a laboratory setting with the help of linear optical elements and homodyne detection (see Chaps. 9 and 11).

where z belongs to the real $2N$-dimensional phase space and

$$s = \begin{cases} 0, & \text{for symmetric ordering;} \\ -1, & \text{for anti-normal ordering;} \\ 1, & \text{for normal ordering} \end{cases} \tag{2.173}$$

(relate with the discussion in Sect. 2.3.2). One can see from such a representation that, in the phase space picture, the tensor product can be replaced by a direct sum, and then emerges the conception of N-mode phase space.

It might also be helpful recalling that the characteristic function contains all the necessary information to reconstruct the density matrix including the photon number statistics and the moments. The nth order moment of the quadrature for instance can be obtained by applying

$$\langle \hat{q}^n \rangle = \left(i \frac{\partial}{\partial z} \right)^n \Phi_s(z)|_{z=0}, \tag{2.174}$$

where the phase is incorporated in z-parameter (relate with Eq. (2.65)). The family of the characteristic functions can also be related by using a complex Fourier transform to the pertinent quasi-statistical distributions that constitute another set of a complete description of the quantum states (see also Eq. (2.74));

$$P(\alpha; s) = \frac{1}{\pi^2} \int \Phi_s(z) e^{iz^* \Omega \alpha} d^{2N} z. \tag{2.175}$$

Description of electromagnetic radiation as a continuous variable is based on the fact that these variables are Gaussian by nature; with the understanding that Gaussian states are a set of states with a Gaussian characteristic or quasi-statistical distribution function on the multi-mode phase space. So the discussion on Gaussian optical states such as coherent, squeezed and thermal is expected to be an important area of quantum optics and quantum information processing in the context of continuous variable (see Chaps. 9, 11 and 12).

In line with this, existing experience shows that a Gaussian state can be completely described by the first and second-order moments of the quadrature operators depicted in a covariance matrix

$$\sigma_{lj} = \frac{1}{2} [\langle \hat{R}_l \hat{R}_j \rangle + \langle \hat{R}_j \hat{R}_l \rangle] - \langle \hat{R}_l \rangle \langle \hat{R}_j \rangle, \tag{2.176}$$

where $\hat{R}_{l,j}$ stand for the real phase space operators. Notice that the off-diagonal elements of this matrix represents intermodal correlations and the diagonal elements inter-correlations associated with intensity or mean photon number.

The Gaussian nature of the pertinent distribution function implies that the first-order moments can be arbitrarily adjusted by local unitary operations: displacement in a phase space or locally re-centering the reduced Gaussian of each mode where the reduced state that can be obtained from the Gaussian state by a partial tracing over the subset of the modes would still be Gaussian—which leaves the relevant information including entropy and entanglement unchanged.[11] So the Wigner function of the Gaussian state, which should be a Gaussian function, can be expressed in terms of the phase space quadrature variables as

$$W(R) = \frac{1}{\sqrt{\det\sigma_{lj}}} e^{-\frac{1}{2}R\sigma_{lj}^{-1}R^T}. \tag{2.177}$$

Despite the infinite dimension of the associated Hilbert space, a complete description up to a local unitary operation of arbitrary Gaussian state can be obtained from $2N \times 2N$-covariance matrix (2.176).

Pertaining to the proposal that a real covariance matrix contains a complete locally invariant information of the Gaussian state, one should expect some constraints to be imposed on the pertinent covariance matrix such as positive semi-definiteness of the associated density matrix. Such consideration together with the canonical commutation relations lead to the bona fide condition

$$\sigma_{lj} + i\Omega_{lj} \geq 0, \tag{2.178}$$

which is the necessary and sufficient constraint a given matrix has to fulfill to be a covariant matrix that represents a Gaussian state.

A Gaussian variable is also designated as symplectic if there is a linear transformation S that preserves the symplectic form of Ω, that is,

$$S^T \Omega S = \Omega. \tag{2.179}$$

So it might be good noting that ideal beam splitter, phase shifter and squeezer can be described by some sort of symplectic transformation. For example, a two-mode squeezed state,

$$|r, r\rangle = \sqrt{1 - \tanh^2(r)} \sum_{n=0}^{\infty} (-1)^n \tanh^n(r) |n\rangle_a \otimes |n\rangle_b, \tag{2.180}$$

can be described by symplectic covariance matrix of the form

$$\sigma_{sq} = \begin{pmatrix} \cosh(2r) & 0 & \sinh(2r) & 0 \\ 0 & \cosh(2r) & 0 & -\sinh(2r) \\ \sinh(2r) & 0 & \cosh(2r) & 0 \\ 0 & -\sinh(2r) & 0 & \cosh(2r) \end{pmatrix}. \tag{2.181}$$

---

[11]Detailed discussion on various kinds of photon measurement including homodyne is provided in Sect. 4.5 (see also [14]). Some of the issues discussed here would be valuable when one seeks to explore the strength of the quantum feature of the radiation and its application (see Chaps. 7 and 11, and also [15]).

Despite the vastness of the dimension, a Gaussian state of two bosonic modes can be characterized by a simple analytical formulae, and represents the simplest possible states for exploring continuous variable quantum features such as entanglement (see Chap. 7). For a two-mode Gaussian state, the covariance matrix can be constructed from Eq. (2.176) and rewritten in a block form as

$$\sigma = \begin{pmatrix} A & C^T \\ C & B \end{pmatrix}, \tag{2.182}$$

where A, B and C are 2×2-matrices. One may note that a diagonal form of this matrix can be expressed as $\sigma^{\oplus} = v_- \mathbf{I} \oplus v_+ \mathbf{I}$, where the symplectic eigenvalues or spectrum $\{v_-, v_+\}$ is found to be [16]

$$v_{\pm} = \sqrt{\frac{\Delta \pm \sqrt{\Delta^2 - 4 \det \sigma}}{2}}, \tag{2.183}$$

with $\Delta = \det A + \det B - 2 \det C$. The uncertainty relation in this case would look like

$$\det \sigma \geq 1. \tag{2.184}$$

The covariance matrix in the context of information processing is required to be positive definite;[12] which entails that its symplectic eigenvalues should satisfy

$$v_k \geq 1. \tag{2.185}$$

Consistent with such a proposal, von Neumann entropy $S(\hat{\rho})$ can be expressed as

$$S(\hat{\rho}) = \sum_{k=1}^{N} g(v_k), \tag{2.186}$$

where $g(x) = \frac{x+1}{2} \log\left(\frac{x+1}{2}\right) - \frac{x-1}{2} \log\left(\frac{x-1}{2}\right)$.

The quadrature operators can be measured in a laboratory setting with the help of homodyning that can be implemented by combining the target mode with a local oscillator on a balanced beam splitter, and then measuring the intensity of the out going modes using two photon detectors. The subtraction of the signal of both photon detectors gives rise to a signal proportional to $\hat{q}$ while $\hat{p}$ quadrature is measured by applying $\pi/2$ phase shift on the local oscillator. It should be possible to construct the covariance matrix of the two-mode Gaussian state by making use of

---

[12]This observation will be used to explain quantum features of light in view of the information content such as quantum discord and entanglement steering (see Sects. 7.4 and 7.5).

a single homodyne operator. In the same way, a heterodyne detection can be carried out by projection onto a coherent state via combining the measured bosonic mode with an ancillary vacuum mode on a balanced beam splitter, and then carrying out a homodyne measurement on the quadratures of the output mode (see Chap. 4). It may then be reasonable to propose that the relative simplicity and the possibility of achieving high efficiency while measuring and manipulating continuous quadratures are the main reasons for schemes based on continuous variable to be more attractive in application than those based on discrete variables.

Appendix

Beam Splitter Transformation

Transformation on pairs of modes j and k can be expressed with the aid of the Hamiltonian

$$\hat{B}(\theta) = \exp\left[\theta(\hat{a}_j^\dagger \hat{a}_k - \hat{a}_j \hat{a}_k^\dagger)\right] \tag{2.187}$$

for a phase free beam splitter as

$$\hat{a}_j \to \hat{a}_j \cos\theta + \hat{a}_k \sin\theta, \qquad \hat{a}_k \to \hat{a}_j \sin\theta - \hat{a}_k \cos\theta, \tag{2.188}$$

where θ signifies the transmissivity of the beam splitter. For instance, for a 50:50 symmetric beam splitter ($\theta = \pi/4$; see also Sect. 4.3.2), these expressions reduce to

$$\hat{a}_j \to \frac{1}{\sqrt{2}}(\hat{a}_j + \hat{a}_k), \qquad \hat{a}_k \to \frac{1}{\sqrt{2}}(\hat{a}_j - \hat{a}_k). \tag{2.189}$$

In order to detect the quadrature as in the homodyne detection (see Sect. 4.5), the mode must be combined with an intense local oscillator at a 50:50 beam splitter. In this case, the intended oscillator would be assumed to be in a coherent state with a large photon number and represented by a classical complex amplitude rather than the pertinent annihilation operator. With this description, the two output modes of the beam splitter may be approximated as

$$\hat{a}_1^{out} = \frac{1}{\sqrt{2}}[\alpha + \hat{a}], \qquad \hat{a}_2^{out} = \frac{1}{\sqrt{2}}[\alpha - \hat{a}]. \tag{2.190}$$

References

1. A. Einstein, Ann. D. Phys. **17**, 132 (1905) (English translation: A.B. Arons, M.B. Peppard, Am. J. Phys. **33**, 367 (1965)); G.N. Lewis, Nature **118**, 874 (1926)
2. R. Loudon, *The Quantum Theory of Light*, 3rd edn. (Oxford University Press, New York, 2000)

3. C.R. Stroud, E.T. Jaynes, Phys. Rev. A **1**, 106 (1970); R.K. Nesbet, Phys. Rev. A **4**, 259 (1971); J. Barwick, Phys. Rev. A **17**, 1912 (1978)
4. R.J. Glauber, Phys. Rev. **131**, 2766 (1963)
5. W. Vogel, D.G. Welsch, *Quantum Optics*, 3rd revised and extended edition (Wiley-VCH Verlag GmbH and Co. KGaA, Weinheim, 2006)
6. D.F. Walls, G.J. Milburn, *Quantum Optics*, 2nd edn. (Springer, Berlin, 2008)
7. H.P. Yuen, Phys. Rev. A **13**, 2226 (1976); C.M. Cave, Phys. Rev. D **23**, 1693 (1981); C.T. Lee, Phys. Rev. A **42**, 4193 (1990); G.S. Agarwal, K. Tara, Phys. Rev. A **43**, 492 (1991)
8. L.D. Ming, Int. J. Theor. Phys. **54**, 2239 (2015)
9. H. Weyl, Z. Phys. **46**, 1 (1927); E.P. Winger, Phys. Rev. **40**, 749 (1932); E.C.G. Sudarshan, Phys. Rev. Lett. **10**, 277 (1963); R.J. Glauber, Phys. Rev. **131**, 2766 (1963); H.W. Lee, Phys. Rep. **259**, 147 (1995); W.P. Scleich, *Quantum Optics in Phase Space* (Wiley-VCH Verlag Berlin GmbH, Berlin, 2001)
10. R. Loudon, *The Quantum Theory of Light*, 3rd edn. (Oxford University Press, New York, 2000)
11. K.E. Cahill, R.J. Glauber, Phys. Rev. **177**, 1882 (1969)
12. S. Tesfa, A nondegenerate three-level cascade laser coupled to a two-mode squeezed vacuum reservoir. Ph.D. thesis, Addis Ababa University, 2008
13. G. Adesso, F. Illuminati, J. Phys. A **40**, 7821 (2007); P. Kok, B.W. Lovett, *Introduction to Optical Quantum Information Processing* (Cambridge University Press, New York, 2010); C. Weedbrook et al., Rev. Mod. Phys. **84**, 621 (2012); G. Adesso, S. Ragy, A.R. Lee, Open Syst. Info. Dynam. **21**, 140001 (2014)
14. V. D'Auria et al., Phys. Rev. Lett. **102**, 020502 (2009)
15. R. Simon, N. Mukunda, B. Dutta, Phys. Rev. A **49**, 1567 (1994); S.L. Braunstein, P. van Loock, Rev. Mod. Phys. **77**, 513 (2005)
16. A. Serafini, Phys. Rev. Lett. **96**, 110402 (2006); S. Pirandola, A. Serafini, S. Lloyd, Phys. Rev. A **79**, 052327 (2009)

Mathematical Descriptions of Quantized Light 3

The discussion of quantized light in terms of number state, coherent state and squeezed state may not be adequate to analyze the statistical properties and quantum features of the radiation that can be generated from sources such as correlated emission laser and parametric oscillator. In light of this, one may be compelled to extend the conversation on the quantized light beyond mere associated quantum states and probability distributions, and to formulate the mathematical tools that can be utilized to designate and solve the emerging problems. This chapter is thus meant to expand the scope of Chap. 2 so that one can figure out the nature of light in terms of statistical interpretation specially when the radiation is presumed to be trapped in a reflective geometrical structure, and the dynamics is sought for. To grasp the dynamics in quantum optics, one may begin with the presumption that quantum optical systems do evolve in time in the same way as other systems.

The program in this respect is about being able to explain how the properties of the system evolve in time as a result of the involved interaction.[1] This can be done in Schrödinger picture by assuming the state vector to evolve in time while the dynamical operators are treated as constant of time;

$$\frac{d}{dt}|\psi(t)\rangle = -i\hat{H}|\psi(t)\rangle, \qquad (3.1)$$

where $|\psi(t)\rangle$ is the state vector, $\hat{H}$ is the Hamiltonian that describes the system and $\hbar$ is set to 1 throughout. But in Heisenberg picture, the operators are assumed to

---

[1]The overall approach is limited to introducing the basics required to understand and follow the topics to be presented rather than going into details of the covered issues (for detail; see for instance [1]).

© The Author(s), under exclusive license to Springer Nature Switzerland AG 2020
S. Tesfa, *Quantum Optical Processes*, Lecture Notes in Physics 976,
https://doi.org/10.1007/978-3-030-62348-7_3

evolve in time while the state vector is taken to be constant of time;

$$\frac{d}{dt}\hat{O}(t) = -i[\hat{O}(t), \hat{H}]$$

(3.2)

in which $\hat{O}(t)$ is the operator that depends on time.[2]

So as to understand the fascinating field of quantum optics, and also be able to exploit it in the emerging technologies as summarized in Chap. 12, it might be crucial having the knowhow of the underlying mathematical landscape since the fruitfulness of the study on the theoretical aspects of quantum optics is heavily reliant on the complexity of the pertinent mathematical rigor one chooses to adopt (see Sect. 3.2 and also [2]). For a system exposed to damping mechanism, the approaches such as quantum Langevin equation and/or master equation can be applied. With this in mind, different ways of deriving the master equation would be outlined; after that, the pertinent quantum Langevin equation would be derived; and then, the input-output mechanism including the causality effect would be addressed (see Sect. 3.1 and also [3]). Various formulations commonly applicable in obtaining the sought for correlations such as propagator formulation, Feynman path integral, characteristic function and stochastic differential equations are also highlighted. Special techniques such as the ways of solving the Fokker-Planck equation and linearization procedure are included.

3.1   Cavity Dynamics

The quantum system one seeks to study is often enclosed in a region between walls that have known optical properties: the structure dubbed as cavity where the system enclosed in it interacts with the external environment due to the radiation exchange in which the ensuing change of the properties of the quantum system is referred to as cavity dynamics. Even though addressing the involved interaction process at atomic level is challenging, the cavity scheme is expected to offer a considerable simplification in some respect. For example, the light enclosed in the cavity has the highest tendency to be continuously reflected from the walls,[3] whereby the system has a better chance of interacting with the radiation in equal proportion. This characteristic of the reflection is frequently exploited in establishing mathematically

---

[2]The final results should be independent of the approach one opts to follow: adherence to a given quantum approach is mainly a preference based on the depth of the required rigor or the acclimatization to the accompanying procedure.

[3]The bath or reservoir in a conceivable experimental setting can impact the system only over small angle. This makes equally exposing the quantum system with the modes of the reservoir in the same way from all directions an intractable task since one may need to place the detector all over 4π solid angle in a free space scenario which is not possible from practical point of view [4].

Fig. 3.1 Schematic representation of the interaction of the atoms of the system in the cavity with the intra-cavity radiation mode denoted by operator $\hat{a}$, where γ is the atomic damping rate, κ is the cavity damping constant and $\hat{a}_k$'s are the radiation modes of the reservoir that can have a chance to enter the cavity

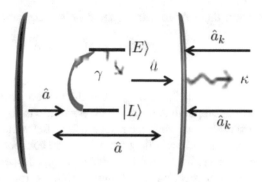

manageable approach and generating experimentally measurable data. It is also possible to close one of the mirrors, and then tactically direct the output radiation in a specified direction so that a fewer number of detectors are required during photon measurement (see Fig. 3.1).

The interaction that involves the cavity can be characterized by certain parameters: the atomic dipole coupling constant that designates the energy exchange between the atoms of the system in the cavity and the cavity radiation; the cavity damping constant that measures the rate at which the radiation enters or leaves the cavity, and can also be recognized as an indicator of the degree by which the cavity radiation is influenced by the environment; and the atomic decay rate that portrays the rate at which the atoms spontaneously emit to modes other than the privileged cavity modes, and measures the degree of the noise associated with the cavity radiation. One may note that the cavity can also interact with the environment via the exchange of fluctuations that requires the cavity to be at equilibrium with the surrounding enclosure before the commencement of the interaction.

The reservoir on the other hand can be visualized as a large number of fictitious harmonic oscillators where the pertinent modes are taken to be very closely spaced in frequency and virtually constitute a continuum from which only modes with frequency closer to the cavity frequency are expected to significantly interact with the cavity. Since there should not be a major difference between the reservoir modes and cavity radiation except where they exist, the reservoir can also be epitomized by Hamiltonian of the form

$$\hat{R} = \sum_k \omega_k \hat{a}_k^\dagger \hat{a}_k, \tag{3.3}$$

where the operators are taken to satisfy the commutation relation:

$$[\hat{a}_j, \hat{a}_k] = [\hat{a}_j^\dagger, \hat{a}_k^\dagger] = 0, \qquad [\hat{a}_j, \hat{a}_k^\dagger] = \hat{I}\delta_{j,k}. \tag{3.4}$$

The radiation outside the cavity is designated as the reservoir mode and presumed to be known. So some kind of fluctuation of the cavity mode, though very small when compared to the reservoir fluctuations, is expected due to its coupling with the reservoir even when the reservoir is in the vacuum state. In light of the exchange of radiation and noise fluctuation, one can conceive a meaningful interaction that has a potential to lead to modification in optical properties of the system to emerge. Owing to such perception, developing various formulations by which such a modification can be studied specially different ways of deriving the master equation in the context of Born-Markov approximation would be outlined. It might be required sometime to strictly assume that the coupling between the quantum system and the cavity radiation is quite small, which amounts to invoking the perturbation technique. Besides, on account of the relative size of the reservoir, the density operator formalism is adopted. The standard procedure by which the master equation can be derived once the relevant Hamiltonian is known along with its c-number equivalent would be provided.

3.1.1 Master Equation

The dynamics of a quantum system can be portrayed by time evolution of the reduced density operator that often dubbed as master equation [5–7]. While studying the ensuing interaction the system and the reservoir are denoted by S and R, and the combined system by $\hat{\rho}_{SR}$. The reduced density operator for the system can be obtained by taking the trace over the reservoir variables: $\hat{\rho} = Tr_R(\hat{\rho}_{SR})$ that amounts to regarding the effect of the reservoir via its overall mean contribution rather than taking the whole effect since the reservoir is too big to be accounted for. The tracing operation is then applied with the assumption that one is interested in the quantum dynamics of the system alone.

To begin with, the system coupled to the reservoir can be designated in the interaction picture by the Hamiltonian of the form

$$\hat{H} = \hat{H}_S + \hat{H}_{SR}, \tag{3.5}$$

where $\hat{H}_S$ is the system Hamiltonian and $\hat{H}_{SR}$ is the Hamiltonian that denotes the interaction between the system and reservoir. Then in view of Liouville or Neumann relation,

$$\frac{\partial \hat{\rho}}{\partial t} = -i[\hat{H}, \hat{\rho}(t)], \tag{3.6}$$

the combined density operator for the system and reservoir evolves according to

$$\frac{\partial \hat{\rho}_{SR}}{\partial t} = i\left[\hat{H}_S + \hat{H}_{SR}, \hat{\rho}_{SR}\right] \tag{3.7}$$

which can be integrated as

$$\hat{\rho}_{SR}(t) = \hat{\rho}_{SR}(t_0) - i \int_{t_0}^{t} \left[\hat{H}_S(t') + \hat{H}_{SR}(t'), \ \hat{\rho}_{SR}(t')\right] dt', \tag{3.8}$$

where t_0 is the initial time at which the interaction is taken to be commenced. Once the involved Hamiltonian and the initial state of the system are known, the time evolution of the reduced density operator is believed to be straightforwardly obtained. However, since the reduced density operator is also part of the integrand, the iterative integration turns out to be complicated.

It then appears convenient and appealing resorting to some approximation schemes such as when interaction Hamiltonian is zero, that is, when the reservoir and system do not interact: when the system and reservoir are separable. The combined density operator in this case would be a direct product of the two density operators: $\hat{\rho}_{SR}(t) = \hat{R}(t_0) \otimes \hat{\rho}(t)$ in which the reservoir is assumed to be initially at equilibrium, and so would not be appreciably disturbed from this state by the interaction, that is, the density operator for the reservoir modes remains almost constant of time.

For the case when the interaction Hamiltonian is small, which is often true, the solution of Eq. (3.7) can then be proposed as

$$\hat{\rho}_{SR}(t) = \hat{R}(t_0) \otimes \hat{\rho}(t) + \hat{\rho}_c(t), \tag{3.9}$$

where the additional term $\hat{\rho}_c(t)$ is presumed to be of the higher-order in the interaction Hamiltonian. Based on the facts that

$$Tr_R(\hat{R}(t_0)) = 1, \qquad Tr_R(\hat{\rho}_{SR}(t)) = \hat{\rho}, \tag{3.10}$$

one can see that $Tr_R(\hat{\rho}_c(t)) = 0$.

Application of the proposed solution of the combined density operator along with the operation of tracing over reservoir variables then lead to

$$\frac{d\hat{\rho}(t)}{dt} = -i Tr_R\left(\left[\hat{H}_S(t) + \hat{H}_{SR}(t), \ \hat{R}(t_0) \otimes \hat{\rho}(t_0)\right]\right)$$

$$- Tr_R\left(\int_{t_0}^{t} \left[\hat{H}_S(t) + \hat{H}_{SR}(t), \ \left[\hat{H}_S(t') + \hat{H}_{SR}(t'), \ \hat{R}(t_0) \otimes \hat{\rho}(t')\right]\right]\right) dt'. \tag{3.11}$$

One can see from Eq. (3.11) that the density operator for the system depends on a prior history. But owing to the fact that the damping or decaying process destroys the previous history—assuming the process to be Markovian in which the system does not have memory mechanism—one can replace $\hat{\rho}(t')$ by $\hat{\rho}(t)$ in Eq. (3.11);

$$\frac{d\hat{\rho}(t)}{dt} = -iTr_R\left([\hat{H}_S(t) + \hat{H}_{SR}(t), \ \hat{R}(t_0) \otimes \hat{\rho}(t_0)]\right)$$

$$- Tr_R\left(\int_{t_0}^t [\hat{H}_S(t) + \hat{H}_{SR}(t), \ [\hat{H}_S(t') + \hat{H}_{SR}(t'), \ \hat{R}(t') \otimes \hat{\rho}(t)]]\right)dt'. \tag{3.12}$$

In addition, since the interaction can hardly affect the state of the reservoir, it is possible to assume that its density operator remains constant of time: $\hat{R}(t) = \hat{R}$. Assuming also the strength of the interaction to be quite small in magnitude, it is possible to decouple the combined density operator as $\hat{\rho}_{SR}(t) = \hat{\rho}(t)\hat{R}$: Born approximation. After that, upon noting that

$$Tr_R(\hat{\rho}(t)\hat{R}) = \hat{\rho}(t), \qquad Tr_R(\hat{H}_S\hat{R}) = \hat{H}_S(t), \tag{3.13}$$

$$Tr_R(\hat{H}_{SR}(t)\hat{R}) = \langle\hat{H}_{SR}(t)\rangle_R, \tag{3.14}$$

Eq. (3.12) can be rewritten in case there is interaction between the system and reservoir, that is, $\langle\hat{H}_{SR}(t')\hat{H}_{SR}(t')\rangle_R \neq \langle\hat{H}_{SR}(t')\rangle\langle\hat{H}_{SR}(t')\rangle_R$ as

$$\frac{d\hat{\rho}(t)}{dt} = -i[\hat{H}_S(t), \ \hat{\rho}(t)] - i[\langle\hat{H}_{SR}(t)\rangle_R, \ \hat{\rho}(t)]$$

$$- \int_{t_0}^t Tr_R[\hat{H}_{SR}(t), \ [\hat{H}_{SR}(t'), \ \hat{\rho}(t')\hat{R}]]dt', \tag{3.15}$$

which can be christened as master equation.

To attest to the proposal that the same master equation can also be expressed in various ways, one can begin with the fact that these Hamiltonian's commute: $[\hat{H}_S, \hat{H}_R] = 0$. One can assert in view of this observation that if the system and reservoir are decoupled before the interaction, they will remain decoupled at equal time and the same picture during the interaction, that is, in case the system and reservoir operators commute at $t = t_0$, they should also commute whenever they are in the same picture and time.

Even though the presented approach is often adopted in theoretical quantum optics, there are also some competing tactics specially when the interaction Hamiltonian is defined over certain time span. In this regard, it might be appealing to suggest that

$$\hat{\rho}_{SR}(t) = \hat{\rho}(t_0)\hat{R} - i\int_{t_0}^t [\hat{H}_I(t' - t_0), \ \hat{\rho}(t_0)\hat{R}]dt'$$

$$- \int_{t_0}^t dt' \int_{t_0}^{t'} dt''[\hat{H}_I(t' - t_0), \ [\hat{H}_I(t'' - t_0), \ \hat{\rho}(t)\hat{R}]] + \cdots, \tag{3.16}$$

where $\hat{H}_I$ is the interaction Hamiltonian. Limiting the consideration to only up to second-order perturbation in interaction Hamiltonian, one gets

$$\hat{\rho}(t) = \hat{\rho}(t_0) - i \int_{t_0}^{t} Tr_R[H_I(t' - t_0), \ \rho(t_0)\hat{R}]dt'$$

$$- \int_{t_0}^{t} dt' \int_{t_0}^{t'} dt'' Tr_R[\hat{H}_I(t' - t_0), \ [\hat{H}_I(t'' - t_0), \ \hat{\rho}(t)\hat{R}]], \qquad (3.17)$$

where the Markovian approximation is inferred.

Critical scrutiny may reveal that there is no essential difference between the implications of Eqs. (3.15) and (3.17). To get a more general master equation, the interaction Hamiltonian is assumed to have the form

$$\hat{H}_I = \sum_i \hat{Q}_i \hat{F}_i, \qquad (3.18)$$

where $\hat{Q}_i$ is a function of system operators while $\hat{F}_i$ is a function of reservoir operators. It may not be difficult to see that the same equation and operators can be expressed in the interaction picture as

$$\hat{H}_I(t - t_0) = \sum_i \hat{Q}_i(t - t_0) \hat{F}_i(t - t_0), \qquad (3.19)$$

where

$$\hat{Q}_i(t - t_0) = e^{i\hat{H}_S(t-t_0)} \hat{Q}_i e^{-i\hat{H}_S(t-t_0)}, \qquad (3.20)$$

$$\hat{F}_i(t - t_0) = e^{i\hat{H}_R(t-t_0)} \hat{F}_i e^{-i\hat{H}_R(t-t_0)}. \qquad (3.21)$$

Based on the proposal that the system and reservoir operators commute whenever they are in the same picture and time, one can then rewrite Eq. (3.17) as

$$\hat{\rho}(t) = \hat{\rho}(t_0) - i \sum_i \int_{t_0}^{t} [\hat{Q}'_i(t'), \ \hat{\rho}(t_0)]\langle F'\hat{(t')}\rangle_R dt'$$

$$- \sum_{i,j} \int_{t_0}^{t} dt' \int_{t_0}^{t'} dt'' \{[\hat{Q}'_i(t')\hat{Q}''_j(t'')\hat{\rho}(t_0)$$

$$- \hat{Q}''_j(t'')\hat{\rho}(t_0)\hat{Q}'_i(t')]\langle \hat{F}'_i(t')\hat{F}''_j(t'')\rangle_R$$

$$- [\hat{Q}'_i(t')\hat{\rho}(t_0)\hat{Q}''_j(t'') - \hat{\rho}(t)\hat{Q}''_j(t'')\hat{Q}'_i(t')]\langle \hat{F}''_j(t'')\hat{F}'_i(t')\rangle_R\}.$$

$$(3.22)$$

After that, in view of the fact that the system operator $\hat{Q}_i$ can be expressed in the interaction picture as $\hat{Q}_i = e^{i\omega_i \tau} \hat{Q}_i^S$, Eq. (3.22) can be rewritten by applying the Markovian approximation scheme as

$$\frac{d\hat{\rho}(t)}{dt} = -i[\hat{H}_S, \ \hat{\rho}(t)] - \sum_i \delta(\omega_i - \omega_j)$$

$$\times \{[\hat{Q}_i \hat{Q}_j \hat{\rho}(t) - \hat{Q}_j \hat{\rho}(t) \hat{Q}_i] C_{ij}^+ - [\hat{Q}_i \hat{\rho}(t) \hat{Q}_j - \hat{\rho}(t) \hat{Q}_j \hat{Q}_i] C_{ji}^-\}, \tag{3.23}$$

where

$$C_{ij}^+ = \int_0^\infty e^{i\omega_i \tau} \langle \hat{F}_i(\tau) \hat{F}_j \rangle_R \, d\tau, \qquad C_{ji}^- = \int_0^\infty e^{i\omega_j \tau} \langle \hat{F}_j \hat{F}_i(\tau) \rangle_R \, d\tau. \tag{3.24}$$

The desired master equation when all operators are in the Schrödinger picture is obtained; the approach often dubbed as the standard method.

Equation (3.23) can also be modified when the system Hamiltonian contains other terms that depend on the reservoir variables as

$$\frac{d\hat{\rho}(t)}{dt} = -i[\hat{H}_S + \hat{H}_e, \ \hat{\rho}(t)] - \sum_i \delta(\omega_i - \omega_j)$$

$$\times \{[\hat{Q}_i \hat{Q}_j \hat{\rho}(t) - \hat{Q}_j \hat{\rho}(t) \hat{Q}_i] C_{ij}^+ - [\hat{Q}_i \hat{\rho}(t) \hat{Q}_j - \hat{\rho}(t) \hat{Q}_j \hat{Q}_i] C_{ji}^-\}. \tag{3.25}$$

Owing to the need for going from one picture to the other, Eq. (3.25) can also be put in the interaction picture as

$$\frac{d\hat{\rho}^I(t)}{dt} = -i[\hat{H}_e^I, \ \hat{\rho}^I(t)] - \sum_i \delta(\omega_i - \omega_j)$$

$$\times \{[\hat{Q}_i^I \hat{Q}_j^I \hat{\rho}^I(t) - \hat{Q}_j^I \hat{\rho}^I(t) \hat{Q}_i^I] C_{ij}^+ - [\hat{Q}_i^I \hat{\rho}^I(t) \hat{Q}_j^I - \hat{\rho}^I(t) \hat{Q}_j^I \hat{Q}_i^I] C_{ji}^-\}, \tag{3.26}$$

where

$$\hat{\rho}^I(t) = e^{-i\hat{H}_S t} \hat{\rho} e^{i\hat{H}_S t}, \qquad Q_i^I = e^{-i\hat{H}_S t} Q_i e^{i\hat{H}_S t}, \qquad \hat{H}_e^I = e^{-i\hat{H}_S t} \hat{H}_e e^{i\hat{H}_S t}. \tag{3.27}$$

For now, imagine the situation in which a single-mode squeezed reservoir is coupled to a single-port mirror[4] that contains the cavity radiation. The interaction of the cavity mode with the reservoir in this case can be expressed in the interaction picture by the Hamiltonian of the form

$$\hat{H}_{SR}(t) = i \sum_j \lambda_j \left(\hat{a}^\dagger \hat{a}_j e^{i(\omega_c - \omega_j)t} - \hat{a}\hat{a}_j^\dagger e^{-i(\omega_c - \omega_j)t} \right), \tag{3.28}$$

where $\hat{a}$ ($\hat{a}_j$) are annihilation operators for the cavity (reservoir) modes, λ_j's are real constants that account for the strength of the coupling, ω_j's are the frequencies related to the reservoir modes and ω_c is the atomic transition frequency given by

$$\omega_c = E_e - E_l \tag{3.29}$$

in which E_e and E_l represent the energy of the excited and lower energey states of the atom (see Sect. 5.1).

To obtain the master equation, the expectation values in Eq. (3.15) need to be determined;

$$\langle \hat{H}_{SR}(t) \rangle_R = i \sum_j \lambda_j \left[\hat{a}^\dagger \langle \hat{a}_j \rangle_R e^{i(\omega_c - \omega_j)t} - \hat{a} \langle \hat{a}_j^\dagger \rangle_R e^{-i(\omega_c - \omega_j)t} \right]. \tag{3.30}$$

To get the explicit form of the master equation, one needs to specify the type of the relevant reservoir beforehand. In this regard, a single-mode squeezed vacuum reservoir is chosen as the toy model. It can be asserted based on the discussion in Sect. 2.2.1 that $\langle \hat{a}_j \rangle_R = \langle \hat{a}_j^\dagger \rangle_R = 0$, and thus $\langle \hat{H}_{SR} \rangle_R = 0$. In light of this, Eq. (3.15) can be put in the form

$$\frac{d\hat{\rho}(t)}{dt} = -i \left[\hat{H}_S(t), \hat{\rho} \right]$$

$$- \int_{t_0}^t Tr_R(\hat{R}\hat{H}_{SR}(t')\hat{H}_{SR}(t)\hat{\rho}(t))dt' - \int_0^t \hat{\rho}(t)Tr_R(\hat{R}\hat{H}_{SR}(t')\hat{H}_{SR}(t))dt'$$

$$+ \int_{t_0}^t Tr_R(\hat{H}_{SR}(t')\hat{\rho}(t)\hat{R}\hat{H}_{SR}(t))dt' + \int_{t_0}^t Tr_R(\hat{H}_{SR}(t)\hat{\rho}(t)\hat{R}\hat{H}_{SR}(t'))dt'. \tag{3.31}$$

---

[4]A single-port mirror practically refers to the cavity formed from two mirrors in which the light can enter and leave through one of the sides while it may enter but cannot leave through the other. It is also possible to imagine a two-port mirror in which the light is allowed to enter and leave through both mirrors.

For Hamiltonian (3.28), it may not be difficult to verify that

$$Tr_R(\hat{R}\hat{H}_{SR}(t')\hat{H}_{SR}(t)) = I_1\hat{a}\hat{a}^\dagger + I_2\hat{a}^\dagger\hat{a} + I_3\hat{a}^2 + I_4\hat{a}^{\dagger 2}, \tag{3.32}$$

where

$$I_1 = \sum_{j,k} \lambda_k\lambda_j \langle\hat{a}_j^\dagger\hat{a}_k\rangle_R e^{i(\omega_c-\omega_k)t'-i(\omega_c-\omega_j)t}, \tag{3.33}$$

$$I_2 = \sum_{j,k} \lambda_k\lambda_j \langle\hat{a}_j\hat{a}_k^\dagger\rangle_R e^{i(\omega_c-\omega_j)t-i(\omega_c-\omega_k)t'}, \tag{3.34}$$

$$I_3 = I_4^* = -\sum_{j,k} \lambda_k\lambda_j \langle\hat{a}_j^\dagger\hat{a}_k^\dagger\rangle_R e^{-i(\omega_c-\omega_k)t'-i(\omega_c-\omega_j)t}, \tag{3.35}$$

in which

$$\langle\hat{a}_j^\dagger\hat{a}_k\rangle_R = N\delta_{j,k}, \qquad \langle\hat{a}_j\hat{a}_k^\dagger\rangle_R = (N+1)\delta_{j,k}, \tag{3.36}$$

$$\langle\hat{a}_j\hat{a}_k\rangle_R = \langle\hat{a}_j^\dagger\hat{a}_k^\dagger\rangle_R = M\delta_{j,2k_c-k}. \tag{3.37}$$

One may note that $N = \sinh^2 r$ is the mean photon number corresponding to squeezed vacuum reservoir modes and $M = \sinh r \cosh r = \sqrt{N(N+1)}$ is the measure of the correlation among the reservoir modes (see also Appendix 2).

In the same token, assuming the reservoir mode frequencies to be closely spaced, a summation over k can be converted into integration over ω:

$$\sum_k \lambda_k^2 e^{\pm i(\omega_c-\omega_k)(t-t')} = \int_0^\infty g(\omega)\lambda^2(\omega)e^{\pm i(\omega_c-\omega)(t-t')}d\omega, \tag{3.38}$$

where $g(\omega)$ is the spectral density of the reservoir modes that emulates the density of the number of modes when the frequency lies between ω and $\omega + d\omega$. Owing to the observation that the reservoir modes that have frequency very close to the cavity modes do significantly interact with the system, it is assumed that ω varies very little around ω_c, that means, we can replace $g(\omega)$ and $\lambda^2(\omega)$ by $g(\omega_c)$ and $\lambda^2(\omega_c)$, and extend the lower limit of the integration to $-\infty$ without substantially altering the result:

$$\sum_k \lambda_k^2 e^{\pm i(\omega_c-\omega_k)(t-t')} = g(\omega_c)\lambda^2(\omega_c)\int_{-\infty}^\infty e^{\pm i(\omega_c-\omega)(t-t')}d\omega, \tag{3.39}$$

from which follows

$$\sum_k \lambda_k^2 e^{\pm i(\omega_c - \omega_k)(t-t')} = \kappa \delta(t - t'),$$ (3.40)

where

$$\kappa = 2\pi g(\omega_c) \lambda^2(\omega_c)$$ (3.41)

is referred to as cavity damping constant.

With this information, the master equation that epitomizes the quantum system in the cavity coupled to a single-mode squeezed vacuum reservoir is found[5] to be

$$\frac{\partial \hat{\rho}}{\partial t} = -i[\hat{H}_S, \hat{\rho}] + \frac{\kappa}{2} N [2\hat{a}^\dagger \hat{\rho}\hat{a} - \hat{a}\hat{a}^\dagger \hat{\rho} - \hat{\rho}\hat{a}\hat{a}^\dagger]$$
$$+ \frac{\kappa}{2}(N+1)[2\hat{a}\hat{\rho}\hat{a}^\dagger - \hat{a}^\dagger \hat{a}\hat{\rho} - \hat{\rho}\hat{a}\hat{a}^\dagger]$$
$$- \frac{\kappa}{2} M [2\hat{a}\hat{\rho}\hat{a} - \hat{a}^2\hat{\rho} - \hat{\rho}\hat{a}^2 + 2\hat{a}^\dagger \hat{\rho}\hat{a}^\dagger - \hat{a}^{\dagger 2}\hat{\rho} - \hat{\rho}\hat{a}^{\dagger 2}].$$ (3.42)

The contribution of the reservoir is incorporated via N and M that can be modified based on the properties of the external environment. For example, the time evolution of the density operator for a quantum system coupled to a single-mode reservoir can be expressed by setting $M = 0$ and replacing N by $\bar{n}$ for thermal reservoir, and setting $N = 0$ for vacuum reservoir:

$$\frac{\partial \hat{\rho}}{\partial t} = -i[\hat{H}_S, \hat{\rho}] + \frac{\kappa}{2}(\bar{n}+1)[2\hat{a}\hat{\rho}\hat{a}^\dagger - \hat{a}^\dagger \hat{a}\hat{\rho} - \hat{\rho}\hat{a}\hat{a}^\dagger]$$
$$+ \frac{\kappa}{2}\bar{n}[2\hat{a}^\dagger \hat{\rho}\hat{a} - \hat{a}\hat{a}^\dagger \hat{\rho} - \hat{\rho}\hat{a}\hat{a}^\dagger],$$ (3.43)

$$\frac{\partial \hat{\rho}}{\partial t} = -i[\hat{H}_S, \hat{\rho}] + \frac{\kappa}{2}[2\hat{a}\hat{\rho}\hat{a}^\dagger - \hat{a}^\dagger \hat{a}\hat{\rho} - \hat{\rho}\hat{a}\hat{a}^\dagger].$$ (3.44)

It also appears straightforward to generalize the earlier discussion to a two-mode case by extending the interaction Hamiltonian to

$$\hat{H}_{SR}(t) = i \sum_k \lambda_k [\hat{a}^\dagger \hat{a}_k e^{i(\omega_a - \omega_k)t} - \hat{a}\hat{a}_k^\dagger e^{-i(\omega_a - \omega_k)t}$$
$$+ \hat{b}^\dagger \hat{b}_k e^{i(\omega_b - \omega_k)t} - \hat{b}\hat{b}_k^\dagger e^{-i(\omega_b - \omega_k)t}].$$ (3.45)

---

[5]The same result can be obtained if one sets $\hat{Q}_1 = \hat{a}$, $\hat{Q}_2 = \hat{a}^\dagger$, $\hat{F}_1 = \hat{F}$ and $\hat{F}_2 = \hat{F}^\dagger$ in Eq. (3.23), and evaluate the integrals, where $\langle \hat{F}_i \hat{F}_j \rangle_R$, $\langle \hat{F}_i^\dagger \hat{F}_j \rangle_R$ and $\langle \hat{F}_i \hat{F}_j^\dagger \rangle_R$ are correlations of the reservoir operators (see Eq. (3.36)).

With the help of Eqs. (3.15) and (3.45) along with the claim that the cavity and reservoir operators commute, one can obtain

$$Tr_R(\hat{R}\hat{H}_{SR}(t)\hat{H}_{SR}(t')) = I_1\hat{a}^{\dagger^2} + I_2\hat{a}^\dagger\hat{a} + I_3 2\hat{a}^\dagger\hat{b}^\dagger + I_4 2\hat{a}^\dagger\hat{b} + I_5\hat{a}\hat{a}^\dagger + I_6\hat{a}^2$$

$$+ I_7 2\hat{a}\hat{b}^\dagger + I_8 2\hat{a}\hat{b} + I_9\hat{b}^{\dagger^2} + I_{10}\hat{b}^\dagger\hat{b} + I_{11}\hat{b}\hat{b}^\dagger + I_{12}\hat{b}^2,$$

$$(3.46)$$

where the I_i's can be inferred from the detailed provided in Appendix 2. So following the same approach,

$$\sum_k \lambda_k^2 e^{\pm i(\omega_{a,b}-\omega_k)(t-t')} = \kappa_{a,b}\delta(t-t') \tag{3.47}$$

$$\sum_k \lambda_k\lambda_{k_a+k_b-k}e^{\pm i(\omega_a-\omega_k)(t-t')} = \sqrt{\kappa_a\kappa_b}\delta(t-t'), \tag{3.48}$$

where N and M are taken to be the same for both modes and $k_{a,b}$ is the cavity damping constant for the respective modes that can be different.

Upon making use of the provided information, it is then possible to verify that

$$\frac{d\hat{\rho}(t)}{dt} = -i[\hat{H}_S(t), \hat{\rho}(t)]$$

$$+ \frac{\kappa_a(N+1)}{2}[2\hat{a}\hat{\rho}\hat{a}^\dagger - \hat{a}^\dagger\hat{a}\hat{\rho} - \hat{\rho}\hat{a}^\dagger\hat{a}] + \frac{\kappa_a N}{2}[2\hat{a}^\dagger\hat{\rho}\hat{a} - \hat{a}\hat{a}^\dagger\hat{\rho} - \hat{\rho}\hat{a}\hat{a}^\dagger]$$

$$+ \frac{\kappa_b(N+1)}{2}[2\hat{b}\hat{\rho}\hat{b}^\dagger - \hat{b}^\dagger\hat{b}\hat{\rho} - \hat{\rho}\hat{b}^\dagger\hat{b}] + \frac{\kappa_b N}{2}[2\hat{b}^\dagger\hat{\rho}\hat{b} - \hat{b}\hat{b}^\dagger\hat{\rho} - \hat{\rho}\hat{b}\hat{b}^\dagger]$$

$$- M\sqrt{\kappa_a\kappa_b}[\hat{a}^\dagger\hat{\rho}\hat{b}^\dagger + \hat{b}^\dagger\hat{\rho}\hat{a}^\dagger + \hat{a}\hat{\rho}\hat{b} + \hat{b}\hat{\rho}\hat{a} - \hat{a}^\dagger\hat{b}^\dagger\hat{\rho} - \hat{a}\hat{b}\hat{\rho}$$

$$- \hat{\rho}\hat{a}^\dagger\hat{b}^\dagger - \hat{\rho}\hat{a}\hat{b}], \tag{3.49}$$

which is the master equation for a two-mode cavity radiation coupled to a two-mode squeezed vacuum reservoir. It may not be hard to see that Eq. (3.49) can be specialized for the vacuum and thermal reservoirs following the reasoning provided for the single-mode.

This discussion may evince that an explicit form of the master equation that represents a given quantum system can be obtained once the system Hamiltonian is specified. It should also be clear by now that the main difference between the single-mode and two-mode cases is the cross correlation among the two kinds of the reservoir modes in the latter, which stems from the coherent superposition of the states of the source of the reservoir that enables the reservoir to contribute unbalanced noise fluctuations, and so may have the capacity to enhance the nonclassical features.

3.1.2 Quantum Langevin Equation

The dynamics of the cavity radiation coupled to external environment can also be explored with the aid of the pertinent quantum Langevin equation. The idea of quantum Langevin equation presupposes the existence of the way by which the reservoir can be replaced by the damping term in the Heisenberg equation. This would be mainly done by inserting a random force that adds fluctuation to the system, and chosen in accordance with the criterion that the outcomes to be evinced should be consistent with the results that could have been obtained by using other approaches.

Even though the time dependence in the master equation would be carried by the density operator, suppose for the time being the situation in which the system operator carry the time factor. With this in mind, for a quantum system coupled to the reservoir,

$$\frac{d\hat{a}}{dt} = -i[\hat{a}(t), \; \hat{H}_S(t)] - i[\hat{a}(t), \; \hat{H}_{SR}(t)], \tag{3.50}$$

where the interaction Hamiltonian is given by Eq. (3.28). One can then see with a minimum effort that

$$\frac{d\hat{a}}{dt} = -i[\hat{a}(t), \; \hat{H}_S(t)] + \sum_j \lambda_j \hat{a}_j(t) e^{i(\omega_c - \omega_j)t}. \tag{3.51}$$

One can also write for the annihilation operator of the reservoir mode that

$$\frac{d\hat{a}_j}{dt} = -i[\hat{a}_j(t), \; \hat{H}_S(t)] + \sum_k \lambda_k \left\{ \hat{a}^\dagger [\hat{a}_j, \; \hat{a}_k] e^{i(\omega_c - \omega_k)t} - \hat{a}[\hat{a}_j, \; \hat{a}_k^\dagger] e^{i(\omega_c - \omega_k)t} \right\}. \tag{3.52}$$

Based on the proposal that the system and reservoir operators commute at equal time and picture, it is possible to demand that

$$[\hat{a}_j, \; \hat{a}] = [\hat{a}_j, \; \hat{a}^\dagger] = [\hat{a}_j, \; \hat{H}_S] = 0. \tag{3.53}$$

One can thereafter see from Eq. (3.52) that

$$\frac{d\hat{a}_j}{dt} = -\lambda_j \hat{a} e^{i(\omega_c - \omega_j)t}, \tag{3.54}$$

whose solution can be written as

$$\hat{a}_j(t) = \hat{a}_j(0) - \lambda_j \int_0^t \hat{a}(t') e^{i(\omega_c - \omega_j)(t - t')} dt'. \tag{3.55}$$

Substitution of Eq. (3.55) into (3.51) after that results

$$\frac{d\hat{a}}{dt} = -i[\hat{a}(t), \hat{H}_S(t)] - \sum_j \lambda_j^2 \int_0^t \hat{a}(t')e^{i(\omega_c - \omega_j)(t-t')}dt' + \hat{F}(t), \qquad (3.56)$$

where

$$\hat{F}(t) = \sum_j^\infty \lambda_j \hat{a}_j(0)e^{i(\omega_c - \omega_j)t} \qquad (3.57)$$

is the Langevin noise force that accounts for the contribution of the reservoir. It might be appealing at some point to envision damping as a resistance and the noise operator as noise generator that injects fluctuations into the system so that the uncertainty principle remains valid even when there is damping. As a result, one of the restrictions imposed on this operator is that its statistical mean should be zero since the generated noise is completely random.

Following the procedure outlined to determine the cavity damping constant, and upon taking the positive time in the integration, it is possible to ascertain that

$$\frac{d\hat{a}}{dt} = -i[\hat{a}(t), \hat{H}_S(t)] - \frac{\kappa}{2}\hat{a}(t) + \hat{F}(t). \qquad (3.58)$$

Equation (3.58) constitutes the sought for quantum Langevin equation for annihilation operator, and often referred to as quantum stochastic differential equation for historical reason. Besides, in view of the boson commutation relation at $t = 0$, one can verify that

$$[\hat{F}(t), \hat{F}^\dagger(t')] = \kappa\delta(t - t'), \qquad [\hat{F}(t), \hat{F}(t')] = [\hat{F}^\dagger(t), \hat{F}^\dagger(t')] = 0 \qquad (3.59)$$

and with the aid of Eqs. (3.36) and (3.37) along with Eq. (3.57),

$$\langle\hat{F}^\dagger(t)\hat{F}(t')\rangle = \kappa N\delta(t - t'), \qquad \langle\hat{F}(t)\hat{F}^\dagger(t')\rangle = \kappa(N + 1)\delta(t - t'), (3.60)$$

$$\langle\hat{F}(t)\hat{F}(t')\rangle = \langle\hat{F}^\dagger(t)\hat{F}^\dagger(t')\rangle = \kappa M\delta(t - t'), \qquad (3.61)$$

which are the correlations of the noise forces.

3.1.3 Input-Output Relation

Imagine that the dynamics of the reservoir can also be inferred by reversing the sequence of the time order in Eq. (3.58). Such effort hopefully legitimizes the possibility of expressing the radiation in the cavity in terms of the radiation leaking into or going out of the cavity: the phenomenon usually dubbed as input-output

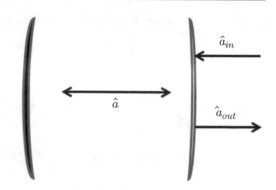

Fig. 3.2 Schematic representation of the input-output scheme, where $\hat{a}$ designates the radiation in the cavity, $\hat{a}_{in}$ part of the reservoir modes that able to leak into the cavity and $\hat{a}_{out}$ the out going radiation modes

mechanism. Let us begin with the fact that there are three operators that corresponds to the accompanying physical process that describe: the system in the cavity, the radiation that enters the cavity and the radiation going out of the cavity (see Fig. 3.2). It might be worthwhile stressing that the going away radiation should not be taken as a part of the reservoir since there is no way for this radiation to enter the cavity again as the reservoir modes are expected to do. Since a measuring device is often placed outside the cavity, it is the output operator that carries the information about the physical process inside the cavity by virtue of its accessibility to the experimenter. Markedly, the output operator $\hat{a}_{out}$ encompasses the superposition of a part of $\hat{a}$ that leaks to the environment and reflected reservoir modes.

If the properties of the radiation entering the cavity and the quantum system in the cavity are known, one may able to systematically predict what will be the result of the interaction and also the properties of the radiation that can escape from the cavity. It can also be argued based on the nature of the quantum interaction theory that there are only some modes that can go into the cavity from the available continuum of the reservoir modes and possibly superimpose where the amount of the radiation that enters the cavity depends on the characteristics of the interaction, properties of the cavity and type of the reservoir.

Equation (3.57) on the other hand may entail as the noise operator can be expressed in terms of the operators of the reservoir modes at initial time. Since the noise fluctuation is taken as an effective contribution of the reservoir that alters the cavity dynamics, the annihilation operator for the input mode can be expressed in terms of the noise operator and cavity damping constant as [8]

$$\hat{a}_{in}(t) = \frac{1}{\sqrt{\kappa}} \hat{F}(t). \tag{3.62}$$

The time development of the annihilation operator for the cavity radiation can thus be modified as

$$\frac{d\hat{a}}{dt} = -i\left[\hat{a}(t), \hat{H}_S(t)\right] - \frac{\kappa}{2}\hat{a}(t) + \sqrt{\kappa}\hat{a}_{in}(t) \tag{3.63}$$

(see Eq. (3.58)). Notice that $\hat{a}_{in}$ characterizes the reservoir modes with properties:

$$[\hat{a}_{in}(t), \hat{a}_{in}^{\dagger}(t')] = \hat{I}\delta(t - t'), \qquad [\hat{a}_{in}(t), \hat{a}_{in}(t')] = [\hat{a}_{in}^{\dagger}(t), \hat{a}_{in}^{\dagger}(t')] = 0, \tag{3.64}$$

$$\langle \hat{a}_{in}^{\dagger}(t)\hat{a}(t')_{in}\rangle = N\delta(t - t'), \qquad \langle \hat{a}_{in}(t)\hat{a}_{in}^{\dagger}(t')\rangle = (N + 1)\delta(t - t'), \tag{3.65}$$

$$\langle \hat{a}_{in}(t)\hat{a}_{in}(t')\rangle = \langle \hat{a}_{in}^{\dagger}(t)\hat{a}_{in}^{\dagger}(t')\rangle = M\delta(t - t'), \tag{3.66}$$

$$\langle \hat{a}_{in}(t)\rangle = \langle \hat{a}_{in}^{\dagger}(t)\rangle = 0, \tag{3.67}$$

The boson operator for the reservoir modes for any time $t \leq T$ can also be examined by replacing 0 by T, and then interchanging the limit of integration from T to t to from t to T in Eq. (3.55):

$$\hat{a}_j(t) = \hat{a}_j(T) + \lambda_j \int_t^T \hat{a}(t')e^{i(\omega_c - \omega_j)(t - t')}dt'. \tag{3.68}$$

It is possible to rewrite Eq. (3.51) after that with the aid of Eq. (3.68) as

$$\frac{d\hat{a}}{dt} = -i[\hat{a}(t), \hat{H}_S(t)] + \sum_j \lambda_j \left[\hat{a}_j(T) + \lambda_j \int_t^T \hat{a}(t')e^{i(\omega_c - \omega_j)t'}dt'\right]e^{i(\omega_c - \omega_j)t}, \tag{3.69}$$

from which follows

$$\frac{d\hat{a}}{dt} = -i[\hat{a}(t), \hat{H}_S(t)] + \frac{\kappa}{2}\hat{a}(t) + \sum_j \lambda_j \hat{a}_j(T)e^{i(\omega_c - \omega_j)t}. \tag{3.70}$$

It might not elude one's perception that the output radiation at any time is significantly influenced by the cavity and reservoir dynamics at earlier times: the output radiation at a given time depends on the nature of the interaction started in the cavity prior to the time when one seeks to consider. On account of this, the annihilation operator for the output radiation can be defined as

$$\hat{a}_{out}(t) = -\frac{1}{\sqrt{\kappa}}\hat{F}(T) = -\frac{1}{\sqrt{\kappa}}\sum_j \lambda_j \hat{a}_j(T)e^{i(\omega_c - \omega_j)t}, \tag{3.71}$$

where the minus sign is inserted to ensure consistency with the prevailing situation that the output radiation is going away taking energy and momentum along with it.

This outcome may suggest that the cavity mode can be expressed in terms of the output variables as

$$\frac{d\hat{a}}{dt} = -i[\hat{a}(t), \hat{H}_S(t)] + \frac{\kappa}{2}\hat{a}(t) - \sqrt{\kappa}\hat{a}_{out}(t). \tag{3.72}$$

Comparison of Eq. (3.63) with (3.72) reveals that

$$\hat{a}(t) = \frac{1}{\sqrt{\kappa}}[\hat{a}_{in}(t) + \hat{a}_{out}(t)], \tag{3.73}$$

that is,

$$\hat{a}_{out}(t) = \sqrt{\kappa}\hat{a}(t) - \hat{a}_{in}(t). \tag{3.74}$$

Equation (3.74) can be interpreted as optical boundary condition that leads to treating the output radiation as the measure of the balance between the input and cavity radiation.

Despite the availability of such a condition, the relationship between the cavity mode and radiation field outside the cavity is yet to be trivial specially when one is interested in the anti-normally ordered expectation values. Since the partially reflecting coupler mirror of the cavity is capable of not only letting the cavity modes out, but also allows the outside radiation field to leak into the cavity, the cavity and input fields are expected to eventually correlate. The emerging correlation oblige the residual fluctuations in the spectrum of the transmitted cavity modes to cancel out with fluctuations in the reflected input radiation field. Such a process can lead for instance to unusually perfect output spectral squeezing at appropriate frequency for systems such as degenerate parametric oscillator.

One may next seek to determine the intensity of the radiation accessible to the experimenter. To do so, the mean photon number of the output radiation can be put in the form

$$\langle \hat{a}_{out}^\dagger(t)\hat{a}_{out}(t)\rangle = \kappa\langle\hat{a}^\dagger(t)\hat{a}(t)\rangle + \langle\hat{a}_{in}^\dagger(t)\hat{a}_{in}(t)\rangle$$
$$- \sqrt{\kappa}[\langle\hat{a}^\dagger(t)\hat{a}_{in}(t)\rangle + \langle\hat{a}_{in}^\dagger(t)\hat{a}(t)\rangle], \tag{3.75}$$

where the first term is the cavity damping constant times the mean photon number of the cavity mode and the second term is the mean photon number of the input radiation, whereas the remaining terms are mixed terms that account for the correlation between the cavity and input modes.

To obtain these expectation values, one needs to resort to the definition of the noise operator;

$$\sqrt{\kappa}\langle\hat{a}^\dagger(t)\hat{a}_{in}(t)\rangle = \langle\hat{a}^\dagger(t)\hat{F}(t)\rangle. \tag{3.76}$$

To evaluate this correlation, one may require the explicit form of $\hat{a}^\dagger(t)$ that can be obtained from the quantum Langevin equation;

$$\hat{a}^\dagger(t) = \hat{a}^\dagger(0)e^{\frac{-\kappa t}{2}} + i \int_0^t e^{-\frac{\kappa(t-t')}{2}} [\hat{a}^\dagger(t'), \hat{H}_S(t')]dt' + \int_0^t e^{-\frac{\kappa(t-t')}{2}} \hat{F}^\dagger(t')dt',$$

(3.77)

from which follows

$$\sqrt{\kappa}\langle\hat{a}^\dagger(t)\hat{a}_{in}(t)\rangle = \langle\hat{a}^\dagger(0)\hat{F}(t)\rangle e^{\frac{-\kappa t}{2}} + i \int_0^t e^{-\frac{\kappa(t-t')}{2}} \big[\langle\hat{a}^\dagger(t')\hat{H}_S(t')\hat{F}(t)\rangle$$

$$- \langle\hat{H}_S(t')\hat{a}^\dagger(t')\hat{F}(t)\rangle\big]dt' + \int_0^t e^{-\frac{\kappa(t-t')}{2}} \langle\hat{F}^\dagger(t')\hat{F}(t)\rangle dt'.$$

(3.78)

Since the reservoir can affect the system after some part of it manages to enter the cavity, it is claimed that the reservoir variables at earlier times do not affect the system variables;

$$\langle\hat{A}(0)\hat{F}(t)\rangle = \langle\hat{A}(t')\hat{F}(t)\rangle_{t'\leq t} = 0,$$

(3.79)

where $\hat{A}$ is arbitrary operator of the system. One can hence obtain by applying Eq. (3.60) that

$$\langle\hat{a}^\dagger(t)\hat{a}_{in}(t)\rangle = \frac{\sqrt{\kappa}N}{2}.$$

(3.80)

So upon using the fact that $\langle\hat{a}^\dagger_{in}(t)\hat{a}_{in}(t)\rangle = N$, it might be straightforward to see that

$$\langle\hat{a}^\dagger_{out}(t)\hat{a}_{out}(t)\rangle = \kappa\langle\hat{a}^\dagger(t)\hat{a}(t)\rangle + N(1 - \kappa).$$

(3.81)

Equation (3.81) indicates that the mean photon number of the output radiation can be described in terms of the mean photon number of the cavity radiation, mean photon number of the reservoir modes and cavity damping constant in which the first term stands for the mean of the radiation leaking out of the cavity and the second term for the mean photon number of the reservoir minus the radiation that enters the cavity.

3.1.4 Two-Time Optical Boundary Condition

It should be clear by now that the cavity system would be in eternal contact with the bath that consists of at least the vacuum modes of continuous spectral distribution that can be described by the Hamiltonian of the form

$$\hat{H} = \hat{H}_{sys} + \int_{-\infty}^{\infty} \omega \hat{b}^{\dagger}(\omega)\hat{b}(\omega)d\omega + i\int_{-\infty}^{\infty} \kappa(\omega)\big[\hat{b}^{\dagger}(\omega)\hat{c} - \hat{c}^{\dagger}\hat{b}(\omega)\big]d\omega, \tag{3.82}$$

where the second and the last terms represent the bath and interaction Hamiltonian with commutation relation $[\hat{b}(\omega),\ \hat{b}^{\dagger}(\omega)] = \hat{I}\delta(\omega - \omega')$, and $\hat{c}$ is one of the several possible system operators. The range from $-\Omega$ to ∞ would be utilized in practice when the frame is presumed to rotate with angular frequency Ω as commonly taken in quantum optics. This assumption certainly corresponds to the case for which Ω is very large when compared to a typical bandwidth the reservoir modes can encompass.

Upon making use of Heisenberg equation (3.2), it may not be difficult to come by with

$$\frac{d\hat{b}(\omega)}{dt} = -i\omega\hat{b}(\omega) + \kappa(\omega)\hat{c}, \tag{3.83}$$

$$\frac{d\hat{a}}{dt} = -i\big[\hat{a},\ \hat{H}_{sys}\big] + \int \kappa(\omega)\big[\hat{b}^{\dagger}(\omega)[\hat{a},\ \hat{c}] - [\hat{a},\ \hat{c}^{\dagger}]\hat{b}(\omega)\big]d\omega, \tag{3.84}$$

where $\hat{a}$ is the system operator that commutes with bath operator. The solution of Eq. (3.83) can then be proposed as

$$\hat{b}(\omega) = \hat{b}_0(\omega)e^{-i\omega(t-t_0)} + \kappa(\omega)\int_{t_0}^{t} e^{-i\omega(t-t')}\hat{c}(t')dt', \tag{3.85}$$

which leads to

$$\frac{d\hat{a}}{dt} = -i\big[\hat{a},\ \hat{H}_{sys}\big] + \int \kappa(\omega)\big[e^{i\omega(t-t_0)}\hat{b}_0^{\dagger}(\omega)[\hat{a},\ \hat{c}] - [\hat{a},\ \hat{c}^{\dagger}]\hat{b}_0(\omega)e^{-i\omega(t-t_0)}\big]d\omega$$
$$+ \int d\omega\kappa^2(\omega)\int_{t_0}^{t} \big[e^{i\omega(t-t')}\hat{c}^{\dagger}(t')[\hat{a},\ \hat{c}] - [\hat{a},\ \hat{c}^{\dagger}]\hat{c}(t')e^{-i\omega(t-t')}\big]dt'. \tag{3.86}$$

It is found appealing introducing Markovian approximation in which the coupling constant is independent of the frequency, that is, $\kappa(\omega) = \sqrt{\gamma/2\pi}$ to expedite the involved integration.

It is also possible to define the field entering the cavity by

$$\hat{b}_{in}(t) = \frac{1}{\sqrt{2\pi}}\int d\omega\hat{b}_0(\omega)e^{-i\omega(t-t')}, \tag{3.87}$$

which satisfies the boson commutation relation $[\hat{b}_{in}(t), \hat{b}_{in}^\dagger(t')] = \hat{I}\delta(t - t')$. In connection to this, it might be easy to obtain

$$\frac{d\hat{a}}{dt} = -i[\hat{a}, \hat{H}_{sys}] - [\hat{a}, \hat{c}^\dagger]\left(\frac{\gamma}{2}\hat{c} + \sqrt{\gamma}\hat{b}_{in}(t)\right) + \left(\frac{\gamma}{2}\hat{c}^\dagger + \sqrt{\gamma}\hat{b}_{in}^\dagger(t)\right)[\hat{a}, \hat{c}]. \tag{3.88}$$

For a harmonic oscillator, $\hat{H}_{sys} = \omega_0\hat{a}^\dagger\hat{a}$ in which $\hat{c} = \hat{a}$;

$$\frac{d\hat{a}}{dt} = -i\omega_0\hat{a} - \frac{\gamma}{2}\hat{a} - \sqrt{\gamma}\hat{b}_{in}(t). \tag{3.89}$$

Even though we begin with frequency dependence of the reservoir operators, one may note that the time evolution of the system operator turns out to have the same form as quantum Langevin equation.

One can also write analogous to Eq. (3.85) for $t_1 > t$ that

$$\hat{b}(\omega) = \hat{b}_1(\omega)e^{-i\omega(t-t_1)} - \kappa(\omega)\int_t^{t_1} e^{-i\omega(t-t')}\hat{c}(t')dt', \tag{3.90}$$

and accordingly define

$$\hat{b}_{out}(t) = \sqrt{\frac{1}{2\pi}}\int \hat{b}_1(\omega)e^{-i\omega(t-t')}d\omega, \tag{3.91}$$

from which the time reversed Langevin equation follows

$$\frac{d\hat{a}}{dt} = -i[\hat{a}, \hat{H}_{sys}] - [\hat{a}, \hat{c}^\dagger]\left(-\frac{\gamma}{2}\hat{c} + \sqrt{\gamma}\hat{b}_{out}(t)\right)$$
$$+ \left(-\frac{\gamma}{2}\hat{c}^\dagger + \sqrt{\gamma}\hat{b}_{out}^\dagger(t)\right)[\hat{a}, \hat{c}]. \tag{3.92}$$

Comparison of the time evolution of the system operator for separately considered cases such as Eqs. (3.88) and (3.92) reveals that

$$\hat{b}_{out}(t) - \hat{b}_{in}(t) = \sqrt{\gamma}\hat{c}(t). \tag{3.93}$$

Equation (3.93) can also be interpreted as two-time optical boundary condition or as causality effect. Besides, it is not difficult to verify that

$$[\hat{a}(t), \hat{b}_{in}(t')] = -u(t - t')\sqrt{\gamma}[\hat{a}(t), \hat{c}(t')], \tag{3.94}$$

$$[\hat{a}(t), \hat{b}_{out}(t')] = u(t' - t)\sqrt{\gamma}[\hat{a}(t), \hat{c}(t')], \tag{3.95}$$

where

$$u(x) \begin{cases} 1, & x > 0; \\ \frac{1}{2}, & x = 0; \\ 0, & x < 0. \end{cases}$$

Equation (3.94) may hint a clue to how the corresponding causality works:[6] the future properties of the system would be affected by the present input while the future value of the output would be affected by the present values of the system operators. Such an idea can be utilized when one seeks to determine for instance the squeezing spectrum of the output radiation (see Sect. 4.2.3). Although this discussion may be relevant for the system driven by noisy input, it does not represent stochastic process since no attempt has been made to take the density operator of the bath into consideration.

3.2 Mathematical Approaches

Even for the cases in which the pertinent differential equation can be solved, the ensuing calculation of the corresponding correlations may require operator ordering. The radiation field one wishes to study can also be in contact with the environment where the prevailing fluctuation incites noise—the phenomenon that makes obtaining the solution of the dynamical operator a formidable task. It might be imperative in such cases to have acclimatization with certain specialized mathematical approaches in quantum optics, although the final result should be independent of the approach one follows. The available formulations mainly differ in the way the relevant differential equations are set. The principal task in theoretical quantum optical research is usually geared towards solving the resulting differential equations with the help of one (or more) of the available suitable approaches such as characteristic function, stochastic differential equation and linearization technique.

In line with the underlying mathematical rigor and physical relevance, a degenerate parametric oscillator is often chosen as a test toy to illustrate the approaches we opt to consider. In a degenerate parametric oscillation, a strong pump photon with frequency 2ω would be down converted into a pair of signal photons each with

---

[6]In deriving Eq. (3.94), the general mathematical relations,

$$\int_{-\infty}^{\infty} e^{-i\omega(t-t')} = 2\pi \delta(t - t')d\omega, \qquad \int_{t_0}^{t} \hat{c}(t')\delta(t - t')dt' = \frac{\hat{c}(t)}{2},$$

are utilized [9]. It is also equally possible to use the definitions:

$$\int \hat{b}(\omega)d\omega = \hat{b}_{in} + \frac{\gamma}{2}\hat{c}(t), \qquad \int \hat{b}(\omega)d\omega = \hat{b}_{out} - \frac{\gamma}{2}\hat{c}(t).$$

Fig. 3.3 Schematic representation of a degenerate parametric oscillator in which the pumping radiation $\hat{b}$ is down converted by a nonlinear crystal placed in a cavity to two identical modes designated by an operator $\hat{a}$ (see also Figs. 3.6 and 8.1)

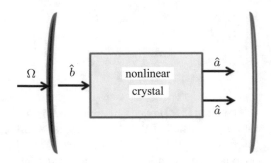

frequency ω that can be described in the interaction picture by the Hamiltonian of the form

$$\hat{H} = \frac{i\lambda}{2} \left(\hat{b}^\dagger \hat{a}^2 - \hat{b}\hat{a}^{\dagger 2} \right), \tag{3.96}$$

where λ is the coupling constant taken to be real and $\hat{a}$ ($\hat{b}$) is the annihilation operator for the signal (pump) mode (see Chap. 8 and Fig. 3.3). Differential equations that follow from such Hamiltonian are nonlinear; and are so difficult to directly solve.

To evade the emerging mathematical hurdle, one may need to resort to some sort of approximation scheme such as taking the pump radiation to be strong. This approach is referred to as parametric approximation in which the operator that describes the pump field would be replaced by the c-number variable or treated classically; which entails that the amplitude of the pump radiation does not significantly change or deplete during the conversion process, that is, Eq. (3.96) can be rewritten in the form

$$\hat{H} = \frac{i\varepsilon}{2} \left(\hat{a}^2 - \hat{a}^{\dagger 2} \right), \tag{3.97}$$

where ε is the product of the coupling constant and the c-number equivalence of the amplitude of the pump mode, and taken to be a real positive constant proportional to the amplitude of the driving radiation.

Then following the standard approach (see Sect. 3.1.1), the master equation when the cavity is assumed to be coupled to a squeezed vacuum reservoir is found to be

$$\frac{\partial \hat{\rho}}{\partial t} = \frac{\varepsilon}{2} \left(\hat{a}^2 \hat{\rho} - \hat{a}^{\dagger 2} \hat{\rho} - \hat{\rho}\hat{a}^2 + \hat{\rho}\hat{a}^{\dagger 2} \right) + \frac{\kappa}{2} (N+1) \left(2\hat{a}\hat{\rho}\hat{a}^\dagger - \hat{a}^\dagger \hat{a}\hat{\rho} - \hat{\rho}\hat{a}^\dagger \hat{a} \right)$$
$$+ \frac{\kappa}{2} N \left(2\hat{a}^\dagger \hat{\rho}\hat{a} - \hat{a}\hat{a}^\dagger \hat{\rho} - \hat{\rho}\hat{a}\hat{a}^\dagger \right)$$
$$- \frac{\kappa}{2} M \left(2\hat{a}\hat{\rho}\hat{a} - \hat{a}^2 \hat{\rho} - \hat{\rho}\hat{a}^2 + 2\hat{a}^\dagger \hat{\rho}\hat{a}^\dagger - \hat{a}^{\dagger 2} \hat{\rho} - \hat{\rho}\hat{a}^{\dagger 2} \right). \tag{3.98}$$

3.2.1 Characteristic Function

Calculating the required moments directly using the master equation might be tedious and even can be tricky. Aiming at avoiding such predicament, the approach we intend to introduce is geared towards obtaining the Q-function from the dynamical equations via relevant characteristic function (see Sect. 2.3.1) in which the time development of the signal mode can be expressed upon employing Eq. (3.58) as

$$\frac{d\hat{a}(t)}{dt} = -\varepsilon\hat{a}^\dagger(t) - \frac{\kappa}{2}\hat{a}(t) + \hat{F}(t), \tag{3.99}$$

where $\hat{F}(t)$ is the noise operator. One may see that Eq. (3.99) is nonlinear, and so difficult to solve directly.

To unravel this challenge, the quadrature operators (2.47) can be utilized whereupon taking the complex conjugate of Eq. (3.99) one finds

$$\frac{d\hat{a}_+}{dt} = -\frac{\lambda_+}{2}\hat{a}_+ + \hat{F}^\dagger(t) + \hat{F}(t), \tag{3.100}$$

$$\frac{d\hat{a}_-}{dt} = -\frac{\lambda_-}{2}\hat{a}_- + i(\hat{F}^\dagger(t) - \hat{F}(t)) \tag{3.101}$$

in which $\lambda_\pm = \kappa \pm 2\varepsilon$.

Then one can propose the solution of the form

$$\hat{a}_+(t) = \hat{a}_+(0)e^{-\frac{\lambda_+}{2}t} + \int_0^t e^{\frac{-\lambda_+}{2}(t-t')}\big[\hat{F}(t') + \hat{F}^\dagger(t')\big]dt', \tag{3.102}$$

$$\hat{a}_-(t) = \hat{a}_-(0)e^{-\frac{\lambda_-}{2}t} + i\int_0^t e^{\frac{-\lambda_-}{2}(t-t')}\big[\hat{F}^\dagger(t') - \hat{F}(t')\big]dt', \tag{3.103}$$

from which follows

$$\hat{a}(t) = C_+\hat{a}(0) + C_-\hat{a}^\dagger(0) + \hat{D}_+(t) + \hat{D}_-(t), \tag{3.104}$$

$$\hat{a}^\dagger(t) = C_+\hat{a}^\dagger(0) + C_-\hat{a}(0) + \hat{D}_+(t) - \hat{D}_-(t), \tag{3.105}$$

where

$$C_\pm = \frac{1}{2}\left(e^{-\frac{\lambda_\pm}{2}t} \pm e^{-\frac{\lambda_\mp}{2}t}\right), \tag{3.106}$$

$$\hat{D}_\pm(t) = \frac{1}{2}\int_0^t e^{-\frac{\lambda_\pm}{2}(t-t')}\big[\hat{F}(t') \pm \hat{F}^\dagger(t')\big]dt'. \tag{3.107}$$

It may worth noting that the dynamics of the radiation can be analyzed by exploiting Eqs. (3.104) and (3.105) directly without necessarily applying the char-

acteristic function formulation. Nonetheless, the correlations between the operators
that involve the noise force and system operators at different times with associated
operator ordering might be required to determine various moments where some of
the suggested moments can be obtained with the help of the correlations of the noise
operators of the reservoir.

Even then, since the c-number phase space distribution functions such as Q-
function can be derived from the characteristic function, calculating the characteristic function might be more economical specially when a higher-order correlations
are sought for. So with the aid of Eqs. (2.83), (3.104) and (3.105), the anti-normally
ordered characteristic function for a single-mode cavity radiation can be expressed
as

$$\phi_a(z, t) = e^{-\frac{z^* z}{2}} Tr(\hat{\rho}(0) \exp[\hat{a}^\dagger(zC_+ - z^*C_-) + \hat{a}(zC_- - z^*C_+)$$
$$+ \hat{D}_+(t)(z - z^*) - \hat{D}_-(t)(z + z^*)]). \tag{3.108}$$

The separation of the operators into the system and reservoir parts may envisage
that the system and reservoir operators commute at equal time; and hence, can be
solved separately.[7] Note that one can write by citing the Baker-Hausdroff's relation
[5] (see also Appendix 1) that

$$Trs\big(\hat{\rho}_S(0) \exp[\hat{a}(zC_+ - z^*C_-) + \hat{a}^\dagger(zC_- - z^*C_+)]\big)$$
$$= Trs\big(\hat{\rho}_S(0) \exp[-\frac{1}{2}(zC_+ - z^*C_-)(zC_- - z^*C_+)]$$
$$\times \exp[\hat{a}(zC_+ - z^*C_-) \exp\hat{a}^\dagger(zC_- - z^*C_+)]\big). \tag{3.109}$$

After that, one can proceed with evaluating the contribution of the reservoir
alone. It might be helpful noting that

$$Tr_R\big(\hat{\rho}_R(0) \exp[\hat{D}_+(t)(z - z^*) - \hat{D}_-(t)(z + z^*)]\big)$$
$$= \big\langle \exp\big[\hat{D}_+(t)(z - z^*) - \hat{D}_-(t)(z + z^*)\big]\big\rangle_R \tag{3.110}$$

and for a Gaussian variable of zero mean [10];

$$\big\langle \exp\big[\hat{D}_+(t)(z - z^*) - \hat{D}_-(t)(z + z^*)\big]\big\rangle_R = \exp\Big[\frac{1}{2}\big[(z - z^*)^2\langle\hat{D}_+^2(t)\rangle_R$$
$$+(z + z^*)^2\langle\hat{D}_-^2(t)\rangle_R - (z^2 - z^{*2})\big(\langle\hat{D}_+(t)\hat{D}_-(t)\rangle_R + \langle\hat{D}_-(t)\hat{D}_+(t)\rangle_R\big)\big]\Big].$$
$$\tag{3.111}$$

---

[7]Extending the assumption that at initial time the system and reservoir variables are not correlated
to the later stages of the interaction process implies that the system and reservoir operators remain
to be uncorrelated at equal times since the reservoir needs some time to perceivably alter the
properties of the system.

Then applying Eqs. (3.60) and (3.61), one can find

$$
\psi_a(z, t) = \exp\left[\frac{z^*z}{2} - zz^*\left(\frac{\kappa e^{2r}}{4\lambda_+}(1 - e^{-\lambda_+ t}) + \frac{\kappa e^{-2r}}{4\lambda_-}(1 - e^{-\lambda_- t})\right)\right.
$$
$$
\left. + \frac{1}{2}(z^2 + z^{*2})\left(\frac{\kappa e^{2r}}{4\lambda_+}(1 - e^{-\lambda_+ t}) - \frac{\kappa e^{-2r}}{4\lambda_-}(1 - e^{-\lambda_- t})\right)\right]
$$
$$
\times Trs\left(\hat{\rho}_S(0) \exp\left[-\frac{1}{2}(zC_+ - z^*C_-)(zC_- - z^*C_+)\right]\right.
$$
$$
\times \exp\left[\hat{a}(zC_+ - z^*C_-)\right]\exp\left[\hat{a}^\dagger(zC_- - z^*C_+)\right]). \qquad (3.112)
$$

It should be possible to obtain the Q-function once the characteristic function is known. Markedly, the intended approach may be advantageous when directly finding the Q-function from the master equation is complicated or when including the effects of the noise force is required.

Contrary to the discussion in Sect. 2.3, the emphasis here is geared towards establishing the way how the initial properties of the system would be included. To begin with, imagine the signal mode to be initially in vacuum state. In this case, upon inserting the completeness relation for a coherent state in Eq. (3.109); then making use of the equivalence that $e^{c\hat{a}}|\alpha\rangle \longrightarrow e^{c\alpha}$ along with its complex conjugate; after that, upon integrating the resulting expression, one can verify that

$$
Trs\left(\hat{\rho}_S(0) \exp[\hat{a}(zC_+ - z^*C_-) + \hat{a}^\dagger(zC_- - z^*C_+)]\right)
$$
$$
= \exp\left[\frac{1}{2}(z^2 + z^{*2})C_-C_- - \frac{z^*z}{2}(C_+^2 + C_-^2)\right]. \qquad (3.113)
$$

The anti-normally ordered characteristic function as a result can be put in the form

$$
\psi_a(z, t) = \exp\left[-az^*z + \frac{b}{2}(z^2 + z^{*2})\right], \qquad (3.114)
$$

where

$$
a = \frac{1}{2} + \frac{1}{4}(e^{-\lambda_+ t} + e^{-\lambda_- t}) + \frac{\kappa e^{2r}}{4\lambda_+}(1 - e^{-\lambda_+ t}) + \frac{\kappa e^{-2r}}{4\lambda_-}(1 - e^{-\lambda_- t}), \quad (3.115)
$$

$$
b = \frac{1}{4}(e^{-\lambda_+ t} + e^{-\lambda_- t}) + \frac{\kappa e^{2r}}{4\lambda_+}(1 - e^{-\lambda_+ t}) - \frac{\kappa e^{-2r}}{4\lambda_-}(1 - e^{-\lambda_- t}). \qquad (3.116)
$$

The Q-function thus can be expressed upon applying the anti-normally ordered characteristic function as

$$Q(\alpha, t) = \frac{1}{\pi} \left[\frac{1}{a^2 - b^2} \right]^{1/2} \exp \left[\frac{-a\alpha^*\alpha + \frac{b}{2}(\alpha^{*2} + \alpha^2)}{a^2 - b^2} \right] \tag{3.117}$$

(compare with Eq. (3.216) at steady state). It might also be straightforward to see when the signal mode is initially in coherent state that

$$Q(\alpha, t) = \frac{1}{\pi} \left[\frac{1}{a^2 - b^2} \right]^{1/2} \exp \left[\frac{b(c_+^2 + c_-^2) - 2ac_+c_-}{2(a^2 - b^2)} \right]$$

$$\times \exp \left[\frac{-a\alpha^*\alpha + \frac{b}{2}(\alpha^{*2} + \alpha^2) + (bc_+ - ac_-)\alpha^* + (bc_- - ac_+)\alpha}{a^2 - b^2} \right], \tag{3.118}$$

where $c_{\pm} = \left[\gamma(e^{-\frac{\lambda_+ t}{2}} \pm e^{-\frac{\lambda_- t}{2}}) + \gamma^*(e^{-\frac{\lambda_+ t}{2}} \mp e^{-\frac{\lambda_- t}{2}}) \right]/2$ and γ stands for a c-number equivalence of the amplitude of the coherent light. Since the resulting amplification can be too strong to overshadow the signature of the initially seeded light, the effect of the coherent light at steady state is not apparent. Besides, if the signal mode is initially in chaotic state, the Q-function would have a similar form except for the modification in the parameters related to the mean photon number.

3.2.2 Stochastic Differential Equation

Existing experience evinces that a theoretical research in quantum optics at some level amounts to solving differential equation. Based on the proposal that mathematical method in quantum optics should encompass the ways of solving the emerging differential equations, we now seek to outline the procedure that can be applied in expressing the operator differential equations in terms of the equivalent c-number variables hoping that the c-number differential equations can be relatively easier to handle. The approach we intend to discuss comprises the method of converting the time evolution of the operators that can be obtained, let us say, from the master equation to the c-number equation associated with the normal ordering.

First of all, stochastic differential equation can be obtained by making use of quantum Langevin equation. In view of Eqs. (3.58) and (3.96), the time evolution of the annihilation operator for the signal-mode turns out in this regard to be

$$\frac{d\hat{a}(t)}{dt} = -\frac{\kappa}{2}\hat{a}(t) - \lambda\hat{b}(t)\hat{a}^{\dagger}(t) + \hat{F}(t), \tag{3.119}$$

whose solution can be written as

$$\hat{a}(t) = \hat{a}(0)e^{-\frac{\kappa}{2}t} + \int_0^t e^{-\frac{\kappa}{2}(t-t')}[\hat{F}(t') - \lambda \hat{b}(t')\hat{a}^\dagger(t')]dt'. \tag{3.120}$$

Since such an iterative integration is not easy to carry out directly, obtaining the explicit solution of Eq. (3.120) may not be a straightforward task. Besides, inferring the time evolution of the noise force directly may not be obvious since the reservoir is quite big when compared to the system, and its extent is presumed to be unknown to the observer. The calculation of the higher-order moments can also be ruined by the required operator ordering. One may hence evade such obstacles by systematically avoiding the direct contribution of the noise force.

Critical observation may expose that the suggested remedy can be expedient by expressing the expectation value of Eq. (3.119) in terms of the c-number variables;

$$\frac{d}{dt}\langle\alpha(t)\rangle = -\frac{\kappa}{2}\langle\alpha(t)\rangle - \lambda\langle\beta(t)\alpha^*(t)\rangle, \tag{3.121}$$

where the mean value of a random Gaussian noise operator is taken to be zero. Without loss of generality, it is also possible to express Eq. (3.121) as

$$\frac{d\alpha(t)}{dt} = -\frac{\kappa}{2}\alpha(t) - \lambda\beta(t)\alpha^*(t) + \eta(t), \tag{3.122}$$

where $\eta(t)$ is the relevant noise force that accounts for the stochastic process and whose properties remain to be determined. Comparison of Eqs. (3.121) and (3.122) reveals that the expectation value of this noise force should also vanish. Since the expectation value of any operator is a c-number, $\alpha(t)$ and $\beta(t)$ can be taken as c-number variables associated with preassigned order (not relevant for present case). One may note that the emerging differential equation looks just like an ordinary stochastic differential equation; and hence, the name.

Even then, one of the main challenges is obtaining the correlation of the noise force with the system operators. In light of this, we intend to outline the way of expressing the correlations of the stochastic noise force with the system variable and itself without employing the explicit solution of the stochastic differential equation. To this end, on the basis of the fact that

$$\frac{d}{dt}\langle\hat{a}^2(t)\rangle = \left\langle\frac{d\hat{a}(t)}{dt}\hat{a}(t)\right\rangle + \left\langle\hat{a}(t)\frac{d\hat{a}(t)}{dt}\right\rangle \tag{3.123}$$

along with Eq. (3.119), one can write

$$\frac{d}{dt}\langle\hat{a}^2(t)\rangle = -\kappa\langle\hat{a}^2(t)\rangle - 2\lambda\langle\hat{b}(t)\hat{a}^\dagger(t)\hat{a}(t)\rangle - \lambda\langle\hat{b}(t)\rangle$$

$$+ \langle\hat{a}(t)\hat{F}(t)\rangle + \langle\hat{F}(t)\hat{a}(t)\rangle. \tag{3.124}$$

Then after, it might be required to calculate the expectation values involved in Eq. (3.124). To do so, one can show with the use of Eq. (3.120) that

$$\langle \hat{a}(t)\hat{F}(t)\rangle = \langle \hat{a}(0)\hat{F}(t)\rangle e^{-\frac{\kappa}{2}t}$$

$$+ \int_0^t e^{-\frac{\kappa}{2}(t-t')}\big[\langle \hat{F}(t')\hat{F}(t)\rangle - \lambda\langle \hat{b}(t')\hat{a}^\dagger(t')\hat{F}(t)\rangle\big]dt'. \qquad (3.125)$$

On the basis of the assertion that the reservoir can affect the dynamics of the system only after some time, note that the noise operator evaluated at a certain time does not correlate with the system operators at earlier times;

$$\langle \hat{a}(t)\hat{F}(t)\rangle = \int_0^t e^{-\frac{\kappa}{2}(t-t')}\langle \hat{F}(t')\hat{F}(t)\rangle dt'. \qquad (3.126)$$

Equation (3.126) indicates that the correlation of the noise operator with the system variable depends on the correlation of the reservoir noise forces at different times. For a single-mode squeezed vacuum reservoir, one can verify upon using Eq. (3.61) that

$$\langle \hat{a}(t)\hat{F}(t)\rangle = \langle \hat{F}(t)\hat{a}(t)\rangle = \frac{\kappa M}{2}, \qquad (3.127)$$

where substitution of Eq. (3.127) into (3.124) results

$$\frac{d}{dt}\langle \hat{a}^2(t)\rangle = -\kappa\langle \hat{a}^2(t)\rangle - 2\lambda\langle \hat{b}(t)\hat{a}^\dagger(t)\hat{a}(t)\rangle - \lambda\langle \hat{b}(t)\rangle + \kappa M. \qquad (3.128)$$

Since the operators are already expressed in the normal order, it is possible to put Eq. (3.128) in terms of the c-number variables associated with the normal ordering (relevant in this case) in the form

$$\frac{d}{dt}\langle \alpha^2(t)\rangle = -\kappa\langle \alpha^2(t)\rangle - 2\lambda\langle \beta(t)\alpha^*(t)\alpha(t)\rangle - \lambda\langle \beta(t)\rangle + \kappa M. \qquad (3.129)$$

Then by assuming that the c-number variables in Eq. (3.122) are associated with normal ordering, one can write

$$\frac{d}{dt}\langle \alpha^2(t)\rangle = -\kappa\langle \alpha^2(t)\rangle - 2\lambda\langle \beta(t)\alpha^*(t)\alpha(t)\rangle + 2\langle \alpha(t)\eta(t)\rangle. \qquad (3.130)$$

Comparison of Eqs. (3.129) and (3.130) hence indicates

$$\kappa M - \lambda\langle \beta(t)\rangle = 2\langle \alpha(t)\eta(t)\rangle. \qquad (3.131)$$

One may note that the correlation of the system or the signal mode with the pertinent stochastic noise force can be expressed in terms of the correlation of the reservoir modes and the expectation value of the pump mode.

To determine the correlation of the stochastic noise force of different times, one may opt to proceed with the general solution of Eq. (3.122):

$$\alpha(t) = \alpha(0)e^{-\frac{\kappa}{2}t} + \int_0^t e^{-\frac{\kappa}{2}(t-t')}\big[\eta(t') - \lambda\beta(t')\alpha^*(t')\big]dt',\tag{3.132}$$

from which follows

$$\langle\alpha(t)\eta(t)\rangle = \langle\alpha(0)\eta(t)\rangle e^{-\frac{\kappa}{2}t} + \int_0^t e^{-\frac{\kappa}{2}(t-t')}\big[\langle\eta(t')\eta(t)\rangle - \lambda\langle\beta(t')\alpha^*(t')\eta(t)\rangle\big]dt'.\tag{3.133}$$

With the observation that the system variables at earlier time and the noise force at later time do not correlate, one gets

$$\int_0^t e^{-\frac{\kappa}{2}(t-t')}\langle\eta(t')\eta(t)\rangle dt' = -\frac{1}{2}\lambda\langle\beta(t)\rangle + \frac{\kappa M}{2},\tag{3.134}$$

from which follows[8]

$$\langle\eta(t)\eta(t')\rangle = \big[\kappa M - \lambda\langle\beta(t)\rangle\big]\delta(t - t').\tag{3.135}$$

Comparison with the correlation of the Langevin noise operator (3.61) indicates that there is a tangible distinction between the two. Such a difference mainly stems from what they designed to represent: the Langevin noise operator strictly stands for the reservoir mode fluctuations while the stochastic noise force for all sources of the noise related to random fluctuations in the system. It is hence worthwhile to observe that the quantum features of the signal radiation would be affected by the available noise that includes the fluctuation arising from pumping mechanism.

In the same way, to obtain other forms of the correlations of the stochastic noise force, one can start from Eq. (3.122);

$$\frac{d}{dt}\langle\alpha^*(t)\alpha(t)\rangle = -\kappa\langle\alpha^*(t)\alpha(t)\rangle - \lambda\langle\beta^*(t)\alpha^2(t)\rangle - \lambda\langle\beta(t)\alpha^{*2}(t)\rangle$$

$$+ \langle\alpha^*(t)\eta(t)\rangle + \langle\alpha(t)\eta^*(t)\rangle,\tag{3.136}$$

---

[8]The principal aim in the present approach is to characterize the correlation properties of the stochastic noise force. To achieve this goal, it might be appealing to begin with the general mathematical condition that if $\int_0^t e^{-a(t-t')}\langle f(t')g(t)\rangle dt' = \sigma$, then $\langle f(t')g(t)\rangle = 2\sigma\delta(t-t')$, where a is taken to be constant of time.

where following a similar approach leads to

$$\langle\alpha^*(t)\eta(t)\rangle = \int_0^t e^{-\frac{\kappa}{2}(t-t')}\langle\eta(t)\eta^*(t')\rangle dt', \tag{3.137}$$

$$\langle\alpha(t)\eta^*(t)\rangle = \int_0^t e^{-\frac{\kappa}{2}(t-t')}\langle\eta^*(t)\eta(t')\rangle dt'. \tag{3.138}$$

One can ascertained from Eqs. (3.137) and (3.138) that $\langle\alpha^*(t)\eta(t')\rangle = \langle\alpha(t)\eta^*(t')\rangle$. In the same manner, it may not be hard to show by using Eq. (3.119) that

$$\frac{d}{dt}\langle\hat{a}^\dagger(t)\hat{a}(t)\rangle = -\kappa\langle\hat{a}^\dagger(t)\hat{a}(t)\rangle - \lambda\left[\langle\hat{a}^{\dagger^2}(t)\hat{b}(t)\rangle + \langle\hat{a}^2(t)\hat{b}^\dagger(t)\rangle\right]$$
$$+ \langle\hat{a}^\dagger(t)\hat{F}(t)\rangle + \langle\hat{F}^\dagger(t)\hat{a}(t)\rangle \tag{3.139}$$

in which

$$\langle\hat{F}^\dagger(t)\hat{a}(t)\rangle = \langle\hat{F}^\dagger(t)\hat{a}(0)\rangle e^{-\frac{\kappa}{2}t}$$
$$+ \int_0^t e^{-\frac{\kappa}{2}(t-t')}\left[\langle\hat{F}^\dagger(t)\hat{F}(t')\rangle - \lambda\langle\hat{F}^\dagger(t)\hat{b}(t')\hat{a}^\dagger(t')\rangle\right]dt'. \tag{3.140}$$

Since the noise force at later time does not correlate with the system variables at earlier times, one can verify that

$$\langle\hat{F}^\dagger(t)\hat{a}(t)\rangle = \frac{\kappa N}{2}. \tag{3.141}$$

Equation (3.139) can therefore be expressed in terms of the c-number variables associated with the normal ordering as

$$\frac{d}{dt}\langle\alpha^*(t)\alpha(t)\rangle = -\kappa\langle\alpha^*(t)\alpha(t)\rangle - \lambda\left[\langle\alpha^{*^2}(t)\beta(t)\rangle + \langle\alpha^2(t)\beta^*(t)\rangle\right] + \kappa N. \tag{3.142}$$

After that, comparison of Eq. (3.136) and (3.142) indicates that $\langle\alpha(t)\eta^*(t)\rangle = \kappa N/2$. One can then attest following the approach already outlined to

$$\langle\eta(t)\eta^*(t')\rangle = \kappa N\delta(t - t'), \tag{3.143}$$

which looks the same as the correlation of the Langevin noise operator (see Eq. (3.60)). This can be connoted as the correlation of the sources of random fluctuations other than the reservoir do not contribute directly to the mean photon number or intensity of the signal mode.[9]

Stochastic differential equation can also be obtained from the master equation with the aid of the definition

$$\frac{d}{dt}\langle \hat{a}(t)\rangle = Tr\left(\frac{d\hat{\rho}}{dt}\hat{a}\right). \tag{3.144}$$

The corresponding expressions in terms of the c-number variables associated with the normal ordering so can be put upon using Eq. (3.98) in the form

$$\frac{d}{dt}\langle \alpha(t)\rangle = -\frac{\kappa}{2}\langle \alpha(t)\rangle - \varepsilon\langle \alpha^*(t)\rangle, \tag{3.145}$$

$$\frac{d}{dt}\langle \alpha^*(t)\alpha(t)\rangle = -\kappa\langle \alpha^*(t)\alpha(t)\rangle - \varepsilon\left[\langle \alpha^2(t)\rangle + \langle \alpha^{*^2}(t)\rangle\right] + \kappa N, \tag{3.146}$$

$$\frac{d}{dt}\langle \alpha^2(t)\rangle = -\kappa\langle \alpha^2(t)\rangle - 2\varepsilon\langle \alpha^*(t)\alpha(t)\rangle - \varepsilon + \kappa M. \tag{3.147}$$

One may note that it is admissible to express Eq. (3.145) as

$$\frac{d\alpha(t)}{dt} = -\frac{\kappa}{2}\alpha(t) - \varepsilon\alpha^*(t) + f(t), \tag{3.148}$$

where $f(t)$ is the relevant stochastic noise force whose properties remain to be determined (compare with Eq. (3.122). Markedly, Eq. (3.145) would be equal to the expectation value of Eq. (3.148) provided that $\langle f(t)\rangle = 0$. It is hence desirable to designate Eq. (3.148) as the sought for stochastic differential equation whose property is the same as what is already obtained.

In addition to calculating the correlations of the stochastic noise force, one may intend to attempt to obtain the time evolution of the c-number dynamical variable. To this effect, the solution of Eq. (3.148) can be written as

$$\alpha(t) = \alpha(0)e^{-\frac{\kappa}{2}t} + \int_0^\infty e^{-\frac{\kappa}{2}(t-t')}\left[f(t') - \varepsilon\alpha^*(t')\right]dt'. \tag{3.149}$$

---

[9]Parametric approximation can also be invoked in which the amplitude of the pump mode $\beta(t)$ is treated as a real positive constant: $\varepsilon = \lambda\beta(t)$. Once the properties of the stochastic noise force are known, one should be able to evaluate all important correlations, that is, the system one wishes to consider could be analyzed directly by employing the correlations resulting from stochastic differential equation.

One may now seek to determine the correlations of the pertinent noise force following the outlined procedure. To do so, it may not be hard to see upon using Eq. (3.148) that

$$\frac{d}{dt}\langle\alpha^*(t)\alpha(t)\rangle = -\kappa\langle\alpha^*(t)\alpha(t)\rangle - \varepsilon\big[\langle\alpha^2(t)\rangle + \langle\alpha^{*2}(t)\rangle\big]$$

$$+ \langle\alpha^*(t)f(t)\rangle + \langle f^*(t)\alpha(t)\rangle, \tag{3.150}$$

where comparison with Eq. (3.146) indicates

$$\langle\alpha^*(t)f(t)\rangle + \langle f^*(t)\alpha(t)\rangle = \kappa N. \tag{3.151}$$

And application of Eq. (3.148) once again signifies

$$\frac{d}{dt}\langle\alpha^2(t)\rangle = -\kappa\langle\alpha^2(t)\rangle - 2\varepsilon\langle\alpha^*(t)\alpha(t)\rangle + 2\langle\alpha(t)f(t)\rangle, \tag{3.152}$$

from which follows

$$\langle\alpha(t)f(t)\rangle = \frac{1}{2}(\kappa M - \varepsilon). \tag{3.153}$$

It is also possible to delve into the process of obtaining the correlation properties of the stochastic noise force, that is, in view of Eq. (3.149), one can write

$$\langle\alpha(t)f(t)\rangle = \langle\alpha(0)f(t)\rangle e^{-\frac{\kappa}{2}t} + \int_0^t e^{-\frac{\kappa}{2}(t-t')}\big[\langle f(t')f(t)\rangle - \varepsilon\langle\alpha^*(t')f(t)\rangle\big]dt'. \tag{3.154}$$

Based on the proposal that the system variables at sometime and the noise force at any later time do not correlate, it may not be hard to get

$$\langle\alpha(t)f(t)\rangle = \int_0^t e^{-\frac{\kappa}{2}(t-t')}\langle f(t')f(t)\rangle dt', \tag{3.155}$$

from which follows

$$\langle f(t')f(t)\rangle = (\kappa M - \varepsilon)\delta(t-t'), \qquad \langle f^*(t)f(t')\rangle = \kappa N\delta(t-t'). \tag{3.156}$$

In the interest of shifting the attention to finding the evolution of the dynamical variables, it might be desirable to introduce a new variable since the emerging stochastic differential equation may not be directly solvable due to the required lengthy iterative integrations. It thus appears appealing to put Eq. (3.149) and its complex conjugate in a more convenient way with the help of two c-number variables defined by $\alpha_\pm(t) = \alpha^*(t) \pm \alpha(t)$ as

$$\frac{d\alpha_\pm}{dt} = -\lambda_\pm \alpha_\pm + f_\pm(t), \tag{3.157}$$

where $\lambda_\pm = k \pm 2\varepsilon$ and $f_\pm(t) = f^*(t) \pm f(t)$, and the solution of Eq. (3.157) can be written as

$$\alpha_\pm(t) = \alpha_\pm(0)e^{-\frac{\lambda_\pm}{2}t} + \int_0^t e^{-\frac{\lambda_\pm}{2}(t-t')} f_\pm(t')dt', \tag{3.158}$$

from which follows[10]

$$\alpha(t) = A_{+(t)}\alpha(0) + A_-(t)\alpha^*(0) + B_+(t) - B_-(t), \tag{3.159}$$

where

$$A_\pm(t) = \frac{1}{2}\left(e^{-\frac{\lambda_+}{2}t} \pm e^{-\frac{\lambda_-}{2}t}\right), \tag{3.160}$$

$$B_\pm(t) = \frac{1}{2}\int_0^t e^{-\frac{\lambda_\pm}{2}(t-t')} f_\pm(t')dt'. \tag{3.161}$$

It might also be required to obtain the Q-function even when the pertinent differential equation is solvable specially when one wishes to study the photon statistics in terms of the relevant photon distribution function. The anti-normal ordered characteristic function (2.83) can thus be rewritten in terms of the c-number variables associated with the normal ordering as

$$\phi(z, t) = e^{-z^*z} \exp\frac{1}{2}\left[z^2\langle\alpha^{*^2}(t)\rangle + z^{*^2}\langle\alpha^2(t)\rangle - 2z^*z\langle\alpha^*(t)\alpha(t)\rangle\right] \tag{3.162}$$

since $\alpha(t)$ is a Gaussian variable of zero mean epitomized by Eq. (3.111).

It should be clear by now that the correlation functions in Eq. (3.162) can be obtained from Eq. (3.159) and its complex conjugate. By assuming the signal mode to be initially in the vacuum state, and making use of Eqs. (3.156) and (3.161), one can ascertain that

$$\langle B_-(t)B_+(t)\rangle = \langle B_+(t)B_-(t)\rangle = 0, \tag{3.163}$$

$$\langle B_\pm^2(t)\rangle = \frac{\kappa N \pm \kappa M - \varepsilon}{2\lambda_\pm}\left[1 - e^{-\lambda_\pm t}\right]. \tag{3.164}$$

---

[10]One can also verify that the c-number equation associated with the normal ordering pertinent to Eq. (3.104) is the same as Eq. (3.159).

With the aid of Eq. (3.163), it may not be hard to reach at

$$\langle \alpha^2(t) \rangle = \frac{\kappa N + \kappa M - \varepsilon}{2\lambda_+} \left[1 - e^{-\lambda_+ t} \right] + \frac{\kappa M - \kappa N - \varepsilon}{2\lambda_-} \left[1 - e^{-\lambda_- t} \right], \qquad (3.165)$$

$$\langle \alpha^*(t)\alpha(t) \rangle = \frac{\kappa N + \kappa M - \varepsilon}{2\lambda_+} \left[1 - e^{-\lambda_+ t} \right] - \frac{\kappa M - \kappa N - \varepsilon}{2\lambda_-} \left[1 - e^{-\lambda_- t} \right].$$
$$(3.166)$$

Then upon using Eqs. (3.165) and (3.166), the characteristic function can be expressed as (compare with Eq. (3.114))

$$\phi_a(z, t) = \exp \left[-a_1 z^* z + \frac{a_2}{2} (z^{*2} + z^2) \right], \qquad (3.167)$$

with

$$a_1 = 1 + \frac{\kappa N + \kappa M - \varepsilon}{2\lambda_+} (1 - e^{-\lambda_+ t}) - \frac{\kappa M - \kappa N - \varepsilon}{2\lambda_-} (1 - e^{-\lambda_- t}), (3.168)$$

$$a_2 = \frac{\kappa N + \kappa M - \varepsilon}{2\lambda_+} (1 - e^{-\lambda_+ t}) + \frac{\kappa M - \kappa N - \varepsilon}{2\lambda_-} (1 - e^{-\lambda_- t}). \qquad (3.169)$$

3.2.3 Fokker-Planck Equation

Fokker-Planck equation is a c-number equivalent of the master equation put in a certain operator ordering, and happens to be a second-order differential equation for quantum optical systems frequently studied. Hoping that devising the way of solving the Fokker-Planck equation would be desirable in theoretical research, the operators in Eq. (3.98) are required to be put in the normal order [6, 11]. Despite this observation, note that some of the terms in the master equation are not put in the normal order and so should be put in that way by applying the commutation relations:

$$\left[\hat{a}, \ f(\hat{a}, \hat{a}^\dagger) \right] = \frac{\partial f}{\partial \hat{a}^\dagger}, \qquad \left[f(\hat{a}, \hat{a}^\dagger), \ \hat{a}^\dagger \right] = \frac{\partial f}{\partial \hat{a}}. \qquad (3.170)$$

In case the cavity is assumed to be in contact with vacuum reservoir, one can obtain

$$\frac{\partial Q(\alpha)}{\partial t} = \left[\frac{\varepsilon}{2} \left(2\frac{\partial}{\partial \alpha^*}\alpha + 2\frac{\partial}{\partial \alpha}\alpha^* + \frac{\partial^2}{\partial \alpha^2} + \frac{\partial^2}{\partial \alpha^{*2}} \right) \right.$$
$$\left. + \frac{\kappa}{2} \left(\frac{\partial}{\partial \alpha}\alpha + \frac{\partial}{\partial \alpha^*}\alpha^* + \frac{2\partial^2}{\partial \alpha \partial \alpha^*} \right) \right] Q(\alpha); \qquad (3.171)$$

which is the Fokker-Planck equation.

It might also be appealing to describe Eq. (3.171) in terms of Cartesian coordinates by exploiting the variable transformations:

$$\alpha = x + iy \tag{3.172}$$

$$\frac{\partial}{\partial\alpha} = \frac{1}{2}\left[\frac{\partial}{\partial x} - i\frac{\partial}{\partial y}\right], \tag{3.173}$$

$$\frac{\partial}{\partial\alpha}\alpha = \frac{1}{2}\left[\frac{\partial}{\partial x}x + i\frac{\partial}{\partial x}y - i\frac{\partial}{\partial y}x + \frac{\partial}{\partial y}y\right] \tag{3.174}$$

as

$$\frac{\partial}{\partial t}Q(x, y, t) = \left[\frac{\kappa + 2\varepsilon}{2}\frac{\partial}{\partial x}x + \frac{\kappa - 2\varepsilon}{2}\frac{\partial}{\partial y}y\right.$$
$$\left. + \frac{\kappa + \varepsilon}{4}\frac{\partial^2}{\partial x^2} + \frac{\kappa - \varepsilon}{4}\frac{\partial^2}{\partial y^2}\right]Q(x, y, t), \tag{3.175}$$

which denotes Fokker-Planck equation in phase space representation. The Fokker-Planck equation can also be expressed when the system is placed in thermal reservoir as

$$\frac{\partial}{\partial t}Q(x, y, t) = \left[\frac{\kappa + 2\varepsilon}{2}\frac{\partial}{\partial x}x + \frac{\kappa - 2\varepsilon}{2}\frac{\partial}{\partial y}y\right.$$
$$\left. + \frac{\kappa(\bar{n} + 1) + \varepsilon}{4}\frac{\partial^2}{\partial x^2} + \frac{\kappa(\bar{n} + 1) - \varepsilon}{4}\frac{\partial^2}{\partial y^2}\right]Q(x, y, t). \tag{3.176}$$

While devising the means of solving the pertinent differential equation, one may strive to look into some of the techniques used to solve the emerging differential equations. Details of the approach explicably comprises techniques such as converting the operator equation into c-number equation, solving the resulting differential equations using classical approaches, and then expressing the Q-function in terms of c-number variables. One of the methods that can be applied to solve the emerging second-order differential equation is the coherent state propagator formulation. In this approach, the Fokker-Planck equation would be expressed in the same way as Schrödinger wave equation.

It might be helpful in this case to begin with a common experience that the time development of the operator and the state vector can be expressed in terms of the evolution operator as

$$|\alpha(t)\rangle = \hat{U}(t)|\alpha(0)\rangle, \qquad \hat{x}(t) = \hat{U}^\dagger(t)\hat{x}\hat{U}(t), \tag{3.177}$$

where the corresponding wave function can be put as

$$\langle x | a'; t \rangle = \langle x' | a' \rangle \exp\left(-iEt\right), \qquad \Phi(x, t) = \Phi(x(0)) \exp\left(-iEt\right),$$
$$(3.178)$$

which can be expressed in a semiclassical approximation as

$$\Phi(x, t) = \frac{\text{constant}}{\left[V(x) - E\right]^{\frac{1}{4}}} \exp\left[\pm \int dx' [2m(V(x') - E)]^{\frac{1}{2}} - iEt\right], \qquad (3.179)$$

where E is the total energy and $V(x)$ is the potential.

In the same way, the corresponding state vector evolves according to

$$|\alpha; t\rangle = \exp\left[-i\hat{H}t\right]|\alpha; t = 0\rangle, \qquad (3.180)$$

whereupon introducing the completeness relation; and then, multiplying the resulting expression by $\langle x''|$ from left indicate

$$\langle x'' | \alpha; t \rangle = \sum_{a'} \langle x'' | a' \rangle \langle a' | \alpha, t = 0 \rangle \exp\left[-iEt\right]. \qquad (3.181)$$

With the aid of the completeness relation once again, it is not then hard to see that

$$\Phi(x'', t) = \int d^3x' \sum_{a'} \langle x'' | a' \rangle \langle a' | x' \rangle \langle x' | \alpha, t = 0 \rangle \exp\left(-iEt\right), \qquad (3.182)$$

which can be put in a more appealing form as

$$\Phi(x'', t) = \int d^3x' K(x'', t; x', t = 0) \Phi(x', t = 0), \qquad (3.183)$$

where

$$K(x'', t; x', t = 0) = \sum_{a'} \langle x'' | a' \rangle \langle a' | x' \rangle \exp\left(-iEt\right) \qquad (3.184)$$

is the propagator that can also be defined as

$$K(x'', t; x', t = 0) = \langle x'' | \exp\left(-i\hat{H}t\right) | x' \rangle. \qquad (3.185)$$

Equation (3.185) can also be rewritten as

$$K(x'', t; x', t = 0) = \sum_{a'} \langle x'' | \exp\left(-i\hat{H}t\right) | a' \rangle \langle a' | \exp\left(i\hat{H}t\right) | x' \rangle, \qquad (3.186)$$

from which follows $K(x'', t; x', t = 0) = \langle x''; t | x'; t_0 \rangle$. Critical observation may reveal that the notion of propagator evokes the possibility of expressing the transition through composite in time and position steps.

In the concurrent state manipulation formulation, it is required to convert the Fokker Planck equation into the Schrödinger wave equation type by using the recipe that [12]

$$\frac{\partial}{\partial t} |Q(t)\rangle = -i\hat{H}|Q(t)\rangle. \tag{3.187}$$

In line with the usual transformation from classical to quantum,

$$\left(x, y, \frac{\partial}{\partial x}, \frac{\partial}{\partial y}, Q(x, y, t) \right) \longrightarrow (\hat{x}, \hat{y}, i\hat{p}_x, i\hat{p}_y, |Q(t)\rangle), \tag{3.188}$$

the corresponding quantum Hamiltonian would look like

$$\hat{H} = -i\frac{\kappa + \varepsilon}{4}\hat{p}_x^2 - i\frac{\kappa - \varepsilon}{4}\hat{p}_y^2 - \frac{\kappa + 2\varepsilon}{2}\hat{p}_x x - \frac{\kappa - 2\varepsilon}{2}\hat{p}_y y \tag{3.189}$$

(compare with Eq. (3.175)).

Note that the solution of Eq. (3.187) can be written as

$$|Q(t)\rangle = e^{-i\hat{H}t}|Q(0)\rangle, \tag{3.190}$$

where $|Q(0)\rangle$ stands for the initial state and the associated wave function can be expressed as $Q(x, y, t) = \langle x, y | Q(t) \rangle$, which is equivalent to the relevant Q-function in the phase space representation. On account of this, Eq. (3.190) can be put in the form

$$Q(x, y, t) = \langle x, y | e^{-i\hat{H}t} | Q(0) \rangle. \tag{3.191}$$

With the application of a two-dimensional completeness relation for a position eigenstate, one gets

$$Q(x, y, t) = \int dx' dy' \langle x, y | e^{-i\hat{H}t} | x', y' \rangle \langle x', y' | Q(0) \rangle. \tag{3.192}$$

One may note that it is also possible to define

$$\langle x, y | e^{-i\hat{H}t} | x', y' \rangle = Q(x, y, t | x', y', 0) \tag{3.193}$$

as Q-function propagator. Notice that

$$Q(x, y, t | x', y', 0)|_{t=0} = \delta(x - x')\delta(y - y') \tag{3.194}$$

since the position eigenstate is assumed to be complete. It may not be difficult to see that $Q(x', y', 0) = \langle x', y' | Q(0) \rangle$ is the Q-function taken as the initial wave function.

It is also good to highlight that the Q-function propagator related to the quadratic Hamiltonian of the form

$$\hat{H} = a\hat{p}_x^2 + b(t)\hat{p}_x\hat{x} + c(t)\hat{x}^2, \tag{3.195}$$

for $a \neq 0$ can be defined as

$$Q(x'', T | x', 0) = \left[\frac{i\partial^2 S_c}{2\pi \partial x' \partial x''} \right]^{1/2} \exp\left[-\frac{1}{2} \int_0^T b(t)dt + iS_c \right] \tag{3.196}$$

for anti-standard ordering, where $S_c(x', x'', T) = \int_0^T \mathcal{L}(x, \dot{x}, t)dt$ is the classical action with $x' = x(0)$ and $x'' = x(T)$.

To obtain the classical action, one first needs to determine the c-number equation related to the Hamiltonian, and then apply Hamilton's equations. To this effect, the c-number expression that corresponds to Eq. (3.189) turns out to be

$$H = -i\frac{\kappa + \varepsilon}{4}p_x^2 - i\frac{\kappa - \varepsilon}{4}p_y^2 - \frac{\kappa + 2\varepsilon}{2}p_x x - \frac{\kappa - 2\varepsilon}{2}p_y y \tag{3.197}$$

in which application of the Hamilton's equation, $\dot{q}_i = \frac{\partial H}{\partial p_i}$, gives rise to

$$\dot{x} = -i\frac{\kappa + \varepsilon}{2}p_x - \frac{\kappa + 2\varepsilon}{2}x, \qquad \dot{y} = -i\frac{\kappa - \varepsilon}{2}p_y - \frac{\kappa - 2\varepsilon}{2}y, \tag{3.198}$$

and solving Eq. (3.198) separately yields

$$p_x = \frac{2i}{\kappa + \varepsilon}\left[\dot{x} + \frac{\kappa + 2\varepsilon}{2}x \right], \qquad p_y = \frac{2i}{\kappa - \varepsilon}\left[\dot{y} + \frac{\kappa - 2\varepsilon}{2}y \right]. \tag{3.199}$$

With the help of $\mathcal{L} = \sum_i p_i\dot{q}_i - H$, the classical Lagrange function is found after that to be

$$\mathcal{L}_c(\dot{x}, x, \dot{y}, y, t) = \frac{i}{\kappa + \varepsilon}\left[\dot{x} + \frac{\kappa + 2\varepsilon}{2}x \right]^2 + \frac{i}{\kappa - \varepsilon}\left[\dot{y} + \frac{\kappa - 2\varepsilon}{2}y \right]^2. \tag{3.200}$$

Besides, employing the Euler-Lagrange equation, $\frac{d}{dt}\left(\frac{\partial \mathcal{L}}{\partial \dot{q}_i} \right) - \frac{\partial \mathcal{L}}{\partial q_i} = 0$, one can see that

$$\ddot{x} - \left[\frac{\kappa + 2\varepsilon}{2} \right]^2 x = 0, \qquad \ddot{y} - \left[\frac{\kappa - 2\varepsilon}{2} \right]^2 y = 0, \tag{3.201}$$

where the solution of these expressions can be put in the form

$$x(t) = E e^{\frac{\kappa+2\varepsilon}{2}t} + F e^{-\frac{\kappa+2\varepsilon}{2}t}, \qquad y(t) = G e^{\frac{\kappa-2\varepsilon}{2}t} + H e^{-\frac{\kappa-2\varepsilon}{2}t}, \qquad (3.202)$$

in which E, F, G and H are constants to be determined from the initial conditions. Then substitution of Eq. (3.202) and the first derivatives into Eq. (3.200) leads to

$$\mathcal{L}_c(\dot{x}, x, \dot{y}, y, t) = i \frac{(\kappa+2\varepsilon)^2}{\kappa+\varepsilon} E^2 e^{(\kappa+2\varepsilon)t} + i \frac{(\kappa-2\varepsilon)^2}{\kappa-\varepsilon} G^2 e^{(\kappa-2\varepsilon)t}, \qquad (3.203)$$

in which

$$E = -\frac{x' - x'' e^{\frac{\kappa+2\varepsilon}{2}T}}{e^{(\kappa+2\varepsilon)T} - 1}, \qquad G = -\frac{y' - y'' e^{\frac{(\kappa-2\varepsilon)}{2}T}}{e^{(\kappa-2\varepsilon)T} - 1}, \qquad (3.204)$$

where $y' = y(0)$ and $y'' = y(T)$.

Once the Lagrange function is obtained, the classical action follows

$$S_c(x', x'', y', y'', T) = i \frac{\kappa+2\varepsilon}{\kappa+\varepsilon} \left(\frac{x' - x'' e^{\frac{\kappa+2\varepsilon}{2}T}}{e^{(\kappa+2\varepsilon)T} - 1} \right) + i \frac{\kappa-2\varepsilon}{\kappa-\varepsilon} \left(\frac{y' - y'' e^{\frac{\kappa-2\varepsilon}{2}T}}{e^{(\kappa-2\varepsilon)T} - 1} \right). \qquad (3.205)$$

Equation (3.196) obviously can be generalized for a two-mode quantum system as

$$Q(x'', y'', T|x', y', 0) = \left[\frac{i\partial^2 S_c}{2\pi \partial x' \partial x''} \right]^{1/2} \left[\frac{i\partial^2 S_c}{2\pi \partial y' \partial y''} \right]^{1/2}$$

$$\times \exp\left[-\frac{1}{2} \int_0^T b'(t)dt - \frac{1}{2} \int_0^T c'(t)dt + i S_c \right], \qquad (3.206)$$

where $b'(t) = (\kappa + 2\varepsilon)/2$ and $c'(t) = (\kappa - 2)\varepsilon/2$ for a degenerate parametric oscillator.

On the basis of Eqs. (3.205) and (3.206), one can reach at

$$Q(x, y, t|x', y', 0) = \frac{1}{\pi} \left[\frac{\kappa^2 - 4\varepsilon^2}{(\kappa^2 - \varepsilon^2)(1 - e^{-(\kappa+2\varepsilon)t})(1 - e^{-(\kappa-2\varepsilon)t})} \right]$$

$$\times \exp\left[-\frac{\kappa+2\varepsilon}{\kappa+\varepsilon} \left(\frac{(x' e^{-\frac{\kappa+2\varepsilon}{2}t} - x)^2}{1 - e^{-(\kappa+2\varepsilon)t}} \right) \right.$$

$$\left. -\frac{\kappa-2\varepsilon}{\kappa-\varepsilon} \left(\frac{(y' e^{-\frac{\kappa-2\varepsilon}{2}t} - y)^2}{1 - e^{-(\kappa-2\varepsilon)t}} \right) \right], \qquad (3.207)$$

where the term corresponding to $t = 0$ is ignored and x'' is replaced by x, T by t and y'' by y. If the signal mode is assumed to be initially in vacuum state, the Q-function that describes the initial state in a position space representation would have the form

$$Q(x', y', 0) = \frac{1}{\pi} \exp\left[-x'^2 - y'^2\right].\tag{3.208}$$

Next upon applying Eqs. (3.192), (3.193) and (3.207), it is possible to show that

$$Q(x, y, t) = \frac{A(t)}{\pi} \sqrt{\frac{1}{(1 + B(t)e^{-(\kappa+2\varepsilon)t})(1 + C(t)e^{-(\kappa-2\varepsilon)t})}}$$
$$\times \exp\left[x^2 B(t)\left(\frac{B(t)e^{-(\kappa+2\varepsilon)t}}{1 + B(t)e^{-(\kappa+2\varepsilon)t}} - 1\right)\right.$$
$$\left. + y^2 C(t)\left(\frac{C(t)e^{-(\kappa-2\varepsilon)t}}{1 + C(t)e^{-(\kappa-2\varepsilon)t}} - 1\right)\right],\tag{3.209}$$

in which

$$A(t) = \sqrt{\frac{\kappa^2 - 4\varepsilon^2}{(\kappa^2 - \varepsilon^2)(1 - e^{-(\kappa+2\varepsilon)t})(1 - e^{-(\kappa-2\varepsilon)t})}},\tag{3.210}$$

$$B(t) = \frac{\kappa + 2\varepsilon}{(\kappa + \varepsilon)(1 - e^{-(\kappa+2\varepsilon)t})},\tag{3.211}$$

$$C(t) = \frac{\kappa - 2\varepsilon}{(\kappa - \varepsilon)(1 - e^{(\kappa-2\varepsilon)})}.\tag{3.212}$$

In light of the introduced variable transformations, the Q-function can then be put in terms of coherent state representation in the form

$$Q(\alpha^*, \alpha, t) = \frac{D(t)}{\pi} \exp[-E_+(t)\alpha^*\alpha + E_-(t)(\alpha^{*2} + \alpha^2)],\tag{3.213}$$

with

$$D(t) = \sqrt{\frac{\kappa^2 - 4\varepsilon^2}{(\kappa + \varepsilon(1 + e^{-(\kappa+2\varepsilon)t}))(\kappa - \varepsilon(1 + e^{-(\kappa-2\varepsilon)t}))}},\tag{3.214}$$

$$E_\pm(t) = \pm\frac{1}{2}\left[\frac{\kappa + 2\varepsilon}{\kappa + \varepsilon(1 + e^{-(\kappa+2\varepsilon)t})} \pm \frac{\kappa - 2\varepsilon}{\kappa - \varepsilon(1 + e^{-(\kappa-2\varepsilon)t})}\right]\tag{3.215}$$

(compare with Eq. (3.118)).

It should also be straightforward to verify that at $t = 0$ the Q-function for the vacuum state would be regained, and at steady state Eq. (3.213) reduces to

$$Q(\alpha^*, \alpha) = \frac{1}{\pi} \left(\frac{\kappa^4 - 4\varepsilon^4}{\kappa^2 - \varepsilon^2} \right)^{1/2}$$

$$\times \exp\left[-\alpha^*\alpha \left(\frac{\kappa^2 - 2\varepsilon^2}{\kappa^2 - \varepsilon^2} \right) - (\alpha^2 + \alpha^{*2}) \left(\frac{\kappa\varepsilon}{\kappa^2 - \varepsilon^2} \right) \right], \qquad (3.216)$$

which is the Q-function for a degenerate parametric oscillator when the signal mode is initially in vacuum state. One can see from Figs. 3.4 and 3.5 that the Q-function for this case shows noticeable Gaussian behavior with the characteristics that near threshold it is stretched along y-axis while shirking along x, which might be the reminiscent of the inherent squeezing feature (relate with the interpretation in Sect. 2.3).

When the cavity containing the system is coupled to thermal reservoir, the pertinent Hamiltonian likewise turns out to be

$$\hat{H} = -i\frac{\kappa(\bar{n}+1)+\varepsilon}{4}\hat{p}_x^2 - i\frac{\kappa(\bar{n}+1)-\varepsilon}{4}\hat{p}_y^2 - \frac{\kappa+2\varepsilon}{2}\hat{p}_x x - \frac{\kappa-2\varepsilon}{2}\hat{p}_y y,$$
$$(3.217)$$

from which follows using the same approach

$$Q(\alpha^*, \alpha) = \frac{1}{\pi}\sqrt{\frac{\kappa^2 - 4\varepsilon^2}{(\kappa(\bar{n}+1))^2 - \varepsilon^2}} \exp\left[-\alpha^*\alpha \left(\frac{\kappa^2(\bar{n}+1) - 2\varepsilon^2}{(\kappa(\bar{n}+1))^2 - \varepsilon^2} \right) \right.$$

$$\left. -(\alpha^2 + \alpha^{*2}) \left(\frac{\kappa\varepsilon(2\bar{n}+1)}{(\kappa(\bar{n}+1))^2 - \varepsilon^2} \right) \right]. \qquad (3.218)$$

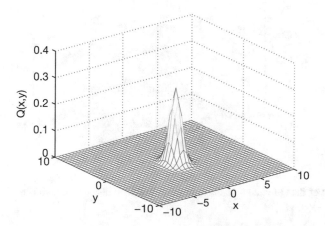

Fig. 3.4 Plot of the Q-function of the light generated by a degenerate parametric oscillator far from threshold ($\varepsilon = 0.25\kappa$)

Fokker-Planck equation can also be put in the form

$$\frac{\partial F(x_i)}{\partial t} = \left[-\sum_i^n a_i \frac{\partial}{\partial x_i} + \frac{1}{2} \sum_{ij}^n D_{ij} \frac{\partial^2}{\partial x_i \partial x_j} \right] F(x_i), \tag{3.219}$$

where a_i's are the drift terms and D_{ij}'s are the diffusion coefficients. To solve such a problem, it might be required to impose initial condition such as assuming the system to be initially in vacuum state and the quantum states as complete: $F(x_i, t)|_{t=0} = \prod_i \delta(x_i - x_i')$, where $\prod_i$ stands for multiple products and the primed variables are the values fixed at initial time.

According to Wang-Uhlenbeck method, the solution of Eq. (3.219) can be written as

$$F(x_i, t) = \frac{1}{\sqrt{\pi^n \det \sigma_{ij}(t)}} \exp \left[-\sum_{ij}^n \sigma_{ij}^{-1}(t)(x_i - x_i' e^{a_i t})(x_j - x_j' e^{a_j t}) \right],$$
$$\tag{3.220}$$

where n represents the number of the variables involved in describing the system and

$$\sigma_{ij}(t) = -\frac{2D_{ij}}{a_i + a_j} \left(1 - e^{(a_i + a_j)t} \right). \tag{3.221}$$

One of the advantages of this method is that knowing the drift terms and diffusion coefficients is sufficient.

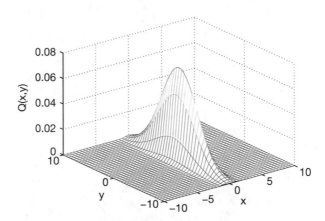

Fig. 3.5 Plot of the Q-function of the light generated by a degenerate parametric oscillator near threshold ($\varepsilon = 0.49\kappa$)

As illustration, imagine a differential equation of the form

$$\frac{\partial}{\partial t} Q(x, y, t|x', y', 0) = \lambda \left[\frac{\partial}{\partial x} + \frac{\partial}{\partial y} + \frac{1}{4} \left(\frac{\partial^2}{\partial^2 x} - \frac{\partial^2}{\partial^2 y} \right) \right] Q(x, y, t|x', y', 0).$$
(3.222)

To express Eq. (3.222) in the same way as Eq. (3.219), it is desirable setting $a_1 = a_2 = -\lambda$, $D_{11} = \lambda/2$, $D_{22} = -\lambda/2$ and $D_{12} = D_{21} = 0$, that is,

$$\sigma_{11} = \frac{1}{2} \left(1 - e^{-2\lambda t} \right), \qquad \sigma_{22} = \frac{1}{2} \left(1 - e^{2\lambda t} \right), \qquad \sigma_{12} = \sigma_{21} = 0. \quad (3.223)$$

So with the help of Eq. (3.223) and using the fact that

$$\det \sigma_{ij} = \begin{vmatrix} \sigma_{11} & 0 \\ 0 & \sigma_{22} \end{vmatrix} = \sigma_{11}\sigma_{22},$$
(3.224)

one gets

$$\sqrt{\det \sigma_{ij}} = \left[\frac{1}{4} (e^{\lambda t} - e^{-\lambda t})(e^{\lambda t} - e^{-\lambda t}) \right]^{\frac{1}{2}} = i \sinh \lambda t.$$
(3.225)

Applying the variable transformations $x_1 = x$ and $x_2 = y$, the solution of Eq. (3.222) can then be put in view of Eq. (3.220) in the form

$$Q(x, y, t|x', y', 0) = \frac{1}{i\pi \sinh \lambda t} \exp \left[-\frac{2(x - x'e^{-\lambda t})^2}{1 - e^{-2\lambda t}} - \frac{2(y - y'e^{\lambda t})^2}{1 - e^{2\lambda t}} \right].$$
(3.226)

One may realize that the Q-function at later time can be obtained by integrating Eq. (3.226) over the initial variables:

$$Q(x, y, t) = \int dx'dy' Q(x, y, t|x', y', 0)Q(x', y', 0).$$
(3.227)

Assuming that the system is initially in vacuum state, and making use of Eqs. (3.226) and (3.227), one can obtain[11]

$$Q(x, y, t) = \frac{\sec h\lambda t}{\pi} \exp \left[\frac{1}{\sinh^2 \lambda t} (2(x^2 + y^2) - x^2 e^{2\lambda t} - y^2 e^{-2\lambda t}) \right],$$
(3.228)

which stands for the Q-function related to Eq. (3.222).

---

[11] The integral identity

$$\int dx'dy' e^{-ax'^2 + bx' - cy'^2 + dy'} = \frac{\pi}{\sqrt{ac}} \exp \left[\frac{b^2}{4a} + \frac{d^2}{4c} \right]$$

is applied.

One may also need to outline the formulation for solving a general Fokker-Planck equation at steady state. Imagine the Fokker-Planck equation of the type given by Eq. (3.171) that can be expressed in phase space representation using the usual variable transformation as

$$\frac{\partial}{\partial t} F(x_i) = \sum_{i=1}^{2} \frac{\partial}{\partial x_i} \left[A_i x_i + \sum_{j=1}^{2} B_{ij} \frac{\partial}{\partial x_j} \right] F(x_i), \tag{3.229}$$

where $A_1 = (\kappa + 2\varepsilon)/2$, $A_2 = (\kappa - 2\varepsilon)/2$, $B_{11} = (\kappa + \varepsilon)/2$, $B_{22} = (\kappa - \varepsilon)/2$ and $B_{12} = B_{21} = $ constant. It may worth stressing that Eq. (3.229) is different from the Wang-Uhlenbeck type of differential equation specially in the drift part. At steady state, $\frac{\partial F(x_i)}{\partial t} = 0$; as result

$$\frac{\partial F}{\partial x_1} = -\left[\frac{A_1 B_{22} x_1 - A_2 B_{12} x_2}{B_{12} B_{21} - B_{22} B_{11}} \right] F, \qquad \frac{\partial F}{\partial x_2} = -\left[\frac{A_2 B_{11} x_1 - A_2 B_{12} x_2}{B_{21} B_{12} - B_{11} B_{22}} \right] F. \tag{3.230}$$

The solution of Eq. (3.229) can thus be written as

$$F(x_1, x_2) = \frac{1}{N} \exp\left[-\beta_1 x_1^2 - \beta_2 x_2^2 + (\beta_3 + \beta_4) x_1 x_2 \right], \tag{3.231}$$

where

$$\beta_1 = -\frac{A_1 B_{22}}{2(B_{12} B_{21} - B_{22} B_{11})}, \qquad \beta_2 = -\frac{A_2 B_{11}}{2(B_{21} B_{12} - B_{22} B_{11})},$$

$$\beta_3 = \frac{A_2 B_{12}}{2(B_{12} B_{21} - B_{22} B_{11})}, \qquad \beta_4 = \frac{A_1 B_{21}}{2(B_{21} B_{12} - B_{22} B_{11})}. \tag{3.232}$$

In present case, ($B_{12} = B_{21} = 0$ in which $\beta_3 = \beta_4 = 0$), Eq. (3.231) can be put in the form

$$F(x_1, x_2) = \frac{\sqrt{\beta_1 \beta_2}}{\pi} \exp\left[-\beta_1 x_1^2 - \beta_2 x_2^2 \right]. \tag{3.233}$$

On account of the variable transformation $\alpha = x_1 + i x_2$, one straightforwardly gets Eq. (3.216). Even though the quantum system can be explored by using one of these approaches, it is often presumed to be sufficient applying the steady state consideration that potentially reduce the rigor.

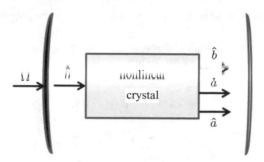

Fig. 3.6 Schematic representation of a degenerate parametric oscillation in which an external coherent radiation of frequency 2ω would be partially down converted to two subharmonic modes of frequency ω by a nonlinear crystal, where the converted radiation is denoted by a boson operator $\hat{a}$ and the light that passes unconverted by $\hat{b}$ (see also Figs. 3.3 and 8.1)

3.2.4 Linearization Procedure

It may not be difficult to conceive that in the process of degenerate parametric oscillation only some part of the pump radiation could be down converted into a pair of signal photons. The cavity as a result can be presumed to contain the down converted radiation commonly known as the fundamental or subharmonic mode and the unchanged radiation designated as the second-harmonic or residual pump mode (see Fig. 3.6). Although treating the second-harmonic mode classically as highlighted earlier appears to be mathematically friendly, the prevailing physical circumstance indicates that the residual pump mode can correlate with the fundamental mode, and so can incite a nonclassical aspect of light.[12] Hoping to adhere to a more realistic scenario, we opt to abandon the customary assumption of treating the residual pump mode classically. In this venture, owing to the difficulty associated with analytically solving the emerging nonlinear coupled differential equations, we intended to use the linearization procedure that amounts to taking the dynamical properties of the system to vary slightly around the steady state values.

Interaction of a coherent radiation with a nonlinear crystal can be described by interaction Hamiltonian of the form

$$\hat{H}_I = i\Omega \left[\hat{b}^\dagger - \hat{b}\right] + \frac{i\lambda}{2}\left[\hat{a}^{\dagger 2}\hat{b} - \hat{a}^2\hat{b}^\dagger\right], \tag{3.234}$$

where Ω is a real positive constant proportional to the amplitude of the coherent input, $\hat{a}$ and $\hat{b}$ are the time independent annihilation operators for the subharmonic

---

[12]Note that assuming the pumping radiation as the classical light may compromise the description of the quantum features and classical properties of the generated signal mode, residual pump mode and the superposed two-mode light specially when a linearization procedure is applied [13].

and second-harmonic modes, and λ is a real positive coupling constant (see also Eqs. (2.84) and (8.30)).

Since $\hat{a}$ and $\hat{b}$ are mutually commuting operators, the pertinent differential equations are found using Eq. (3.58) to be

$$\frac{d\hat{a}}{dt} = \lambda \hat{a}^\dagger \hat{b} - \frac{\kappa}{2}\hat{a} + \hat{F}_1(t), \tag{3.235}$$

$$\frac{d\hat{b}}{dt} = \Omega - \frac{\lambda}{2}\hat{a}^2 - \frac{\kappa}{2}\hat{b} + \hat{F}_2(t), \tag{3.236}$$

where κ is the cavity damping constant taken to be the same for both modes and $\hat{F}_i(t)$ with $i = 1, 2$ are the Langevin noise operators that satisfy:

$$\langle \hat{F}_j^\dagger(t)\hat{F}_i(t')\rangle_{i\neq j} = \langle \hat{F}_j(t)\hat{F}_i(t')\rangle_{i\neq j} = \langle \hat{F}_i(t)\rangle = 0, \tag{3.237}$$

$$\langle \hat{F}_i^\dagger(t)\hat{F}_i^\dagger(t')\rangle = \langle \hat{F}_i(t)\hat{F}_i(t')\rangle = \kappa M_i \delta(t - t'), \tag{3.238}$$

$$\langle \hat{F}_i^\dagger(t)\hat{F}_i(t')\rangle = \kappa N_i \delta(t - t'), \qquad \langle \hat{F}_i(t)\hat{F}_i^\dagger(t')\rangle = \kappa(N_i + 1)\delta(t - t'), \tag{3.239}$$

in which $N_i = \sinh^2(r_i)$, $M_i = \sinh(r_i)\cosh(r_i)$ and r_i's are the squeeze parameters of the respective independent modes (see Sect. 3.1.2).

Since directly solving coupled nonlinear differential equations of this kind is a formidable task, one may utilize the linearization procedure starting with the approximation that

$$\hat{a}(t) = \alpha + \hat{A}(t), \tag{3.240}$$

$$\hat{b}(t) = \beta + \hat{B}(t), \tag{3.241}$$

where $\hat{A}(t)$ and $\hat{B}(t)$ are taken to be very small variations about the mean values at steady state: $\alpha = \langle \hat{a}(t)\rangle_{ss}$ and $\beta = \langle \hat{b}(t)\rangle_{ss}$. After that, upon calculating the statistical averages of Eqs. (3.235) and (3.236) along with the semiclassical approximation[13] in which at steady state the modes are assumed to be uncorrelated, and using the classical de-correlation in which the involved operators are assumed to be factorized, one can obtain

$$\lambda \alpha^* \beta - \frac{\kappa}{2}\alpha = 0, \tag{3.242}$$

$$\lambda \alpha^2 + \kappa \beta = 2\Omega. \tag{3.243}$$

---

[13]The semiclassical assumption is found to work for a weak nonlinearity or weak coupling between the external radiation and nonlinear crystal, that is, when the mean photon number at the threshold is presumed to be very large.

With the aid of Eqs. (3.235), (3.236), (3.240)–(3.243), and the fact that $\hat{A}$ and $\hat{B}$ are small perturbations about the steady state values, it is possible to write

$$\frac{d\hat{A}(t)}{dt} = \varepsilon_1^* \hat{B}(t) + \varepsilon_2 \hat{A}^\dagger(t) - \frac{\kappa}{2} \hat{A}(t) + \hat{F}_1(t), \tag{3.244}$$

$$\frac{d\hat{B}(t)}{dt} = -\varepsilon_1 \hat{A}(t) - \frac{\kappa}{2} \hat{B}(t) + \hat{F}_2(t), \tag{3.245}$$

where

$$\varepsilon_1^* = \lambda \alpha^*, \qquad \varepsilon_2 = \lambda \beta. \tag{3.246}$$

Multiplication of Eq. (3.242) by α indicates that the phase of β is the same as that of α^2. This suggestion would be consistent with Eq. (3.243) provided that both α and β are real (see also Appendix 5). It hence follows from Eqs. (3.242) and (3.243) that

$$\varepsilon_1 = \pm\sqrt{2\lambda\Omega - \kappa\varepsilon_2}, \qquad \varepsilon_1 = \frac{2\varepsilon_1^* \varepsilon_2}{k}, \tag{3.247}$$

in which $\varepsilon_2 = \kappa/2$ for $\varepsilon_1 \neq 0$.

To solve the linearized coupled differential equations, it turns out to be convenient to introduce new operators defined by

$$\hat{A}_\pm = \hat{A}^\dagger \pm \hat{A}, \qquad \hat{B}_\pm = \hat{B}^\dagger \pm \hat{B}. \tag{3.248}$$

In light of such consideration, one can show that

$$\frac{d}{dt}\hat{A}_\pm(t) = -a_\pm \hat{A}_\pm(t) + \varepsilon_1 \hat{B}_\pm(t) + \hat{E}_\pm(t), \tag{3.249}$$

$$\frac{d}{dt}\hat{B}_\pm(t) = -\varepsilon_1 \hat{A}_\pm(t) - \frac{\kappa}{2}\hat{B}_\pm(t) + \hat{F}_\pm(t), \tag{3.250}$$

where $a_\pm = (\kappa \mp 2\varepsilon_2)/2$ and the noise operators are defined by

$$\hat{E}_\pm(t) = \hat{F}_1^\dagger(t) \pm \hat{F}_1(t), \qquad \hat{F}_\pm(t) = \hat{F}_2^\dagger(t) \pm \hat{F}_2(t). \tag{3.251}$$

One may note that Eqs. (3.249) and (3.250) are linearized coupled differential equations that can be solved by introducing a matrix equation of the form

$$\frac{d}{dt}\mathcal{A}(t) = -\mathcal{B}\mathcal{A}(t) + \mathcal{C}(t), \tag{3.252}$$

where

$$\mathcal{A}(t) = \begin{pmatrix} \hat{A}_\pm(t) \\ \hat{B}_\pm(t) \end{pmatrix}, \qquad \mathcal{B} = \begin{pmatrix} a_\pm & -\varepsilon_1 \\ \varepsilon_1 & \frac{\kappa}{2} \end{pmatrix}, \qquad \mathcal{C}(t) = \begin{pmatrix} \hat{E}_\pm(t) \\ \hat{F}_\pm(t) \end{pmatrix}. \tag{3.253}$$

The solution of such a differential equation is provided in Appendix 2 following a straightforward algebra where some terms are found to take complex values when $\varepsilon_2 > 2\varepsilon_1$ in which case the obtained solution is relegated to oscillate rapidly at steady state. Besides, the condition for ε_1 to be real requires that $2\lambda\Omega \geq \kappa\varepsilon_2$, and so the case for which $2\lambda\Omega = \kappa\varepsilon_2$ is denoted as threshold condition.

Once the dynamics of the system operators is known, the squeezing properties of the second-harmonic mode can be studied with the help of the quadrature operators defined by

$$\hat{b}_+ = \hat{b}^\dagger + \hat{b}, \qquad \hat{b}_- = i[\hat{b}^\dagger - \hat{b}] \tag{3.254}$$

that can be rewritten applying Eq. (3.241) as

$$\hat{b}_+ = 2\beta + \hat{B}_+, \qquad \hat{b}_- = i\hat{B}_-, \tag{3.255}$$

where the squeezing properties can be captured by the variances of these quadrature operators

$$\Delta b_\pm^2 = \pm\left[\langle \hat{B}_\pm^2(t) \rangle - \langle \hat{B}_\pm(t) \rangle^2\right]. \tag{3.256}$$

Upon taking the subharmonic and second-harmonic modes to be initially in vacuum state, and applying the fact that the expectation value of the noise force is zero, one can find with the aid of Eq. (3.332) that

$$\langle \hat{B}_\pm(t) \rangle = \left(\frac{1 \pm 1}{\lambda}\right)\left[\varepsilon_1 c_\pm(t) - \varepsilon_2 d_\pm(t)\right],$$

which reduces at steady state to

$$\langle \hat{B}_\pm(t) \rangle_{ss} = 0. \tag{3.257}$$

In view of the derivation presented in Appendix 4 along with the proposal that the noise force at later time is not correlated with the system variables at the earlier times, one can obtain

$$\langle \hat{B}_\pm^2(t) \rangle = \left(\frac{1 \pm 1}{\lambda}\right)\left[\varepsilon_1 c_\pm(t) - \varepsilon_2 d_\pm(t)\right]\langle \hat{B}_\pm(t) \rangle \pm \left[d_\pm^2(t) + c_\pm^2(t)\right]$$

$$+ \langle \hat{h}_\pm^2(t) \rangle + \langle \hat{k}_\pm^2(t) \rangle + 2\langle \hat{k}_\pm(t)\hat{h}_\pm(t) \rangle. \tag{3.258}$$

With the aid of Eqs. (3.238), (3.239), (3.338), (3.339), and upon taking the squeezing parameters of the two reservoir modes to be the same ($r_a = r_b$), it should be possible to verify that

$$\langle \hat{B}_{\pm}^2(t) \rangle = \left(\frac{1 \pm 1}{\lambda} \right) \left[\varepsilon_1 c_{\pm}(t) - \varepsilon_2 d_{\pm}(t) \right] \langle \hat{B}_{\pm}(t) \rangle \pm \left[d_{\pm}^2(t) + c_{\pm}^2(t) \right]$$

$$+ \frac{\kappa (2M \pm 2N \pm 1)}{4} \left[\frac{1 + p^2 + q^2 \mp 2p}{2\eta_{\pm}} \left(1 - e^{-2\eta_{\pm}t} \right) \right.$$

$$+ \frac{1 + p^2 + q^2 \pm 2p}{2\mu_{\pm}} \left(1 - e^{-2\mu_{\pm}t} \right) + \frac{2(1 - p^2 - q^2)}{\eta_{\pm} + \mu_{\pm}} \left(1 - e^{-(\eta_{\pm} + \mu_{\pm})t} \right) \Bigg]$$

$$(3.259)$$

(see also Sect. 8.1.2).

On account of the parameters provided in the appendix, the quadrature variances of the second-harmonic mode at steady state then turn out to be

$$\Delta b_{\pm}^2 = \left[\frac{\kappa^2(\kappa \mp 3\varepsilon_2) + 2\kappa\varepsilon_2^2 + 4\kappa\varepsilon_1^2}{(\kappa \mp \varepsilon_2)[\kappa(\kappa \mp 2\varepsilon_2) + 4\varepsilon_1^2]} \right] e^{\pm 2r}, \tag{3.260}$$

which reduces for $\lambda \neq 0$ to

$$\Delta b_{\pm}^2 = \left(\frac{16\lambda\Omega - \kappa^2(1 \pm 3)}{(2 \mp 1)[8\lambda\Omega - \kappa^2(1 \pm 1)]} \right) e^{\pm 2r} \tag{3.261}$$

and for $\lambda = 0$ to

$$\Delta b_{\pm}^2 = e^{\pm 2r}.$$

In the same way, the quadrature variances of the subharmonic mode are found to be

$$\Delta a_{\pm}^2 = e^{\pm 2r} \begin{cases} 1 & \lambda = 0, \\ \frac{16\lambda\Omega - \kappa^2(2\pm 1)}{(2\mp 1)(8\lambda\Omega - \kappa^2(1\pm 1))} & \text{otherwise.} \end{cases} \tag{3.262}$$

As one can see from Figs. 3.7 and 3.8, the second-harmonic and subharmonic modes exhibit quadrature squeezing even when there is only vacuum fluctuation. Nonetheless, comparison may reveal that the values of the parameters for which the maximum squeezing is obtained is different, but the degree of squeezing appears the same for large $\lambda\Omega$. In view of the obtained result, it does not seem convincing treating the second-harmonic mode as classical light for the benefit of mathematical simplicity (compare also with the discussion in Sect. 8.1.2).

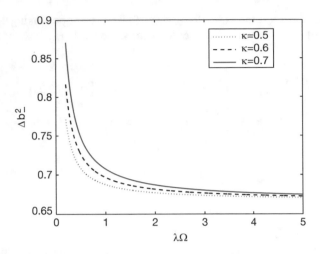

Fig. 3.7 Plots of the minus quadrature variance of the second-harmonic mode at steady state for $r = 0$ and different values of κ

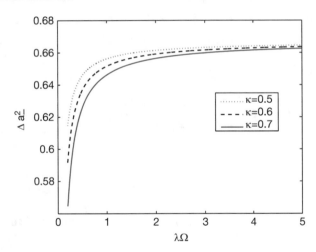

Fig. 3.8 Plots of the minus quadrature variance of the subharmonic mode at steady state for $r = 0$ and different values of κ

   In addition, due to the coherence associated with the driving mechanism, it might be reasonable expecting a nonclassical correlation between the subharmonic and second-harmonic modes as the emerging correlations can lead to a two-mode squeezing and entanglement due to the down conversion of a single high frequency photon into a pair of correlated lower frequency photons that forges harmonically induced superposition.[14] It is, therefore, compelling to expect the coherently driven degenerate parametric down conversion phenomenon to be a source of two-mode squeezed light characterized by a strong correlation between the subharmonic and second-harmonic modes.

As noted in Sect. 2.2.3, the pertinent two-mode radiation can be epitomized by annihilation operator

$$\hat{c} = \frac{1}{\sqrt{2}}(\hat{a} + \hat{b}).$$
(3.263)

In view of boson commutation relation, one can verify that $[\hat{c}, \hat{c}^\dagger] = \hat{I}$ and $[\hat{c}, \hat{c}] = 0$. The squeezing properties of the two-mode radiation can hence be explored with the help of the quadrature operators $\hat{c}_+ = \hat{c}^\dagger + \hat{c}$ and $\hat{c}_- = i(\hat{c}^\dagger - \hat{c})$ that can be put by employing Eqs. (3.240) and (3.241) in the form

$$\hat{c}_+ = \frac{1}{\sqrt{2}}(2\alpha + 2\beta + \hat{A}_+ + \hat{B}_+), \qquad \hat{c}_- = \frac{i}{\sqrt{2}}(\hat{A}_- + \hat{B}_-).$$
(3.264)

The variances of these quadrature operators are found to have the form

$$\Delta c_\pm^2 = \pm\frac{1}{2}\big[\langle\hat{A}_\pm^2(t)\rangle + \langle\hat{B}_\pm^2(t)\rangle + \langle\hat{A}_\pm(t)\hat{B}_\pm(t)\rangle + \langle\hat{B}_\pm(t)\hat{A}_\pm(t)\rangle$$
$$- \langle\hat{A}_\pm(t)\rangle^2 - \langle\hat{B}_\pm(t)\rangle^2 - 2\langle\hat{B}_\pm(t)\rangle\langle\hat{A}_\pm(t)\rangle\big],$$
(3.265)

where various expectation values can be determined by taking both the subharmonic and second-harmonic modes to be initially in vacuum state and employing the fact that the expectation values of the noise forces are zero. With this consideration, one can see at steady state that $\langle\hat{A}_\pm(t)\rangle_{ss} = 0$.

The rest of the correlations in Eq. (3.265) can also be calculated in the same way [13];

$$\langle\hat{A}_\pm^2(t)\rangle = -\left(\frac{1\pm 1}{\lambda}\right)\big[\varepsilon_1 b_\pm(t) + \varepsilon_2 c_\pm(t)\big]\langle\hat{A}_\pm(t)\rangle \pm \big[b_\pm^2(t) + c_\pm^2(t)\big]$$
$$+ \frac{\kappa(2M \pm 2N \pm 1)}{4}\left[\frac{1 + p^2 + q^2 \pm 2p}{2\eta_\pm}\left(1 - e^{-2\eta_\pm t}\right)\right.$$
$$+ \frac{1 + p^2 + q^2 \mp 2p}{2\mu_\pm}\left(1 - e^{-2\mu_\pm t}\right)$$
$$\left. + \frac{2(1 - p^2 - q^2)}{\eta_\pm + \mu_\pm}\left(1 - e^{-(\eta_\pm + \mu_\pm)t}\right)\right],$$
(3.266)

---

[14]Based on a similar kind of argument, the existence of a strong correlation between the fundamental and second-harmonic modes in the opposite up parametric conversion process has been reported [14] and the inclusion of the quantum properties of the pump mode in the nondegenerate parametric oscillator results modification in the quantum features of the cavity radiation [15]. The manifestation of nonclassical properties including harmonic entanglement between the fundamental and residual pump modes for a degenerate parametric oscillator has also been explored [16], even though we opt not to pursue this issue further.

$$\langle \hat{B}_\pm(t)\hat{A}_\pm(t)\rangle = \left(\frac{1\pm1}{\lambda}\right)\left[\varepsilon_1 c_\pm(t) - \varepsilon_2 d_\pm(t)\right]\langle \hat{A}_\pm(t)\rangle + c_\pm(t)\left[d_\pm(t) - b_\pm(t)\right]$$

$$- \frac{\kappa(1 + 2N \pm 2M)}{4}qp\left[\frac{1}{\eta_\pm}\left(1 - e^{-2\eta_\pm t}\right) + \frac{1}{\mu_\pm}\left(1 - e^{-2\mu_\pm t}\right)\right.$$

$$\left. - \frac{2}{(\eta_\pm + \mu_\pm)}\left(1 - e^{-(\eta_\pm + \mu_\pm)t}\right)\right]. \tag{3.267}$$

Thereafter, it should be possible to see at steady state that

$$\Delta c_\pm^2 = e^{\pm 2r}\begin{cases} 1 & \lambda = 0, \\ \frac{32\lambda\Omega - \kappa^2(3\pm4) - 4\kappa\sqrt{2\lambda\Omega - \kappa^2/2}}{2(2\mp1)[8\lambda\Omega - \kappa^2(1\pm1)]} & \text{otherwise.} \end{cases} \tag{3.268}$$

As one can see from Fig. 3.9, the superposed two-mode radiation also exhibits a considerable degree of squeezing even when the oscillator is coupled to vacuum reservoir. It is expected that the coherence in the external radiation prior to the down conversion process is responsible for the correlation between the subharmonic and second-harmonic modes that leads to a two-mode squeezing. This observation can be understood as the down conversion process is unable to destroy the existing coherence despite the realization of splitting. This could also be perceived as if the phase sensitive down conversion process allocates the arising quantum noise disproportionately to the quadrature components, which is presumed to account for the emerging quantum aspects of the radiation.

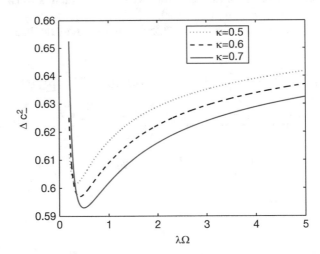

Fig. 3.9 Plots of the minus quadrature variance of the superposed two-mode radiation at steady state for $r = 0$ and different values of κ

Appendix 1: Important Mathematical Identities

We provide some very important mathematical identities without proof:

1. Many of the integrations in this book can be carried out by using a standard integral identity of the form

$$\int \frac{d^2 z}{\pi} \exp\left[-az^* z + bz + cz^* + Az^2 + Bz^{*2}\right]$$
$$= \sqrt{\frac{1}{a^2 - 4AB}} \exp\left[\frac{abc + Ab^2 + Bc^2}{a^2 - 4AB}\right], \tag{3.269}$$

where $a > 0$.

2. In case α is a c-number parameter, and $\hat{a}$ is an operator, it is admissible to write a power series expansion of the form

$$e^{\alpha \hat{a}} = \sum_{n=0}^{\infty} \frac{(\alpha \hat{a})^n}{n!}. \tag{3.270}$$

3. If $\hat{A}$ and $\hat{B}$ are noncommuting operators, and α is a constant, it is possible to verify upon expanding about α that

$$e^{-\alpha \hat{A}} \hat{B} e^{\alpha \hat{A}} = \hat{B} - \alpha[\hat{A}, \hat{B}] + \frac{\alpha^2}{2!}[\hat{A}, [\hat{A}, \hat{B}]] + \dots. \tag{3.271}$$

4. In case $\hat{A}$ and $\hat{B}$ are noncommuting operators, but satisfy the commutation relations

$$[[\hat{A}, \hat{B}], \hat{A}] = [[\hat{A}, \hat{B}], \hat{B}] = 0, \tag{3.272}$$

the Baker-Hausdroff relation holds;

$$e^{\hat{A}+\hat{B}} = e^{-\frac{1}{2}[\hat{A}, \hat{B}]} e^{\hat{A}} e^{\hat{B}}, \qquad e^{\hat{A}} e^{\hat{B}} = e^{\hat{B}} e^{\hat{A}} e^{[\hat{A}, \hat{B}]}, \qquad e^{\hat{A}+\hat{B}} = e^{\frac{1}{2}[\hat{A}, \hat{B}]} e^{\hat{B}} e^{\hat{A}}. \tag{3.273}$$

5. If $f(\hat{a}^\dagger, \hat{a})$ can be expanded in a power series of $\hat{a}^\dagger$ and $\hat{a}$,

$$[\hat{a}, f(\hat{a}^\dagger, \hat{a})] = \frac{\partial f}{\partial \hat{a}^\dagger}, \qquad [\hat{a}^\dagger, f(\hat{a}^\dagger, \hat{a})] = -\frac{\partial f}{\partial \hat{a}}, \tag{3.274}$$

$$e^{c\hat{a}} \hat{f}(\hat{a}, \hat{a}^\dagger) e^{-c\hat{a}} = \hat{f}(\hat{a}, \hat{a}^\dagger + c), \qquad e^{-c\hat{a}^\dagger} \hat{f}(\hat{a}, \hat{a}^\dagger) e^{c\hat{a}^\dagger} = \hat{f}(\hat{a} + c, \hat{a}^\dagger), \tag{3.275}$$

where c is a real constant parameter. One may note that these relations hold regardless of the order of the operators in the function. Based on these definitions, it is thus possible to verify that

$$\hat{a}^n \hat{a}^{\dagger n} = \sum_{m=0}^{n} \binom{n}{m} \frac{n!}{(n-m)!} \hat{a}^{\dagger n-m} \hat{a}^{n-m}. \tag{3.276}$$

6. The number operator $\hat{a}^\dagger \hat{a}$ exhibits the following important properties:

$$[\hat{a}^\dagger \hat{a}, \hat{a}^m] = -m\hat{a}^m, \qquad [\hat{a}^\dagger \hat{a}, \hat{a}^{\dagger m}] = m\hat{a}^{\dagger m}, \tag{3.277}$$

and so for a constant parameter c

$$e^{c\hat{a}^\dagger \hat{a}} \hat{a} e^{-c\hat{a}^\dagger \hat{a}} = \hat{a} e^{-c}, \qquad e^{c\hat{a}^\dagger \hat{a}} \hat{a}^\dagger e^{-c\hat{a}^\dagger \hat{a}} = \hat{a}^\dagger e^{c}. \tag{3.278}$$

Appendix 2: Correlation of the Reservoir Modes

We now seek to obtain various expectation values of the squeezed vacuum reservoir modes. To this end, we consider the case in which a single-mode squeezed vacuum radiation is impinged onto a single-port mirror. The reservoir modes in this case can be described by

$$|r\rangle_k = \hat{S}_k(r)|0\rangle, \tag{3.279}$$

where

$$\hat{S}_k(r) = e^{-r(\hat{a}_k \hat{a}_k - \hat{a}_k^\dagger \hat{a}_k^\dagger)} \tag{3.280}$$

denotes the single-mode squeeze operator. Owing to this description, the pertinent density operator can be expressed as

$$\hat{\rho}_k = |r\rangle\langle r|_k = \hat{S}_k(r)|0\rangle\langle 0|\hat{S}_k^\dagger(r). \tag{3.281}$$

Then upon applying Eq. (3.281), one can write $\langle \hat{a}_k \rangle = \langle 0|\hat{a}_k(r)|0\rangle$, where

$$\hat{a}_k(r) = \hat{S}_k^\dagger(r)\hat{a}_k \hat{S}_k(r). \tag{3.282}$$

After that, with the aid of Eqs. (3.280) and (3.282), one can verify that

$$\hat{a}_k(r) = \hat{a}_k \cosh r + \hat{a}_k^\dagger \sinh r, \tag{3.283}$$

in which $\hat{a}_k = \hat{a}_k(0)$. So one can see that

$$\langle \hat{a}_k \rangle = \cosh r \langle 0|\hat{a}_k|0 \rangle + \sinh r \langle 0|\hat{a}_k^\dagger|0 \rangle, \tag{3.284}$$

from which follows $\langle \hat{a}_k \rangle = 0$.

Upon making use of the fact that $\hat{S}_k^\dagger(r)\hat{S}_k(r) = \hat{I}$ along with Eqs. (3.281) and (3.283), one can then obtain

$$\langle \hat{a}_k^\dagger \hat{a}_{k'} \rangle = \langle 0|\hat{a}_k^\dagger(r)\hat{a}_{k'}(r)|0 \rangle, \tag{3.285}$$

which can be put in the form

$$\langle \hat{a}_k^\dagger \hat{a}_{k'} \rangle = \cosh^2 r \langle 0|\hat{a}_k^\dagger \hat{a}_{k'}|0 \rangle + \sinh^2 r \langle 0|\hat{a}_k \hat{a}_{k'}^\dagger|0 \rangle$$
$$+ \cosh r \sinh r \langle 0|\hat{a}_k \hat{a}_{k'}|0 \rangle + \cosh r \sinh r \langle 0|\hat{a}_k^\dagger \hat{a}_{k'}^\dagger|0 \rangle, \tag{3.286}$$

from which follows

$$\langle \hat{a}_k^\dagger \hat{a}_{k'} \rangle = N \delta_{k,k'}, \tag{3.287}$$

where $N = \sinh^2 r$.

On the basis of the commutation relation $[\hat{a}_k, \hat{a}_{k'}^\dagger] = \hat{I}\delta_{k,k'}$ and Eq. (3.287), one can also see that

$$\langle \hat{a}_k \hat{a}_{k'}^\dagger \rangle = (N+1)\delta_{k,k'}. \tag{3.288}$$

In the same way, one can verify that

$$\langle \hat{a}_k \hat{a}_{k'} \rangle = M \delta_{k,k'}, \tag{3.289}$$

where $M = \cosh r \sinh r$. Assuming that k varies very little around k_c (which stands for cavity radiation), one can write $k \approx 2k_c - k$, and so Eq. (3.289) can be put in the form

$$\langle \hat{a}_k \hat{a}_{k'} \rangle = M \delta_{2k_c - k, k'}. \tag{3.290}$$

We next consider the case in which a two-mode squeezed vacuum radiation is impinged upon a single-port mirror. The reservoir modes in this case can be described by

$$|r\rangle_k = \hat{S}_k(r)|0, 0\rangle, \tag{3.291}$$

where

$$\hat{S}_k(r) = e^{-r(\hat{a}_k \hat{b}_k - \hat{a}_k^\dagger \hat{b}_k^\dagger)} \tag{3.292}$$

is the two-mode squeeze operator in which the corresponding density operator can be expressed as

$$\hat{\rho}_k = \hat{S}_k(r)|0,0\rangle\langle 0,0|\hat{S}_k^\dagger(r).$$

(3.293)

So one can verify that

$$\hat{a}_k(r) = \hat{a}_k \cosh r + \hat{b}_k^\dagger \sinh r,$$

(3.294)

$$\hat{b}_k(r) = \hat{b}_k \cosh r + \hat{a}_k^\dagger \sinh r.$$

(3.295)

It can also be established following the same approach that

$$\langle \hat{b}_k \rangle = \langle \hat{b}_k \hat{b}_{k'} \rangle = \langle \hat{a}_k \hat{b}_{k'}^\dagger \rangle = 0,$$

(3.296)

$$\langle \hat{b}_k^\dagger \hat{b}_{k'} \rangle = N\delta_{kk'}, \qquad \langle \hat{b}_k \hat{b}_{k'}^\dagger \rangle = (N+1)\delta_{kk'}.$$

(3.297)

Upon assuming that k varies very little around $(k_a + k_b)/2$, one can write

$$k \approx k_a + k_b - k,$$

(3.298)

from which follows

$$\langle \hat{a}_k \hat{b}_{k'} \rangle = M\delta_{k_a+k_b-k,k'}.$$

(3.299)

Appendix 3: Correlation Between the Noise Force Operators

The noise force operators that correspond to a single-mode squeezed vacuum reservoir that can be designated in terms of the annihilation operators $\hat{a}_k$ can be expressed as

$$\hat{F}_{Ra}(t) = \sum_k \lambda_k \hat{a}_k e^{i(\omega_a - \omega_k)t},$$

(3.300)

where λ_k is the coupling constants (see Eq. (3.57)). One can then write with the aid of Eq. (3.300) that

$$\langle \hat{F}_{Ra}^\dagger(t) \hat{F}_{Ra}(t') \rangle = \sum_{k,k'} \lambda_k \lambda_{k'} \langle \hat{a}_k^\dagger \hat{a}_{k'} \rangle e^{-i(\omega_a - \omega_k)t + i(\omega_a - \omega_k)t'}.$$

(3.301)

Then by making use of Eq. (3.287), one can see that

$$\langle \hat{F}_{Ra}^\dagger(t) \hat{F}_{Ra}(t') \rangle = N \sum_k \lambda_k^2 e^{-i(\omega_a - \omega_k)(t-t')},$$

(3.302)

which reduces in light of Eq. (3.40) to

$$\langle \hat{F}_{Ra}^{\dagger}(t)\hat{F}_{Ra}(t')\rangle = N\kappa\delta(t - t'). \tag{3.303}$$

One can also establish in a similar manner that

$$\langle \hat{F}_{Ra}(t)\hat{F}_{Ra}^{\dagger}(t')\rangle = (N + 1)\kappa\delta(t - t'), \qquad \langle \hat{F}_{Ra}(t)\hat{F}_{Ra}(t')\rangle = 0. \tag{3.304}$$

Upon introducing

$$\hat{F}_{Rb}(t) = \sum_{j}\lambda_{j}\hat{b}_{j}e^{i(\omega_{b}-\omega_{j})t} \tag{3.305}$$

for the other mode, it is possible to establish in the same way that

$$\langle \hat{F}_{Rb}(t)\hat{F}_{Rb}(t')\rangle = \langle \hat{F}_{Rb}^{\dagger}(t')\hat{F}_{Ra}(t)\rangle = 0, \tag{3.306}$$

$$\langle \hat{F}_{Rb}^{\dagger}(t)\hat{F}_{Rb}(t')\rangle = N\kappa\delta(t - t'), \qquad \langle \hat{F}_{Rb}(t)\hat{F}_{Rb}^{\dagger}(t')\rangle = (N + 1)\kappa\delta(t - t'), \tag{3.307}$$

$$\langle \hat{F}_{Rb}(t')\hat{F}_{Ra}(t)\rangle = M\kappa\delta(t - t'). \tag{3.308}$$

Appendix 4: Solution of Coupled Differential Equations

It turns out to be important outlining the way of solving the differential equation that can be represented by matrix equation (3.252) [17]. It is straightforward to note that the eigenvalues and eigenvectors of the matrix $\mathcal{B}$ can be obtained by applying the relation

$$\mathcal{B}\mathcal{V}_{i} = \lambda_{i}\mathcal{V}_{i}, \tag{3.309}$$

in which

$$\mathcal{V}_{i} = \begin{pmatrix} x_{i} \\ y_{i} \end{pmatrix}, \tag{3.310}$$

with the normalization condition $x_{i}^{2} + y_{i}^{2} = 1$, where $i = 1, 2$. Equation (3.309) would have a nontrivial solution provided that

$$\begin{vmatrix} a_{\pm} - \lambda & -\varepsilon_{1} \\ \varepsilon_{1} & \frac{\kappa}{2} - \lambda \end{vmatrix} = 0, \tag{3.311}$$

from which follows

$$\eta_\pm = \frac{\kappa \mp \varepsilon_2}{2} - \frac{1}{2}\sqrt{\varepsilon_2^2 - 4\varepsilon_1^2}, \qquad \mu_\pm = \frac{\kappa \mp \varepsilon_2}{2} + \frac{1}{2}\sqrt{\varepsilon_2^2 - 4\varepsilon_1^2}, \qquad (3.312)$$

where $\eta_\pm$ and $\mu_\pm$ are the eigenvalues.

One then needs to proceed to evaluate the eigenvectors. To this end, with the aid of Eqs. (3.309) and (3.310), it is possible to see that

$$a_\pm x_1 - \varepsilon_1 y_1 = \mu_\pm x_1, \qquad (3.313)$$

$$\varepsilon_1 x_1 + \frac{\kappa}{2} y_1 = \mu_\pm y_1. \qquad (3.314)$$

After that, upon solving Eqs. (3.313) and (3.314) simultaneously, one gets

$$x_1 = \frac{-\varepsilon_1}{\sqrt{\varepsilon_1^2 + (\mu_\pm - a_\pm)^2}}, \qquad y_1 = \frac{\mu_\pm - a_\pm}{\sqrt{\varepsilon_1^2 + (\mu_\pm - a_\pm)^2}}, \qquad (3.315)$$

$$x_2 = \frac{-\varepsilon_1}{\sqrt{\varepsilon_1^2 + (\eta_\pm - a_\pm)^2}}, \qquad y_2 = \frac{\eta_\pm - a_\pm}{\sqrt{\varepsilon_1^2 + (\eta_\pm - a_\pm)^2}}. \qquad (3.316)$$

So based on the obtained result, the vector defined by

$$V = \begin{pmatrix} x_1 & x_2 \\ y_1 & y_2 \end{pmatrix},$$

can be expressed as

$$V = \begin{pmatrix} \dfrac{-\varepsilon_1}{\sqrt{\varepsilon_1^2 + (\mu_\pm - a_\pm)^2}} & \dfrac{-\varepsilon_1}{\sqrt{\varepsilon_1^2 + (\eta_\pm - a_\pm)^2}} \\ \dfrac{\mu_\pm - a_\pm}{\sqrt{\varepsilon_1^2 + (\mu_\pm - a_\pm)^2}} & \dfrac{\eta_\pm - a_\pm}{\sqrt{\varepsilon_1^2 + (\eta_\pm - a_\pm)^2}} \end{pmatrix}. \qquad (3.317)$$

To get the inverse of this matrix, one needs to evaluate its characteristic equation:

$$\lambda^2 - \lambda \left(\frac{-\varepsilon_1}{\sqrt{\varepsilon_1^2 + (\mu_\pm - a_\pm)^2}} + \frac{\eta_\pm - a_\pm}{\sqrt{\varepsilon_1^2 + (\eta_\pm - a_\pm)^2}} \right)$$

$$+ \frac{\varepsilon_1(\mu_\pm - \eta_\pm)}{\sqrt{\varepsilon_1^2 + (\mu_\pm - a_\pm)^2}\sqrt{\varepsilon_1^2 + (\eta_\pm - a_\pm)^2}} = 0. \qquad (3.318)$$

Owing to the fact that a matrix behaves in the same way as its characteristic equation, one can write

$$
\mathcal{V}\left[\mathcal{V} - \mathcal{I}\left(\frac{\eta_1}{\sqrt{\varepsilon_1^2 + (\mu_\pm - a_\pm)^2}} + \frac{\eta_\perp \quad \eta_\perp}{\sqrt{\varepsilon_1^2 + (\eta_\pm - a_\pm)^2}}\right)\right]
$$

$$
= \frac{\varepsilon_1(\eta_\pm - \mu_\pm)\mathcal{I}}{\sqrt{\varepsilon_1^2 + (\mu_\pm - a_\pm)^2}\sqrt{\varepsilon_1^2 + (\eta_\pm - a_\pm)^2}}, \tag{3.319}
$$

from which follows using Eq. (3.317)

$$
\mathcal{V}^{-1} = \frac{1}{\eta_\pm - \mu_\pm}\begin{pmatrix} -\frac{\eta_\pm - a_\pm}{\varepsilon_1}\sqrt{\varepsilon_1^2 + (\mu_\pm - a_\pm)^2} & -\sqrt{\varepsilon_1^2 + (\mu_\pm - a_\pm)^2} \\ \frac{\mu_\pm - a_\pm}{\varepsilon_1}\sqrt{\varepsilon_1^2 + (\eta_\pm - a_\pm)^2} & \sqrt{\varepsilon_1^2 + (\eta_\pm - a_\pm)^2} \end{pmatrix}. \tag{3.320}
$$

Then upon employing the fact that $\mathcal{V}^{-1}\mathcal{V} = \mathcal{V}\mathcal{V}^{-1} = \mathcal{I}$, one can rewrite Eq. (3.309) as

$$
\frac{d\mathcal{A}(t)}{dt} = -\mathcal{V}\mathcal{V}^{-1}\mathcal{B}\mathcal{V}\mathcal{V}^{-1}\mathcal{A}(t) + \mathcal{C}(t). \tag{3.321}
$$

Upon multiplying Eq. (3.321) from left by $\mathcal{V}^{-1}$, one gets

$$
\frac{d}{dt}\left(\mathcal{V}^{-1}\mathcal{A}(t)\right) = -\mathcal{R}(\mathcal{V}^{-1}\mathcal{A})(t) + \mathcal{V}^{-1}\mathcal{C}(t), \tag{3.322}
$$

where

$$
\mathcal{R} = \mathcal{V}^{-1}\mathcal{B}\mathcal{V} = \begin{pmatrix} \mu_\pm & 0 \\ 0 & \eta_\pm \end{pmatrix}. \tag{3.323}
$$

Equation (3.322) would have a valid solution at steady state provided that the matrix $\mathcal{R}$ is positive. As we can see from Eq. (3.323), $\mathcal{R}$ would be positive if both $\mu_\pm$ and $\eta_\pm$ are positive; the restriction that constitutes the threshold condition.

One may note that the solution of Eq. (3.322) can be proposed as

$$
\mathcal{V}^{-1}\mathcal{A}(t) = e^{-\mathcal{R}t}\mathcal{V}^{-1}\mathcal{A}(0) + \int_0^t e^{-\mathcal{R}(t-t')}\mathcal{V}^{-1}\mathcal{C}(t')dt' \tag{3.324}
$$

in which multiplying Eq. (3.324) from left by $\mathcal{V}$ yields

$$
\mathcal{A}(t) = \mathcal{V}e^{-\mathcal{R}t}\mathcal{V}^{-1}\mathcal{A}(0) + \int_0^t \mathcal{V}e^{-\mathcal{R}(t-t')}\mathcal{V}^{-1}\mathcal{C}(t')dt' \tag{3.325}
$$

where

$$e^{-\mathcal{R}t} = \begin{pmatrix} e^{-\mu_\pm t} & 0 \\ 0 & e^{-\eta_\pm t} \end{pmatrix}. \tag{3.326}$$

Besides, with the help of Eqs. (3.253), (3.317), (3.320) and (3.326), one can show that

$$\mathcal{V}e^{-\mathcal{R}t}\mathcal{V}^{-1}\mathcal{A}(0) = \frac{1}{\eta_\pm - \mu_\pm}$$

$$\times \begin{pmatrix} [(\eta_\pm - a_\pm)e^{-\mu_\pm t} - (\mu_\pm - a_\pm)e^{-\eta_\pm t}]\hat{A}_\pm(0) + \varepsilon_1(e^{-\mu_\pm t} - e^{-\eta_\pm t})\hat{B}_\pm(0) \\ \frac{(\mu_\pm - a_\pm)(\eta_\pm - a_\pm)}{\varepsilon_1}(e^{-\eta_\pm t} - e^{-\mu_\pm t})\hat{A}_\pm(0) + [(\eta_\pm - a_\pm)e^{-\eta_\pm t} - (\mu_\pm - a_\pm)e^{-\mu_\pm t}]\hat{B}_\pm(0) \end{pmatrix}, \tag{3.327}$$

$$\mathcal{V}e^{-\mathcal{R}(t-t')}\mathcal{V}^{-1}\mathcal{C}(t') = \frac{1}{\eta_\pm - \mu_\pm}\begin{pmatrix} r_\pm \\ s_\pm \end{pmatrix}, \tag{3.328}$$

where

$$r_\pm = \left[(\eta_\pm - a_\pm)e^{-\mu_\pm(t-t')} - (\mu_\pm - a_\pm)e^{-\eta_\pm(t-t')}\right]\hat{E}_\pm(t')$$

$$+ \varepsilon_1[e^{-\mu_\pm(t-t')} - e^{-\eta_\pm(t-t')}]\hat{F}_\pm(t'), \tag{3.329}$$

$$s_\pm = \frac{(\mu_\pm - a_\pm)(\eta_\pm - a_\pm)}{\varepsilon_1}\left(e^{-\eta_\pm(t-t')} - e^{-\mu_\pm(t-t')}\right)\hat{E}_\pm(t')$$

$$+ \left((\eta_\pm - a_\pm)e^{-\eta_\pm(t-t')} - (\mu_\pm - a_\pm)e^{-\mu_\pm(t-t')}\right)\hat{F}_\pm(t'). \tag{3.330}$$

So on account of the obtained results, it is possible to verify that

$$\hat{A}_\pm(t) = -\left(\frac{1\pm 1}{\lambda}\right)\left[\varepsilon_1 b_\pm(t) + \varepsilon_2 c_\pm(t)\right]$$

$$+ b_\pm(t)\hat{A}_\pm(0) + c_\pm(t)\hat{B}_\pm(0) + \hat{g}_\pm(t) + \hat{f}_\pm(t), \tag{3.331}$$

$$\hat{B}_\pm(t) = \left(\frac{1\pm 1}{\lambda}\right)\left[\varepsilon_1 c_\pm(t) - \varepsilon_2 d_\pm(t)\right]$$

$$- c_\pm(t)\hat{A}_\pm(0) + d_\pm(t)\hat{B}_\pm(0) + \hat{k}_\pm(t) + \hat{h}_\pm(t), \tag{3.332}$$

where

$$b_{\pm}(t) = \frac{1}{2}\left[(1 \pm p)e^{-\eta_{\pm}t} + (1 \mp p)e^{-\mu_{\pm}t}\right], \tag{3.333}$$

$$c_{\pm}(t) = \frac{q}{2}\left[e^{-\eta_{\pm}t} - e^{-\mu_{\pm}t}\right], \tag{3.334}$$

$$d_{\pm}(t) = \frac{1}{2}\left[(1 \mp p)e^{-\eta_{\pm}t} + (1 \pm p)e^{-\mu_{\pm}t}\right], \tag{3.335}$$

$$\hat{f}_{\pm}(t) = \frac{q}{2}\int_0^t \left[e^{-\eta_{\pm}(t-t')} - e^{-\mu_{\pm}(t-t')}\right]\hat{F}_{\pm}(t')dt', \tag{3.336}$$

$$\hat{g}_{\pm}(t) = \frac{1}{2}\int_0^t [(1 \pm p)\,e^{-\eta_{\pm}(t-t')} + (1 \mp p)\,e^{-\mu_{\pm}(t-t')}]\hat{E}_{\pm}(t')dt', \tag{3.337}$$

$$\hat{h}_{\pm}(t) = \frac{1}{2}\int_0^t [(1 \mp p)\,e^{-\eta_{\pm}(t-t')} + (1 \pm p)\,e^{-\mu_{\pm}(t-t')}]\hat{F}_{\pm}(t')dt', \tag{3.338}$$

$$\hat{k}_{\pm}(t) = \frac{q}{2}\int_0^t \left[e^{-\mu_{\pm}(t-t')} - e^{-\eta_{\pm}(t-t')}\right]\hat{E}_{\pm}(t')dt', \tag{3.339}$$

with

$$p = \frac{\varepsilon_2}{\sqrt{\varepsilon_2^2 - 4\varepsilon_1^2}}, \tag{3.340}$$

$$q = \frac{2\varepsilon_1}{\sqrt{\varepsilon_2^2 - 4\varepsilon_1^2}}. \tag{3.341}$$

Appendix 5: Proof That ε_1 Is a Positive Constant

In the hope of ascertaining the proposal that ε_1 is a positive constant by making use of mathematical argument, one can multiply Eqs. (3.242) and (3.243) by λ;

$$\varepsilon_1^*\varepsilon_2 - \frac{\kappa}{2}\varepsilon_1 = 0, \tag{3.342}$$

$$\varepsilon_1^2 + \kappa\varepsilon_2 = 2\lambda\Omega, \tag{3.343}$$

from which follows

$$\varepsilon_1 = \frac{2\varepsilon_1^* \varepsilon_2}{\kappa}. \tag{3.344}$$

Multiplying Eq. (3.343) after that by ε_1^*, and then inserting the complex conjugate of Eq. (3.344) into the resulting expression lead to

$$\varepsilon_1^* \varepsilon_1^2 + 2\varepsilon_1 \varepsilon_2^* \varepsilon_2 - 2\lambda\Omega\varepsilon_1^* = 0. \tag{3.345}$$

Then after subtracting Eq. (3.345) from its complex conjugate indicates

$$(\varepsilon_1^* \varepsilon_1 + 2\varepsilon_2^* \varepsilon_2 + 2\lambda\Omega)(\varepsilon_1 - \varepsilon_1^*) = 0, \tag{3.346}$$

which holds true if $\varepsilon_1 = \varepsilon_1^*$. One can also see from Eq. (3.344) that $\varepsilon_2 = \varepsilon_2^*$ since κ is taken to be a real positive constant. On the basis of these facts, it can be noted that $\alpha = \alpha^*$ and $\beta = \beta^*$ since λ is taken to be a positive constant.

References

1. H. Carmichael, *An Open Systems Approach to Quantum Optics* (Springer, Berlin, 1993)
2. R.R. Puri, *Mathematical Methods of Quantum Optics* (Springer, Berlin, 2001)
3. C.W. Gardiner, P. Zoller, *Quantum Noise*, 2nd edn. (Springer, Berlin, 2000)
4. H.A. Bachor, T.C. Ralph, *A Guide to Experiments in Quantum Optics*, second, revised and elarged edition (Wiley-VCH Verlag GmbH and Co. KGaA, Weinheim, 2004)
5. W.H. Louisell, *Quantum Statistical Properties of Radiation* (Wiley, New York, 1973); T.A.B. Kennedy, D.F. Walls, Phys. Rev. A **37**, 152 (1988)
6. H.J. Carmichael, *Statistical Methods in Quantum Optics 1: Master Equations and Fokker-Planck Equations* (Springer, Berlin, 2002)
7. S. Tesfa, A nondegenerate three-level cascade laser coupled to a two-mode squeezed vacuum reservoir. Ph.D. thesis, Addis Ababa University, 2008
8. M.J. Collett, C.W. Gardiner, Phys. Rev. A **30**, 1386, (1984)
9. C.W. Gardiner, M.J. Collett, Phys. Rev. A **31**, 3761 (1985); C. Brukner, Nature Phys. **10**, 259 (2014)
10. S.M. Barnett, P.M. Badmore, *Methods in Theoretical Quantum Optics* (Oxford University Press, New York, 1997)
11. H. Risken, *The Fokker-Planck Equation: Methods of Solution, Applications* (Springer, Berlin, 1984); N.A. Enaki, J. Mod. Opt. **57**, 1397 (2010)
12. K. Fesseha, J. Math. Phys **33**, 2179 (1992); B. Daniel, K. Fesseha, Opt. Commun. **151**, 384 (1998)
13. S. Tesfa, Eur. Phys. J. D **46**, 351 (2008). J. Phys. B **41**, 065506 (2008)
14. M.K. Olsen, Phys. Rev. A **70**, 035801 (2004)
15. A.S. Villar et al., Phys. Rev. Lett. **97**, 140504 (2006); M.K. Olsen, A.S. Bradley, Phys. Rev. A **74**, 063809 (2006)
16. S. Tesfa, J. Phys. B **41**, 065506 (2008)
17. S. Tesfa, A nondegenerate three-level cascade laser coupled to a two-mode squeezed vacuum reservoir. Ph.D. thesis, Addis Ababa University, 2008

Photon Measurement

4

The aim of a given measurement is often geared towards acquiring as much accurate information as possible with the aspiration that the gained information enables a unique state determination. Existing experience however reveals that a unique quantum state determination via a measurement process is not practically attainable. Since interaction of the radiation with the measuring device is often complex,[1] detection of a photon or photon measurement is one of such impairments. In addition to providing as much information about the system as possible, it is desirable for a measurement to have as little imprecision as possible. In perspective, it is required to measure some of the observables to get a viable information about the quantum state, although the outcome of the measurement of one of the observables exhibits only a specific aspect of the physical system. One should also note that a complete description of a quantum system requires a deeper inspection than just measuring specific observable, which demands a considerable effort and resource. As a result, it would be a matter of convenience to go for feasibility that allows to acquire relatively sufficient amount of information, and at the same time stick to selected sets of measurements. Taking this consideration into account, we strive to explain the underlying physical phenomena, techniques and principles required to explore the process of photon measurement, and also to provide the discussion on a more contentious issues as put forward by various school of thoughts.

To this effect, one may need to begin with the observation that the measurement approach in quantum optics relies on a photoelectric effect in which the current or the voltage produced by the photoelectrons induced by the photo-ionization process is measured. Since the measurement is destructive as the photon responsible for the creation of photoelectron disappears, the detectors are assumed to work on absorptive mechanism. To make the interpretation of the measured data comparatively

---

[1] Although the theory of photon detection necessitates a complete knowledge of the interaction of the radiation with matter, with the aim of limiting the involved rigor such consideration is not included in the present discussion (for some practical account; see for instance [1]).

© The Author(s), under exclusive license to Springer Nature Switzerland AG 2020
S. Tesfa, *Quantum Optical Processes*, Lecture Notes in Physics 976,
https://doi.org/10.1007/978-3-030-62348-7_4

manageable, it might therefore be compelling to assume each absorbed photon to give rise to no more than one electron; and conversely, each electron to be ejected only by one photon.

Upon taking the detectors to be insensitive to spontaneous emission, the annihilation operator of the electric field $\hat{\mathbf{E}}^{(\dagger)}$ is presumed to engage in the measurement process. In case the field makes a transition from an initial state $|i\rangle$ to a final state $|f\rangle$ during measurement in which a single photon is absorbed, the element of the transition matrix can be expressed as

$$T_{if} = \langle f|\hat{\mathbf{E}}^{(\dagger)}|i\rangle. \tag{4.1}$$

The probability per unit time at which the photon is absorbed at position $\mathbf{r}$ in space and at time t for an ideal photon detector would be proportional to

$$W_{if} = \langle i \mid \hat{\mathbf{E}}^{(-)}(\mathbf{r}, t) \mid f \rangle \langle f \mid \hat{\mathbf{E}}^{(\dagger)}(\mathbf{r}, t) \mid i \rangle, \tag{4.2}$$

since $(\hat{\mathbf{E}}^{(\dagger)})^{\dagger} = \hat{\mathbf{E}}^{(-)}$ (see Sect. 2.1).

One may note that the final state of the field cannot be interrogated, and so cannot be measured; rather, it is the total counting rate that can be inferred. To obtain the total counting rate or average field intensity, it should be reasonable to sum overall states of the field that can be reached from the initial state via the absorption process, that is,

$$I(\mathbf{r}, t) = \langle i \mid \hat{\mathbf{E}}^{(-)}(\mathbf{r}, t)\hat{\mathbf{E}}^{(\dagger)}(\mathbf{r}, t) \mid i \rangle, \tag{4.3}$$

where the expectation value in Eq. (4.3) is evaluated over the initial state and when the operator is put in the normal ordering [2].

There is still another glaring setback related to the inherent conundrum in quantum measurement theory resulting from the incompatibility of a pair of conjugate canonical variables [3]. Particularly, since the photon number and the phase are related by the uncertainty relation, it is not feasible to precisely measure the intensity and the phase of the light at the same time. Photon measurement scheme so suffers in two ways: the measurement process invariably induces noise as prescribed by quantum theory; and at the same time, the measuring device has inherent inefficiency. As a way out, quantum nondemolition measurement scheme in which the photon intensity can be measured without necessarily altering the photon number or the phase has been introduced (see Sect. 4.1 and also [4]).

In addition to measuring the intensity or the phase, it might be required to know the interrelation among various beams of light to better understand the properties of the radiation. In view of the fact that the photons produced at different positions and times can fall on a detector at the same time, the way of evaluating the correlations or coincidence measurement among photons generated from different sources and at different times is forwarded, where the emphasis is given to distinguish quantum from classical correlations (see Sect. 4.2 and also [5]). The approaches and physical apparatuses such as wave mixer and beam splitter that can be utilized to separate the beam of light based on its different characteristics, and then recombine after certain delay in time are also presented (see Sects. 4.3 and 4.5). In doing so, the number of photons is presumed to be counted by a suitable detector; and the pertinent distribution in space or time or frequency is then plotted. Besides, aiming at facilitating the process of photon number manipulation required in application, the technique of photon addition and subtraction is highlighted (see Sect. 4.4).

4.1 Basics of Quantum Measurement Theory

When some apparatus and procedure are envisioned to prepare the quantum state, the experimental setup that includes the properties of the apparatus should reflect the pertinent mathematical structure. In case the experiment is repeated so as to measure the same property of a quantum state prepared in the same way, the outcome of successive measurements can often be different. The other important aspect of quantum measurement theory is related to the presumption that the result of the measurement can be described by the probability distribution in which the process can be perceived as random and indeterministic.[2] Regardless of such perception, if a measurement is repeated without re-preparing the state, one finds the same result as the earlier; which entails that after measuring some aspects of the quantum state, one should update the quantum state to recap the result of the measurement. It is the updating of the quantum state model that is actually designated as wave function collapse.

Notably, the reality of the wave function even after a century of debate remains mysterious and contentious. The elusiveness of the quantum state stems from the fact that the state itself is not observable entity: the quantum is not directly accessible. Even then, measurement outcomes are classical like click in the detector or the positioning of the meter or fringes on the film. This reflection may suggest that it is the classical measurement that eventually reveals the consequences of the underneath quantum layer and somehow leads to the determination of the quantum state of the system. The process of reading the information even in this case can be relatively simpler when the set of the possible output states is known and the states in the set are mutually orthogonal; otherwise, discriminating perfectly among the involved states would not be tractable. Due to such inherent inconsistencies in the theory of quantum measurement, there are various philosophical and interpretational

---

[2]There has been a considerable disagreement over this observation ranging from the result merely appears random and indeterministic to indeterminism is core and irreducible, which might have stemmed from the assumption that one of the important features of quantum measurement theory is the wave function collapse the nature and interpretation of which vary according to the adopted approach [6].

contentions[3] that emanate from the basic understanding in quantum theory which encompasses that all measurements have associated operator with properties: the observable is epitomized by a Hermitian operator that maps the Hilbert space into itself; the eigenvalues of the observable are real; for each eigenvalue there is one (or more) eigenvector or eigenstate that makes up the state of the system after the measurement; and the observable has a set of eigenvectors that span the space of the state, that is, a quantum state can be represented as a superposition of the eigenstates.

4.1.1 Interpretation of Quantum Measurement

Orthodox interpretation of quantum measurement is based on the perceived probabilistic nature of the wave function: the measurement process is assumed to randomly pick exactly one out of the many possibilities allowed by the wave function of the system in a manner consistent with the well defined probabilities assigned to each accessible state. It is an attempt to explain the results of the experiment and the corresponding mathematical formulation in terms of the principles of quantum theory in such a way that the act of measurement is believed to cause the calculated set of the probabilities to collapse to the value defined by the measurement. To gain some insight of the approaches in the standard quantum measurement theory, one can begin with the assumption that the Hilbert space is finite and discrete while the initial state is pure.

On account of the basic understanding of quantum theory, the process of the measurement and the accompanying mathematical expression hence can be put forward:

1. to every observable corresponds a Hermitian operator $\hat{X}$ with a spectral representation

$$\hat{X} = \sum_j \lambda_j |j\rangle\langle j|, \tag{4.4}$$

   where from the Hermiticity of $\hat{X}$ follows that the eigenvalues λ_j are real, and in case the eigenvalues are nondegenerate, the pertinent eigenvectors form a complete orthonormal basis set, that is, $\langle i|j\rangle = \delta_{ij}$;
2. the projectors $P_j = |j\rangle\langle j|$ span the entire Hilbert space, that is,

$$\sum_j P_j = 1; \tag{4.5}$$

---

[3]The discussion on measurement theories and the accompanying controversies as coined originally by the proponents of pertinent interpretation is included to provide the theoretical background for the photon measurement approaches and to foresee the application of the quantum features (see Chaps. 10, 11 and 12).

3. from the orthogonality of the states, one can have $P_i P_j = P_{ij} \delta_{ij}$, which implies that $P_i^2 = P_i$: the eigenvalues of any projector are 0 and 1, and so measurement of $\hat{X}$ yields only one of the eigenvalues;

4. in line with the presumption of the collapse of the wave function, the state of the system after the measurement would be

$$|\phi_j\rangle = \frac{P_j|\Phi\rangle}{\sqrt{\langle\Phi|P_j|\Phi\rangle}},$$ (4.6)

if the outcome is one of the eigenvalues;

5. the probability for finding this particular outcome as measurement result is

$$p_j = \langle\Phi|P_j|\Phi\rangle;$$ (4.7)

6. if one performs the measurement, but does not record the results, the post measurement state can be described by the density operator of the form

$$\hat{\rho} = \sum_j p_j|\phi_j\rangle\langle\phi_j| = \sum_j P_j|\Phi\rangle\langle\Phi|P_j.$$ (4.8)

Such description may entail that the measurement process is random; so one cannot tell about the outcome for sure except predicting the spectrum of the possible outcomes and the probability that a particular outcome can appear. This categorically leads to the ensemble interpretation of quantum theory, which affirms the impossibility of predicting the exact outcome of an individual measurement except when the probability is 0 or 1.

Once the main ingredients are in place, one should be able to calculate the moments of the probability distribution generated by the measurement. The first moment in this respect is taken to be the average of a large number of identical measurements performed on the initial ensemble, which is usually designated as expectation value of $\hat{X}$ denoted by $\langle\hat{X}\rangle$ and defined as

$$\langle\hat{X}\rangle = \sum_j \lambda_j p_j = Tr(X\hat{\rho}),$$ (4.9)

whereas the second moment[4] is related to the variance and defined as

$$\langle(\hat{X} - \langle\hat{X}\rangle)^2\rangle = Tr(\hat{X}^2\hat{\rho}) - \left[Tr(\hat{X}\hat{\rho})\right]^2.$$ (4.10)

One may recap from earlier discussion that the final eigenstate emerges randomly with probability that equals to the square of its overlap with the original state due

---

[4]Higher-order moments can also be calculated in the same way, although the first and second moments are found to be the most important to look for in practice (see Chap. 3).

to the inherent wave function collapse. The notion of the wave function collapse however leads to serious questions on the theory of quantum measurement, and the conception of determinism and locality. Concerted effort over the years indicates that while the measurement theory resulting from orthodox interpretation correctly predicts the form and probability distribution of the final eigenstates, it does not explain the randomness inherent in the choice of the final state; and so calls for contention that stems from the apparent conflict between the linear dynamics of quantum theory and nonlinear collapse of the wave function. The presumption of the collapse particularly seems to be right about what happens when one makes measurements, but about the dynamics whenever one is not measuring. It might be worthy noting that it is the dichotomy between the wave function as mathematical model and observed macroscopic events that happens at fundamental level which is the source of controversies.

Since the mathematical model never directly pertains to the accompanying physical observations, one needs to interpret the outcome of measurement; which is perhaps the most difficult part of the game. One also often models a given system based on the superposition of two or more possible outcomes; and yet, the question of how the superposition of different possibilities evolves into some particular observation should be dealt with properly. So to explore some of the contentions, it might be essential starting with the suggestion that the orthodox interpretation denies the wave function to be anything more than a theoretical concept. It might thus be desirable to expose the nature of the Copenhagen interpretation with the aid of quite few experiments and inconsistencies:

1. **Schrödinger's cat**
   This thought experiment may show the implication of accepting uncertainty at the microscopic level has on the description of the properties of the macroscopic objects. Imagine that a cat is put in a sealed box with its life or death situation made dependent on the state of a subatomic particle put in the box along with it. The description of the cat during the course of the experiment becomes a blur of living and dead. But this cannot be accurate interpretation when viewed from our everyday intuition since it implies that the cat is both dead and alive until the box is opened. But the cat, if it survives, will remember only being alive. How can the living cat be both alive and dead at the same time remains the source of contention and debate.

2. **Wigner's friend**
   To push the same argument further, let us enclose Wigner's friend in the box along with the cat. The external observer (Wigner) believes that the cat is in the entangled state $(|\text{dead}\rangle + |\text{alive}\rangle)/\sqrt{2}$ (see Chaps. 7 and 11). His friend however is convinced that the cat is alive (if it is actually alive): for him, the cat is in the state $|\text{alive}\rangle$. It might hence be imperative inquiring how can Wigner and his friend see different wave functions. It might be insightful noting at this juncture that each observer has different information; and therefore, acquires different wave functions. One should note that the distinction between the objective nature

of reality and the subjective nature of probability can lead to a great deal of controversy.

3. **Double slit diffraction pattern**

It is an experimental fact that the light that passes through double slits can cast a diffraction pattern on screen—which leads to the controversy surrounding whether a light is a particle or wave. According to Copenhagen interpretation, it is neither. A particular experiment can demonstrate a particle (photon) or wave property but not both at the same time. Due to the smallness of Planck's constant, it is impossible to realize experiments that directly reveal the wave nature for a system bigger than a few atoms despite prediction of quantum theory. To save face, the bigger systems are usually considered as classical object but where the classical-quantum boundary lies is a genuine question.

4. **Einstein-Podolsky-Rosen (EPR) contention**

Since entangled particles are emitted in a single event, the conservation laws ensure that the measured spin of one particle must be the opposite of the other so that if the spin of one particle is measured, the spin of the other would be instantaneously known provided that the spin of the combined system is known to the experimenter. The most discomforting aspect of this phenomenon is that its effect is instantaneous: something that happens in one galaxy could cause instantaneous change in another galaxy. However, according to Einstein's theory of special relativity, no information bearing signal or entity can travel at or faster than the speed of light, which is known to be finite. It thus seems as if Copenhagen interpretation is not consistent with special relativity (for EPR contention's significance; see Chaps. 7 and 11).

As far as the measurement process is concerned, standard interpretation is perceived as the interplay between the quantum system and its classical counterpart. It recognizes the coexistence of two mutually opposite laws of state determination via state vector changes (or the unitary evolution) and the law of discrete projection postulate (or collapse of the wave function). Even though the interaction mechanism between the quantum system and measuring device remains unclear, the interpretation in this sense seems dualistic.

4.1.2 von Neumann Measurement Scheme

The way out of such a conundrum appears to be self evident in case the measurement process is also regarded as quantum phenomenon; which suggests that one should treat the measuring device as quantum system and the measurement process as interaction between two quantum systems: the to be measured and measuring [7]. It seems that the basic conception of this scheme is so natural, simple and convincing, but even according to von Neumann, the undesirable projection postulate persists, and it should also be applied to the measuring device. One might not wonder why then von Neumann came to the speculation that the collapse of the state vector should occur in the consciousness of the observer.

In line with this, imagine the quantum state to be in the superposed state $|\psi\rangle = \sum_n c_n|\psi_n\rangle$, where $|\psi_n\rangle$'s are eigenstates of the operator that needs to be measured. Assume also that the to be measured system with state $|\psi\rangle$ to interact with the measuring apparatus with state $|\phi\rangle$ so that the total wave function before the interaction be the product state $|\psi\rangle \otimes |\phi\rangle$. During the interaction of the system and measuring instrument, the unitary evolution is supposed to initiate the transition

$$|\psi\rangle \otimes |\phi\rangle \longrightarrow \sum_n c_n|\psi_n\rangle \otimes |\phi_n\rangle, \tag{4.11}$$

which designates the measurement of the first kind, where $|\phi_n\rangle$'s are the orthonormal states of the measuring apparatus: $\langle\phi_m|\phi_n\rangle = \delta_{nm}$. Such a unitary evolution is referred to as pre-measurement process.

One may note that the relation with the wave function collapse would be established by calculating the final density operator of the system that would be interpreted as the ensemble of the systems that describes the measurement with probability $|c_n|^2$ in the state $|\psi_n\rangle$. With this in mind, the transition

$$|\psi\rangle \rightarrow \sum_n |c_n|^2|\psi_n\rangle \otimes \langle\psi_n| \tag{4.12}$$

is often referred to as weak von Neumann projection[5] while the wave function collapse related to a strong von Neumann projection can be portrayed by

$$|\psi\rangle \rightarrow \sum_n |c_n|^2|\psi_n\rangle\langle\psi_n| \rightarrow |\psi_n\rangle, \tag{4.13}$$

which corresponds to additional selection of the involved subensemble by a means of observation.

In the measurement of the second kind on the other hand the unitary evolution during the interaction of the system and measuring instrument is assumed to have the form

$$|\psi\rangle \otimes |\phi\rangle \rightarrow \sum_n c_n|\chi_n\rangle \otimes |\phi_n\rangle \tag{4.14}$$

---

[5]In case the measured observable has a degenerate spectrum, the weak von Neumann projection can be generalized to Löders projection [8]:

$$|\psi\rangle \rightarrow \sum_n |c_n|^2 P_n; \quad P_n = \sum_i |\psi_{ni}\rangle\langle\psi_{ni}|$$

in which the vectors $|\psi_{ni}\rangle$'s for fixed n are the degenerate eigenvectors of the measured observable. For arbitrary state describable by the density operator ρ, Löders projection is given by $\hat{\rho} \rightarrow \sum_n P_n\hat{\rho}P_n$.

in which the states $|\chi_n\rangle$'s of the system are determined by specific properties of the interaction between the system and measuring instrument. These states are normalized but not necessarily mutually orthogonal. The wave function collapse thus can be linked to the idea that the final state of the system would be $|\chi_n\rangle$ with probability $|c_n|^2$. One may note that many present day measurement procedures are predominantly measurement of the second kind; some are even working correctly only as a consequence of being the second kind. For example, a photon counter detects a photon via absorbing (annihilating) it, and so leaving the electromagnetic field in a lesser number.

4.1.3 Positive Operator Valued Measure

One may observe that the last three postulates of the standard theory provide an algorithm to generate the probabilities in which the number of possible outcomes is bounded by the number of terms in the orthogonal decomposition of the identity operator of the Hilbert space. This may indicate that one cannot have more orthogonal projections than the dimensionality of the Hilbert space of the system. Nonetheless, it might be desirable to have more outcomes than the dimensionality of the system while keeping the positivity and normalization of the probabilities. It should be clear from the outset that this proposition is acceptable in case one manages to relax the above restrictive postulates, and replace them with more flexible ones. Let us hence begin with postulate 5 that embodies the prescription for the generation of probabilities.

To get a positive probability, it might be sufficient if P_j^2 is a positive operator—the positivity of the underlying P_j operator may not be required. To modify this postulate in the context of less restrictive condition, one can introduce a positive operator: $\hat{\Pi}_j \geq 0$, which is a generalization of P_j^2 with

$$p_j = Tr(\hat{\Pi}_j \hat{\rho}). \tag{4.15}$$

One should ensure the pertinent probability distribution to be still normalized. In light of this, inspection of the postulates that have been discussed so far reveals that normalization is the consequence of postulate 1; and so requires $\sum_j \hat{\Pi}_j = \hat{I}$. This envisages that the positive operator—which represents the decomposition of the identity—is denoted as positive operator valued measure, where $\{\hat{\Pi}_j\} \geq 0$ are its elements.

One may note that for the positive operator valued measure to exist, the orthogonality and positivity of P_j operators are not required. The operators that determine the post measurement via postulates 4 and 6 can hence be any operator including non-Hermitian operators. Recall that for the projective measurement, orthogonality is a consequence of postulate 3, which is the most constraining postulate since it restricts the number of terms in the decomposition of the identity to utmost the dimensionality of the system.

In case one abandons postulate 3, the operators that generate the probability distribution are no longer the same as that generate the post measurement states; and hence, one can enjoy a considerable amount of freedom in choosing them:

1. the decomposition of the identity $\hat{I} = \sum_j \hat{\Pi}_j$ in terms of the positive operators $\{\hat{\Pi}_j\} \geq 0$ is epitomized as the positive operator valued measure, where $\{\hat{\Pi}_j\}$ are the pertinent elements;
2. the elements of the positive operator valued measure can be expressed in terms of the detection operators $\hat{A}_j$'s, which are non-Hermitian in general and restricted only by the requirement that $\sum_j \hat{A}_j^\dagger \hat{A}_j = \hat{I}$ as $\hat{\Pi}_j = \hat{A}_j^\dagger \hat{A}_j$;
3. detection yields one of the alternatives corresponding to the element of the positive operator valued measure;
4. the state of the system after the measurement is

$$|\phi\rangle = \frac{\hat{A}_j|\Phi\rangle}{\sqrt{\langle\Phi|\hat{A}_j^\dagger\hat{A}_j|\Phi\rangle}}, \tag{4.16}$$

   if it is initially in the pure state $|\Phi\rangle$ and

$$\rho_j = \frac{\hat{A}_j\hat{\rho}\hat{A}_j^\dagger}{Tr(\hat{A}_j\hat{\rho}\hat{A}_j^\dagger)} = \frac{\hat{A}_j\hat{\rho}\hat{A}_j^\dagger}{Tr(\hat{A}_j^\dagger\hat{A}_j\hat{\rho})}, \tag{4.17}$$

   if it is initially in the mixed state $\hat{\rho}$;
5. the probability that this particular alternative is found as the measurement outcome is

$$p_j = Tr(\hat{A}_j\hat{\rho}\hat{A}_j^\dagger) = Tr(\hat{\Pi}_j\hat{\rho}); \tag{4.18}$$

6. if one performs the measurement, but do not record the results, the post measurement state would be described by the density operator

$$\hat{\bar{\rho}} = \sum_j p_j\hat{\rho}_j = \sum_j \hat{A}_j\hat{\rho}\hat{A}_j^\dagger. \tag{4.19}$$

Very often one is concerned with the resulting probability distribution rather than the state of the system after such an operation is performed. It may suffice in this case to consider postulates 1 and 5 that define the probability of finding an alternative j as the detection result. It is also good to note that any operation that satisfies postulates 1 and 2 can be taken as a legitimate operation that generates valid probability distribution.

4.1.4   Viewing Measurement as Interaction Process

In a double slit experiment, one sends a stray of particles through two narrow slits, and then tricks them to impinge upon a common screen. In this framework, looking for the probability distribution of detecting the particles over the surface of the screen is perhaps the most rational thing to do. One should not just take the probabilities of the passage through the slits; multiply with the probabilities of detection at the screen conditional on passage through either slit; and sum over the contributions of the two slits; rather, one should look into the additional interference term to expose the underlying subtlety. The acceptable description of the particle in terms of quantum wave function is the one in which the wave is presumed to pass through both slits at the same time. There are situations even in this context when the interference term would not be observed. The classical probability formula would be enough to describe the properties of the interacting systems in this case such as when one seeks to carry out the detection measurement at the slits, and when another system happens to interact with the particles in their way between the slits and screen.

The suppression of interference with the help of a suitable spontaneous interaction with the environment is often dubbed as decoherence; a phenomenon anticipated to have a significant implication in the understanding of the theory of quantum information processing (see Chaps. 10 and 11). It might not be difficult to comprehend that the actual issue of quantum measurement is more challenging than the entanglement of measuring apparatus with the system to be measured since everything is in interaction with everything else. Decoherence theory is thus alluded to provide an explanation for the transition of the system to a mixture of states, where the ensued mixture is taken as a proper quantum ensemble in case the measurement leads to the realization of precisely one state in the ensemble. Besides, the notion that interference can be suppressed between localized states of the macroscopic objects justifies why macroscopic objects appear to be in localized states and why one never observes the pointer of the apparatus pointing to two different results as the conventional quantum theory might suggest.

Quantum theory treats the issue of measurement problem that arises from the evolution of quantum state according to Schrödinger equation into a linear superposition of different states and leads to a paradoxical predictions as one to be resolved by the nonunitary transformation of the state vector at the time of measurement into a definite state. Quantum theory also tries to provide accurate ways of predicting the value of the definite state that will be measured in the form of probability for each possible measurement outcome. The nature of the transition from the quantum superposition of the states to a definite classical state during measurement process however is not explained by the orthodox quantum theory; rather, assumed as an axiom.

The notion of quantum Darwinism may on the other hand able to explain how the classical world emerges from the quantum world in the sense that the transition of quantum features from the vast superimposed states to the greatly reduced set of pointer states (the states to which quantum system decoheres) could be perceived as a selection process imposed on the quantum system via its continuous interaction with the environment [9]. As a result, it should be admissible to construe that the available quantum states collapse into a single state due to the interaction with the environment. Decoherence can thus be taken as a selection process that leads to the final stable state usually perceived as pointer state. Even though pointer states are quantum states, they can be fit enough to be transmitted through the environment without collapsing; and can then make copies of themselves contrary to the principle of quantum no-cloning (see Chap. 10).

It might also be judicious to argue that what the observer knows is inseparable from what the observer is, that is, a distinct memory or identity states of the observer—which are also his states of knowledge—cannot be superimposed. It is the persistence of correlations between the records or data in the possession of the observer and the recorded states of the macroscopic system which is required to recover familiar reality. At some point, it might suffice to consider that the states of the decohering system quickly evolve into mixtures of preferred or pointer states, that is, all that can be known about the system is its decoherence resistant identity tag. The observer thus can monitor the system indirectly by intercepting a small fraction of the environment such as a fraction of photons that have been reflected or emitted. Since the principle of superposition admits arbitrary linear combinations of any state as possible quantum state, it may be legitimate to argue that the interpretational problem in the quantum theory stems from the vastness of the Hilbert space. Consequently, one may conceive the available quantum superpositions to be treated equally by decoherence, although the interaction with the environment singles out the preferred set of states. Therefore, it might be valid to propose that these pointer states remain untouched in spite of the nature of the environment while their superposition may lose phase coherence or information, and so decohere. Critical scrutiny may also reveal that it is the environment that imposes the selection rules by preserving part of the information that resides in the correlations between the system and measuring apparatus.

Existing experience indicates that the observer applies the environment or measuring device as communication channel to learn about the system. One can therefore propose that the spreading of information about the system throughout the environment might be responsible for the emergence of the objective reality. The transition from the quantum indeterminism of the global state vector to the classical definiteness of the states of individual systems can be rephrased as the outcome of a generic measurement of the state of quantum system is not deterministic. The randomness is blamed on the collapse of the wave function, which is invoked whenever the quantum system comes into contact with classical apparatus. This issue still arises in fully quantum version in spite of the overall deterministic quantum evolution of the state vector of the universe.

To illustrate such subtlety, imagine a quantum system S initially in state $|\Phi_0\rangle$ to interact with a quantum apparatus A initially in state $|\alpha_0\rangle$, that is,

$$|\Phi\rangle = |\Phi_0\rangle \otimes |\alpha_0\rangle = \left(\sum_j a_j |\upsilon_j\rangle \right) \otimes |\alpha_0\rangle. \qquad (4.20)$$

Owing to the perceived interaction, the quantum apparatus in the initial state is expected to change and the intended change can be captured by the process that can be represented as

$$\left(\sum_j a_j |\upsilon_j\rangle \right) \otimes |\alpha_0\rangle \longrightarrow \sum_j a_j |\upsilon_j\rangle \otimes |\alpha_j\rangle = |\Phi_t\rangle, \qquad (4.21)$$

where $\{|\alpha_j\rangle\}$ and $\{|\upsilon_j\rangle\}$ are states of the apparatus and system, and a_j are complex coefficients.

The conditional dynamics of such pre-measurement process can be accomplished by means of a unitary Schrödinger evolution where $|\Phi_t\rangle$ would be EPR type entangled state. The EPR nature of the state that emerges from the pre-measurement can be made more explicit by rewriting it in a different chosen basis;

$$|\Phi_t\rangle = \sum_j a_j |\upsilon_j\rangle \otimes |\alpha_j\rangle = \sum_r b_j |\mu_j\rangle \otimes |\beta_j\rangle. \qquad (4.22)$$

Such freedom of basis choice can be considered as the source of the ambiguity guaranteed by the principle of superposition. If one were to associate the states of the apparatus (or the observer) with decompositions of $|\Phi_t\rangle$—even before inquiring about a specific outcome of the measurement—one would have to decide on the decomposition of $|\Phi_t\rangle$. This assertion might be understood as the change of the basis can redefine the measured quantity. It is in such an analysis that the interpretational problem in quantum measurement takes refuge.

4.1.5   Quantum Nondemolition Measurement

As already highlighted, the process of the measurement of the observable induces noise, that is, successive measurements of the same observable may yield different results where the uncertainty in the measurement is linked to the Heisenberg's uncertainty relation. It may not be difficult to infer that one of the central principles of quantum theory ruins the repeatability of successive measurements of the physical system, which is the basic foundation of science since consecutive measurements of a given system is expected to yield the same result as required by the reproducibility criteria. The detection of the radiation field as done by photon counting techniques for instance is field destructive; and hence, makes repetitive measurement unattainable.

It can thus be asserted that if there is at all a way of measuring the observable successively without altering the subsequent results, it would be of a considerable importance. Such insight hopefully unveils the need for the quantum nondemolition (QND) measurement scheme in which one monitors the observable that can be measured repeatedly with the readout of each measurement being the same as the initial trial [10]. This conception entails that a genuine QND experiment can be performed several times without necessarily affecting the quality of the quadrature to be interrogated.

Consider a free particle that has the Hamiltonian of the form $\hat{H} = \hat{p}^2/2m$. It may not be hard to imagine that the measurement of the position of the particle along the x-axis perturbs the momentum in the same direction by an amount of $\Delta p_x \geq \hbar/2\Delta x$. Such a change in the momentum during successive measurements can alter the subsequent values of the position:

$$\Delta x^2(t) = \Delta x^2(t = 0) + \left(\frac{\hbar}{2m\,\Delta x(t = 0)} \right)^2 t^2. \tag{4.23}$$

One can therefore infer from Eq. (4.23) that a precise measurement of the position for the second time would be spoiled by the noise arising from the measurement. To overcome such a challenge, one may need to delve into the discussion of the notion of QND measurement. To this effect, one can begin with the presumption that the signal observable $\hat{A}_s$ of the quantum system to be measured by detecting changes in observable $\hat{A}_p$ of the probe system coupled to the quantum system during the measurement time T without necessarily perturbing the subsequent evolution of $\hat{A}_s$ as usually expected.

Imagine that the total Hamiltonian of the system-probe coupled system can be expressed as

$$\hat{H} = \hat{H}_s + \hat{H}_p + \hat{H}_I, \tag{4.24}$$

where the equations of motion for the operators $\hat{A}_s$ and $\hat{A}_p$ would be

$$\frac{d\hat{A}_s}{dt} = -i[\hat{A}_s, \, \hat{H}_s + \hat{H}_I], \tag{4.25}$$

$$\frac{d\hat{A}_p}{dt} = -i[\hat{A}_p, \, \hat{H}_p + \hat{H}_I]. \tag{4.26}$$

Since the two systems are engaged in the QND measurement in which one of them is measured by varying the other, some constraints should be imposed on both systems so that the necessary coupling can be forged. The main idea is to invoke a connection between the quadratures via a nonlinear interaction; and then, to arrange the experimental setup in such a way that the back action noise is transferred to the quadrature that would be measured as in squeezing.

In view of this requirements, certain conditions are expected to be satisfied:

1. since $\hat{A}_r$ is the quantity to be measured by varying the observable $\hat{A}_p$, the interaction Hamiltonian $\hat{H}_I$ should be a function of $\hat{A}_s$:

$$\frac{\partial \hat{H}_I}{\partial \hat{A}_s} \neq 0; \qquad (4.27)$$

2. since the probe is used to measure the dynamics of the quantum system, it cannot be constant of time:

$$[\hat{A}_p, \ \hat{H}_I] \neq 0; \qquad (4.28)$$

3. based on the claim that the system would not be altered during the measurement, it is possible to demand that

$$[\hat{A}_s, \ \hat{H}_I] = 0, \qquad (4.29)$$

which can be referred to as back action invasion criterion;
4. the dynamics of $\hat{A}_s$ will be predictable despite the uncertainty imposed by the conjugate variable $\hat{A}_s^c$ due to the measurement process: the measurement of $\hat{A}_s$ will not affect $\hat{A}_s$ itself,

$$\frac{\partial \hat{H}_s}{\partial \hat{A}_s^c} = 0. \qquad (4.30)$$

Owing to the fact that the interaction Hamiltonian restricted by these conditions is the combined contribution of the system and the probe, the outlined conditions are believed to determine the kind of the probe system or variables one can use. To ascertain whether a QND measurement can be viable, it might be imperative looking for the QND operator that should satisfy certain criteria.[6] Such a proposal is tantamount to calculating various normalized correlations between two quadratures of input and output fields.

To do so, one can apply two real Hermitian operators such as $\hat{p}$ and $\hat{q}$ for which the correlation in frequency domain can be defined as

$$C(\omega) = \frac{\left| \int_{-\infty}^{\infty} e^{-i\omega\tau} \langle \hat{p}(\tau)\hat{q} + \hat{q}(\tau)\hat{p} \rangle d\tau \right|^2}{\int_{-\infty}^{\infty} e^{-i\omega(\tau_p + \tau_q)} \langle \hat{p}(\tau_p)\hat{p} \rangle \langle \hat{q}(\tau_q)\hat{q} \rangle d\tau_q d\tau_p}, \qquad (4.31)$$

---

[6]If quantum observable that can be represented by a self-adjoint operator in the Heisenberg picture can be made to commute with itself at time t and t', $[\hat{O}(t), \hat{O}(t')] = 0$, the observable O can be measured with no limit on the repeatability. The operator that obeys such a general condition is designated as QND operator.

and can also be put in a more appealing manner as

$$C = \frac{|\langle \hat{p}\hat{q}\rangle|^2}{\langle \hat{p}^2\rangle\langle \hat{q}^2\rangle}. \tag{4.32}$$

On the basis of this definition and the requirement associated with nondemolition measurement, the pertinent variable should also satisfy certain conditions:

1. the coefficient that represents how accurately one can determine the signal input by measuring the probe output can be quantified by the correlation coefficient denoted by C_1 and defined as

$$C_1 = \frac{|\langle \hat{S}_{in}\hat{P}_{out}\rangle|^2}{\langle \hat{S}_{in}^2\rangle\langle \hat{P}_{out}^2\rangle}, \tag{4.33}$$

where $\hat{S}_{in}$ is the noise part of the quadrature of the signal one seeks to measure and $\hat{P}_{out}$ is the noise part of the quadrature of the probe one reads—perfect QND measurement entails $C_1 = 1$;
2. a nondestructive measurement scheme can be quantified by the correlation coefficient C_2 defined by

$$C_2 = \frac{|\langle \hat{S}_{in}\hat{S}_{out}\rangle|^2}{\langle \hat{S}_{in}^2\rangle\langle \hat{S}_{out}^2\rangle}, \tag{4.34}$$

where $\hat{S}_{out}$ is the noise part of the quadrature of the signal one reads;[7]
3. the coefficient that used to determine how good the scheme is as state preparation device is the variance in the signal output in relation to the measured value for the probe field, which can be denoted by C_3 and given by

$$C_3 = \frac{|\langle \hat{S}_{out}\hat{P}_{out}\rangle|^2}{\langle \hat{S}_{out}^2\rangle\langle \hat{P}_{out}^2\rangle}. \tag{4.35}$$

One may note that the conditional variance of the signal output defined in terms of the measured value of the probe field can be expressed as

$$V_C(\hat{S}_{in}|\hat{P}_{out}) = \langle \hat{S}_{out}^2\rangle(1 - C_3). \tag{4.36}$$

---

[7] A measurement is presumed as perfectly nondestructive as possible if $C_2 = 1$. It is also required that QND scheme to have a better performance than a simple beam splitter; in which case, the sum $C_1 + C_2$ is envisioned to give a good indication of the ability of the scheme to perform QND measurement. For example, one has $C_1 + C_2 = 1$ for a simple beam splitter and $C_1 + C_2 = 2$ for a perfect QND measurement scheme; realistic quantum system should exhibit the condition that lies within this interval.

A perfect state preparation device is feasible provided that $V_C(\hat{S}_{in}|\hat{P}_{out}) = 0$, which implies that $C_3 = 1$.

Existing experience reveals that the phase shift can be introduced in one arm of the Mach-Zehnder type interferometer by coaxing the light into passing across a nonlinear medium in which the accurate measurement of the phase of the probe field destroys the interference pattern generated by the interferometer (see Sect. 4.3.2). The measurement of the phase shift of the probe beam can provide information about the photon number whose exact knowledge in one of the pathes of a single-photon interferometer determines which path the photon follows. It is such knowledge that subsequently annihilates the interference. To explain such a phenomenon in optical Kerr medium, one may begin with the refractive index that can be described by the relation of the form

$$\eta = \eta_0 + \eta_2 E^2, \tag{4.37}$$

where η_0 represents the weak field refractive index and η_2 the strength of the rate of change of the refractive index with variation in optical intensity.

If a strong signal wave is incident on such nonlinear material, it causes change in the refractive index; the weak probe wave consequently experiences a phase shift while propagating through the material, and the phase shift would be proportional to the intensity of the signal wave. Suppose the reflectivity of the mirrors M_1 and M_2 is zero for the signal frequency ω_s and unity for the probe frequency ω_p so that the interferometer is taken to be active for the probe field. One can begin to explore this idea with the measurement scheme in which the photon number of the signal wave is measured via the phase of the probe wave upon using an optical Kerr effect (see Fig. 4.1 and also [11]). The probe wave is expected to undergo a phase shift due to the change in the refractive index inside the Kerr medium while the signal wave passes through it without experiencing change in the photon number. The observation that the phase of the probe field would be modulated by signal photon

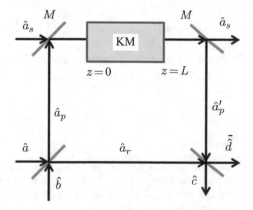

Fig. 4.1 Schematic representation of optical Kerr effect, which is also known as Match-Zehnder interferometer. The radiation propagates in one of the arms in a free space, and in the other through a Kerr medium (KM) that invokes the phase shift. Then after, the mixture of these two different waves would be captured on the screen

number indicates that a phase shift for the probe wave can be measured in terms of a photon current of a balanced mixed detector.

The Hamiltonian that describes such a process can be proposed to be

$$\hat{H} = \omega_s \hat{a}_s^\dagger \hat{a}_s + \omega_p \hat{a}_p^\dagger \hat{a}_p + \kappa \hat{a}_s^\dagger \hat{a}_s \hat{a}_p^\dagger \hat{a}_p \tag{4.38}$$

in which κ is the coupling constant that depends on the third-order nonlinear susceptibility related to the optical Kerr effect. Upon taking the observable to be measured as photon number of the signal wave $\hat{A}_s = \hat{a}_s^\dagger \hat{a}_s$ and the probe or readout observable as suitable phase operator that satisfies the conditions of the QND variable,[8] the probe operator can be defined by

$$\hat{a}_p = \sqrt{\hat{a}_p^\dagger \hat{a}_p} e^{i\phi_p}. \tag{4.39}$$

One should note that $e^{i\phi_p}$ and $e^{-i\phi_p}$ are non-Hermitian operators from which one generates Hermitian operators

$$\cos\phi_p = \frac{1}{2}\left[e^{i\phi_p} + e^{-i\phi_p}\right], \qquad \sin\phi_p = \frac{1}{2i}\left[e^{i\phi_p} - e^{-i\phi_p}\right] \tag{4.40}$$

that can be regarded as operators that stand for the observable related to the measurement of phase properties. It then should follow from choosing $\hat{A}_p = \sin\phi_p$ that

$$\hat{A}_p = \frac{1}{2i}\left[\frac{1}{\sqrt{\hat{a}_p^\dagger \hat{a}_p}}\hat{a}_p - \hat{a}_p^\dagger \frac{1}{\sqrt{\hat{a}_p^\dagger \hat{a}_p}}\right]. \tag{4.41}$$

One can verify that the previous four conditions set for a QND operator hold true.

In this regard, the Heisenberg equation for the probe operator $\hat{a}_p$ inside the Kerr medium yields

$$\frac{d\hat{a}_p}{dt} = -i\kappa \hat{A}_s \hat{a}_p. \tag{4.42}$$

It should be possible to express Eq. (4.42) as spatial evolution for a traveling wave by replacing t by $-\frac{z}{v}$ when the propagation is towards $+z$ direction with a constant

---

[8]The phase problem is one of the oldest issues initiated by Dirac who introduced commutation relation of the form $[\hat{n}, \hat{\phi}] = \hat{I}$ that leads to the Heisenberg inequality $\Delta\hat{n}\Delta\hat{\phi} \geq 1$, where $\hat{n}$ ($\hat{\phi}$) is the operator for the photon number (phase), which limits the sensitivity in the precision phase measurements [12].

velocity v. When the spatial evolution of the traveling wave is temporally stationary: $\frac{d}{dt} \to -v\frac{d}{dz}$;

$$\frac{d}{dz}\hat{a}_p(z) = \frac{i\kappa}{v}\hat{A}_s\hat{a}_p(z),\qquad (4.43)$$

where $\hat{A}_s$ is taken to be constant, and the integration of Eq. (4.43) results

$$\hat{a}_p(L) = \hat{a}_p(0)e^{i\frac{\kappa L}{v}\hat{A}_s} \qquad (4.44)$$

in which the operator quantity $\kappa L\hat{A}_s/v$ corresponds to the phase shift in the probe $\hat{a}_p$. In this case, the operator $\hat{A}_p$ at $z = L$ can be expressed as

$$\hat{A}_p(L) = \frac{1}{2i}\left[\frac{1}{\sqrt{\hat{a}_p^\dagger(0)\hat{a}_p(0)}}\hat{a}_p(0)e^{i\frac{\kappa L}{v}\hat{A}_s} - e^{-i\frac{\kappa L}{v}\hat{A}_s}\hat{a}_p^\dagger(0)\frac{1}{\sqrt{\hat{a}_p^\dagger(0)\hat{a}_p(0)}}\right].$$
$$(4.45)$$

Since the main task is to measure the phase shift in the probe field and the signal photon number with the help of a balanced homodyne detector, the photon number difference can be expressed in view of the discussion in Sect. 4.3.2 as

$$\hat{n}_{cd} = \hat{d}^\dagger\hat{d} - \hat{c}^\dagger\hat{c}. \qquad (4.46)$$

Upon making use of the properties of a symmetric 50:50 beam splitter, one can also realize that

$$\hat{a}_p = \frac{1}{\sqrt{2}}(i\hat{a} + \hat{b}), \qquad \hat{a}_r = \frac{1}{\sqrt{2}}(\hat{a} + i\hat{b}). \qquad (4.47)$$

It is a well known fact that the Kerr medium shifts the phase of the probe wave, that is,

$$\hat{a}_p' = \exp\left[i\left(\frac{\kappa L}{v}\hat{A}_s + \frac{\pi}{2}\right)\right]\hat{a}_p(0), \qquad (4.48)$$

where $\frac{\pi}{2}$ can be added by adjusting the interferometer.

Upon taking the second beam splitter to be also symmetric, one can see that

$$\hat{c} = \frac{1}{\sqrt{2}}(i\hat{a}_r + \hat{a}_p'), \qquad \hat{d} = \frac{1}{\sqrt{2}}(\hat{a}_r + i\hat{a}_p'), \qquad (4.49)$$

as a result follows

$$\hat{n}_{cd} = -i\big[\hat{a'}_p^\dagger \hat{a}_r - \hat{a}_r^\dagger \hat{a'}_p\big].$$ (4.50)

One can also infer from Eq. (4.47) that

$$\hat{a}_p^\dagger \hat{a}_r - \hat{a}_r^\dagger \hat{a}_p = -i(\hat{a}^\dagger \hat{a} - \hat{b}^\dagger \hat{b}).$$ (4.51)

One can obtain after that

$$\hat{n}_{cd} = (\hat{a}^\dagger \hat{a} - \hat{b}^\dagger \hat{b}) \sin\left(\frac{\kappa L}{v}\hat{A}_s\right) - i\sqrt{2}(\hat{a}^\dagger \hat{b} - \hat{b}^\dagger \hat{a}) \sin\left(\frac{\kappa L}{v}\hat{A}_s\right)$$
$$- \left[\hat{b}^\dagger \hat{a} \exp\left(i\frac{\kappa L}{v}A_s\right) + \hat{a}^\dagger \hat{b} \exp\left(-i\frac{\kappa L}{v}A_s\right)\right].$$ (4.52)

Assuming the field $\hat{b}$ to be in vacuum state, one gets

$$\langle \hat{A}_s \rangle = \langle \hat{a}_s^\dagger \hat{a}_s \rangle = \frac{v}{\kappa L} \frac{\langle \hat{n}_{cd} \rangle}{\langle \hat{a}^\dagger \hat{a} \rangle},$$ (4.53)

where the approximation $\kappa L A_s \ll v$ is employed. Equation (4.53) may entail that the mean photon number of the signal mode can be measured directly from the mean photon number of the probe laser and mean of the photon difference between the reference and the light coming through the Kerr medium; and so does not alter the state of the signal mode.

One may also seek to measure the number of photons based on the detection of a dispersive phase shift (energy shift that is not accompanied by the absorption of photons) induced by the cavity field on the wave function of a non-resonant atoms crossing the cavity [13]. The probe is no longer a field as previously assumed; rather, it is a beam of atoms that interact nonlinearly with the signal mode. Imagine the atomic beam to comprise a three-level atomic system in the cascade configuration in which the cavity mode is detuned from the $|a\rangle \to |b\rangle$ transition by $\Delta = \omega - \omega_{ab}$, where $|a\rangle$ and $|b\rangle$ are the upper and intermediate energy levels, and ω_{ab} is the transition (atomic) frequency (see Sect. 6.5). Detuning is usually taken to be very small when compared to ω_{ab} and other transition parameters in the atomic spectrum, which implies that it is the upper and intermediate energy levels that engage in the interaction. One may note that the main idea originates from the fact that the interaction of the non-resonant field with the atom induces a dispersive energy shift in the state $|b, n\rangle$, which is proportional to the number of photons in the field. So the energy shift should be detected by measuring the dephasing accumulated between $|b\rangle$ and the lower energy level $|c\rangle$. The nondemolition character would be related to the energy shift depending on the number of photons in the field due to the dispersion effect that does not affect the number of photons.

The interaction Hamiltonian that describes the intended system can be captured by

$$\hat{H} = \omega \hat{a}^\dagger \hat{a} + \frac{\omega_{ab}}{2}\left[|a\rangle\langle a| - |b\rangle\langle b|\right] + g\left[|a\rangle\langle b|\hat{a}^\dagger + \hat{a}|b\rangle\langle a|\right], \qquad (4.54)$$

where the interaction between the field and lower energy level is neglected (relate with Eq. (8.71)). One may note that the atomic dressed state which diagonalizes Eq. (4.54) can be expressed as

$$|+\rangle = \cos\theta_n|a, n-1\rangle - \sin\theta_n|b, n\rangle, \qquad (4.55)$$

$$|-\rangle = \sin\theta_n|a, n-1\rangle + \cos\theta_n|b, n\rangle, \qquad (4.56)$$

with the corresponding eigenvalues

$$E_{+n} = \left[(n-1)\omega + \frac{\omega_{ab}}{2}\right] - \frac{1}{2}(\Omega_n - \Delta), \qquad (4.57)$$

$$E_{-n} = \left[n\omega - \frac{\omega_{ab}}{2}\right] + \frac{1}{2}(\Omega_n - \Delta), \qquad (4.58)$$

where

$$\sin\theta_n = \frac{\Omega_n - \Delta}{\sqrt{(\Omega_n - \Delta)^2 + 4g^2 n}}, \qquad \cos\theta_n = \frac{2g\sqrt{n}}{\sqrt{(\Omega_n - \Delta)^2 + 4g^2 n}}, \qquad (4.59)$$

in which $\Omega_n = \sqrt{\Delta^2 + 4g^2 n}$ (see Sect. 6.1.1).

In case Δ is assumed to be large when compared to $2g\sqrt{n}$, that is, when $4g^2 n \ll \Delta^2$, one can see that Eqs. (4.55) and (4.56) reduce to

$$|+\rangle = |a, n-1\rangle, \qquad |-\rangle = |b, n\rangle. \qquad (4.60)$$

Besides, expanding Ω_n, and then keeping the first term indicate that

$$E_{+n} = \left[(n-1)\omega + \frac{\omega_{ab}}{2}\right] - \frac{g^2}{\Delta}n, \qquad (4.61)$$

$$E_{-n} = \left[n\omega - \frac{\omega_{ab}}{2}\right] + \frac{g^2}{\Delta}n. \qquad (4.62)$$

One can observe that there is no net change in the photon number but the state $|b, n\rangle$ whose unperturbed energy is $n\omega - \omega_{ab}/2$ experiences the energy shift of $g^2 n/\Delta$, which is proportional to the number of signal photons.

The effective Hamiltonian that describes the atom-radiation interaction can also be put in the form

$$\hat{H} = \frac{\omega_{bc}}{2}\hat{\sigma}_z + \omega\hat{a}_s^\dagger\hat{a}_s + \frac{g^2}{\Delta}\hat{a}_s^\dagger\hat{a}_s\hat{\sigma}_+\hat{\sigma}_-,$$ (4.63)

where

$$\hat{\sigma}_z = |b\rangle\langle b| - |c\rangle\langle c|, \qquad \hat{\sigma}_+ = |b\rangle\langle c|, \qquad \hat{\sigma}_- = |c\rangle\langle b|.$$ (4.64)

Hamiltonian (4.63) can also induce a phase shift proportional to $g^2 n/\Delta$ in the energy level $|b\rangle$ and leaves the energy level $|c\rangle$ unperturbed.

Note that the photon number n is the eigenvalue of the signal operator $\hat{A}_s = \hat{a}_s^\dagger\hat{a}_s$, and the probe is chosen to be the atomic dipole operator

$$\hat{A}_p = \frac{1}{2i}(\hat{\sigma}_+ - \hat{\sigma}_-).$$ (4.65)

It can be verified that operators $\hat{A}_s$ and $\hat{A}_p$ satisfy the criteria set for QND variable where the entire system resembles an optical Kerr effect.

4.2   Coincidence Measurement

Since recording photon intensity with single detector does not encompass all kinds of measurements on all aspects of the field, it should be necessary and straightforward at some point to think of other possible approaches such as coincidence measurement in which two or more beams of light are created from the same initial wave packet, and after interweaving their optical properties then send to a coincidence counter. The resulting outcome can be understood as correlation; the notion that can be loosely perceived as the expectation value of the involved product states [14]. While studying a random process in which each sample is presumed to be at different points and times, the correlation is expected to depend on how quickly they change and how far they are separated. The correlation function hence can be linked to the interrelation between such random variables at two different points in space and/or time, and can be taken as a useful indicator of the interdependencies. In case one considers the interrelation between random variables that represents the same quantity measured at two different points, the emerging correlation is often referred to as autocorrelation function, whereas the intra-relation between different random variables is often designated as cross correlation function.

Imagine that there are two electromagnetic fields that emerge from position $\mathbf{r}$ and detected at separate times t_1 and t_2 in which the correlation between the two photons can be captured by

$$G^{(1)}(\mathbf{r}; t_1, t_2) = Tr(\hat{\rho}\hat{\mathbf{E}}^{(-)}(\mathbf{r}, t_1)\hat{\mathbf{E}}^{(\dagger)}(\mathbf{r}, t_2)).$$ (4.66)

Equation (4.66) stands for the first-order two-time correlation function that turns out to be sufficient to capture the result of classical interference experiments, and interpreted as the transition probability for the detector atom while absorbing the photon from the field at position **r** in time between t and $t + dt$. Since a precise knowledge of the state of the field is almost absent, one may resort to statistical formulations where the density operator corresponds to the initial state of the radiation field. One should also note that stationary fields are common interest in the field of experimental quantum optics in which the correlation function is taken to be invariant under the displacement of the time variable.

The correlation function is thus presumed to depend on the difference of t_1 and t_2: $\tau = t_2 - t_1$, that is,

$$G^{(1)}(\mathbf{r}; t_1, t_2) = G^{(1)}(\mathbf{r}; \tau). \tag{4.67}$$

The joint probability for detecting one photon at position $\mathbf{r}_1$ between t_1 and $t_1 + dt_1$ and another at $\mathbf{r}_2$ between t_2 and $t_2 + dt_2$ with $t_1 < t_2$ can also be designated by the second-order quantum correlation function

$$G^{(2)}(\mathbf{r}_1, \mathbf{r}_2; t_1, t_2) = Tr(\hat{\rho}\hat{\mathbf{E}}^{(-)}(\mathbf{r}_1, t_1)\hat{\mathbf{E}}^{(-)}(\mathbf{r}_2, t_2)\hat{\mathbf{E}}^{(\dagger)}(\mathbf{r}_2, t_2)\hat{\mathbf{E}}^{(\dagger)}(\mathbf{r}_1, t_1)), \tag{4.68}$$

where the right side of Eq. (4.68) is time and normally ordered, and can be interpreted as the photon coincidences or delayed coincidences between the two photons.

A normalized correlation function of the radiation in terms of the boson operators may also be required where the first-order two-time normalized correlation function can be expressed as

$$g^{(1)}(\tau) = \frac{\langle \hat{a}^\dagger(t)\hat{a}(t + \tau) \rangle}{\langle \hat{a}^\dagger(t)\hat{a}(t) \rangle} \tag{4.69}$$

and the second-order normalized two-time correlation function as

$$g^{(2)}(\tau) = \frac{\langle \hat{a}^\dagger(t)\hat{a}^\dagger(t + \tau)\hat{a}(t + \tau)\hat{a}(t) \rangle}{\langle \hat{a}^\dagger(t)\hat{a}(t) \rangle^2} \tag{4.70}$$

in which the field is taken to be statistically stationary and the operators are put in the normal order. The main theme is to devise a viable means of obtaining these expectation values.

The quantum nature of the radiation imposes the condition that the photon correlation experiments should be consistent with the theory of quantum electrodynamics: once the photon is absorbed, the state of the radiation field and the atom should change in the way provided in Chaps. 5 and 6. The subsequent absorption events should then occur in accordance with the immediate initial state of the system that would be substantially altered from the one before it. As a result, the state with

n available photons can have correlations up to nth order that can be perceived as subsequent absorption events led to different states of the field; and the pertinent interaction process may not be describable in terms of a classical correlation alone.

Let us now expose the merits of the normalized correlation function in identifying the photon statistics in the sense of revealing the nature of the radiation. For example, when the radiation field satisfies the inequality

$$g^{(2)}(\tau) < g^{(2)}(0) \qquad (4.71)$$

for all τ's less than some critical time τ_c commonly dubbed as correlation time, the photon would exhibit excess correlation for times less than the correlation time. This phenomenon is designated as photon bunching, that is, as the photons tend to distribute themselves in bunches rather than at random since the correlation for the photons arriving at the same time ($\tau = 0$) is greater than the one's coming at separate time τ. When such a light falls on the photon detector, more pairs of photons are detected closer together than further apart. It might be good to note that thermal and coherent lights exhibit a phenomenon of photon bunching.

In addition, one may observe in a certain quantum optical system the situation in which

$$g^{(2)}(\tau) > g^{(2)}(0) \qquad (4.72)$$

that corresponds to the phenomenon of photon anti-bunching: fewer photon pairs are detected closer together, that is, the photons tend to reach the detector separately which can be considered as one of the manifestations of the quantum features of light. For example, the light emitted by a two-level atom confined in a cavity and externally pumped by coherent radiation exhibits photon anti-bunching [15]. The photon statistics of a given light can also be identified by using the second-order correlation function defined overtime: if $g^{(2)}(\tau) = 1$, the corresponding photon statistics is dubbed as Poissonian; for $g^{(2)}(\tau) > 1$ super-Poissonian; for $g^{(2)}(\tau) < 1$ sub-Poissonian.

The second-order correlation function can also be expressed for a single-mode light in terms of the mean and variance of the photon number as

$$g^{(2)}(0) = 1 + \frac{\Delta n^2 - \langle \hat{n} \rangle}{\langle \hat{n} \rangle^2}, \qquad (4.73)$$

which is found for coherent, chaotic and squeezed light respectively to be

$$g^{(2)}(0) = 1, \qquad g^{(2)}(0) = 2, \qquad g^{(2)}(0) = 2 + \frac{\cosh^2 r}{\sinh^2 r}. \qquad (4.74)$$

Moreover, to connect the photon correlation function with the conception of the nonclassical behavior of the radiation field, one may begin with a general statistical condition that

$$Tr(\hat{\rho}\hat{A}^{\dagger}\hat{A}) \geq 0. \tag{4.75}$$

The first-order autocorrelation is thus expected to satisfy the condition that $G^{(1)}(\mathbf{r}; \tau) \geq 0$: the first-order correlation function should be a positive quantity, and in the same manner nth order correlation function should also be positive.

The cross correlation of the same dynamical variable measured at different positions and/or times is hence expected to satisfy

$$\sum_{i,j} \lambda_i^* \lambda_j G^{(1)}(x_i; x_j) \geq 0, \tag{4.76}$$

where $\lambda_{i,j}$'s are arbitrary sets of complex numbers that designate the amplitude of the radiation field. Such a matrix would have a positive definite quadratic form provided that

$$\det[G^{(1)}(x_i; x_j)] \geq 0, \tag{4.77}$$

which implies that

$$G^{(1)}(x_1; x_1)G^{(1)}(x_2; x_2) \geq |G^{(1)}(x_1; x_2)|^2. \tag{4.78}$$

The violation of such inequalities is tantamount to witnessing a nonclassical feature (see Chap. 7).

4.2.1 Optical Coherence

To illustrate the underlying physical process required in establishing photon correlation, it might be necessary to connect the optical coherence with the inequalities introduced earlier. To do so, one can begin with the classical interpretation of coherence which is linked to the existence of fringe in conventional experiment such as Young's double slit interference where the appearance of the fringe is associated with the correlation of the beams that travel over a significant; and yet, short distance in relation to each other. Optical coherence hence refers to the ability of the light wave to cast interference pattern: a light generated by a conventional source can interfere if it is coherent. The notion of interference (coherence) can also be construed as a measure of the correlation of the amplitudes of the mixed beams, and so amounts to adding two waves to create a wave with greater amplitude than the constituents or subtracting one from the other to create a wave of lesser amplitude than at least either of the two. The nature of the interference depends on the relative phase, that is, two waves exhibit coherence when they have a constant relative phase, and the degree of coherence is quantified by the interference visibility—how much perfectly the waves can cancel each other due to destructive interference.

In case two light waves are brought together and failed to produce interference pattern, they are regarded as incoherent, and as fully coherent if they produce a perfect interference pattern in the sense that regions with a complete destructive interference exist. Since the interference visibility of a given wave lies between these extreme cases, the light generated by an optical source is believed to possess only partial coherence. A radiation field may not interfere in practical setting with the time delayed version of itself in case the time delay is sufficiently large or the distance one of the beams travels is sufficiently long. The delayed distance more than which there is no interference is taken as coherence length, and the corresponding time beyond which the optical coherence disappears as coherence time. This kind of interpretation would remain satisfactory as long as optical experiments are limited to measuring field intensities, that is, quantities quadratic in the electric field. But with the advent of laser, which leads to the feasibility of experiments such as Hanbury, Brown and Twiss, sharpening the classical interpretation of the coherence is found to be necessary. It then turns out to be convenient applying the concept of the normalized correlation function that can be regarded as either classical or nonclassical.[9]

With this in mind, let us now begin with the normalized first-order correlation function

$$g^{(1)}(\mathbf{r}, t; \mathbf{r}, t') = \frac{G^{(1)}(\mathbf{r}, t; \mathbf{r}, t')}{[G^{(1)}(\mathbf{r}, t; \mathbf{r}, t)G^{(1)}(\mathbf{r}, t'; \mathbf{r}, t')]^{\frac{1}{2}}}. \tag{4.79}$$

Based on the condition typified by Eq. (4.78), one may note that $g^{(1)}(\mathbf{r}, t; \mathbf{r}, t') \le 1$. In the same way, the normalized nth order correlation function can be expressed as

$$g^{(n)}(x_1, \ldots, x_{2n}) = \frac{G^{(n)}(x_1, \ldots, x_{2n})}{\left[\prod_{j=1}^{2n} G^{(1)}(x_j, x_j)\right]^{\frac{1}{2}}}, \tag{4.80}$$

where $g^{(n)}(x_1, \ldots, x_n) \le 1$.

It might be therefore imperative stressing that various degrees of coherence can be manifested by different orders of the normalized nth order correlation function with the condition that for the light to be coherent

$$g^{(n)}(x_1, \ldots, x_n) = 1 \tag{4.81}$$

---

[9]Due to the inherent difference between classical and quantum coherence, even with the advent of new techniques, gadgets and optical sources of light such as laser, it is demanding to perform experiments such as interferences, photon counting or coincidences of photons on fields that cannot be classically described [16].

for all orders: the nth order coherence thus heralds that the admissibility of condition (4.81) up to the nth order.

A stable laser is believed to produce a coherent light, which is coherent in all orders (see Sect. 2.2). Since a coherent light possesses both amplitude and phase as a classical signal, it should be enormously suitable for performing certain quantum optical realizations even at the level of small intensity when the particle nature of light is expected to influence the outcome of the observation. It can also be taken as a convenient tool for addressing the classical limit while selecting a given light for application (see Chaps. 11 and 12, and also [17]).

4.2.2 Theoretical Approaches

Even though counting the number of photons generated by a single source with the help of a single detector appears most obvious way of accounting for photon properties, over the years, the photons delayed in time and space are counted when the light generated by a single source is orchestrated to travel over unequal distances aiming at casting the interference pattern on the screen. The understanding that the photon separated in time and space is often counted by more than one detector when the nature of the emerging correlation is required makes such an approach the basis of research in quantum optics [18]. The correlation function in prescribed order that can be obtained once the time evolution of the dynamical operator is known is thus deemed to be important. Obtaining the transition probability distribution in practice might also be required. The mechanism that can be utilized to determine the correlation among the variables of the system separated by space and time is recognized in the nutshell as space-time formulation, which can be evaluated with the help of the explicit form of the pertinent one-time correlation function.

Once the time evolution of the pertinent operator is known, it should be possible to exploit the utility offered by the quantum regression theorem to evaluate the two-time expectation values. One can begin to this effect with a Markovian approximation in which the correlation between the system and reservoir operators at equal time is taken to be irrelevant: $\hat{\rho}_{SR}(t) = \hat{\rho}_S(t) \otimes \hat{\rho}_R(t)$.

The time evolution of a single-time expectation value in this approximation can be expressed as

$$\langle \hat{A}(t + \tau) \rangle = Tr_S[\hat{A}(t) Tr_R(\hat{U}(\tau)\hat{\rho}_S(t) \otimes \hat{\rho}_R(t)\hat{U}^\dagger(\tau))]. \tag{4.82}$$

Since $\hat{\rho}_S(t) \otimes \hat{\rho}_R(t)$ represents the combined system-reservoir system;

$$\langle \hat{A}(t + \tau) \rangle = Tr_S(\hat{A}(t) Tr_R(\hat{\rho}_S(t + \tau) \otimes \hat{\rho}_R(t + \tau))), \tag{4.83}$$

which can also be rewritten as

$$\langle \hat{A}(t+\tau)\rangle = \sum_j G_j(\tau)\langle \hat{A}_j(t)\rangle, \tag{4.84}$$

where $G_j(\tau)$'s are coefficients that depend on τ.

The two-time correlation function can be put in the same way in the form

$$\langle \hat{A}(t+\tau)\hat{B}(t)\rangle = Tr_S\big(\hat{A}(t)\hat{B}(t)Tr_R(\hat{U}(\tau)\hat{\rho}_S(t)\otimes\hat{\rho}_R(t)\hat{U}^\dagger(\tau))\big). \tag{4.85}$$

Comparison of Eqs. (4.84) and (4.85) shows that

$$\langle \hat{A}(t+\tau)\hat{B}(t)\rangle = \sum_j G_j(\tau)\langle \hat{A}_j(t)\hat{B}_j(t)\rangle. \tag{4.86}$$

It is such a subtle way of transferring the time τ from the dynamical operator to a parameter $G(\tau)$ that is commonly perceived as quantum regression or Onsager-Lax theorem.

What remains to determine the two-time correlation is obtaining the parameter $G(\tau)$ from the time evolution of the system. For example, the first-order correlation function can be expressed as

$$g^{(1)}(\tau) = G(\tau)\langle \hat{a}^\dagger(t)\hat{a}(t)\rangle, \tag{4.87}$$

where the normalization coefficient is ignored for brevity. So in the spirit that once the dynamical equation for the operator is known, it is possible to determine $G(\tau)$ in straightforward manner, we turn the attention to uncover the ways of obtaining the two-time correlation function with the aid of the pertinent coherent state propagator defined in Sect. 3.2.3 (see also [19]).

To do so, imagine arbitrary correlation function of the form $g(\tau) = \langle \hat{a}^\dagger(t+\tau)\hat{a}(t)\rangle$ that can be expressed in the Heisenberg picture in terms of the initial density operator as

$$g(\tau) = Tr\big(\hat{\rho}(0)\hat{a}^\dagger(t+\tau)\hat{a}(t)\big). \tag{4.88}$$

So by using $Tr(\hat{\rho}(0)\hat{A}(t)) = Tr\big(\hat{\rho}(t)\hat{A}(0)\big)$, one can write

$$g(\tau) = Tr(\hat{\rho}(t)\hat{a}^\dagger(\tau)\hat{a}), \tag{4.89}$$

where $\hat{\rho}(t)$ can be generated from $\hat{\rho}(0)$ by using $\hat{\rho}(t) = \hat{U}(t)\hat{\rho}(0)\hat{U}^\dagger(t)$. Introduction of the completeness relation for a coherent state in Eq. (4.89) afterward indicates

$$g(\tau) = \int \frac{d^2\alpha}{\pi}\, \alpha\langle\alpha|\hat{U}(t)\hat{\rho}(0)\hat{U}^\dagger(t)\hat{a}^\dagger(\tau)|\alpha\rangle, \tag{4.90}$$

where setting the initial state as $|\alpha_0\rangle$ leads to

$$g(\tau) = \int \frac{d^2\alpha}{\pi}\,\alpha\,\langle\alpha|\hat{U}(t)|\alpha_0\rangle\langle\alpha_0|\hat{U}^\dagger(t)\hat{a}^\dagger(\tau)|\alpha\rangle. \tag{4.91}$$

With the aid of the definition of the coherent state propagator for a single-mode radiation, $K(\alpha, t|\beta, 0) = \langle\alpha|\hat{U}(t)|\beta\rangle$, one can obtain (see also Eq. (3.184))

$$\langle\alpha|\hat{U}(t)|\alpha_0\rangle = \int \frac{d^2\alpha_1}{\pi} K(\alpha, t|\alpha_1, 0)\,\langle\alpha_1|\alpha_0\rangle \tag{4.92}$$

in which by inserting the completeness relation once again, it is possible to verify that

$$\langle\alpha_0|\hat{U}^\dagger(t)\hat{a}^\dagger(\tau)|\alpha\rangle = \int \frac{d^2\alpha_2}{\pi}\frac{d^2\alpha_3}{\pi} K^*(\alpha_2, t|\alpha_3, 0)\,\langle\alpha_0|\alpha_3\rangle\langle\alpha_2|\hat{a}^\dagger(\tau)|\alpha\rangle, \tag{4.93}$$

where $K^*(\alpha_2, t|\alpha_3, 0) = \langle\alpha_3|\hat{U}^\dagger(t)|\alpha_2\rangle$.

Besides, it is not difficult to notice that

$$\langle\alpha_2|\hat{a}^\dagger(\tau)|\alpha\rangle = Tr\big(|\alpha\rangle\langle\alpha_2|\hat{a}^\dagger(\tau)\big).$$

One can also shift the time dependence in the same way;

$$\langle\alpha_2|\hat{a}^\dagger(\tau)|\alpha\rangle = Tr\big(\hat{U}(\tau)|\alpha\rangle\langle\alpha_2|\hat{U}^\dagger(\tau)\hat{a}^\dagger\big). \tag{4.94}$$

Inserting the completeness relation for a coherent state into Eq. (4.94), one then after obtains

$$\langle\alpha_2|\hat{a}^\dagger(\tau)|\alpha\rangle = \int \frac{d^2\alpha_4}{\pi}\alpha_4^* K(\alpha_4, \tau; \alpha, 0) K^*(\alpha_4, \tau; \alpha_2, 0). \tag{4.95}$$

On account of Eqs. (4.91), (4.93) and (4.94), one then gets

$$g(\tau) = \int \frac{d^2\alpha}{\pi}\frac{d^2\alpha_1}{\pi}\frac{d^2\alpha_2}{\pi}\frac{d^2\alpha_3}{\pi}\frac{d^2\alpha_4}{\pi}\alpha\alpha_4^*\langle\alpha_1|\alpha_0\rangle\langle\alpha_0|\alpha_3\rangle$$
$$\times K(\alpha_4, \tau|\alpha, \tau) K^*(\alpha_4, \tau|\alpha_2, 0) K(\alpha, t|\alpha_1, 0) K^*(\alpha_2, t|\alpha_3, 0). \tag{4.96}$$

The two-time second-order correlation function can also be evaluated by using the coherent state propagator in the same manner. What remains to do is to extrapolate this derivation to the case when there are four operators instead of two, adapt the coherent state propagator for different variables, and then carry out the integration. Since the propagator associated with physically realizable optical systems can be expressed in an exponential form, note that solving the involved mathematical task is usually simple despite the number of integrations to be performed.

The Q-function that corresponds to the time-dependent density operator can be defined in the same spirit as

$$Q(\beta, t) = \frac{1}{\pi} \langle \beta | \hat{U}(t) | \beta_0 \rangle \langle \beta_0 | \hat{U}^\dagger(t) | \beta \rangle \qquad (4.97)$$

(see Eq. (2.80)), where introducing the completeness relation for a coherent state in Eq. (4.97) leads to

$$Q(\beta, t) = \frac{1}{\pi} \int \frac{d^2\beta_1}{\pi} \frac{d^2\beta_2}{\pi} \langle \beta | \hat{U}(\tau) | \beta_1 \rangle \langle \beta_1 | \beta_0 \rangle \langle \beta_0 | \beta_2 \rangle \langle \beta_2 | \hat{U}^\dagger(t) | \beta \rangle, \qquad (4.98)$$

from which follows

$$Q(\beta, t) = \frac{1}{\pi} \int \frac{d^2\beta_1}{\pi} \frac{d^2\beta_2}{\pi} K(\beta, \tau | \beta_1, 0) K^*(\beta, t | \beta_2, 0) \langle \beta_1 | \beta_0 \rangle \langle \beta_0 | \beta_2 \rangle. \qquad (4.99)$$

Taking Eq. (4.99) as a definition of the Q-function in terms of the coherent state propagator, one can see that

$$g(\tau) = \int \frac{d^2\beta d^2\eta d^2\gamma}{\pi} Q(\gamma, \gamma^*, \tau) Q(\beta, \eta^*, t) \beta \gamma^*. \qquad (4.100)$$

Since the time evolution can be directly obtained from the density operator, it is possible to assume that $g(\tau) = Tr(\hat{a}^\dagger \hat{a}(\tau) \hat{\rho}(t))$. So upon expanding the density operator in the normal order, and introducing the coherent state completeness relation, one can see that

$$g(\tau) = \int \frac{d^2\alpha}{\pi} \sum_{l,m} C_{lm}(t) Tr(\hat{a}^\dagger \hat{a}(\tau) | \alpha \rangle \langle \alpha | \hat{a}^{\dagger l} \hat{a}^m). \qquad (4.101)$$

In view of the fact that $\langle \alpha | \hat{a}^{\dagger l} = \alpha^{*l} \langle \alpha |$ and $\langle \alpha | \hat{a}^m = \left(\alpha + \frac{\partial}{\partial \alpha^*} \right)^m \langle \alpha |$, one may see that

$$g(\tau) = \int \frac{d^2\alpha}{\pi} \sum_{l,m} C_{lm}(t) \alpha^{*l} \left(\alpha + \frac{\partial}{\partial \alpha^*} \right)^m Tr(\hat{a}^\dagger \hat{a}(\tau) | \alpha \rangle \langle \alpha |). \qquad (4.102)$$

On the basis of the fact that $Q(\alpha, \alpha^*, t) = \frac{1}{\pi} \sum_{l,m} C_{lm}(t) \alpha^{*l} \alpha^m$, when the operators are initially put in the normal order, one finds

$$g(\tau) = \int d^2\alpha \, Q\left(\alpha^*, \alpha + \frac{\partial}{\partial \alpha^*}, t \right) Tr(\hat{a}^\dagger \hat{a}(\tau) | \alpha \rangle \langle \alpha |). \qquad (4.103)$$

In light of a cyclic property of the trace operation, note that it is not hard to see

$$Tr(\hat{a}^\dagger \hat{a}(\tau)|\alpha\rangle\langle\alpha|) = \alpha^* \langle\alpha|\hat{a}(\tau)|\alpha\rangle. \tag{4.104}$$

It is also possible to write for $\hat{\rho} = |\alpha\rangle\langle\alpha|$ that $\langle\alpha|\hat{a}(\tau)|\alpha\rangle = Tr(\hat{a}(\tau)\hat{\rho})$, where the time factor can be transferred to the density operator: $\langle\alpha|\hat{a}(\tau)|\alpha\rangle = Tr(\hat{a}\hat{\rho}(\tau))$. After that, upon introducing the coherent state completeness relation, one can write

$$\langle\alpha|\hat{a}(\tau)|\alpha\rangle = \int d^2\beta \ Q(\beta, \beta^*, \tau)\beta. \tag{4.105}$$

Therefore, on account of earlier discussion, one can see that

$$g(\tau) = \int d^2\alpha d^2\beta \ Q\left(\alpha^*, \alpha + \frac{\partial}{\partial\alpha^*}, t\right) Q(\beta^*, \beta, \tau)\alpha^*\beta. \tag{4.106}$$

Even though the stationarity condition prevails in a wide variety of optical experiments, the focus in earlier discussion is on circumstances relevant to the space-time domain. The space-time formulation however may not be suitable for analyzing some problems involving propagation of light in a material media or scattering. The space-frequency formulation in this case is found to lead to discoveries and understanding of some new physical aspects of the radiation such as correlation induced by spectral modifications and changes in the polarization properties of light while propagating. It is also expected that this consideration can be of a great help in studying the structure of objects by applying the inverse scattering approaches [20].

On account of the assumption that statistical stationarity cannot be always guaranteed specially when emission of the light takes place over a finite time interval, one may be compelled to consider the case of non-stationary light. To explore a quantum theoretical approach that can be applied in obtaining the first-order optical coherence, it hence appears appealing starting with the assertion that the quantized electromagnetic fields can be expressed as

$$\hat{\mathbf{E}}(\mathbf{r}, t) = i\sqrt{\frac{\omega}{2\varepsilon_0 V}} \sum_{k,s} \hat{a}_{k,s}\hat{e}_{k,s}e^{i(\mathbf{k}.\mathbf{r}-\omega t)} - i\sqrt{\frac{\omega}{2\varepsilon_0 V}} \sum_{k,s} \hat{a}_{k,s}^\dagger \hat{e}_{k,s}^* e^{-i(\mathbf{k}.\mathbf{r}-\omega t)},$$

$$\tag{4.107}$$

where the wave vector $\mathbf{k}$ represents the plane wave modes and $\hat{e}_{k,s}$ ($s = 1, 2$) are basis vectors that satisfy (see Sect. 2.1):

$$\mathbf{k}.\hat{e}_{k,s} = 0, \qquad \hat{e}_{k,s}^*.\hat{e}_{k,s'} = \delta_{ss'}, \qquad \hat{e}_{k,1} X \hat{e}_{k,2} = \mathbf{k}/k. \tag{4.108}$$

In view of the fact that a discrete mode representation is appropriate when the electric field is assumed to be confined in the cavity, the field operator can be designated by using a Fourier integral transform as

$$\hat{\mathbf{E}}(\mathbf{r}, t) = \int_{-\infty}^{\infty} \hat{\mathbf{e}}(\mathbf{r}, \omega) e^{-i\omega t} d\omega, \tag{4.109}$$

where the positive part of the electric field can be expressed as

$$\hat{\mathbf{e}}^{(+)}(\mathbf{r}, \omega) = \frac{1}{2\pi} \int_{-\infty}^{\infty} \hat{\mathbf{E}}^{(+)}(\mathbf{r}, t) e^{i\omega t} dt. \tag{4.110}$$

With this characterization, upon neglecting the pertinent polarization, the first-order correlation function can be defined in space-frequency domain as

$$\Gamma^{(1)}(\mathbf{r}, \omega; \mathbf{r}', \omega') = Tr\left(\hat{\rho}\hat{\mathbf{e}}^{(-)}(\mathbf{r}, -\omega)\hat{\mathbf{e}}^{(+)}(\mathbf{r}', \omega')\right). \tag{4.111}$$

Then upon invoking a Fourier transformation, Eq. (4.111) can be put in terms of the first-order correlation function in space-time domain in the form

$$\Gamma^{(1)}(\mathbf{r}, \omega; \mathbf{r}', \omega') = \frac{1}{4\pi^2} \int \int_{-\infty}^{\infty} G^{(1)}(\mathbf{r}, t; \mathbf{r}', t') e^{i(\omega' t' - \omega t)} dt dt', \tag{4.112}$$

where $\Gamma^{(1)}(\mathbf{r}, \omega; \mathbf{r}', \omega')$ can be referred to as two-frequency cross-spectral density function.

4.2.3 Practical Applications

Various ways of counting the number of photons having specific frequency by using ideal detector are already available despite the fact that the distribution of the number of photon spans over a range of frequency or energy. If the measurement can be done overall possible frequencies, even in this case, one may come up with unique and reliable distribution function for the photon number. In light of this, the distribution of the photons on the surface of detector and evaluated against the frequency is referred to as spectrum. So hoping to avoid misunderstanding in interpreting the data that comes out of a measuring device, we seek to present different spectra.

4.2.3.1 Power Spectrum
To begin with, the power spectrum can be defined as

$$S(\omega) = \int_{-\infty}^{\infty} d\tau e^{i\omega\tau} \langle \hat{\mathbf{E}}^{(-)}(t)\hat{\mathbf{E}}^{+}(t+\tau)\rangle_{ss}, \tag{4.113}$$

where ss stands for steady or stationary state. In this situation, the photons are required to be collected over a long period of time to establish the intended description. The power spectrum can hence be perceived as the frequency domain interpretation of the first-order two-time correlation function, that is, as the number of photons falling on the ideal detector over a long period of time and weighted against the frequency parameter

The power spectrum can also be expressed as

$$S(\omega) = \int_{-\infty}^{0} d\tau e^{i\omega\tau} \langle \hat{a}^{\dagger}(t)\hat{a}(t+\tau)\rangle_{ss} + \int_{0}^{\infty} d\tau e^{i\omega\tau} \langle \hat{a}^{\dagger}(t)\hat{a}(t+\tau)\rangle_{ss}. \tag{4.114}$$

In line with the presumption that the correlation function at steady state depends only on the time difference τ, one can replace t by $t - \tau$ in the first integral. Besides, upon changing the variable of integration from τ to $-\tau$, it is possible to see that

$$S(\omega) = \int_{0}^{\infty} d\tau e^{-i\omega\tau} \langle \hat{a}^{\dagger}(t+\tau)\hat{a}(t)\rangle_{ss} + \int_{0}^{\infty} d\tau e^{i\omega\tau} \langle \hat{a}^{\dagger}(t)\hat{a}(t+\tau)\rangle_{ss}. \tag{4.115}$$

One may note that one of the right side terms is the complex conjugate of the other;[10]

$$S(\omega) = 2\mathrm{Re} \int_{0}^{\infty} d\tau e^{i\omega\tau} \langle \hat{a}^{\dagger}(t)\hat{a}(t+\tau)\rangle_{ss} = 2\mathrm{Re} \int_{0}^{\infty} d\tau e^{-i\omega\tau} \langle \hat{a}^{\dagger}(t+\tau)\hat{a}(t)\rangle_{ss}. \tag{4.116}$$

The spectrum that corresponds to the normalized two-time first-order correlation function or the normalized power spectrum is more often required. The power spectrum is so integrated overall possible frequencies, where the final result is set to unity to ensure normalization: $\int_{-\infty}^{\infty} S(\omega)d\omega = 1$.

As illustration, imagine a system whose cavity mode can be expressed as

$$\hat{a}(t) = \hat{a}(t_0)e^{-\frac{\kappa t}{2}} + \int_{t_0}^{t} e^{-\frac{\kappa}{2}(t-t')} \hat{F}(t')dt', \tag{4.117}$$

where κ is the usual cavity damping constant and t_0 is the time at which the interaction is assumed to be commenced. Upon replacing t by $t + \tau$, and letting t_0 to be t in Eq. (4.117), one can see that

$$\hat{a}(t+\tau) = \hat{a}(t)e^{-\frac{\kappa(t+\tau)}{2}} + \int_{t}^{t+\tau} e^{-\frac{\kappa}{2}(\tau+t-t')} \hat{F}(t')dt', \tag{4.118}$$

---

[10]It should be clear by now that $\langle \hat{a}^{\dagger}(t)\hat{a}(t+\tau)\rangle_{ss}$ is the first-order two-time correlation function, which can be obtained by applying quantum regression theorem or the pertinent Q-function or the coherent state propagator. Nonetheless, it appears legitimate arguing that in order to obtain the power spectrum, it might be more suitable to solve the Heisenberg equation or quantum Langevin equation than the master equation.

where setting $t' = t + \tau'$ leads to

$$\hat{a}(t + \tau) = \hat{a}(t)e^{-\frac{\kappa(t+\tau)}{2}} + \int_t^{t+\tau} e^{-\frac{\kappa}{2}(\tau-\tau')} \hat{F}(t + \tau')d\tau'. \tag{4.119}$$

Markedly, with the help of the proposal that the noise force at time t does not correlate with the system variable at earlier time along with the assumption that initially the cavity is in a chaotic state, one can verify at steady state that

$$\langle \hat{a}^\dagger(t)\hat{a}(t + \tau) \rangle = Ne^{-\frac{\kappa}{2}\tau}, \tag{4.120}$$

where N is the mean photon number of the reservoir modes. The power spectrum can thus be expressed as

$$S(\omega) = 4N\mathrm{Re}\left(\frac{1}{\kappa + i2\omega}\right), \tag{4.121}$$

where taking the real value of Eq. (4.121) indicates

$$S(\omega) = \frac{4N\kappa}{\kappa^2 + 4\omega^2}, \tag{4.122}$$

which turns out to be a Lorenzian of width κ (note that quantum regression theorem should also lead to the same result).

In addition, since it is the output radiation that is accessible to measurement, it might be imperative looking into the output quantities such as the output power spectrum that stands for the spectrum of the radiation outside the cavity:

$$S^{out}(\omega) = 2\mathrm{Re}\int_0^\infty d\tau e^{i\omega\tau} \langle \hat{a}_{out}^\dagger(t)\hat{a}_{out}(t + \tau)\rangle_{ss}. \tag{4.123}$$

So in view of Eq. (3.74), it is straightforward to see that

$$\langle \hat{a}_{out}^\dagger(t)\hat{a}_{out}(t + \tau) \rangle = \kappa \langle \hat{a}^\dagger(t)\hat{a}(t + \tau)\rangle - \sqrt{\kappa}\langle \hat{a}_{in}^\dagger(t)\hat{a}(t + \tau)\rangle$$
$$- \sqrt{\kappa}\langle \hat{a}^\dagger(t)\hat{a}_{in}(t + \tau)\rangle + \langle \hat{a}_{in}^\dagger(t)\hat{a}_{in}(t + \tau)\rangle. \tag{4.124}$$

Then on account of the fact that the input variables can be expressed for a normally ordered operators in terms of the associated noise force as $\hat{a}_{in}(t) = \hat{F}(t)/\sqrt{\kappa}$, one can write

$$\langle \hat{a}_{out}^\dagger(t)\hat{a}_{out}(t + \tau) \rangle = \kappa \langle \hat{a}^\dagger(t)\hat{a}(t + \tau)\rangle - \langle \hat{a}^\dagger(t)\hat{F}(t + \tau)\rangle + \frac{1}{\kappa}\langle \hat{F}^\dagger(t)\hat{F}(t + \tau)\rangle. \tag{4.125}$$

Since the reservoir noise force operates outside the cavity, the noise force at time t can correlate with the system variables only after it enters the cavity.

Owing to the fact that such a correlation depends on the time evolution of the dynamical variables, it is possible to show that (see the appendix)

$$\langle \hat{a}^{\dagger}(t)\hat{a}(t+\tau)\rangle \quad \frac{\kappa N}{2} \quad \tau,$$

(4.126)

After that, upon making use of the property of the noise operator for the squeezed vacuum reservoir, one can see that

$$\langle \hat{a}_{out}^{\dagger}(t)\hat{a}_{out}(t+\tau)\rangle = \kappa \langle \hat{a}^{\dagger}(t)\hat{a}(t+\tau)\rangle - \frac{\kappa N}{2}e^{\frac{-\kappa\tau}{2}} + N\delta(\tau).$$

(4.127)

With such information, it would not be hard to ascertain that

$$2\mathrm{Re}\int_0^{\infty} d\tau\, e^{(i\omega - \frac{\kappa}{2})\tau}(-\frac{\kappa N}{2}) = -\frac{N\kappa^2}{\kappa^2 + 4\omega^2},$$

(4.128)

$$2\mathrm{Re}\int_0^{\infty} d\tau\, e^{i\omega\tau} N\delta(\tau) = N,$$

(4.129)

and then

$$S^{out}(\omega) = \frac{4N\omega^2}{\kappa^2 + 4\omega^2} + \kappa S(\omega) = \frac{4N(\omega^2 + \kappa^2)}{\kappa^2 + 4\omega^2},$$

(4.130)

where $S(\omega) = 2\mathrm{Re}\int_0^{\infty} d\tau\, e^{i\omega\tau}\langle\hat{a}^{\dagger}(t)\hat{a}(t+\tau)\rangle_{ss}$ is the power spectrum of the radiation inside the cavity.

It might not be difficult to infer from Fig. 4.2 that the output power spectrum is almost independent of the initial cavity radiation since the effect of the reservoir

Fig. 4.2 Plots of the cavity and output power spectra for $N = 0.2$ and $\kappa = 0.8$

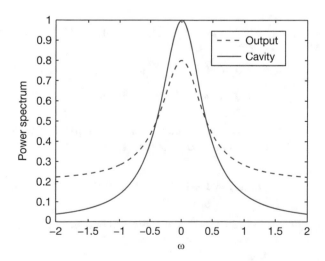

would be quite significant overtime when compared to the initial condition that assumed to remain constant throughout the observation time. One can also see for both cavity and output cases that the height increases with the intensity of the reservoir and the width broaden with the cavity damping constant.

4.2.3.2 Squeezing Spectrum

It should be clear by now that to describe the phenomenon of squeezing, two quadrature components are required; and hence, the squeezing spectrum needs to be calculated for each quadrature component (see also Sect. 2.2). So the degree of squeezing that can be inferred from the calculation of the quadrature variance and squeezing spectrum should not be directly compared since the squeezing spectrum corresponds to the correlation of the quadrature operators evaluated at different times while the quadrature variance is at equal time [21].

Taking this into consideration, the squeezing spectrum of the cavity radiation for the quadrature components

$$\hat{\chi}_+ = \hat{a}^\dagger + \hat{a}, \qquad \hat{\chi}_- = i(\hat{a}^\dagger - \hat{a}) \tag{4.131}$$

can be defined as

$$V_j(\omega) = \int_{-\infty}^{\infty} d\tau e^{i\omega\tau} \langle \hat{\chi}_j(t), \ \hat{\chi}_j(t+\tau) \rangle_{ss}, \tag{4.132}$$

where j stands for the index of the quadrature operators and

$$\langle \hat{\chi}_j(t), \ \hat{\chi}_j(t+\tau) \rangle = \langle \hat{\chi}_j(t)\hat{\chi}_j(t+\tau) \rangle - \langle \hat{\chi}_j(t) \rangle \langle \hat{\chi}_j(t+\tau) \rangle. \tag{4.133}$$

As demonstration, the squeezing spectrum of the plus quadrature component can be expressed as

$$V_+(\omega) = \int_{-\infty}^{\infty} d\tau e^{i\omega\tau} [\langle \hat{a}^\dagger(t)\hat{a}(t+\tau) \rangle_{ss} + \langle \hat{a}^\dagger(t)\hat{a}^\dagger(t+\tau) \rangle_{ss} + \langle \hat{a}(t)\hat{a}(t+\tau) \rangle_{ss}$$
$$+ \langle \hat{a}(t)\hat{a}^\dagger(t+\tau) \rangle_{ss} - \langle \hat{a}^\dagger(t) \rangle_{ss} \langle \hat{a}(t+\tau) \rangle_{ss} - \langle \hat{a}^\dagger(t) \rangle_{ss} \langle \hat{a}^\dagger(t+\tau) \rangle_{ss}$$
$$- \langle \hat{a}(t) \rangle_{ss} \langle \hat{a}(t+\tau) \rangle_{ss} - \langle \hat{a}(t) \rangle_{ss} \langle \hat{a}^\dagger(t+\tau) \rangle_{ss}]. \tag{4.134}$$

With the help of the commutation relation, $[\hat{a}(t), \hat{a}^\dagger(t+\tau)] = \hat{I}\delta(\tau)$, one can thus put Eq. (4.134) in a more compact form as

$$V_+(\omega) = 1 + 2\text{Re} \int_0^{\infty} d\tau e^{i\omega\tau} \langle: \hat{\chi}_+(t), \ \hat{\chi}_+(t+\tau) :\rangle_{ss}, \tag{4.135}$$

where :: indicates that the operators in between are already put in the normal order.

In the same way, the squeezing spectrum for the minus quadrature would be

$$V_-(\omega) - 1 = 2\text{Re} \int_0^\infty d\tau e^{i\omega\tau} \langle: \hat{\chi}_-(t), \hat{\chi}_-(t + \tau) :\rangle_{ss}, \tag{4.136}$$

where combination of these results reveal

$$V_\pm(\omega) = 1 \pm 2\text{Re} \int_0^\infty d\tau e^{i\omega\tau} \langle: \hat{\chi}_\pm(t), \hat{\chi}_\pm(t + \tau) :\rangle_{ss}. \tag{4.137}$$

Equation (4.137) denotes the squeezing spectrum in which the strength of the squeezing can be attributed to the degree by which one of $V_\pm(\omega)$ is close to zero.

When the squeezing spectrum that involves a two-time correlation function is found to be difficult to directly obtain, the squeezing spectrum can be calculated in the frequency space. It is possible to write Eq. (4.137) in this case in terms of c-number variables associated with the normal ordering as

$$V_\pm^{fd}(\omega) = 1 \pm \kappa \int_{-\infty}^\infty \langle \alpha_\pm(t)\alpha_\pm(t + \tau) \rangle_{ss} e^{i\omega\tau} d\tau, \tag{4.138}$$

where $\alpha_\pm(t) = \alpha^*(t) \pm \alpha(t)$ and when $\alpha(t)$ is taken to be Gaussian variable of zero mean.

To obtain a suitable expression for Eq. (4.138) in the frequency space, one can begin with Fourier transforms of the form

$$\alpha_\pm(\omega) = \frac{1}{\sqrt{2\pi}} \int_{-\infty}^\infty \alpha_\pm(t) e^{-i\omega t} dt, \qquad \alpha_\pm(\omega') = \frac{1}{\sqrt{2\pi}} \int_{-\infty}^\infty \alpha_\pm(t') e^{-i\omega' t'} dt', \tag{4.139}$$

with the aid of which one can write

$$\langle \alpha_\pm(\omega)\alpha_\pm(\omega') \rangle = \frac{1}{2\pi} \int_\infty^\infty \int_{-\infty}^\infty \langle \alpha_\pm(t)\alpha_\pm(t') \rangle e^{-i(\omega+\omega')t - i\omega'(t'-t)} dt' dt. \tag{4.140}$$

Then setting $\tau = t' - t$ in Eq. (4.140) leads to

$$\langle \alpha_\pm(\omega)\alpha_\pm(\omega') \rangle = \frac{1}{2\pi} \int_{-\infty}^\infty \int_{-\infty}^\infty \langle \alpha_\pm(t)\alpha_\pm(t + \tau) \rangle e^{-i(\omega+\omega')t - i\omega'\tau} d\tau dt, \tag{4.141}$$

where assuming that the correlation function at steady state depends on the time difference τ, and carrying out the integration over t yield

$$\langle \alpha_\pm(\omega)\alpha_\pm(\omega') \rangle = \int_{-\infty}^\infty \langle \alpha_\pm(t)\alpha_\pm(t + \tau) \rangle_{ss} e^{-i\omega'\tau} \delta(\omega + \omega') d\tau. \tag{4.142}$$

Now integrating both sides of Eq. (4.142) over ω' indicates

$$\int_{-\infty}^{\infty} d\omega' \langle \alpha_{\pm}(\omega)\alpha_{\pm}(\omega') \rangle = \int_{-\infty}^{\infty} \langle \alpha_{\pm}(t)\alpha_{\pm}(t+\tau) \rangle_{ss} e^{i\omega\tau} d\tau. \tag{4.143}$$

As a result, in frequency domain, the squeezing spectrum turns out to be

$$V_{\pm}^{fd}(\omega) = 1 \pm \kappa \int_{-\infty}^{\infty} \langle \alpha_{\pm}(\omega)\alpha_{\pm}(\omega') \rangle d\omega'. \tag{4.144}$$

4.2.3.3 Output Squeezing Spectrum

One may also seek to proceed with obtaining the spectrum that describes the squeezing properties of the output radiation, which is often designated simply as squeezing spectrum. Recall that the input and output quadrature operators can be expressed in interaction picture as

$$\chi_{\theta}^{in}(t) = \hat{a}_{in}(t)e^{-i(\phi_{in}+\theta)t} + \hat{a}_{in}^{\dagger}(t)e^{i(\phi_{in}+\theta)t}, \tag{4.145}$$

$$\chi_{\theta}^{out}(t) = \hat{a}_{out}(t)e^{-i(\phi_{out}+\theta)t} + \hat{a}_{out}^{\dagger}(t)e^{i(\phi_{out}+\theta)t}, \tag{4.146}$$

where ϕ_{in} and ϕ_{out} are the input and output phases, and θ is the phase between the quadrature components (see also Sect. 3.1.3). It is possible to set $\theta = 0$ or $\theta = \pi/2$ for the amplitude and phase quadratures of the input and output fields while the phase difference between the two quadrature components is maintained at $\pi/2$.

With this description, the squeezing spectrum for the quadrature fluctuations can be defined in a more appealing form as

$$V(\theta, \omega) = \int_{-\infty}^{\infty} d\tau e^{i\omega\tau} \langle \hat{\chi}_{\theta}(t+\tau), \hat{\chi}_{\theta}(t) \rangle_{ss}, \tag{4.147}$$

where θ distinguishes the quadrature of interest. Upon using the usual boson commutation relation, it is possible to put Eq. (4.147) phenomenologically as

$$V(\theta, \omega) = 1 + \xi(+\omega) + \xi(-\omega) + 2\chi(\omega; \theta), \tag{4.148}$$

where the constant 1 is the vacuum or the shot noise level, $\xi(\pm\omega)$ is the optical spectrum independent of the quadrature phase angle θ and $\chi(\omega; \theta)$ is the phase sensitive contribution responsible for squeezing.

It is not as such difficult to characterize:

$$\xi(\pm\omega) = \int_{-\infty}^{\infty} d\tau e^{\pm i\omega\tau} \langle \hat{a}_{out}^{\dagger}(t), \hat{a}_{out}(t+\tau) \rangle, \tag{4.149}$$

$$\chi(\omega; \theta) = 2\text{Re}\left[e^{-i2(\theta+\phi_{out})} \int_{0}^{\infty} d\tau e^{i\omega\tau} \langle \hat{a}_{out}(t+\tau), \hat{a}_{out}(t) \rangle\right], \tag{4.150}$$

where

$$\langle \hat{a}^\dagger_{out}(t+\tau), \hat{a}_{out}(t)\rangle = \langle \hat{a}^\dagger_{in}(t+\tau), \hat{a}_{in}(t)\rangle + \kappa\big\{(N+1)\langle \hat{a}^\dagger(t+\tau),\, \hat{a}(t)\rangle$$
$$+ N\langle[\hat{a}^\dagger(t+\tau),\, \hat{a}(t)]\rangle - Mu(\tau)\langle[\hat{a}^\dagger(t+\tau),\, \hat{a}^\dagger(t)]\rangle$$
$$- M^* u(-\tau)\langle[\hat{a}(t+\tau),\, \hat{a}(t)]\rangle\big\}, \tag{4.151}$$

$$\langle \hat{a}_{out}(t+\tau), \hat{a}_{out}(t)\rangle = \langle \hat{a}_{in}(t+\tau)\hat{a}_{in}(t)\rangle + \kappa\big\{(N+1)\langle \hat{T}(\hat{a}^\dagger(t+\tau)\hat{a}(t))\rangle$$
$$- N\langle\tilde{\hat{T}}\hat{a}(t+\tau)\hat{a}(t)\rangle + Mu(-\tau)\langle[\hat{a}^\dagger(t+\tau),\, \hat{a}^\dagger(t)]\rangle$$
$$- M^* u(\tau)\langle[\hat{a}^\dagger(t),\, \hat{a}(t+\tau)]\rangle\big\} \tag{4.152}$$

in which $u(\tau)$ is the Heaviside step function[11] and $\hat{T}$ $(\tilde{\hat{T}})$ is time (anti-time) ordering that amounts to putting earlier times to the right (left) (compare with causality effect provided in Sect. 3.1.4).

The squeezing spectrum that represents the output signal can thus be expressed as

$$V_j^{out}(\omega) = 1 + \kappa \int_{-\infty}^{\infty} d\tau e^{-i\omega\tau}\, \hat{T}\langle : \hat{\chi}_j(t+\tau),\, \hat{\chi}_j(t) :\rangle_{ss}, \tag{4.153}$$

where $\hat{T}$ stands for time ordering and j for respective quadrature component. As already observed, the output field accounts for the superposition of the reservoir modes reflected from the mirror and the light that passes through the cavity via the same mirror. Since both contributions are significantly influenced by the squeezed input overtime, the output radiation is expected to show a higher degree of squeezing as captured by the pertinent spectrum.

4.3 Techniques of Wave Mixing

During the interaction of light with a material medium, the external light impinged onto the medium or sample scatters where the result of the measurement can be inferred from the properties of the scattered light that can be captured via collecting, collimating, and after that guiding the emerging beam to the detecting device. The

---

[11]The Heaviside step function for continuous variable x can be defined as

$$H(x) = \begin{cases} 0, & x < 0; \\ \frac{1}{2}, & x = 0; \\ 1, & x > 1. \end{cases}$$

beam of light in the process is made to split, travel via different paths, and then combined at the mixing device. When a light of known frequency is impinged onto an active medium such as optical Kerr medium, the light with different frequency from the incident can emerge from the interaction, that is, a superposition between the indistinguishable amplitudes of the engaged waves can occur. For example, in a double slit experiment, a light wave is presumed to have two indistinguishable alternative amplitudes that instigate the photon events at space-time point.

The quantum interpretation of superposition on the other hand can be linked to the inability to know for sure which path the light followed. Even though quantum theory may never able to resolve through which slit the light passes, it has the capacity to accurately predict the counting rate as a function of the relative delay between the two amplitudes. It is with such a conviction that we seek to focus on the description of the statistical properties of the light ensuing from various wave mixing processes rather than contemplating on the which-path conundrum. The experimental observation of Hanbury, Brown and Twiss type may also initiate the need for looking into the second-order interferometry that epitomizes an alternative physical description of the joint photon events at different space-time points created by two light quanta with or without distinguishable amplitudes.[12]

One may thus need to explore some possible scenarios that able to address the physical processes involved in wave mixing. To this effect, the mechanism of direct wave mixing where two light waves with the same frequency but different properties are injected to an empty cavity is considered. The main task in this mixing scheme encompasses the process of pumping the light to the cavity that contains another light, and then study the emerging properties of the mixture. The emphasis is geared towards expressing the quasi-statistical distribution of the mixture in terms of the individual distribution functions. The idea of mixing the waves with the aid of the beam splitter that can be construed as impinging two lights onto the beam splitter can sanction the propagation of the emerging wave as a mixture would also be discussed. In both schemes, the involved process is passive in a sense that the frequency or energy of the wave is unchanged. But later, active wave mixing process in which the frequency of the original light changes due to the involved exchange of excitation in a nonlinear medium would be addressed (for nonlinear multi-wave mixing; see Chap. 8).

4.3.1 Direct Wave Mixing

Suppose a light whose property is well known is pumped into an empty lossless cavity. In this case, the light in the cavity after the commencement of the intended

---

[12] Two-photon interference may not be the interference between two photons but the result of the superposition among indistinguishable amplitudes of the two photons. Contrary to the appealing nature of such a prospect, the concept of a two-photon amplitude has somehow been confined to the study of entangled photons.

interaction would be in the superposition of the vacuum mode and the light pumped into the cavity. Once the the mixing process is established, it may appear appropriate and appealing to look for the density operator that describes the mixture. It should also be admissible to utilize the quasi statistical distribution functions such as P function, Q-function and Wigner function to explore properties of the mixture. We hence seek to outline the mathematical approaches that can ease the required detail for understanding the properties of the radiation generated by different sources placed in the cavity. To entertain such an idea, one can employ the definition of the density operator in terms of one of the quasi-statistical distribution functions.[13]

In view of Eq. (2.70), the density operator after the radiation is pumped into the cavity can be expressed as

$$\hat{\rho}' = \int d^2\alpha' \, P(\alpha')|\alpha'\rangle\langle\alpha'|, \tag{4.154}$$

where α' signifies the coherent state representation of the radiation in the cavity. Note that this kind of description should be possible for all types of P-functions even when they are not expected to be a well behaved distribution functions. With the aid of Eqs. (2.24) and (4.154), one can then write

$$\hat{\rho}' = \int d^2\alpha' \, P(\alpha')D(\alpha')|0\rangle\langle0|\hat{D}^\dagger(\alpha'). \tag{4.155}$$

Since the cavity radiation is initially in ordinary vacuum state, denoted by density operator $\hat{\rho}_0$ one can see that

$$\hat{\rho}' = \int d^2\alpha' \, P(\alpha')\hat{D}(\alpha')\hat{\rho}_0\hat{D}^\dagger(\alpha'). \tag{4.156}$$

One may note that the density operator $\hat{\rho}'$ accounts for the light in the cavity after the known light is pumped to the cavity containing the vacuum mode.

If another light that can be described by a coherent state $|\alpha''\rangle$ is assumed to be injected into the cavity containing the radiation designated by the new density operator (4.156), it is possible to intuitively propose that the density operator for the mixture of the two lights can be expressed as

$$\hat{\rho}'' = \int d^2\alpha'' \, P(\alpha'')\hat{D}(\alpha'')\hat{\rho}'\hat{D}^\dagger(\alpha''). \tag{4.157}$$

One can hence rewrite Eq. (4.157) by making use of Eqs. (4.155) and (4.156) as

$$\hat{\rho}'' = \int d^2\alpha' d^2\alpha'' \, P(\alpha')P(\alpha'')\hat{D}(\alpha'')|\alpha'\rangle\langle\alpha'|\hat{D}^\dagger(\alpha''). \tag{4.158}$$

---

[13]Even though the procedure of coupled system approach and interaction theory can also be employed to achieve similar goal, the rigor involved in incorporating the contribution of each constituting system may not be as handy as one wishes it to be.

It may also be required to obtain the explicit form of the action of the displacement operator on the coherent state:

$$\hat{D}(\alpha)|\beta\rangle = \exp[\alpha \hat{a}^\dagger - \alpha^* \hat{a}] \exp[\beta \hat{a}^\dagger - \beta^* \hat{a}]|0\rangle, \tag{4.159}$$

which can be rewritten upon applying the operator identity, $e^{\hat{A}} e^{\hat{B}} = e^{\hat{A}+\hat{B}+\frac{1}{2}[\hat{A},\,\hat{B}]}$, as

$$\hat{D}(\alpha)|\beta\rangle = \exp\left[\frac{1}{2}(\alpha \beta^* - \alpha^* \beta)\right]|\alpha + \beta\rangle. \tag{4.160}$$

It is therefore found advantageous expressing equation (4.158) as

$$\hat{\rho}'' = \int d^2\alpha' d^2\alpha'' P(\alpha') P(\alpha'')|\alpha'' + \alpha'\rangle\langle\alpha' + \alpha''|. \tag{4.161}$$

With the notion that the superposed light can undergo interference due to the involved indistinguishability of the light quanta, the density operator corresponding to the superposition of light beams as in the wave-mixing can be exemplified by such an expression. One can also see from Eq. (4.161) that the order of injection of the light beams into the cavity is immaterial, which may endorse the intuitive understanding that the phenomenon of superposition, if there is any, should be independent of the order one chooses to follow.

With the help of Eqs. (2.67), (2.73), (2.74) and (4.161), the P-function that represents the mixed light can be put in the form

$$P(\alpha) = \frac{1}{\pi^2} \int d^2z \int d^2\alpha' d^2\alpha'' P(\alpha') P(\alpha'') Tr(|\alpha'' + \alpha'\rangle\langle\alpha' + \alpha''|e^{z\hat{a}^\dagger} e^{-z^*\hat{a}})$$

$$\times \exp[z^*\alpha - z\alpha^*], \tag{4.162}$$

from which follows

$$P(\alpha) = \int d^2\alpha' P(\alpha') P(\alpha - \alpha'). \tag{4.163}$$

Equation (4.163) can be utilized in exploring the properties of light resulting from the mixture of the two beams for which their P-functions are well behaved. Due to the nature of the P-function, the application of Eq. (4.163) is unfortunately limited to the mixture of the coherent and chaotic light beams.

To obtain the Q-function in the same way, it turns out to be advisable making use of Eq. (4.161);

$$Q(\alpha) = \frac{1}{\pi^2} \int d^2z \, d^2\alpha' d^2\alpha'' P(\alpha') P(\alpha'')$$

$$\times \exp[-z^*z + z(\alpha'^* + \alpha''^* - \alpha^*) - z^*(\alpha' + \alpha'' - \alpha)], \tag{4.164}$$

from which follows

$$Q(\alpha) = \frac{1}{\pi} \int d^2\alpha' d^2\alpha'' P(\alpha')P(\alpha'') \exp[-\alpha'^*\alpha' + \alpha'^*\alpha - \alpha''^*\alpha' - \alpha'_{/*}\alpha''$$
$$+ \alpha\alpha''^* + \alpha^*\alpha' + \alpha''\alpha^* - \alpha^*\alpha - \alpha'^*\alpha'']. \tag{4.165}$$

Equation (4.165) clearly indicates that once the P-function for each light is known, and verified to be well behaved, the Q-function that describes the mixture of the two light beams can be obtained. Nonetheless, this approach may not resolve the problem associated with the P-function of light with nonclassical features as anticipated.

To overcome such impediment, it appears appealing expressing the Q-function of the mixture in terms of the Q-function of the individual lights by using Eqs. (2.80), (2.106) and (4.154) as

$$Q(\alpha) = \frac{e^{\alpha^*\alpha}}{\pi} \int d^2\beta d^2\gamma \, Q(-\beta^*, \beta) Q(-\gamma^*, \gamma) \exp[-\beta^*\beta - \gamma^*\gamma$$
$$+ \alpha\gamma^* - \beta\gamma^* - \beta\alpha^* - \beta^*\gamma + \beta^*\alpha - \alpha^*\gamma]. \tag{4.166}$$

It should be possible to obtain the Q-function for the mixture of light beams described by the coherent state variables β and γ once the Q-functions pertinent to the individual light beams are known. The Q-function designated by Eq. (4.166) is symmetric about β and γ variables—which might suggest that one can use either of them as pumping light beam.

The mixture of two light beams can also be studied by applying the Wigner function defined by Eq. (2.97);

$$W(\alpha) = \frac{1}{\pi^2} \int d^2z \phi_s(z) \exp[z^*\alpha - z\alpha^*].$$

On account of Eq. (4.161), one can see that

$$W(\alpha) = \frac{2e^{\frac{\alpha^*\alpha}{2}}}{\pi} \int d^2\alpha' d^2\alpha'' P(\alpha')P(\alpha'') \exp\frac{1}{2}[-\alpha'^*\alpha' - \alpha''^*\alpha''$$
$$+ \alpha^*\alpha' + \alpha^*\alpha'' + \alpha\alpha'^* + \alpha\alpha''^* - \alpha'\alpha''^* - \alpha''\alpha'^*]. \tag{4.167}$$

So taking into consideration

$$P(\beta) = \frac{2}{\pi} \int d^2\alpha W(\alpha) \exp\left[2|\beta - \alpha|^2\right], \tag{4.168}$$

it might not be difficult to verify that

$$W(\alpha) = \frac{2e^{2\alpha^*\alpha}}{\pi} \int d^2\beta d^2\gamma \, W(\beta) W(\gamma) \exp[2(\beta^*\beta + \gamma^*\gamma)]$$

$$\times \exp[2(\gamma^*\beta + \gamma\beta^* - \beta^*\alpha - \gamma^*\alpha - \gamma\alpha^* - \beta\alpha^*)]. \qquad (4.169)$$

It might worth noting that these kinds of procedures can help in understanding the mixture of light beams in the cavity scenario.

4.3.2 Device Supported Wave Mixing

It is customary sending the light in an interferometric experiment via different paths, recombining them on the beam splitter, and then sending the mixed light to the detector by using optical elements so that the casted interference pattern is applied to visualize the involved correlations—the process that encompasses the notion of device supported wave mixing. To do so, a device that has a capacity to split the incoming light beam with predetermined proportions into two without necessarily interacting with it is required. One of the passive devices that boasts such a potential is the optical beam splitter.[14] Hoping to get an insight to the theoretical techniques that can help us to capture the physics of the radiation coming out of one of the arms in terms of the incident light, we try to describe the process of radiation splitting in classical and quantum realms.

Markedly, a beam splitter can be considered in classical realm as a linear dispersive medium with loss that splits the input beam into reflected and transmitted beams that would be characterized by a complex transmission and reflection coefficients (see Fig. 4.3). One may note that the output electric fields E_3 and E_4 can be related to the input electric fields E_1 and E_2 by

$$E_3 = R_{31}E_1 + T_{32}E_2, \qquad E_4 = R_{42}E_2 + T_{41}E_1, \qquad (4.170)$$

where R_{ij} and T_{ij} denote the reflection and transmission coefficients with the indices showing the connection between the input and output electric fields, and they are complex quantities that depend on the optical frequency. But one may choose to restrict the discussion to the non-dispersive case in which the reflection and transmission coefficients are presumed to be constant of frequency.

On account of the idea that the energy of the lossless beam splitter should be conserved, it is possible to propose that $|E_1|^2 + |E_2|^2 = |E_3|^2 + |E_4|^2$, which is

---

[14]An optical beam splitter can be made of a glass (two triangular glass prisms glued together at their base) or half silvered mirror (sheet of glass or plastic with transparently thin coating of metal) having a partial transmittance (reflectance) in a sense that the beam of light impinged onto the beam splitter is partly transmitted (reflected), and in the process splits it into two.

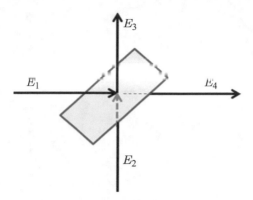

Fig. 4.3 Schematic representation of beams of light engaged in the splitting process in terms of the classical electric field, where E_1 and E_2 are the two input electric fields, and E_3 and E_4 are the output electric fields. E_3 denotes the mixture of the electric field E_1 partly reflected from the glass and E_2 partly transmitted through the same glass, whereas E_4 designates the sum of the electric field partly transmitted from E_1 and partly reflected from E_2

true for arbitrary input electric field provided that

$$|R_{31}|^2 + |T_{41}|^2 = |R_{42}|^2 + |T_{32}|^2 = 1, \qquad R_{31}T_{32}^* + R_{42}^*T_{41} = 0. \qquad (4.171)$$

So with the aid of Eq. (4.171), one can assert that

$$\frac{|R_{31}|}{|T_{41}|} = \frac{|R_{42}|}{|T_{32}|}, \qquad (4.172)$$

from which the magnitude of the transmission and reflection coefficients that correspond to each input electric field are inferred to be the same:

$$|R_{31}| = |R_{42}| = |R|, \qquad |T_{32}| = |T_{41}| = |T|. \qquad (4.173)$$

These requirements impose additional constraints that restrict the beam splitter matrix to be unitary: the matrix is equal with its complex conjugate transpose;

$$\begin{pmatrix} R_{31} & T_{32} \\ T_{41} & R_{42} \end{pmatrix} = \begin{pmatrix} R_{31}^* & T_{41}^* \\ T_{32}^* & R_{42}^* \end{pmatrix}, \qquad (4.174)$$

which should be consistent with the assumption that the transmission and reflection coefficients of the glass or plastic are independent of the frequency of the light falling on it.

The general structure of the beam splitter matrix can be simplified upon dividing the coefficients into the amplitude and phase factors as

$$R_{31} = R_{42} = |R|e^{i\phi_R} = R, \qquad T_{32} = T_{41} = |T|e^{i\phi_T} = T. \qquad (4.175)$$

Fig. 4.4 Schematic representation of optical beam splitter in the regime of quantum theory in which the input radiation modes describable by $\hat{a}_1$ and $\hat{a}_2$ that impinged onto the beam splitter emerge as output modes $\hat{a}_3$ and $\hat{a}_4$

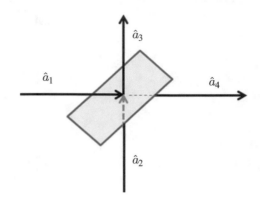

With this introduction, Eq. (4.171) reduces to

$$|R|^2 + |T|^2 = 1, \qquad RT^* + TR^* = 0. \tag{4.176}$$

Then upon using Eqs. (4.173) and (4.175), one can also see that $|R||T|[\cos(\phi_R - \phi_T)] = 0$, which would be true when $|R|$ and $|T|$ are different from zero provided that $\phi_R - \phi_T = \pm\pi/2$. For a 50:50 beam splitter, one may see that the reflection and transmission coefficients can be taken as $|R| = |T| = 1/\sqrt{2}$ with $\phi_R - \phi_T = \frac{\pi}{2}$.

On the other hand, a beam splitter can be perceived in the quantum realm as a device having an ability to convert the incoming photon state into a linear superposition of the output states. So schematic representation 4.3 would be modified by replacing the classical electric fields with the pertinent boson operators (see Fig. 4.4). For a symmetric beam splitter,[15] the output mode operators can be expressed in terms of the input mode operators as

$$\hat{a}_3 = R\hat{a}_1 + T\hat{a}_2, \qquad \hat{a}_4 = T\hat{a}_1 + R\hat{a}_2, \tag{4.177}$$

---

[15]The advent of entanglement based technologies that include quantum computation and communication makes the beam splitter one of the important components of the present day optical experiments (see Chaps. 9 and 11). Particularly, the application of the beam splitter in quantum technologies is usually sought for via the beam splitter interaction Hamiltonian denoted by

$$\hat{H}_b = i\chi\left(\hat{a}_2^\dagger\hat{a}_1 - \hat{a}_1^\dagger\hat{a}_2\right),$$

where χ is the interaction coupling constant. In connection to this Hamiltonian, for equal frequency scenario, the beam splitter operator can be introduced as

$$\hat{b} = e^{\chi t\left(\hat{a}_2^\dagger\hat{a}_1 - \hat{a}_1^\dagger\hat{a}_2\right)},$$

where

$$\hat{b}\hat{a}_1\hat{b}^\dagger = \hat{a}_1\cos\chi t + \hat{a}_2\sin\chi t, \qquad \hat{b}\hat{a}_2\hat{b}^\dagger = \hat{a}_2\cos\chi t - \hat{a}_1\sin\chi t$$

in which $\cos\chi t$ and $\sin\chi t$ can be related to the transmission and reflection coefficients (see also Appendix 5 of Chap. 3).

from which follows

$$\hat{a}_1^\dagger = R\hat{a}_3^\dagger + T\hat{a}_4^\dagger, \qquad \hat{a}_2^\dagger = T\hat{a}_3^\dagger + R\hat{a}_4^\dagger. \tag{4.178}$$

On the basis of the assertion that the boundary conditions related to classical and quantum theoretical approaches should be the same, it is possible to propose that

$$[\hat{a}_3, \hat{a}_3^\dagger] = [\hat{a}_4, \hat{a}_4^\dagger] = 1, \tag{4.179}$$

$$[\hat{a}_3, \hat{a}_3] = [\hat{a}_4, \hat{a}_4] = [\hat{a}_3, \hat{a}_4] = [\hat{a}_3, \hat{a}_4^\dagger] = [\hat{a}_3^\dagger, \hat{a}_4] = 0, \tag{4.180}$$

where the output operators are taken to satisfy the usual boson commutation relation. It is also possible to verify by applying Eqs. (4.177) and (4.179) that the conditions provided in Eq. (4.176) still hold true.

In addition, the photon number operators for the beam splitter for arbitrary arm can be expressed as $\hat{n}_i = \hat{a}_i^\dagger \hat{a}_i$, where $i = 1, 2, 3, 4$. With the help of this definition and Eq. (4.177), one gets

$$\hat{n}_3 = |R|^2 \hat{a}_1^\dagger \hat{a}_1 + |T|^2 \hat{a}_2^\dagger \hat{a}_2 + R^* T \hat{a}_1^\dagger \hat{a}_2 + T^* R \hat{a}_2^\dagger \hat{a}_1, \tag{4.181}$$

$$\hat{n}_4 = |T|^2 \hat{a}_1^\dagger \hat{a}_1 + |R|^2 \hat{a}_2^\dagger \hat{a}_2 + RT^* \hat{a}_1^\dagger \hat{a}_2 + T R^* \hat{a}_2^\dagger \hat{a}_1. \tag{4.182}$$

Summing Eqs. (4.181) and (4.182) in view of Eq. (4.176) leads to

$$\hat{n}_3 + \hat{n}_4 = \hat{n}_1 + \hat{n}_2, \tag{4.183}$$

which entails that the number of the photon would be conserved during beam splitting.

Recall that the density operator for the combined output modes (4.161) can be expressed as

$$\hat{\rho}_{3,4} = \int d^2\alpha \, d^2\beta \, P_1(\alpha) P_2(\beta) |\alpha\rangle_1 |\beta\rangle_2 \langle\beta|_2 \langle\alpha|_1, \tag{4.184}$$

where $|\alpha\rangle_1$ and $|\beta\rangle_2$ are the coherent state representation of the input radiation fields in the separate arms and $\hat{\rho}_{3,4}$ stands for the radiations emerging from the beam splitter along arms 3 and 4 separately. Since the input modes are independent before they made to interact on the beam splitter, the combined state of the input modes can be expressed in a coherent state representation as $|\varphi\rangle_{in} = |\alpha\rangle_1 |\beta\rangle_2$. In line with the definition of the coherent state in terms of the pertinent displacement operator (see Sect. 2.2), it is possible to put this expression by applying Eq. (4.178) as

$$|\varphi\rangle_{in} = |R\alpha + T\beta\rangle_3 \otimes |T\alpha + R\beta\rangle_4. \tag{4.185}$$

Then upon taking Eq. (4.185) into account, it is possible to see that

$$\hat{\rho}_{3,4} = \int d^2\alpha d^2\beta P_1(\alpha) P_2(\beta) |R\alpha + T\beta\rangle_3 |T\alpha + R\beta\rangle_4 \langle R\beta + T\alpha| \langle R\alpha + T\beta|, \tag{4.186}$$

which might envisage the superposition or entanglement of the output modes. With the help of Eq. (4.186), it should be admissible to obtain various quasi-statistical distribution functions including the photon number probability distribution. For instance, the joint probability for finding n_3 photons in mode 3 and n_4 photons in mode 4 can be expressed as

$$P(n_3, n_4) = \langle n_3 | \hat{\rho}_{3,4} | n_4 \rangle. \tag{4.187}$$

Upon exploiting the property of the probability distribution function, the mean photon number in each mode can be defined as

$$\langle \hat{n}_3 \rangle = \sum_{n_3=0} P(n_3, n_4) n_3, \qquad \langle \hat{n}_4 \rangle = \sum_{n_4=0} P(n_3, n_4) n_4. \tag{4.188}$$

Moreover, since the beam splitter induces noise as a consequence of quantum fluctuation while the beam passes through it, the photon number is anticipated to fluctuate. To illustrate such fluctuation, the photon number variances of the two output modes when arm 2 is in vacuum state can be put in the form

$$\Delta n_3^2 = |R|^4 \Delta n_1^2 + |R|^2 |T|^2 \langle \hat{n}_1 \rangle, \tag{4.189}$$

$$\Delta n_4^2 = |T|^4 \Delta n_1^2 + |R|^2 |T|^2 \langle \hat{n}_1 \rangle. \tag{4.190}$$

Equations (4.189) and (4.190) may reveal that the variances of the output photon number in both arms depend on the variance and mean photon number of the input mode. The second term in both expressions in particular can be perceived as the heating of the input field in arm 1 as a result of the vacuum fluctuation in arm 2, that means, the random division of the input photon at the beam splitter causes noise.

In case an arbitrary input is injected to both arms where the photons in arms 1 and 2 are assumed to be uncorrelated, the variance of the photon number of the output modes in each arm turns out upon employing Eqs. (4.181) and (4.182) to be

$$\Delta n_3^2 = |R|^4 \Delta n_1^2 + |T|^4 \Delta n_2^2 + |R|^2 |T|^2 [\langle \hat{n}_1 \rangle + \langle \hat{n}_2 \rangle - 2\langle \hat{n}_1 \rangle \langle \hat{n}_2 \rangle], \tag{4.191}$$

$$\Delta n_4^2 = |T|^4 \Delta n_1^2 + |R|^4 \Delta n_2^2 + |R|^2 |T|^2 [\langle \hat{n}_1 \rangle + \langle \hat{n}_2 \rangle - 2\langle \hat{n}_1 \rangle \langle \hat{n}_2 \rangle]. \tag{4.192}$$

As one can see in this case as well, the variances of the output modes are related to the variance and mean photon numbers of the input radiation. The last term in the right side of Eqs. (4.191) and (4.192) can be interpreted in the same way as the noise caused by the fluctuation due to a random splitting of the beam at the beam splitter.

Now imagine the case when a single photon is incident on arm 1 while arm 2 is in vacuum state. The input state in such a scenario can be taken as $|1\rangle_1|0\rangle_2 = \hat{a}_1^\dagger|0, 0\rangle$, where $|0, 0\rangle$ is the joint vacuum state. The input state then can be converted into the output state with the help of Eq. (4.178) as

$$|1\rangle_1|0\rangle_2 = (R\hat{a}_3^\dagger + T\hat{a}_4^\dagger)|0, 0\rangle, \tag{4.193}$$

from which follows

$$|1\rangle_1|0\rangle_2 = R|1\rangle_3|0\rangle_4 + T|0\rangle_3|1\rangle_4. \tag{4.194}$$

Such conversion of the input state into a linear superposition of the two possible output states signifies the basic quantum optical processes associated with the beam splitting, that is, the splitting process can induce a coherent superposition that leads to entanglement between the two output states.

In light of earlier discussion, the mean photon number of the output mode can be calculated by using the relation of the form $\langle \hat{n}_3 \rangle = \langle 0|\langle 1|\hat{n}_3|1\rangle|0\rangle$:

$$\langle \hat{n}_3 \rangle = \langle 0|\langle 1||R|^2\hat{a}_1^\dagger\hat{a}_1 + |T|^2\hat{a}_2^\dagger\hat{a}_2 + R^*T\hat{a}_1\hat{a}_2 + T^*R\hat{a}_2\hat{a}_2^\dagger|1\rangle|0\rangle, \tag{4.195}$$

from which follows $\langle \hat{n}_3 \rangle = |R|^2$. It is also possible to ascertain that $\langle \hat{n}_4 \rangle = |T|^2$; which implies that $\langle \hat{n}_3 \rangle + \langle \hat{n}_4 \rangle = 1$. Since there is only one photon that impinged onto the beam splitter, it is reasonable expecting only one photon to emerge from the output arms. But for a general case, the photon should be divided between the two output arms according to the magnitude of the reflection and transmission coefficients.

In the same token, the correlation of the photon numbers in the output arms for this case would be

$$\langle \hat{n}_3\hat{n}_4 \rangle = \langle 0|\langle 1|\hat{n}_3\hat{n}_4|1\rangle_1|0\rangle_2. \tag{4.196}$$

One can see with the help of Eqs. (4.181) and (4.182) that $\langle \hat{n}_3\hat{n}_4 \rangle = 0$: the two photon numbers are not correlated, which is a clear manifestation of the particle like aspect of a single photon.

One may also seek to generalize for the case of arbitrary input state in arm 1 while arm 2 still remains in vacuum state. In this case, the combined input state would be denoted by $|\text{arb}\rangle_1|0\rangle_2$; as a result, one may verify employing Eqs. (4.181) and (4.182) that

$$\langle \hat{n}_3 \rangle = |R|^2\langle \text{arb}|\hat{a}_1^\dagger\hat{a}_1|\text{arb}\rangle, \qquad \langle \hat{n}_4 \rangle = |T|^2\langle \text{arb}|\hat{a}_1^\dagger\hat{a}_1|\text{arb}\rangle, \tag{4.197}$$

where $\langle\text{arb}|\hat{a}_1^\dagger\hat{a}_1|\text{arb}\rangle$ represents the mean photon number of the arbitrary input. The correlation of the photon number can also be expressed in the same way as

$$\langle\hat{n}_3\hat{n}_4\rangle = |R|^2|T|^2\langle\hat{n}_1(\hat{n}_1 - 1)\rangle. \tag{4.198}$$

The normalized photon number correlation for the two output modes can thus be put in the form

$$g_{3,4}^{(2)}(0) = \frac{\langle\hat{n}_3\hat{n}_4\rangle}{\langle\hat{n}_3\rangle\langle\hat{n}_4\rangle}, \tag{4.199}$$

which reduces to

$$g_{3,4}^{(2)}(0) = \frac{\langle\hat{n}_1(\hat{n}_1 - 1)\rangle}{\langle\hat{n}_1\rangle^2} = g_{1,1}^{(2)}(0). \tag{4.200}$$

One may see that Eq. (4.200) is identical to the normalized second-order one-time correlation: $g^{(2)}(0) = \langle\hat{a}^\dagger\hat{a}(\hat{a}^\dagger\hat{a} - 1)\rangle/\langle\hat{a}^\dagger\hat{a}\rangle^2$ (see also Sect. 4.2). There is indeed no way in which the photon can simultaneously be assigned to a particular path and at the same time contribute to the interference pattern; and hence, the correlation between the two output photon numbers.

4.3.3 Active Wave Mixing

Optical elements such as mirror, grating, polarizer, filter and lens are capable of altering the direction of the propagation or blocking a portion of the incident light. The interaction leading to such processes can be taken as passive as long as the devices do not exchange energy with the radiation. But the interaction of the external radiation with a nonlinear medium—in which the frequency of the output radiation has a chance to be different from the input—is referred to as active interaction. The notion of active wave mixing hence epitomizes the active participation of the medium in the interaction via the accompanying polarization, which can be construed by assuming that the medium has a frequency dependent refractive index. In this case, if the medium is shined with an external radiation, the refractive index of the medium changes, that means, if the same medium is shined with the same light once more, the medium may respond in a different way since the anticipated successive changes can induce a nonlinearity in the interaction process. The modulation of the refractive index due to the perceived successive interaction can also induce modulation in the output radiation in the sense that the emerging frequency would be different from the input radiation. It is such a modulation of the frequency due to the change in the response of the medium that is interpreted as active process.

In other words, the first input field causes oscillation of the polarization in the dielectric of the medium. As a result of the induced oscillation, the medium then can be conned to release the radiation it absorbs with some phase shift that depends on the damping of the individual dipole. In the same manner, the application of the second field believed to have a different frequency drives the polarization of the dielectric already modified by first radiation. In the process, the interaction of the medium with two different radiations is expected to induce interference. This kind of superposition of the waves can consequently lead to a generation of harmonics in the polarization at the sum and difference frequencies. Then application of the third field may drive the already altered polarization; the mechanism that can lead to the beating in the input fields as well as the sum and difference frequencies and gives rise to the fourth field. Since each of the produced beat frequencies can act as a new source field, a myriad number of interactions and fields with a significant degree of correlation can be expected.

Active wave mixing process can thus be explained in the realm of quantum theory in terms of the underlying absorption-emission events associated with the atoms of the medium. Imagine that the medium has four-level dipole allowed atomic transition roots as in the Raman scheme and there are two external radiations that can drive the atom to upper energy levels (see Sect. 6.5). Assume also that the atom can decay to two lower energy levels other than from which it is driven; and in the process, gives off two radiations with frequency different from the driving radiations. In case one keeps on pumping such a medium for sometime, the resulting absorption-emission events lead to a nonlinear process and the medium gets a chance to actively engage in the energy conversion.

The envisioned nonlinear interaction may also lead to a correlation between the radiations that can be emitted via different transition roots. So the emerging nonlinear interaction is expected to incite nonclassical properties of the radiation where the squeezing property of light is first demonstrated with the help of the mechanism outlined earlier. Other nonlinear optical processes such as parametric oscillation, second-harmonic generation and correlated emission lasing known to be the source of nonclassical light can also be taken as a good example of active wave mixing process (see Sect. 8.1). Owing to the anticipated nonclassical interaction, quite diverse applications have been attached to the process of active wave mixing. One can cite for example the mechanism of phase conjugation that refers to the process of reversing the phase (the phase conjugated wave appears to travel backward in time), real time holographic imaging, real time image processing, photon processing techniques with the aid of Fourier transform optics and Raman spectroscopy (see also [22]).

To expose the process of active wave mixing, it might be appropriate starting with a nonlinear medium that can be represented by an aggregate of atoms (see Sect. 5.6). Suppose the atoms in the medium are described by two energy levels whose energy transition is resonant with the pumping radiation. It is also assumed that, once excited by the external light, the atoms can go over to any dipole allowed lower energy level; the pertinent transition is designated as open (see Sect. 5.3). At microscopic level, the radiation impinged onto this medium interacts with the atoms via the accompanying successive absorption-emission events, where the strength or effectiveness of the mixing process depends on the density of the sample, the geometry of the scheme or phase matching and the response of the medium. In line with the discussion provided in Chap. 6, we thus seek to outline how the response of a two-level atom to a radiation that propagates in it would be addressed.

To this effect, imagine that the pumping radiation propagates in the z-direction; and hence, can be described by the electric field of the form

$$E(z, t) = \tilde{E}(z, t)e^{i(kz-\omega t)} + c.c, \qquad (4.201)$$

where the complex field amplitude is taken to change little in the range of the optical wavelength, ω is the oscillation frequency and k is the wave number (see Sect. 5.6). The polarization that can be induced in the medium can be defined in the same way as

$$P(z, t) = \tilde{P}(z, t)e^{i(kz-\omega t)} + c.c. \qquad (4.202)$$

So in view of the observation that the electric field and polarization can be related via Maxwell equation, one can see that $\frac{d\tilde{E}}{dz} = -\alpha\tilde{E}$, where the complex absorption coefficient α is given by

$$\alpha = -\frac{ik}{\varepsilon}\frac{\tilde{P}}{\tilde{E}} \qquad (4.203)$$

in which ε is the permittivity of the medium.

One can also assume that the polarization for an open two-level atom when the upper (denoted by a) and lower energy levels (denoted by b) are pumped with different rates, that is, when the medium is incoherently broadened. Assume also that the polarization can be defined in terms of the off-diagonal elements of the density matrix: $P(z, t) = \mu_{ab}\rho_{ab}(t) + c.c$, where μ_{ab}'s are the elements of the dipole moment matrix of the atom. With this consideration, the slowly varying terms of the polarization can be written as

$$\tilde{P}(z, t) = 2\mu_{abe}e^{-i(kz-\omega t)}\rho_{ab}(z, t). \qquad (4.204)$$

To understand the response of the medium, one may need to obtain the off-diagonal elements of the dipole moment and atomic density matrix.

To do so, it might be sufficient to see that the evolution of the density operator under the influence of the external electric field takes the form

$$\frac{d\rho(z, t)}{dt} = \sum_i \lambda_i(z, t)\rho(z, t_0) + \sum_i \int_{-\infty}^{\infty} \lambda_i(z, t)\frac{d\rho(z, t, t_0)}{dt}dt, \qquad (4.205)$$

where λ_i is the pump rate per unit volume for each energy level. Note that there is no off-diagonal excitation prior to the interaction: $\rho_{jk}(z, t_0, t_0) = \delta_{ja}\delta_{kb}$. Taking this into account, one can obtain

$$\frac{d\rho_{ab}}{dt} = -(i\omega_0 + \gamma_{ab})\rho_{ab} + iV_{ab}(z,t)(\rho_{aa} - \rho_{bb}), \tag{4.206}$$

$$\frac{d\rho_{aa}}{dt} = \lambda_a - \gamma_a\rho_{aa} - (iV_{ab}\rho_{ba} + c.c), \tag{4.207}$$

$$\frac{d\rho_{bb}}{dt} = \lambda_b - \gamma_b\rho_{bb} + iV_{ab}\rho_{ba} + c.c, \tag{4.208}$$

where V_{ab} is the off-diagonal elements of the interaction potential, γ_{ab} is the dephasing rate and $\gamma_{a,b}$ are the atomic damping rates of respective energy levels. The remaining thing to do in analyzing the response of the medium is to solve such coupled differential equations: then after, find the pertinent polarization (see Sect. 5.6 and also [23]).

4.4   Photon Addition

Current advances in the field of quantum computation require a way of adding and subtracting photons at will so that a light can be used as a resource to carry out computational tasks (see Chaps. 9 and 12). For example, adding a photon to the vacuum state and then removing it can be taken as par with the on and off operations in classical computation. Even then, due to uncertainty and commutation relations, two sequences of photon addition-subtraction and subtraction-addition processes are not expected to lead to identical outcomes since the two operators do not commute. The other subtle aspect in the quantum theory is the possibility of having the superposition of the original state and the state of the photon to be added. As a result of the ensuing induced superposition, by adding just one photon to the system, it is possible to obtain a state that cannot be described by classical theory. Despite such a promise, since the nonclassical features are associated with the superposition of a given state with a finite number of photon states, it might be good to be forewarned that the suggested quantum effect may not be noticeable for states with high intensity.

The photon addition can also be perceived in practical setting as the action of the creation operator on the number state, and as an event of stimulated emission from physical point of view. Although the process of photon addition seems quite simple, leaving alone the technical challenge of adding or subtracting a definite number of photons, the theoretical advance is marred with the involved mathematical rigor.[16] For now, we opt to begin with a well established fact that a

---

[16]The discussion on the photon subtraction is not included with the understanding that it can be achieved by replacing the creation operator $\hat{a}^\dagger$ by the annihilation operator $\hat{a}$, even though the result one obtains for photon subtraction is not necessarily the reverse of the photon addition as the boson commutation relation does not allow to be so [24].

coherent state has properties very close to classical radiation and the action of the creation operator on a number state modifies the number of the photon according to $\hat{a}^\dagger|n\rangle = \sqrt{n+1}|n+1\rangle$.

To describe the photon addition, imagine that initially there is a light that can be represented by a coherent state on which the photon creation operator is successively applied:

$$|\alpha, m\rangle \propto \hat{a}^{\dagger^m}|\alpha, 0\rangle, \tag{4.209}$$

where the normalization constant is yet to be determined and $|\alpha, 0\rangle$ is the superposed state of the vacuum and coherent states (see Chap. 2). Such a process can be perceived henceforth as a successive one photon excitations of a coherent light, and so designated as photon addition. One may also note that this state lies between coherent and number states.

If just a single quanta of the field excitation is added to coherent light, the emerging single photon-added coherent state can be denoted by

$$|\alpha, 1\rangle = \frac{\hat{a}^\dagger|\alpha, 0\rangle}{\sqrt{1 + \alpha^*\alpha}}, \tag{4.210}$$

where the denominator is inserted to ensure normalization. The same state can also be expressed in a number state as

$$|\alpha, 1\rangle = \frac{\exp\left(-\frac{1}{2}\alpha^*\alpha\right)}{\sqrt{1 + \alpha^*\alpha}} \sum_{n=0} \frac{\alpha^n}{\sqrt{n!}}\sqrt{n+1}|n+1\rangle. \tag{4.211}$$

When a large number of photons are intended to be added, the emerging mathematics becomes cumbersome. Besides, the effect of the photon addition process may not be perceived when a single photon is added to a state that has a large number of photons. Photon addition is then thought to be interesting when one wishes to manipulate a finite number of photons. It might be worthy to note that a single photon excitation of the coherent state changes it into something quite different specially for less intense light when the absence of the vacuum term has a strong impact.

In light of this, consider the case when one photon is added to thermal light; the scheme designated as a single photon-added thermal state. Such perception signifies the process in which a completely classical light is excited by a single photon in which a single photon-added thermal state can be epitomized by the density operator in a number state:

$$\hat{\rho} = \frac{1}{\bar{n}(\bar{n}+1)} \sum_{n=0} \left(\frac{\bar{n}}{1+\bar{n}}\right)^n n|n\rangle\langle n|. \tag{4.212}$$

Critical scrutiny may reveal that this state has a nonclassical signature. Since the creation operator has the capacity to excite the vacuum state, the photon addition process can be linked to the removal of the vacuum fluctuations. It is perhaps clear to envisage that the pertinent induced nonclassical behavior is associated with the emerging coherent superposition inherent to the removal of vacuum fluctuations [25]. It is such a subtle way of removing the contribution of the classical noise that makes photon addition a viable candidate to carry out certain quantum information processing tasks.

Based on earlier discussion, a photon-added coherent state can also be defined for arbitrary number of photons as

$$|\alpha, m\rangle = \frac{1}{\sqrt{\langle\alpha, 0|\hat{a}^m \hat{a}^{\dagger m}|\alpha, 0\rangle}} \hat{a}^{\dagger m}|\alpha\rangle, \tag{4.213}$$

where m is a positive integer that represents the number of added photons. This state is expected to exhibit nonclassical features that include phase squeezing and sub-Poissonian photon statistics. One can also verify upon using the number state expansion that

$$|\alpha, m\rangle = \frac{\exp\left(-\frac{1}{2}\alpha^*\alpha\right)}{\sqrt{\mathcal{L}_m(-\alpha^*\alpha)m!}} \sum_{n=0}^{\infty} \frac{\alpha^n \sqrt{(m+n)!}}{n!}|n, m\rangle, \tag{4.214}$$

where $\mathcal{L}_m(x) = \sum_{n=0}^{m} \frac{(-x)^n m!}{(n!)^2 (m-n)!}$.

The photon-added states can generally be expressed in a more compact form as

$$|\psi, m\rangle = N_m \hat{a}^{\dagger m}|\psi, 0\rangle, \tag{4.215}$$

where $|\psi, 0\rangle$ is arbitrary quantum state superimposed with the vacuum state and N_m is the normalization constant. Applying this definition and taking the initial state $|\psi, 0\rangle$ as squeezed vacuum state, one can obtain a photon-added squeezed state. One can define namely the mixed photon-added state as

$$\hat{\rho}_m = N_m \hat{a}^{\dagger m} \hat{\rho} \hat{a}^m. \tag{4.216}$$

To describe the photon addition mechanism, one can take the case when the initial state is in the superposition of the number states, that is, when $|\psi\rangle = \sum_{n=0}^{\infty} c_n|n\rangle$.

With this suggestion, an arbitrary number photon-added state can be expressed as

$$|\psi, m\rangle = N_m \sum_{n=m}^{\infty} c_{n-m} \sqrt{\frac{n!}{(n-m)!}}|n\rangle; \tag{4.217}$$

as a result, the probability of detecting n quanta of photons in this state can be put in terms of the initial probabilities as

$$P_n^{(m)} = \frac{1}{\mathcal{L}_m(-\alpha^*\alpha)m!} \frac{n!}{(n-m)!} P_{n-m}^{(0)}. \tag{4.218}$$

One can see from Eq. (4.218) that the probability of detecting n quanta of light is exactly zero for $m > n$ for all kinds of photon-added states.

It might be good to stress that this discussion can be extended to a multi-mode case. In this regard, just like in the single-mode case (see Sect. 2.2), one can begin with transformation:

$$\hat{D}_{ab}^{-1}\hat{a}\hat{D}_{ab} = \hat{a} + \alpha, \qquad \hat{D}_{ab}^{\dagger}\hat{a}^{\dagger}\hat{D}_{ab} = \hat{a}^{\dagger} + \alpha^*, \tag{4.219}$$

and then find for one of the modes that

$$\hat{S}_{ab}^{\dagger}\hat{a}\hat{S}_{ab} = \hat{a}\cosh r - \hat{b}^{\dagger}e^{i\theta}\sinh r \tag{4.220}$$

(compare with Eqs. (2.56) and (2.57)), where upon replacing $\hat{a}$ with $\hat{b}$ one can write a similar expression for a b-mode.

A photon-added two-mode displaced squeezed vacuum state can then be defined by

$$\hat{a}^{\dagger m}\hat{D}_{ab}\hat{S}_{ab}|0_a, 0_b\rangle = \hat{D}_{ab}\hat{S}_{ab}[\hat{a}^{\dagger}\cosh r - \hat{b}e^{-i\theta}\sinh r + \alpha^*]^m|0_a, 0_b\rangle, \tag{4.221}$$

where taking the action of operator $\hat{b}$ on the two-mode vacuum state yields

$$\hat{a}^{\dagger m}\hat{D}_{ab}\hat{S}_{ab}|0_a, 0_b\rangle = \hat{D}_{ab}\hat{S}_{ab}\sum_{l=0}^{m}\binom{m}{l}\hat{a}^{\dagger l}\cosh^l r \,(\alpha^*)^{m-l}|0_a, 0_b\rangle \tag{4.222}$$

and the action of $\hat{a}^{\dagger}$ on the vacuum state l times leads to

$$\hat{a}^{\dagger m}\hat{D}_{ab}\hat{S}_{ab}|0_a, 0_b\rangle = \hat{D}_{ab}\hat{S}_{ab}\sum_{l=0}^{m}\binom{m}{l}\sqrt{l!}\cosh^l r \,(\alpha^*)^{m-l}|l, 0_b\rangle. \tag{4.223}$$

Generalization of earlier discussion for arbitrary photon number can also be done by introducing an operator function of the form

$$f(\hat{n})\hat{a}|f, \alpha\rangle = \alpha|f, \alpha\rangle, \tag{4.224}$$

where $f(\hat{n})$ is an operator valued function of photon number operator that can be conveniently typified as nonlinear coherent states. One may note that a photon-added coherent state is thought in experimental setting to be generated during the interaction of two-level atom with the cavity field initially prepared in a coherent state in a high fitness cavity (relate with the consideration in Chap. 6). The notion of nonlinear coherent states can then be justified by showing that the state $|\alpha, m\rangle$ obeys the general equation

$$f(\hat{n}, m)\hat{a}|\alpha, m\rangle = \alpha|\alpha, m\rangle, \tag{4.225}$$

with a suitable choice of $f(\hat{n}, m)$.

To assert this proposal, one can multiply the expression $\hat{a}|\alpha\rangle = \alpha|\alpha\rangle$ from left by $\hat{a}^{\dagger m}$:

$$\hat{a}^{\dagger m}\hat{a}|\alpha\rangle = \alpha\hat{a}^{\dagger m}|\alpha\rangle. \tag{4.226}$$

After that, upon making use of the commutation relation

$$[\hat{a}, f(\hat{a}, \hat{a}^{\dagger})] = \frac{\partial f}{\partial \hat{a}^{\dagger}}, \tag{4.227}$$

one can write

$$\left(\hat{a}\hat{a}^{\dagger m} - m\hat{a}^{\dagger m-1}\right)|\alpha\rangle = \alpha\hat{a}^{\dagger m}|\alpha\rangle. \tag{4.228}$$

In relation to the boson commutation relation, and employing the identity

$$\hat{I} = \frac{1}{1 + \hat{a}^{\dagger}\hat{a}}\hat{a}\hat{a}^{\dagger}, \tag{4.229}$$

one can also see that

$$\left[\hat{I} - \frac{m}{1 + \hat{a}^{\dagger}\hat{a}}\right]\hat{a}\hat{a}^{\dagger m}|\alpha\rangle = \alpha\hat{a}^{\dagger m}|\alpha\rangle,$$

where comparison reveals

$$f(\hat{n}, m) = \hat{I} - \frac{m}{1 + \hat{n}}. \tag{4.230}$$

This discussion can be taken as evidence for the photon-added state to represent a class of nonlinear coherent state. One may note that the involved operations annihilate the vacuum state $|0\rangle$ and the definite m-photon state $|m\rangle$ together but not the states in between. This process in this sense is completely different from the m-photon annihilation operator $\hat{a}^m$ that annihilates all number state with photon number between 0 and m.

4.5   Photon Detection Mechanism

One of the mechanisms in the photon detection process encompasses the generation of electric charge via photon absorption, and subsequent amplification of the photon induced electric charges so that the desired electronic signal in the form of electric current or voltage can be produced. The primary task in the photon detection process is thus to incite the interaction of radiation with matter or detector, which is a quantum process by nature. Pertaining to the fact that the number of photons and atoms that participate in such interaction are significantly large, it might be appropriate and desirable emphasizing that a direct photon measurement may not be physically feasible. As a way out, a semiclassical description may suffice to unravel the outcome of most experimental situations involving intensity of the radiation and first-order correlation in which the pertinent photon counting statistics can be captured by Mandel's formula [26]:

$$\rho(n,t,\tau) = \int_0^\infty \frac{\left[\eta_d E(t)\right]^n \left[\exp(-\eta_d E(t))\right]}{n!} P[E(t)]dE(t), \tag{4.231}$$

where $\rho(n,t,\tau)$ is the probability of observing n photoelectron events at time t, η_d is the detection efficiency, $P[E(t)]$ is the distribution of the energy and τ is the measurement time.

The energy of light falling upon the surface of the detector can also be described in terms of the intensity of the light integrated over the measurement time and detector area as

$$E(t) = \int_t^{t+\tau} \int_A I(r,t)dA\,dt. \tag{4.232}$$

By assuming that the measurement time is sufficiently short or the energy fluctuation mimics the intensity fluctuation of interest, it thus is possible to rewrite Eq. (4.231) by replacing the energy with the intensity at the detector as

$$\rho(n) = \int_0^\infty \frac{(\eta I)^n e^{-\eta I}}{n!} P(I)dI, \tag{4.233}$$

where η is the efficiency of the detector.

4.5.1   Direct Photon Detection

Various quantum properties of light are usually appraised without bothering whether or not it is really possible to carry out the measurement intended to quantify the said quantum features. Hoping to change this aspect of the analysis of the quantum features of radiation, we resort to recap the observation that almost all quantum optical measurement processes rely on a photoelectric detection mechanism. Since the number of electrons engaged in the interaction is quite large and the detection device is believed to have associated physical limitations, it might be desirable to look for statistical distribution of the photon count (see also Sect. 2.3), which can be achieved upon counting the librated electrons and then casting the distribution on the screen of the monitor. It is possible as a result to link a direct photon detection to the process of measuring the photon flux of the beam that falls on the phototube or photographic plate in terms of the number of librated electrons.

To illustrate this idea, suppose the light beam that involves in the detection process is narrow band excitation, that is, the associated time dependent operator evolves according to

$$\hat{a}(t) = \frac{1}{\sqrt{2\pi}} \int_{-\infty}^{\infty} \hat{a}(\omega) e^{-i\omega t} d\omega. \tag{4.234}$$

The number of photons that arrive at the detector in a measurement time can then be expressed as

$$\hat{N}(t, \tau) = \int_{t}^{t+\tau} \hat{a}^{\dagger}(t')\hat{a}(t')dt', \tag{4.235}$$

which is related to the actual photon detection experiment. Mark the similarity between $\hat{N}(t, \tau)$ and the photon number operator

$$\hat{n} = \int_{0}^{\infty} \hat{a}^{\dagger}(t')\hat{a}(t')dt' \tag{4.236}$$

with the exception that the expectation value of the former is always finite since the integration is carried over a finite time τ.

Since a device in an experimental setup measures the intensity of the light beam rather than the electric field, its operation depends on the absorption of the portion of the beam whose energy is converted to a detectable form. The process of the amplification of the signal produced by the photon ionized electron in the phototube to generate electric current pulse ensures that such a device is capable of registering the number of individual electrons librated from its surface; from which the number of photons impinged onto the surface of the detecting device is inferred. Unfortunately, no ideal detector that can register every photon falling on its surface is available so far due to the inherent physical constraints. For instance, every detector has a dead time since the detector requires some time to ready itself for subsequent ejection after it has already done so while the photons are kept on arriving without being detected in this time interval. One may thus claim that a detector can register only certain fraction of the photon falling on it. The mechanism of registering the fraction of the photon that arrives at the detector out of the photons that are supposed to reach the detector epitomizes the notion of quantum efficiency.

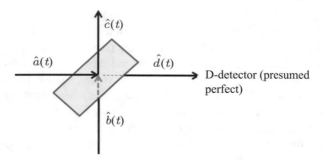

Fig. 4.5 Schematic representation of direct photon detection. The light one seeks to measure is taken to impinge onto one of the arms while the other coupler arm is maintained at a continuous mode vacuum state. The light that reaches the detector in this case is the superposition of the partially transmitted input ($\hat{a}(t)$) and partly reflected local oscillator ($\hat{b}(t)$). If one of this light modes is kept at vacuum state, the detector can capture only part of the incident light

Such an inefficient photon detector can then be described by a hypothetical perfectly efficient detector preceded by a beam splitter[17] (see Fig. 4.5). So with the help of the beam splitter input-output relation, the light falling on the detector can be expressed as

$$\hat{d}(t) = \sqrt{\eta}\hat{a}(t) + i\sqrt{1 - n}\hat{b}(t), \tag{4.237}$$

where η can be taken as efficiency of the detector. With this information, the number of photons that arrive at the detector would be

$$\hat{N}_d(t, \tau) = \int_t^{t+\tau} \hat{d}^\dagger(t')\hat{d}(t')dt', \tag{4.238}$$

where subscript d denotes the quantities to be measured by the inefficient detector, and the mean photon count can be evaluated by taking the average of the photon count operator as

$$\langle \hat{n}_d \rangle = \langle \hat{N}_d(t, \tau) \rangle = \eta \langle \hat{N}(t, \tau) \rangle. \tag{4.239}$$

In addition, the second-moment of the number of photon count can be put in the form

$$\langle \hat{n}_d(\hat{n}_d - 1) \rangle = \eta^2 \langle \hat{N}(t, \tau)[\hat{N}(t, \tau) - 1] \rangle. \tag{4.240}$$

---

[17]The beam splitter can be envisaged as having a complex reflection and transmission coefficients of the form $R = i\sqrt{1 - \eta}$ and $T = \sqrt{\eta}$ (see also Sect. 4.3.2). This kind of interpretation and denotation would be utilized to describe the coherent state based quantum optical processes (see Chap. 11).

Then upon making use of Eqs. (4.237) and (4.239) along with the usual commutation relation of boson operator, the variance of the photon count number is found to be

$$\Delta n_d^2 = \eta^2 \langle \Delta \hat{N}(t, \tau)^2 \rangle + \eta(1 - \eta) \langle \hat{N}(t, \tau) \rangle. \tag{4.241}$$

The first term corresponds to the variance of the photon number of the light beam represented by operator $\hat{N}(t, \tau)$ scaled by the square of quantum efficiency while the second contribution is the partition noise that arises due to the random selection of the fraction of the incident photons by the imperfect photon detector (see also Sect. 4.3.2).

One may note that the photon count distribution can also be expressed when the radiation field is treated quantum mechanically as

$$\varrho_n(t, \tau) = \left\langle : \frac{[\eta \hat{N}(t, \tau)]^n}{n!} \exp\left[-\eta \hat{N}(t, \tau)\right] : \right\rangle. \tag{4.242}$$

Mark that Eq. (4.242) has the same form as the corresponding semiclassical treatment of the same phenomenon as provided in Sect. 2.3.3 (compare specially with Eq. (2.136)).

4.5.2 Homodyne Detection

In the homodyne detection, the expectation value of the electric field or quadrature operators of the incident light as a function of the phase is sought for. The light beam to be measured—denoted by an operator $\hat{a}(t)$—is presumed to be impinged onto one of the arms of the interferometric type setups while another strong coherent light beam commonly known as local oscillator—denoted by operator $\hat{a}_L(t)$—is impinged onto the other coupler arm. The local oscillator is required pertaining to the fact that the signals or the states to be measured have a very low chance to be detected directly by the photon detector against the emerging noise.[18] A passive beam splitter in this mechanism is thus employed to combine the input signal with the local oscillator field where the dominant signal carrying term from each of the two beam splitters turns out to be proportional to the mean of the quadrature of the input signal field.

In a one-port homodyne scheme for example a photon detector is used to monitor the intensity from one output port of the beam splitter, whereas in a two-port homodyne scheme the intensity from both output ports of a 50:50 beam splitter is

---

[18]The local oscillator is assumed to have the carrier frequency ω_c and the signal beam sideband frequencies to be centered around the carrier frequency ($\omega_c \pm \omega_s$). Both the signal and local oscillator are often assumed to have narrow bandwidth so that the time dependent form of the operators can be utilized [27].

monitored. In the latter case, the output can be analyzed via subtracting the outputs of the two detectors. Homodyne detection is hence believed to be a powerful method for measuring a phase sensitive properties of traveling optical fields [28].

In case the mixture at the beam splitter is impinged onto photon detectors, the observed interference fringe that varies with the difference of the phase between the two fields is supposed to reflect the quantum statistics of the signal field, and so can be utilized under certain circumstances to quantify the degree of quantumness of the signal field as in quantum tomography. The homodyne output can therefore be expressed in terms of the joint event probability distribution of the detectors in the output channels in which the difference event statistics measurable by a four-port homodyne detector can be envisaged to yield the quadrature component statistics of the signal field expected to offer a chance for carrying out quantum state measurement. Homodyne detection mechanism can also be realized by a nonlinear mixing of the signal field with the reference field in which the pertinent measurement amounts to taking the reference radiation as derived from the same source as the signal prior to the modulation process in the nonlinear medium. The scattered light modulated by the nonlinear medium afterward would be mixed with the local oscillator on the detector with the help of a passive beam splitter.

The nonclassical features of a radiation can then be captured by taking the intensity of the radiation in one of the arms: the procedure commonly known as ordinary homodyne detection. This conception directly corresponds to the assumption that the signal radiation is readily transmitted through the beam splitter: $T \gg R$. The mean photon number counted by one of the detectors, let us say, in the second arm can thus be expressed as

$$\langle \hat{n}_2 \rangle = \langle \hat{d}_2^\dagger \hat{d}_2 \rangle. \tag{4.243}$$

Then after, upon assuming the intensity of the local oscillator to be high ($\hat{a}_L \to |\beta|$ and $\langle \hat{a}^\dagger \hat{a} \rangle \ll |\beta|^2$), one can write

$$\langle \hat{n}_2 \rangle \simeq (1 - \eta)|\beta|^2 + \sqrt{\eta(1 - \eta)}|\beta|\langle \hat{\chi}(\phi) \rangle, \tag{4.244}$$

where $\hat{\chi}(\phi)$ is the quadrature operator taken to be $\hat{\chi} = i(\hat{a}^\dagger - \hat{a})$.

In the same way, the variance of the photon number can be put in the form

$$\Delta^2 n_2 \simeq (1 - \eta)|\beta|^2 \big[1 - \eta + \eta \Delta^2 \chi(\phi) \big]. \tag{4.245}$$

One can see from Eq. (4.245) that the noise in the intensity of the photon number is related to the noise of the quadrature operator, and so the squeezing can be adjusted by varying the phase of the coherent light.

Following similar reasoning, the variance of the photon difference as in balanced homodyne detection is found to be

$$\Delta^2 n_- \simeq |\beta|^2 \Delta^2 \chi(\phi). \tag{4.246}$$

Since one does not have a privilege of operating a device with $T = 1$, it might be good to note that the oscillator shot noise and the excess noise that enter through the reflectivity of the beam splitter cannot be suppressed in an ordinary homodyne detection. the local oscillator noise hampers the outcome of the measurement. To overcome such setback, one may need to look for the two-port homodyne detection in which the noninterference term can be made to cancel each other. Such a scheme owes its success to the fact that the interference term between the local oscillator and the input signal fields acquires opposite sign to the two outputs of the beam splitter but the noninterference terms take up the same sign. For a 50:50 beam splitter, subtracting the two outputs retains the interference term.

This scheme is also expected to permit a direct observation of one of the quadratures of the input signal without placing much rigorous demand on the local oscillator performance as the conventional one-port homodyning requires. The two-port scheme is, therefore, expected to provide a more viable practical means of capturing the phenomenon of squeezing since the output noise can be made to be insensitive to the local oscillator quadrature noise. It might then be essential looking into the balanced homodyne detection mechanism, which is a two-port homodyne detection where the signal field would be combined with a strong coherent field on a 50:50 beam splitter (see Fig. 4.6). One may recap that the difference of the photon counts in the two outputs of the beam splitter would be proportional to the field quadrature. So upon performing a balanced homodyne detection measurement for different values of the phase of the local oscillator, it is believed that one can get a complete information about the quantum state of the signal via reconstructing the density operator in various representations from which quantum tomography can be inferred.

From technical point of view, a balanced two-port homodyne detection represents a genuine procedure for measuring the amplitude as well as the phase of optical signal by making use of a pair of intensity sensitive photon detectors. The photocurrent difference obtained from the photon detectors placed at both output ports of the beam splitter carries information about the amplitude or phase quadrature of the input signal depending on the relative phase difference between the measured

Fig. 4.6 Schematic representation of balanced homodyne detection in which the detectors are placed in both output arms and the quantity to be measured is the difference between the number of photons arriving at the two detectors during the measurement time

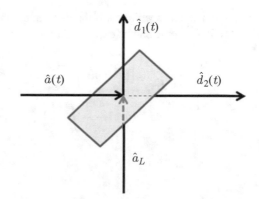

state and local oscillator. It is so envisioned that sampling the desired values of the photocurrent difference during repeated measurements yields the probability distribution of the quadratures. It is hence expected that the required information about the quantum state of the measured optical signal can be inferred from the measured data in this way. Even though a balanced homodyne detection was initially intended to detect squeezing property of light, it is found to be applicable for other experiments such as complete characterization of quantum states by means of quantum tomography, establishment of EPR type quantum correlations, implementation of continuous variable quantum teleportation, and demonstration of the nonclassical features of the electromagnetic fields in the realm of cavity quantum electrodynamics (see Chap. 11).

The difference of the number of photons that arrives at the two detectors can be expressed as

$$\hat{N}_-(t, \tau) = \int_t^{t+\tau} [\hat{a}_1^\dagger(t')\hat{a}_1(t') - \hat{a}_2^\dagger(t')\hat{a}_2(t')]dt', \tag{4.247}$$

where

$$\hat{a}_1(t) = R\hat{a}(t) + T\hat{a}_L(t), \qquad \hat{a}_2(t) = R\hat{a}_L(t) + T\hat{a}(t) \tag{4.248}$$

(compare with Eqs. (4.177)), and the detectors are taken to have the same quantum efficiency that can be modeled by an inefficient beam splitter.

The quantity to be measured thus can be described by the operator of the form

$$\hat{N}_H(t, \tau) = \int_t^{t+\tau} [\hat{d}_1^\dagger(t')\hat{d}_1(t') - \hat{d}_2^\dagger(t')\hat{d}_2(t')]dt' \tag{4.249}$$

in which

$$\hat{d}_1(t) = \sqrt{\eta}\hat{a}_L(t) + i\sqrt{1-\eta}\hat{a}(t), \qquad \hat{d}_2(t) = \sqrt{\eta}\hat{a}(t) + i\sqrt{1-\eta}\hat{a}_L(t). \tag{4.250}$$

So the mean and variance of the difference photon count for a balanced homodyne detection scheme can be expressed as

$$\langle \hat{n}_- \rangle = i \int_t^{t+\tau} \left[\langle \hat{a}^\dagger(t')\hat{a}_L(t') \rangle - \langle \hat{a}_L^\dagger(t')\hat{a}(t') \rangle \right] dt', \tag{4.251}$$

$$\Delta n_-^2 = \frac{1}{4}\Delta N_-^2(t, \tau) - \frac{1}{4}\int_t^{t+\tau} \left[\langle \hat{a}_L^\dagger(t')\hat{a}_L(t') \rangle + \langle \hat{a}^\dagger(t')\hat{a}(t') \rangle \right] dt'. \tag{4.252}$$

Imagine the local oscillator to be a single-mode beam of a coherent light with frequency ω_L chosen to coincide with signal frequency. The operator that describes the local oscillator can be taken as a c-number variable that evolves in time as

$$\alpha_L(t) = |\beta| \exp\left[-i(\omega_L t - \phi_L)\right], \tag{4.253}$$

where ϕ_L is the phase of the local oscillator. In this case, one can write

$$\langle \hat{n}_- \rangle = i|\beta| \int_t^{t+\tau} \left[\langle \hat{a}^\dagger(t') \exp(-i(\omega_L t' - \phi_L)) \rangle - \langle \hat{a}(t') \exp(i(\omega_L t' + \phi_L)) \rangle\right] dt'. \tag{4.254}$$

To connect the homodyne detection mechanism with accessible properties of the radiation, it thus appears useful to define the homodyne electric field operator as

$$\hat{E}_H(\varphi, t, \tau) = \int_t^{t+\tau} \left[\hat{a}^\dagger(t') \exp(-i(\omega_L t' - \varphi)) + \hat{a}(t') \exp(i(\omega_L t' + \varphi))\right] dt', \tag{4.255}$$

where $\varphi = \phi_L + \pi/2$.

In case the balanced homodyne detection is preferred to capture the squeezing properties of the radiation, $\pi/2$ should be introduced to ensure that the quadrature operators are $\pi/2$ out of phase. Critical observation may reveal that the electric field operator can represent the quadrature operators of the radiation field in case an appropriate phase is picked. We then express the mean difference of the photon count with the aid of the definition of the electric field operator as

$$\langle \hat{n}_- \rangle = |\beta|\langle \hat{E}_H(\varphi, t, \tau) \rangle. \tag{4.256}$$

With the introduction of $\pi/2$ in Eq. (4.255), note that the i in Eq. (4.251) can be absorbed. In light of this, the variance is found for a case when the local oscillator flux is much greater than the signal flux approximately to be

$$\Delta n_-^2 = |\beta|^2 \left[\Delta E_H^2(\varphi, t, \tau) + \frac{1}{2}\right]. \tag{4.257}$$

Equation (4.257) indicates that the detection of the squeezed state requires a phase sensitive scheme that measures the variance of the quadrature of the field, and the squeezing properties of the signal radiation can be explored by varying the phase of the oscillator. A balanced homodyne detection was originally designed for measurement in the frequency domain, that is, when both signal to be measured and local oscillator were obtained from a continuous wave sources and the photocurrent difference was analyzed with the help of a spectrum analyzer. Particularly, a narrow frequency window where the experimental noises can be properly minimized as in vacuum, coherent and squeezed states would be selected. In the same spirit, a complete characterization of Fock state containing one or two photons, single-photon added coherent state, single-photon subtracted squeezed state and Schrödinger cat state has been successfully explored by using homodyne detection [29]. One of the challenges related to an optical measurement is the reconstruction of the full information of the quantum state of the light. The maximum amount of information in actual setting that can be inferred from the statistics of the photoelectric count is the photon number distribution. Such a setback in part can be related to the underlying fact that the direct measurement in an optical process requires idealized photon detector with a perfect quantum efficiency, which is not feasible in a realistic experimental setting.

A genuine quest to discover a viable approach in this respect may lead to the unbalanced homodyne detection mechanism. As one can infer from Fig. 4.7, the beam splitter combines the signal field with the local oscillator field in which the superposed light can be expressed as

$$\hat{a}_h = T\hat{a} + R\hat{b}, \tag{4.258}$$

where $\hat{a}$, $\hat{b}$ and $\hat{a}_h$ are the annihilation operators of the signal, local oscillator and superposed or homodyne field. On account of Eq. (4.258), the photon number operator of the homodyne light field takes the form

$$\hat{n}_h = |T|^2 \left(\hat{a} + \frac{R}{T}\hat{b} \right)^{\dagger} \left(\hat{a} + \frac{R}{T}\hat{b} \right). \tag{4.259}$$

The anticipated reconstruction of the quantum states of light can also be linked to the flexibility offered by the s-parameterized quasi-statistical distributions that can be obtained by summing up the measured counting statistics with appropriate weighting factors [30].

Fig. 4.7 Schematic representation of unbalanced homodyne detection in which the detector is placed in front of one of the output arms. Note that this scheme is the same as direct photon detection except that $\hat{b}$ represents a strong local oscillator

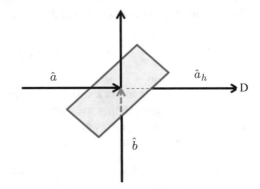

In connection to this, the probability ϱ_m of recording m counts with a photon detector of quantum efficiency ς can be expressed as

$$\varrho_m = \left\langle : \frac{(\varsigma \hat{n})^m e^{-\varsigma \hat{n}}}{m!} : \right\rangle \qquad (4.260)$$

(relate with Eqs. (2.136) and (4.242)). Then assuming the local oscillator to be prepared in coherent state $|\beta\rangle$ (the signal is represented by coherent state $|\alpha\rangle$), it is possible to rewrite the photon counting statistics as

$$\varrho_m(\alpha; \varsigma) = \left\langle : \frac{(\varsigma \hat{N}(\alpha))^m e^{-\varsigma \hat{N}(\alpha)}}{m!} : \right\rangle, \qquad (4.261)$$

where $\hat{N}(\alpha) = \hat{D}(\alpha)\hat{a}^\dagger \hat{a} \hat{D}^\dagger(\alpha)$ is the photon number operator of the displaced signal field. So the coherent amplitude α can be modified as $\alpha = -R\beta/T$ and the overall quantum efficiency η as $\eta = \varsigma |T|^2$. Markedly, in order to keep the overall efficiency η as large as possible, the beam splitter is expected to fulfill the condition that $|R|^2 \ll 1$.

To relate the statistics measured in this scheme to the quantum state of the signal field in terms of the s-parameterized quasi-statistical distributions, one can begin with

$$P(\alpha; s) = \frac{2}{\pi(s'-s)} \int d^2\beta \exp\left[-\frac{2|\alpha - \beta|^2}{s'-s}\right] P(\beta; s'), \qquad (4.262)$$

where the s-parameterized distribution can be expressed in terms of another distribution with parameter s' provided that $s < s'$ and

$$s = \begin{cases} 0, & \text{Winger function;} \\ -1, & Q\text{-function;} \\ 1, & P\text{-function.} \end{cases}$$

For example, taking $s' = 1$ indicates that the right side of Eq. (4.262) can be rewritten as normally ordered expectation value of the form

$$P(\alpha; s) = \frac{2}{\pi(1-s)} \left\langle : \exp\left[-\frac{2\hat{N}(\alpha)}{1-s}\right] : \right\rangle \qquad (4.263)$$

in which a comparison of Eq. (4.261) with (4.263) reveals

$$P(\alpha; s) = \frac{2}{\pi(1-s)} P_0(\alpha; 2/(1-s)), \qquad (4.264)$$

where the efficiency of a virtual photon detector is taken to be $2/(1-s)$ [31].

The exponent in Eq. (4.263) can be decomposed into the sum of the term $-\eta\hat{N}(\alpha)$ that encompasses the physical detector of efficiency η and the remaining term as

$$P(\alpha; s) = \frac{2}{\pi(1-s)} \left\langle : \exp\left[-\frac{(2-\eta(1-s))\hat{N}(\alpha)}{1-s} \right] e^{-\eta\hat{N}(\alpha)} : \right\rangle \qquad (4.265)$$

so that expanding in a power series results

$$P(\alpha; s) = \frac{2}{\pi(1-s)} \sum_{n=0}^{\infty} \left[-\frac{(2-\eta(1-s))\hat{N}(\alpha)}{\eta(1-s)} \right]^2 \rho_n(\alpha; n). \qquad (4.266)$$

This discussion may evince the possibility of obtaining the quasi-statistical distribution functions as a weighted sum over the homodyne counting distributions $\rho_n(\alpha; n)$. This could be taken as one of the simplest schemes that allows to reconstruct the full quantum statistical information from the measured data; and yet, its validity is limited by the involved noise.

4.5.3   Heterodyne Detection

The other commonly used detection mechanism is a heterodyne detection in which the reference local oscillator field consists of sidebands at smaller frequencies with respect to the carrier frequency. In the process of heterodyne detection, a signal of interest at some frequency is nonlinearly mixed with the reference local oscillator set at very close frequency to the signal,[19] and the desired outcome would be inferred from the difference of these frequencies in the same way as in balanced homodyne detection. One may note that the difference in the frequencies carries the information such as amplitude, phase and frequency modulation of the original higher frequency signal but oscillates at a lower carrier frequency that might be easily processed. Since optical frequencies oscillate too rapidly to directly measure the electric field, such a procedure can only reveal the magnitude but not the phase of the electric field. The primary purpose of a heterodyne detection can hence be viewed as a means of converting the signal of the optical band to an experimentally tractable frequency range. Notably, the desired nonlinearity needs to be embedded in the photon absorption process.

In the context that a conventional light detector—which is usually known to be a square law detector—responds to the photon energy of the free electrons, it is straightforward to propose that since the energy flux scales as the square of the

---

[19]The main distinction between the homodyne and heterodyne detections as the names suggest is based on whether the frequency of the signals in the two input ports are the same or different [32].

electric field so does the rate at which the electrons are freed. Prevailing practical experience nonetheless shows that it is the difference frequency that appears in the detector output current when both the local oscillator and the signal fall on the detector at the same time. It may not be hard this to realize that this outcome causes the square of the combined fields to possess a cross term interpreted often as difference frequency modulation. Thus in this case the optical phase of the signal beam is shifted by a certain angle, the phase of the difference frequency is also shifted by the same angle.

To expose the condition required to witness an optical phase shift, one needs to have a common time reference: the signal beam can be generated from the same laser as the local oscillator but can be shifted by some frequency modulator. As long as the modulation source maintains a constant offset phase between the local oscillator and signal source, any added optical phase shifts overtime arising from external modification should be added to the phase of the difference frequency, and thus measurable. So one of the primary sources of the noise is the photon shot noise resulting from the nominally constant level of the local oscillator on the optical detector. Under normal condition, since the shot noise scales as amplitude of the local oscillator electric field, the heterodyne gain also scales in the same way but the ratio of the shot noise to the mixed signal would be constant no matter how large the local oscillator would be. Hence in the heterodyne scheme, an optical field would be combined with the help of a beam splitter on the surface of the photon detector with a strong local oscillator whose frequency ω_0 is offset by an amount $\Delta\omega$ from that of the signal mode ($\Delta\omega \ll \omega_0$). The measured photocurrent would then be filtered in order to select the complex valued component at frequency $\Delta\omega$. The classical statistics of this component corresponds, under certain circumstances, to the quantum statistics of the two-mode operator; $\hat{A} = \hat{a}_S + \hat{a}_I$, where the subscripts S and I are used to denote the signal mode at frequency $\omega_0 + \Delta\omega$ and the imaging mode at frequency $\omega_0 - \Delta\omega$ that can be used to probe the signal mode.

References

1. G.S. Buller, R.J. Collins, Meas. Sci. Technol. **21**, 012002 (2010); M.D. Eisaman, J. Fan, A. Migdall, S.V. Polyakov, Rev. Sci. Instrum. **82**, 071101 (2011)
2. R.J. Glauber, Phys. Rev. **130**, 2529 (1963)
3. M. Born, P. Jordan, Z. Phys. **34**, 858 (1925); W. Heisenberg, Z. Phys. **43**, 172 (1927) (English version: W.A. Fedaka, J.J. Prentis, Am. J. Phys. **77**, 128 (2009))
4. V.B. Braginsky, Y.I. Vorontsov, K.S. Thorne, Science **209**, 547 (1980); C.M. Caves et al., Rev. Mod. Phys. **52**, 341 (1980)
5. C.S. Yu, H. Zhao, Phys. Rev. A **84**, 062123 (2011); G. Adesso, T.R. Bromley, M. Cianciaruso, J. Phys. A **49**, 473001 (2016)
6. M. Schlosshauer, Rev. Mod. Phys. **76**, 1267 (2005)
7. D.J. Starling et al., Phys. Rev. A **80**, 041803 (2009); X.Y. Xu et al., Phys. Rev. Lett. **111**, 033604 (2013); Y. Aharonov, E. Cohen, A.C. Elitzur, Phys. Rev. A **89**, 052105 (2014)
8. J. Dressel et al., Rev. Mod. Phys. **86**, 307 (2014)
9. W.H. Zurek, Phys. Rev. D **24**, 1516 (1981); W.H. Zurek, ibid. **26**, 1862 (1982); A.M. Burke et al., Phys. Rev. Lett. **104**, 176801 (2010)

10. W.G. Unruh, Phys. Rev. D **19**, 2888 (1979); M.J. Holland et al., Phys. Rev. A **42**, 2995 (1990); J.P. Polzat, M.J. Collett, D.F. Walls, ibid. **45**, 5171 (1992); H.A. Bachor, T.C. Ralph, *A Guide to Experiments in Quantum Optics*, second, revised and elarged edition (Wiley-VCH Verlag GmbH and Co. KGaA, Weinheim, 2004); M.F. Bocko, R. Onofrio, Rev. Mod. Phys. **68**, 755 (1996); D.F. Walls, G.J. Milburn, *Quantum Optics*, 2nd edn. (Springer, Berlin, 2008)
11. P. Grangier, J.A. Levenson, J.P. Poizat, Nature **396**, 537 (1998)
12. N. Imoto, H.A. Haus, Y. Yamamoto, Phys. Rev. A **32**, 2287 (1985)
13. G. Nogues et al., Nature **400**, 239 (1999)
14. R. Ghosh, L. Mandel, Phys. Rev. Lett. **59**, 1903 (1987); Z.Y. Ou, L. Mandel, ibid. **62**, 2941 (1989); P.R. Berman, A. Kuzmich, Phys. Rev. A **101**, 033824 (2020)
15. S. Tesfa, J. Mod. Opt. **54**, 1759 (2007)
16. Z. Ficek, S. Swain, *Quantum Interference and Coherence: theory and experiments* (Springer Science+Business Media, New York, 2005); K.H. Kagalwala et al., Nature Photonics **7**, 72 (2013); F. De Zela, Phys. Rev. A **89**, 013845 (2014); X.F. Qian et al., Phys. Rev. Lett. **117**, 153901 (2016)
17. T. Baumgratz, M. Cramer, M.B. Plenio, Phys. Rev. Lett. **113**, 140401 (2014); J. Xu, Phys. Rev. A **93**, 032111 (2016); A. Streltsov, G. Adesso, M.B. Plenio, Rev. Mod. Phys. **89**, 041003 (2017)
18. A.A. Michelson, E.W. Morely, Am. J. Sci. **34**, 333 (1887); L. Onsager, Phys. Rev. **37**, 405 (1931); R.H. Brown, R.Q. Twiss, Phil. Mag. **45**, 663 (1954); R.H. Brown, R.Q. Twiss, Nature **178**, 1046 (1956); M. Lax, Phys. Rev. **129**, 2342 (1963); W.H. Louisell, *Quantum Statistical Properties of Radiation* (Wiley, New York, 1973)
19. M. Hillery, M.S. Zubairy, Phys. Rev. A **26**, 451 (1982); M. Hillery, M.S. Zubairy, ibid. **29**, 1275 (1984)
20. M. Lahiri, E. Wolf, Phys. Rev. A **82**, 043837 (2010)
21. M.J. Collett, C.W. Gardiner, Phys. Rev. A **30**, 1386 (1984); M.J. Collett, D.F. Walls, Phys. Rev. Lett. **61**, 2442 (1988)
22. R.W. Boyd, *Nonlinear Optics*, 3rd edn. (Elsevier Inc., 2008)
23. P. Meystre, M. Sargent III, *Elements of Quantum Optics*, 4th edn. (Springer, Berlin, 2007)
24. R.L. de Matos Filho, W. Vogel, Phys. Rev. A **54**, 4560 (1996); M. Dakna et al., ibid. **55**, 3184 (1997); V.V. Dodonov et al., ibid. **58**, 4087 (1998); M. Bellini, A. Zavatta, Progress Optics **55**, 41 (2010); V. Averchenko et al., New J. Phys. **18**, 083024 (2016); S.M. Barnett et al., Phys. Rev. A **98**, 013809 (2018)
25. P. Malpani et al., Opt. Commun. **459**, 124364 (2019)
26. L. Mandel, Proc. Phys. Soc. **72**, 1037 (1958)
27. B.L. Schumaker, Opt. Lett. **9**, 189 (1984); R. Loudon, *The Quantum Theory of Light*, 3rd edn. (Oxford University Press, New York, 2000)
28. B. Kühn, W. Vogel, Phys. Rev. A **98**, 013832 (2018)
29. G.L. Abbas, V.W.S. Chan, T.R. Yee, Opt. Lett. **8**, 419 (1983); H. Hansen et al., ibid. **26**, 1714 (2001)
30. S. Wallentowitz, W. Vogel, Phys. Rev. A **53**, 4528 (1996)
31. K.E. Cahill, R.J. Glauber, Phys. Rev. **177**, 1882 (1969)
32. H.P. Yuen, V.W.S. Chan, Opt. Lett. **8**, 177 (1983)

Part II
Atom-Radlation Interactions

Atom-Radiation Interaction in Free Space

5

What one observes in a measurement process and subsequent interpretation of the reading of the apparatus are expected to concur with what one knows about the interaction of light with matter where seeing and photosynthesis can be taken as prime illustrations of how such interaction process needs to be perceived. Even though the general understanding of the issue this much is known for quite sometime, a systemic study of the interaction of the atom with radiation in the quantum realm has begun around 1920s in which the interaction is related to the absorption and emission of the photon by the electron of the atom—the perception very closely linked to elastic scattering [1]. The electron in this presumption is taken to radiate at exactly the same frequency as pumping radiation or to undergo a coherent scattering. The incoherently scattered radiation can also emanate from many photon processes where the incoherent part dominates specially when a strong driving laser field is utilized.

One may need to underline from the outset that the understanding drawn from light-matter interaction at the most basic level can differ based on how one treats the light and matter. There can be three alternatives from which a significantly different predictions can follow, that is, when both of them are treated classically, when one of them is treated quantum mechanically (semiclassically) and when both of them are treated quantum mechanically. For instance, in the absence of damping, the semiclassical theory leads to oscillation in the atomic inversion while the quantum theory to collapse and revival. Even then, we opt to restrict the discussion in this chapter to a semiclassical regime (for quantum mechanical treatment; see Chap. 6).

To begin with, one may need to emphasize that objects naturally prefer to go to the lowest energy state accessible to them unless impeded by external agent, that is, if an excited atom is left alone, it emits a photon and returns to the ground state—a process dubbed as de-excitation. An atomic excitation is incidently found to occur instantly but the de-excitation process takes some measurable time that depends on the inherent trait of the excited state and the nature of the interaction. The average

© The Author(s), under exclusive license to Springer Nature Switzerland AG 2020
S. Tesfa, *Quantum Optical Processes*, Lecture Notes in Physics 976,
https://doi.org/10.1007/978-3-030-62348-7_5

time the atom occupies a certain excited energy level which ranges from a billionth of a second to few milliseconds is referred to as atomic lifetime of that level.

The information about the interaction of the atom with radiation can generally be inferred from the pertinent state vector

$$|\Psi(t)\rangle = \sum_j C_j(t)|j\rangle, \tag{5.1}$$

where j stands for the atomic energy level and $C_j(t)$ for the probability amplitude of the occupancy of the jth energy level. Such representation envisages that the dynamics of the emitted radiation can be explored by solving the Schrödinger equation for the wave function or Heisenberg equation for operators (see Sect. 5.2). Once the evolution of the wave function is known, it is expected that the interaction of the atom with external coherent light can be analyzed. We hence seek to examine the radiative properties of the atom and photon statistics of the emitted radiation when the atom is placed in a free space.

Although a given atom can occupy an infinite number of energy levels, it is practically a formidable task to take the contribution of each energy level into consideration. With this in mind, the discussion would be solely confined to a two-level approximation as portrayed in Sect. 5.1. The focus is mainly to describe a free[1] two-level atom in terms of the probability amplitude, density operator, atomic operator, occupancy of each energy level and atomic coherence. Emphasis would also be given to properties of atomic operators that can be characterized by employing the dynamical equations and the possible transition mechanisms along with the physical phenomena that led to different atomic dampings (see Sect. 5.3). Once the evolution of the elements of the density matrix is determined, properties of atomic operator, evolution of the occupancy of each energy level in terms of amplitude, density matrix and Bloch equation, and properties of the emitted radiation in terms of absorption spectrum and squeezing would be discussed (see Sects. 5.4 and 5.5). Aiming at extending the discussion on a single two-level atom to many atoms, the response of dense atomic medium to a light propagating through it via polarization and susceptibility is also highlighted (see Sect. 5.6).

5.1   Description of Two-Level Atom

Owing to the assertions that the strength of the interaction of light with an atom would be significant near resonance and the energy gap between two distinct atomic

---

[1]The notion of a free atom does not mean that the atom is actually free from interaction as it might appear; rather, it is related to the situation in which the atom is placed in a free space and allowed to freely interact with the light that engulfs it. The concept of a free atom thus insinuates the circumstance when the atom is not placed in the cavity, and has no chance of absorbing what it emitted since the emitted light propagates away from the atom [2].

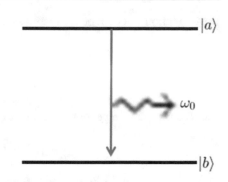

Fig. 5.1 Schematic representation of a two-level atom, where the upper and lower energy levels of the atom are denoted by states $|a\rangle$ and $|b\rangle$, and ω_0 represents the atomic frequency that corresponds to the energy gap between these levels

energy levels can be different from other allowable pairs, it might be admissible to propose that a single-mode radiation can couple only two such distinct levels of the atom meaningfully. One may as a result ignore the contribution of other atomic energy levels. In the two-level approximation scheme, it is therefore imagined as if out of the whole accessible energy levels of the atom two atomic states interact appreciably with the incident optical field, that is, an atom with a large number of allowable atomic energy states can be engineered to absorb and emit radiation with a fixed frequency out of the continuum (see Fig. 5.1). Nonetheless, a number of disparities might be observed while using a real atom to verify theoretical predictions that follow from this approximation in contrast to the need for intricate optical pumping procedure to prepare something close to a true two-level atom. Even though the expected substantial simplification in the involved mathematical rigor ignores the real features of the atomic system, and fails to provide a complete picture of atom-radiation interaction phenomenon, relentless research indicates that there is still enormous richness such as resonance fluorescence, Rabi oscillation and photon anti-bunching in the dynamical process that can be embodied within this approximation.

When a two-level atom is coupled to external field, the interest in quantum optics is often geared towards what happens to the field rather than the atom or the coupled system, which may require the application of the density matrix formalism. With this suggestion, the state vector that designates the unperturbed two-level atom can be defined as

$$|\psi(t)\rangle = C_a(t)|a\rangle + C_b(t)e^{i\varphi}|b\rangle, \tag{5.2}$$

where $C_a(t)$ and $C_b(t)$ are the probability amplitudes for finding the atom in respective quantum states and φ is the relative random phase.[2] The corresponding

---

[2]Equation (5.2) represents the superposition of the upper and lower atomic energy levels as required in quantum information processing in connection with atomic qubits (see Chap. 9).

wave function can then be expressed as

$$\psi(\mathbf{r}, t) = C_a(t)\mathbf{u}_a(\mathbf{r})e^{-i\omega_a t} + C_b(t)\mathbf{u}_b(\mathbf{r})e^{i(\varphi - \omega_b t)}, \tag{5.3}$$

where $\mathbf{u}_a(\mathbf{r})$ and $\mathbf{u}_b(\mathbf{r})$ designate the atomic configurations in space.

The probability of getting the two-level atom in a coherent superposition of the states i and j can be generally defined as

$$\rho_{ij}(t) = C_i(t)C_j^*(t), \tag{5.4}$$

where $i, j = a, b$. In light of the nature of the elements of this density matrix, it is possible to interpret the diagonal element ρ_{ii} as the probability for the two-level atom to occupy the energy eigenstate i and commonly known as atomic population and the off-diagonal element ρ_{ij} as atomic coherence;

$$\begin{pmatrix} \rho_{aa} & \rho_{ab} \\ \rho_{ba} & \rho_{bb} \end{pmatrix} = \begin{pmatrix} C_a C_a^* & C_a C_b^* \\ C_b C_a^* & C_b C_b^* \end{pmatrix}, \tag{5.5}$$

where $\rho_{ab} = \rho_{ba}^*$.

5.1.1   Interaction Hamiltonian

Upon taking the atomic states to be complete and orthonormal, the unperturbed Hamiltonian required to study the atom can be expressed as

$$\hat{H}_0 = \sum_{i,j} |i\rangle\langle i|\hat{H}_0|j\rangle\langle j| = \sum_{i,j} \langle i|\hat{H}_0|j\rangle|i\rangle\langle j|, \tag{5.6}$$

which reduces for a two-level atom to

$$\hat{H}_0 = E_a\hat{\sigma}_{aa} + E_b\hat{\sigma}_{bb}, \tag{5.7}$$

where $\hat{\sigma}_{ij} = |i\rangle\langle j|$ stands for the atomic operator and $\hat{H}_0|i\rangle = E_i|i\rangle$ is the energy eigenvalue equation of respective atomic energy levels with $\langle i|j\rangle = \hat{I}\delta_{ij}$.

Without further assumption, it turns out to be appealing to utilize $\hat{\sigma}_{ii} = (\hat{\sigma}_{ii} + \hat{\sigma}_{ii})/2$, which can also be put with the help of atomic completeness relation,

$$\hat{\sigma}_{aa} + \hat{\sigma}_{bb} = \hat{I}, \tag{5.8}$$

in the form

$$\hat{\sigma}_{aa} = \frac{1}{2}[\hat{\sigma}_{aa} + \hat{I} - \hat{\sigma}_{bb}], \tag{5.9}$$

$$\hat{\sigma}_{bb} = \frac{1}{2}[\hat{\sigma}_{bb} + \hat{I} - \hat{\sigma}_{aa}]. \tag{5.10}$$

The free atom Hamiltonian then can be expressed as

$$\hat{H}_0 = \frac{\omega_0 \hat{\sigma}_z}{2} + \left[\frac{E_b + E_a}{2}\right]\hat{I}, \tag{5.11}$$

where

$$\omega_0 = E_a - E_b, \qquad \hat{\sigma}_z = \hat{\sigma}_{aa} - \hat{\sigma}_{bb} \tag{5.12}$$

are dubbed as atomic transition (natural) frequency and energy operator. In line with the suggested eigenvalue equation with $E_i = \langle i|\hat{H}_0|i\rangle$, one can see that

$$E_a = \frac{\omega_0}{2} + \frac{E_a + E_b}{2}, \qquad E_b = -\frac{\omega_0}{2} + \frac{E_a + E_b}{2}. \tag{5.13}$$

Combination of Eqs. (5.12) and (5.13) leads to $E_a + E_b = 0$, which can be interpreted as the way of shifting the zero value energy to the middle of the two energy levels. As a result,

$$\hat{H}_0 = \frac{1}{2}\omega_0 \hat{\sigma}_z. \tag{5.14}$$

Once the Hamiltonian for a free two-level atom is known, it might be compelling to turn to the interaction of the atom with radiation. To do so, one may start with common knowledge that in case the oscillating string is attached to a pendulum as in the coupled pendulum, the energy can be transferred to the pendulum, and subsequently from the pendulum to the string repeatedly for so many times while the coupled pendulum-string system lose energy overtime; and hence, the oscillation would be damped. The interaction of the atom (pendulum) with radiation field (string) can be perceived metaphorically in this way with the emphasis that the atom-radiation interaction requires quantum treatment. When the atom is in contact with external radiation three distinct physical processes are expected: stimulated absorption, and stimulated and spontaneous emissions. The overall property of or relationship among these phenomena is governed by the conservation of energy and momentum. For example, in stimulated emission, the light should be emitted in the same direction as the absorbed light so that the total net momentum is zero, spontaneously emitted light however should not be directed in any particular direction.[3]

Hoping to ease the rigor, we opt to take into consideration one electron atom for which the contribution of the spin is neglected. Besides, even though the atom should undergo motion due to the accompanying influences, one can still imagine the atom to be stationary or motionless. Even then, the motion of the electron is presumed to be relative to the center of the frame of reference of the mass situated at the center of the atom. One may recall that the Lorentz force for the charged particle moving with velocity $\mathbf{v}$ in the electromagnetic field that can be expressed in terms of the scalar potential $\phi(\mathbf{r}, t)$ and vector potential $\mathbf{A}(\mathbf{r}, t)$ is given by

$$\mathbf{F} = q\left[-\nabla\phi - \frac{\partial\mathbf{A}}{\partial t} + \mathbf{v}X(\nabla X\mathbf{A})\right], \tag{5.15}$$

where q is the charge of the particle. Following a straightforward algebra and upon assuming that the velocity is independent of the position vector, one can find

$$\mathbf{v}X(\nabla X\mathbf{A}) = \nabla(\mathbf{v}.\mathbf{A}) - (\mathbf{v}.\nabla)\mathbf{A}, \qquad \frac{d\mathbf{A}}{dt} = \frac{\partial\mathbf{A}}{\partial t} + (\mathbf{v}.\nabla)\mathbf{A}. \tag{5.16}$$

In this regard, the Lorentz force can be put in the form

$$\mathbf{F} = q\left[-\nabla\phi + \nabla(\mathbf{A}.\mathbf{v}) - \frac{d\mathbf{A}}{dt}\right]. \tag{5.17}$$

Since the scalar and vector potentials do not depend on velocity, one can see that

$$\frac{\partial}{\partial\mathbf{v}}\left[c\phi(\mathbf{r}, t) - \mathbf{v}.\mathbf{A}(\mathbf{r}, t)\right] = -\mathbf{A}(\mathbf{r}, t), \tag{5.18}$$

from which follows

$$\frac{d\mathbf{A}}{dt} = -\frac{d}{dt}\frac{\partial}{\partial\mathbf{v}}\left[c\phi(\mathbf{r}, t) - \mathbf{v}.\mathbf{A}(\mathbf{r}, t)\right]. \tag{5.19}$$

It might not be hard then to verify the generalized electric potential energy to be

$$U = q[\phi - \mathbf{v}.\mathbf{A}], \tag{5.20}$$

---

[3]Even though it is relatively simpler to understand interaction of the atom with radiation field based on Einstein's consideration, such treatment does not necessary lead to light amplification and aid in witnessing the nonclassical properties of the emitted light [3].

and the classical Lagrange function $\mathcal{L} = T - U$ to be expressed afterwards as

$$\mathcal{L} = \frac{mv^2}{\lambda} - q\phi + q\mathbf{v}.\Lambda. \tag{5.21}$$

Consistent with this, the canonical momentum $\mathbf{P} = \frac{\partial \mathcal{L}}{\partial \mathbf{v}}$ can be written in the form $\mathbf{P} = \mathbf{p} + q\mathbf{A}$, where $\mathbf{p} = m\mathbf{v}$ is the usual kinetic momentum.

The Hamiltonian that can be defined in terms of the Lagrange function as $H = \mathbf{v}.\frac{\partial \mathcal{L}}{\partial \mathbf{v}} - \mathcal{L}$ is thus found to be

$$H = \mathbf{v}.\mathbf{P} - \frac{P^2}{2m} + q\phi - q\mathbf{v}.\mathbf{A}, \tag{5.22}$$

where replacing $\mathbf{v}$ by $\mathbf{p}$ leads to

$$H = \frac{1}{2m}[\mathbf{P} - q\mathbf{A}]^2 + q\phi. \tag{5.23}$$

In case one neglects the term quadratic in the vector potential, the procedure usually referred to as minimum coupling approximation, the Hamiltonian takes the form

$$H = \frac{P^2}{2m} - \frac{q}{m}\mathbf{A}.\mathbf{P} + q\phi.$$

Earlier research in quantum electrodynamics unequivocally asserts that the term quadratic in the vector potential contains a product of the annihilation and creation operators of the radiation field, and so related to a two-photon process. But in the two-level system, a two-photon process is much less feasible than the one-photon process; and hence, the rational for the minimum coupling approximation. In another perspective, a one electron atom can be treated as a small electric dipole where the electron on the outer most shell and the proton in the nucleus are the required negative and positive charges separated by the distance in the order of Bohr's radius (see Fig. 5.2). Since the wavelength of the optical field is very large when compared to the size of the atom, it appears acceptable evaluating the interaction potential at the location of the nucleus instead of the electron, and taking the field as a constant over the dimension of the atom—which is the basis for electric dipole approximation.

The Hamiltonian that describes the atom-radiation coupled system for a one electron atom thus can be expressed as

$$H = H_0 + H_I \tag{5.24}$$

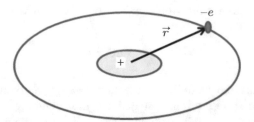

Fig. 5.2 Schematic representation of a one electron atom in which the excited electron $(-e)$ is assumed to be at $\mathbf{r}$ distance from the center of the atom. Treating the positive core of the atom and the electron as a dipole pair, the inherent atomic dipole moment is taken to be $\boldsymbol{\mu} = -e\mathbf{r}$

in which

$$H_0 = \frac{P^2}{2m} - e\phi, \qquad H_I = \frac{e}{m}\mathbf{P}.\mathbf{A}, \tag{5.25}$$

where H_0 designates the unperturbed classical Hamiltonian and H_I interaction Hamiltonian. In the minimum coupling formalism, it is admissible to expand atomic charge density to a multi-polar series and keep the term that corresponds to the dipole as portrayed in Fig. 5.2. The interaction Hamiltonian in this case can be expressed as

$$H_I = -e\mathbf{r}.\mathbf{E}. \tag{5.26}$$

Equation (5.26) so entails as if the electric dipole interacts with the electric field.

In the same token, the interaction between the atom and electromagnetic field can be interpreted in quantum realm as charge-field coupling, and described as exchange of quantized light between the atom and radiation. To pave the way for obtaining the time evolution of the density matrix, it might be interesting to start with quantum Hamiltonian. The interaction of a two-level atom with radiation thus can be represented by the Hamiltonian of the form

$$\hat{H} = \hat{H}_0 + \hat{V}(t), \tag{5.27}$$

where $\hat{V}(t)$ is the interaction Hamiltonian (denoted this way for ease of representation), and the free atom Hamiltonian satisfies

$$H_{0,nm} = E_{nm}\delta_{n,m} \tag{5.28}$$

with $H_{0,nm} = \langle n|\hat{H}_0|m\rangle$.

Quantum interaction Hamiltonian, as in the classical theory, can be expressed in the limit of electric dipole coupling[4] and in the minimum coupling formalism by

$$\hat{V}(t) = -\hat{\boldsymbol{\mu}}.\hat{\tilde{\mathbf{E}}}(t), \qquad (5.29)$$

where $\hat{\boldsymbol{\mu}}$ is the atomic dipole moment operator and $\hat{\tilde{\mathbf{E}}}(t)$ is the operator related to the slowly varying component of the electric field. Since the atom does not possess a permanent electric dipole moment in the present setup, one may see with the aid of Eqs. (5.29) and (5.32) that $V_{aa} = V_{bb} = 0$. The nonvanishing term of the interaction potential, upon taking the electric dipole moment of the electron to point along the direction of the external electric field, would then be

$$V_{ab} = V_{ba}^* = -\mu_{ab}\tilde{E}(t). \qquad (5.30)$$

Equation (5.30) epitomizes the interaction potential expected to substantially affect the details of the solution of the dynamical equations.

5.1.2 Properties of Atomic Operators

In the same way as the density matrix (5.5), a two-level atom can be expressed by the atomic dipole moment matrix as

$$\mu = \begin{pmatrix} \mu_{aa} & \mu_{ab} \\ \mu_{ba} & \mu_{bb} \end{pmatrix}, \qquad (5.31)$$

where

$$\mu_{ij} = \mu_{ji}^* = -e\langle i|\mathbf{r}|j\rangle \qquad (5.32)$$

in which $-e$ is the charge on and $\mathbf{r}$ is the position of the electron (see Fig. 5.2).

Based on the assumption that the atomic energy states $|a\rangle$ and $|b\rangle$ have a definite parity,[5] the diagonal elements of the atomic dipole moment matrix can be set to zero. This proposal envisages that the minimum coupling approximation is not suitable for the description of the atom that possesses a permanent dipole moment, that is, for $\mu_{ii} \neq 0$. It is thus possible to write

$$\langle \hat{\mu} \rangle = Tr(\hat{\rho}\hat{\mu}) = \rho_{ab}\mu_{ba} + \rho_{ba}\mu_{ab}, \qquad (5.33)$$

---

[4] Assuming such interaction Hamiltonian to be responsible for the involved atomic transition, the condition under which $V \neq 0$ is designated as a selection rule.

[5] For a one-electron atom, since the parity of the eigenfunction is even for even values and odd for odd values of the orbital quantum number, and the position vector is an odd function, its expectation value $\langle i|\mathbf{r}|j\rangle$ would be different from zero for atomic states of different parities [4].

from which one can infer that the off-diagonal elements of the density matrix are related to the mean electric dipole moment.

In addition, in view of the condition that the diagonal elements of the density matrix is related to the atomic population in each state and the sum of the probability of the population should be conserved, one can state that

$$\rho_{aa} + \rho_{bb} = 1. \tag{5.34}$$

Once the population of each energy level is known, it should be straightforward to define the atomic population inversion as

$$W = \rho_{aa} - \rho_{bb}, \tag{5.35}$$

which can also be put in the form

$$W = 2\rho_{aa} - 1. \tag{5.36}$$

One can see that if the probability of having the atom in the upper energy level is greater than 50%, Eq. (5.36) would be positive which entails that the population would be reversed from what is expected for a free atom; and hence, the notion of atomic population inversion.

Before directly going into analyzing the interaction of a two-level atom with light, it appears compelling to explore the commutation relation and other properties of the atomic operators. To this end, since the atomic and radiation variables do not correlate before commencement of the interaction, one can begin with the proposal that they do not correlate whenever they are in the same picture and time:

$$[\hat{a}, \hat{\sigma}_i] = [\hat{a}^\dagger, \hat{\sigma}_i] = 0, \tag{5.37}$$

which governs the commutation relation between radiation and atomic operators. But this condition may not be applicable when the operators are kept at different times and when one seeks to envision the correlation between the atomic and radiation operators.

In the same way, upon using the definition that the raising and lowering atomic, and energy operators are

$$\hat{\sigma}_+ = |a\rangle\langle b|, \qquad \hat{\sigma}_- = |b\rangle\langle a|, \qquad \hat{\sigma}_z = \hat{\sigma}_{aa} - \hat{\sigma}_{bb} \tag{5.38}$$

with

$$\left(\hat{\sigma}_\pm\right)^\dagger = \hat{\sigma}_\mp, \qquad \hat{\sigma}_+^2 = \hat{\sigma}_-^2 = 0, \qquad \hat{\sigma}_+\hat{\sigma}_-\hat{\sigma}_+ = \hat{\sigma}_+, \qquad \hat{\sigma}_-\hat{\sigma}_+\hat{\sigma}_- = \hat{\sigma}_-, \tag{5.39}$$

it is possible to verify that

$$[\hat{\sigma}_+, \hat{\sigma}_-] = \hat{\sigma}_z, \qquad [\hat{\sigma}_z, \hat{\sigma}_\pm] = \pm 2\hat{\sigma}_\pm. \tag{5.40}$$

Then after, upon using these commutation relations, one can see that

$$\hat{\sigma}_z\hat{\sigma}_\pm = \pm\hat{\sigma}_\pm, \qquad \hat{\sigma}_\pm\hat{\sigma}_z = \mp\hat{\sigma}_\pm. \tag{5.41}$$

On account of such depictions, the probability of finding a two-level atom in the upper energy level[6] can be expressed in terms of the raising and lower atomic operators as

$$\rho_{aa} = \langle\hat{\sigma}_+\hat{\sigma}_-\rangle, \tag{5.42}$$

from which follows

$$\rho_{aa} = \frac{1}{2}(\langle\hat{\sigma}_z\rangle + 1), \tag{5.43}$$

$$W = \langle\hat{\sigma}_z\rangle. \tag{5.44}$$

The off-diagonal elements of the density matrix or atomic coherence on the other hand can be expressed in terms of the rasing atomic operator as

$$\rho_{ab} = \langle\hat{\sigma}_+\rangle. \tag{5.45}$$

The raising and lowering atomic operators can also be related to the spin operators that stand for the Hermitian dynamical variables and the energy operator as

$$\hat{\sigma}_x = \hat{\sigma}_- + \hat{\sigma}_+, \qquad \hat{\sigma}_y = i(\hat{\sigma}_- - \hat{\sigma}_+), \qquad \hat{\sigma}_z = \hat{\sigma}_+\hat{\sigma}_- - \hat{\sigma}_-\hat{\sigma}_+. \tag{5.46}$$

These operators along with the anti-commutation relation,

$$\hat{\sigma}_-\hat{\sigma}_+ + \hat{\sigma}_+\hat{\sigma}_- = \hat{I}, \tag{5.47}$$

constitute the conception of spin operator. The spatial components of the spin operators along x, y and z directions thus can be represented by Pauli matrices in which

$$\left[\hat{s}_x, \hat{s}_y\right] = \frac{i}{2}\hat{s}_z, \qquad \left[\hat{s}_y, \hat{s}_z\right] = \frac{i}{2}\hat{s}_x, \qquad \left[\hat{s}_z, \hat{s}_x\right] = \frac{i}{2}\hat{s}_y, \qquad \hat{s}_i\hat{s}_j + \hat{s}_j\hat{s}_i = \frac{i}{2}\delta_{ij}\hat{I}, \tag{5.48}$$

where a comparison might reveal $\hat{\sigma} = 2\hat{s}$.

---

[6]Knowing the probability of occupancy of the upper energy level alone should be enough since the population in the lower energy level can be inferred using a more general condition (5.34).

5.2   Dynamical Equations

One may note that the time evolution of the radiative properties of the atom
and statistical properties of the emitted radiation can be examined by solving the
pertinent dynamical equations. This can be done when one seeks to explore the
situation in the absence of damping and the accompanying equation are derived
from either Schrödinger or Heisenberg equation. In light of this, one can express the
density matrix of a two-level system as

$$\hat{\rho} = \begin{pmatrix} \rho_{aa} & \rho_{ba} \\ \rho_{ab} & \rho_{bb} \end{pmatrix}. \tag{5.49}$$

So upon employing the Liouville-Neumann relation (3.6), one can write

$$\frac{d\rho_{nm}}{dt} = -i \sum_i [\hat{H}_{ni}\hat{\rho}_{im} - \hat{\rho}_{ni}\hat{H}_{im}] \tag{5.50}$$

in which one can see using Eq. (5.27)

$$\frac{d\rho_{nm}}{dt} = -i\omega_{nm}\rho_{nm} - i \sum_i (V_{ni}\rho_{im} - \rho_{ni}V_{im}), \tag{5.51}$$

where $\omega_{nm} = E_n - E_m$ is the atomic transition frequency between the energy levels.
It should be straightforward then to write

$$\frac{d\rho_{ab}}{dt} = -i\omega_0\rho_{ab} + i V_{ab}(\rho_{aa} - \rho_{bb}), \tag{5.52}$$

$$\frac{d\rho_{aa}}{dt} = -i(V_{ab}\rho_{ba} - V_{ba}\rho_{ab}), \tag{5.53}$$

$$\frac{d\rho_{bb}}{dt} = -i(V_{ba}\rho_{ab} - V_{ab}\rho_{ba}), \tag{5.54}$$

where $\omega_0 = \omega_{ab} = \omega_a - \omega_b$ is the atomic transition frequency. One may verify that
these equations are consistent with the conservation of the total population since
$\dot{\rho}_{aa} + \dot{\rho}_{bb} = 0$ (compare with Eq. (5.34)).

The atom-radiation coupled system can also be analyzed by employing the
pertinent evolution of the probability amplitudes. When a two-level atom is assumed
to interact with linearly polarized electric field resonantly, it may not be difficult to
see from Schrödinger equation that

$$\frac{d}{dt}C_a(t) = i\Omega C_b(t)e^{i\omega_0 t}, \tag{5.55}$$

$$\frac{d}{dt}C_b(t) = i\Omega C_a(t)e^{-i\omega_0 t}, \tag{5.56}$$

where Ω is the Rabi frequency defined by

$$\Omega = e\mathbf{r}_{ab}.\mathbf{E}. \tag{5.57}$$

These coupled differential equations can be solved by applying iterative integration. Even then, the solutions of these equations can be proposed by assuming the atom to be initially in the upper energy level ($C_a(0) = 1$ and $C_b(0) = 0$) as

$$C_a(t) = \cos(\Omega t), \qquad C_b(t) = \sin(\Omega t). \tag{5.58}$$

The same coupled differential equations can also be rewritten as

$$\frac{d}{dt}C_a(t) = -iC_a(t)V_{aa}(t) - iC_b(t)V_{ab}(t)e^{-i\omega_0 t}, \tag{5.59}$$

$$\frac{d}{dt}C_b(t) = -iC_a(t)V_{ba}(t)e^{i\omega_0 t} - iC_b(t)V_{bb}(t). \tag{5.60}$$

One may note that Eqs. (5.59) and (5.60) reduce when $V_{aa} = V_{bb} = 0$ to

$$\frac{d}{dt}C_a(t) = -iC_b(t)V_{ab}e^{-i\omega_0 t}, \tag{5.61}$$

$$\frac{d}{dt}C_b(t) = -iC_a(t)V_{ba}e^{i\omega_0 t}. \tag{5.62}$$

Equations (5.61) and (5.62) can be solved by using iterative integrations, and perturbation technique in case of weak interaction.

In case the two-level atom is driven externally by a radiation of frequency ω, atomic detuning can be defined as

$$\Delta = \omega - \omega_0, \tag{5.63}$$

which can be perceived as the spread in the spectra of the external radiation. Even when one assumes that the spectrum of the external radiation is sharp, the atomic energy levels are not as rigid as one expects them to be, that is, the uncertainty in the energy of the atomic states can also lead to detuning. Whatever the case, the external radiation cannot be completely resonant with the atomic transitions in realistic setting, and so the need for including the detuning in the analysis. It might not be thus hard to ascertain that when there is detuning, the wave function that denotes the coupled system takes the form

$$\Psi(\mathbf{r}, t) = C_a(t)e^{i(\Delta/2-\omega_a)t}\mathbf{u}_a(\mathbf{r}) + C_b(t)e^{-i(\Delta/2+\omega_b)t}\mathbf{u}_b(\mathbf{r}), \tag{5.64}$$

from which follows

$$\frac{dC_a}{dt} = \frac{i}{2}(\Omega C_b - \Delta C_a), \tag{5.65}$$

$$\frac{dC_b}{dt} = \frac{i}{2}(\Delta C_b + \Omega^* C_a). \tag{5.66}$$

These coupled differential equations can be solved by employing

$$\frac{d\mathcal{C}}{dt} = \frac{i}{2}\mathcal{MC}, \tag{5.67}$$

which can be integrated as

$$\mathcal{C}(t) = \mathcal{C}(0)e^{\frac{i\lambda_\pm t}{2}}, \tag{5.68}$$

where $\lambda_\pm$ is the eigenvalue of matrix $\mathcal{M}$, and

$$\mathcal{M} = \begin{pmatrix} -\Delta & \Omega \\ \Omega^* & \Delta \end{pmatrix}, \qquad \mathcal{C}(t) = \begin{pmatrix} C_a(t) \\ C_b(t) \end{pmatrix}. \tag{5.69}$$

To determine the eigenvalues, one often demands that $\det(\mathcal{M} - \lambda\mathcal{I}) = 0$, from which follows

$$\lambda_\pm = \pm\sqrt{\Delta^2 + \Omega^2}, \tag{5.70}$$

where $\lambda_\pm$ is the generalized Rabi flopping frequency.

5.3   Atomic Transition Mechanisms

Interaction of atom with radiation is characterized by the pertinent atomic transition that can be illustrated by the process in which the atom in the lower energy level jumps to the excited (or upper) energy level upon absorbing or the atom in the upper energy level falls to the lower energy level upon emitting a quanta of light. To explore different features of allowable atomic transition mechanisms, the atom is often assumed to be initially in one of the energy levels so that the initial radiative properties of the atom can be inferred.

5.3.1   Closed Atomic Transition

In case the atom is assumed to spontaneously decay from the upper to the lower accessible energy level at a constant rate, the atomic lifetime of this energy level

Fig. 5.3 Schematic
representation of a closed
two-level atomic transition,
where γ is the spontaneous
atomic decay rate of the
upper energy level (see also
Fig. 5.1)

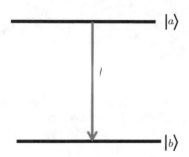

can be defined as

$$T_{sd} = \frac{1}{\gamma},$$ (5.71)

where the subscript sd refers to spontaneous decay, and so γ is known as atomic
decay rate. When the population that leaves the upper energy level directly occupies
the lower energy level as shown in Fig. 5.3, the accompanying atomic transition
is designated as closed ($\dot{\rho}_{bb} + \dot{\rho}_{aa} = 0$ since the population that leaves the upper
energy level is expected to directly go over to the lower energy level).

In the same token, it is legitimate to imagine the atomic dipole moment to
dephase in a characteristic time T_{dd} that can be defined as

$$\gamma_{ab} = \frac{1}{T_{dd}},$$ (5.72)

where the subscript dd stands for atomic dipole moment dephasing, γ_{ab} signifies
the rate at which the atomic coherence would be lost, and referred to as atomic
dephasing.

While studying the interaction of the atom with radiation, the contribution of
damping to environment should also be properly accounted for. In light of this, upon
adding the decay term phenomenologically in Eq. (5.50), the time evolution of the
density matrix can be written as

$$\frac{d\rho_{nm}}{dt} = -i[\hat{H}, \ \hat{\rho}]_{nm} - \frac{1}{2}\left(\hat{\Gamma}\hat{\rho} + \hat{\rho}\hat{\Gamma}\right)_{nm},$$ (5.73)

where $\hat{\Gamma}$ is taken to represent the noise associated with the interaction. The equation
for each element of the density matrix is thus modified as

$$\frac{d\rho_{ab}}{dt} = -\left(i\omega_0 + \gamma_{ab}\right)\rho_{ab} + iV_{ab}(\rho_{bb} - \rho_{aa}),$$ (5.74)

$$\frac{d\rho_{aa}}{dt} = -\gamma\rho_{aa} - i(V_{ab}\rho_{ba} - \rho_{ab}V_{ba}), \tag{5.75}$$

$$\frac{d\rho_{bb}}{dt} = -\gamma\rho_{bb} + i(V_{ab}\rho_{ba} - \rho_{ab}V_{ba}) \tag{5.76}$$

for $\Gamma_{aa} = \Gamma_{bb} = \gamma$ and $\Gamma_{ab} = \gamma_{ab}$. One can see by inspection that condition (5.34) still holds true as perceived.

With the help of Eqs. (5.75) and (5.76), one can verify that the population inversion evolves as

$$\frac{dW}{dt} = -\gamma W - i2(V_{ab}\rho_{ba} - \rho_{ab}V_{ba}). \tag{5.77}$$

A closed atomic system can generally be described by Eqs. (5.74) and (5.77), where formal integration of Eq. (5.74) in the absence of the driving radiation yields

$$\rho_{ab}(t) = \rho_{ab}(0)e^{-(i\omega_0 + \gamma_{ab})t}, \tag{5.78}$$

which represents the time evolution of the atomic coherence. Besides, on account of Eqs. (5.33) and (5.78), the expectation value of the electric dipole moment can be expressed as

$$\langle\mu(t)\rangle = \left[\mu_{ba}\rho_{ab}(0)e^{-i\omega_0 t} + c.c.\right]e^{-\gamma_{ab}t}. \tag{5.79}$$

One may see from Fig. 5.4 (see also Eq. (5.79)) that the mean of the electric dipole moment oscillates with atomic or dipole transition frequency ω_0 and decays with a dipole dephasing time. This result entails that the subsequent study of the dynamics of the population and atomic coherence can also show the reminiscent of this oscillation.

Fig. 5.4 Plots of the mean dipole moment for different values of dephasing in the absence of external driving

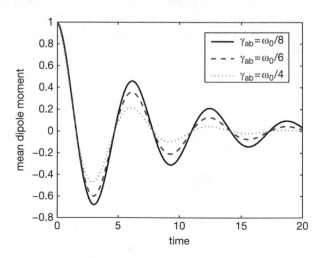

5.3.2 Open Atomic Transition

The other possible atomic transition mechanism is allude to as open, and refers to the situation in which the upper and lower energy levels are allowed to exchange population with the surrounding environment. The rationale for the notion of open transition can be related to the fact that the population that leaves the upper energy level does not necessarily goes over to the prescribed lower energy level since the environment can pump the population to the designated upper and lower energy levels as shown in Fig. 5.5.

In view of this consideration, Eqs. (5.75) and (5.76) are modified due to the exchange of excitation with the surrounding environment as

$$\frac{d\rho_{aa}}{dt} = \lambda_a - \gamma_a \rho_{aa} - i(V_{ab}\rho_{ba} - \rho_{ab}V_{ba}), \tag{5.80}$$

$$\frac{d\rho_{bb}}{dt} = \lambda_b - \gamma_b \rho_{bb} + i(V_{ab}\rho_{ba} - \rho_{ab}V_{ba}), \tag{5.81}$$

where γ_a (γ_b) stands for population decay rate of the upper (lower) energy level and $\lambda_{a,b}$ for the gain of population as in the process of pumping by external agent.

One may seek to integrate Eqs. (5.80) and (5.81) by applying the rate equation approximation in which the dipole dephasing time is assumed to be much smaller than the time for which the population difference can change. To begin with, one can see from Eq. (5.74) that

$$\rho_{ab}(t) = \rho_{ab}(0)e^{-(i\omega_0 + \gamma_{ab})t} + i\int_{-\infty}^{t} dt' \, e^{-(i\omega_0 + \gamma_{ab})(t-t')} V_{ab}(t')(\rho_{aa} - \rho_{bb}). \tag{5.82}$$

If the electric field is taken to be parallel to the dipole moment and the radiation is shined along z-direction, one can observe that

$$V_{ab}(t) = -\frac{1}{2} er_{ab} E \, e^{i(kz - \omega t)} \tag{5.83}$$

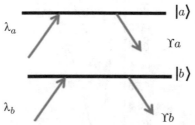

Fig. 5.5 Schematic representation of an open two-level atomic transition mechanism, where $\gamma_{a(b)}$ is the spontaneous atomic decay rate of the upper (lower) energy level to the continuum of the accessible lower energy levels and $\lambda_{a(b)}$ is the external pumping rate

in which case, upon assuming the population inversion to be independent of time and the initial coherence to be negligibly small, and carrying out the integration over t', one can obtain

$$\rho_{ab}(t) = i\frac{(\rho_{aa} - \rho_{bb})er_{ab}E}{2[\gamma_{ab} - i\Delta]}\, e^{i(kz - \omega t)}. \tag{5.84}$$

Then with the aid of Eq. (5.84), it may not be difficult to verify that

$$\frac{d\rho_{bb}}{dt} = \lambda_b - \gamma_b\rho_{bb} + (\rho_{aa} - \rho_{bb})R, \tag{5.85}$$

$$\frac{d\rho_{aa}}{dt} = \lambda_a - \gamma_a\rho_{aa} - (\rho_{aa} - \rho_{bb})R, \tag{5.86}$$

where the rate constant turns out upon keeping the real part to be

$$R = -\frac{\Omega^2\gamma_{ab}}{2[\gamma_{ab}^2 + (\omega_0 - \gamma_{ab})^2]}. \tag{5.87}$$

It is possible to see from Fig. 5.6 that the rate constant depends on the dephasing rate specially when it is less than the natural atomic transition frequency; the situation closer to physically feasible setup. One may also see that the absolute value of the rate constant increases with the natural frequency, and the largest possible value is found near $\gamma_{ab} = \omega_0$.

One can also see that Eqs. (5.85) and (5.86) are the usual rate equations from which follows at steady state

$$\rho_{aa} - \rho_{bb} = \frac{\xi[\gamma_{ab}^2 + (\omega_0 - \gamma_{ab})^2]}{(1 + I)\gamma_{ab}^2 + (\omega_0 - \gamma_{ab})^2}, \tag{5.88}$$

Fig. 5.6 Plots of the rate constant, where the parameters are taken relative to the Rabi Frequency Ω for different values of atomic transition ω_0

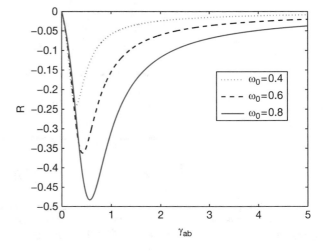

where ξ is the unsaturated population difference $\xi = (\lambda_a\gamma_b - \lambda_b\gamma_a)/\gamma_a\gamma_b$ and I stands for the dimensionless intensity $I = [(\gamma_a + \gamma_b)\Omega^2]/2\gamma_{ab}\gamma_a\gamma_b$. Equation (5.84) can thus be rewritten on account of Eq. (5.88) as

$$\rho_{ab}(t) = -i \left[\frac{\xi\Omega[\gamma_{ab}^2 + (\omega_0 - \gamma_{ab})^2]}{2[\gamma_{ab} - i\Delta][(1 + I)\gamma_{ab}^2 + (\omega_0 - \gamma_{ab})^2]} \right] e^{i(kz - \omega t)}. \tag{5.89}$$

One can see from Fig. 5.7 that when damping constants are taken to be nearly equal to the rate of dephasing, the atom is found to have a better chance to occupy the upper energy level, that is, it is possible to induce population inversion. This result is not the same as what one observes for the atoms placed in a cavity when a closed transition mechanism is adopted (see Chap. 6). It is also possible to see that the population inversion varies substantially with the atomic natural transition frequency specially when $\omega_0 > \gamma_{ab}$. One can also see from Fig. 5.8 that the atomic coherence depends significantly on the atomic natural transition frequency specially near $\omega_0 = \gamma_{ab}$. It turns out that the chance for inducing atomic coherence is greater for an atom with smaller transition frequency but for closed transition mechanism atomic coherence cannot be induced without external pumping. It might also be worthy noting that Eq. (5.88) denotes the population inversion at steady state and Eq. (5.89) the atomic coherence at finite time. The atomic coherence in the open scheme is found to oscillate in time as in the closed transition but does not dephase. This outcome may envisage that the open transition mechanism has a potential to preserve the initially introduced atomic coherence; the attribute that might be interesting in the application of atomic coherence for instance in the phenomenon of electromagnetically induced transparency.

Fig. 5.7 Plots of the population inversion for different values of atomic transition ω_0, $\xi = 0.1$ and $I = \Omega^2/\gamma_{ab}^2$, that is, when $\gamma_a \approx \gamma_b \approx \gamma_{ab}$

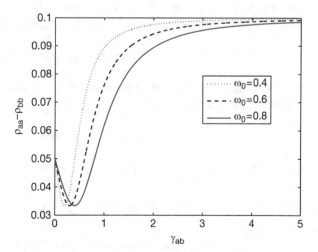

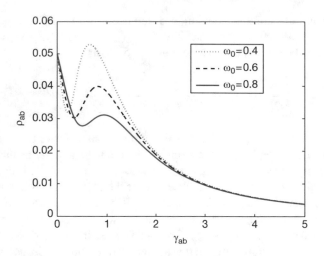

Fig. 5.8 Plots of the atomic coherence for different values of atomic transition ω_0, $\xi = 0.1$, modest detuning, $I = \Omega^2/\gamma_{ab}^2$ and when the exponent is neglected

---

5.4   Atomic Bloch Equations

Atomic Bloch equations are the same as the equations already provided in many respects, although these equations look different within various formalisms, and so one can form differing view points. Hoping to utilize as much as available options to gain versatile understanding of how the populations and coherence of the atomic system evolve, we seek to include a geometrical approach or interpretation to the problem of the interaction of two-level atom with light. To do so, one may begin with the slowly varying off-diagonal density matrix component

$$\tilde{\rho}_{ab}(t) = \sigma_{ab}(t)e^{-i\omega t}. \tag{5.90}$$

In the minimum coupling approximation, one may note that the interaction Hamiltonian takes the form

$$V_{ab}(t) = -\mu_{ab}(t)Ee^{-i\omega t}. \tag{5.91}$$

On the basis of Eqs. (5.90) and (5.91), one can thus rewrite Eqs. (5.74)–(5.76) as

$$\frac{d\sigma_{ab}}{dt} = \left[i(\omega - \omega_0) - \gamma_{ab}\right]\sigma_{ab} - i\mu_{ab}E(\rho_{bb} - \rho_{aa}), \tag{5.92}$$

$$\frac{d}{dt}(\rho_{bb} - \rho_{aa}) = -\gamma(\rho_{bb} - \rho_{aa}) + i2(\mu_{ab}E\sigma_{ba} - \mu_{ba}E^*\sigma_{ab}), \tag{5.93}$$

which can be rewritten upon dropping the subscript ab for atomic parameters as

$$\frac{d\sigma}{dt} = (i\Delta - \gamma_{ab})\sigma - \frac{i}{2}\lambda Ew, \tag{5.94}$$

$$\frac{dw}{dt} = -\gamma w + i(\lambda E\sigma^* - \lambda^* E^*\sigma), \tag{5.95}$$

where λ is the atom-radiation coupling constant taken to be $2\mu_{ab}$ and $w = \rho_{bb} - \rho_{aa}$ (compare with Eq. (5.35)).

One can also look for the equation of the complex amplitude of the induced dipole moment by defining the slowly varying electric dipole moment as

$$\langle \tilde{\mu} \rangle = pe^{-i\omega t} + c.c, \tag{5.96}$$

where $p = \sigma_{ab}\mu_{ba}$ and when the damping term is ignored. The dynamical equations then can be modified as

$$\frac{dp}{dt} = (i\Delta - \gamma_{ab})p - \frac{i}{4}|\lambda|^2 Ew, \tag{5.97}$$

$$\frac{dw}{dt} = -\gamma w - 4\mathrm{Im}(Ep^*). \tag{5.98}$$

In case μ_{ab} and λ are taken to be real quantities, it turns out to be useful introducing two variables that satisfy $\sigma = (u - iv)/2$ with $u = \rho_{ab}e^{i\omega t} + c.c$ (real part of the coherence) and $v = i\rho_{ab}e^{i\omega t} + c.c$ (imaginary part of the coherence). As a result of such variable change, one may see that

$$\frac{du}{dt} = \Delta v - \gamma_{ab}u, \tag{5.99}$$

$$\frac{dv}{dt} = -\Delta u - \gamma_{ab}v + \lambda Ew, \tag{5.100}$$

$$\frac{dw}{dt} = -\gamma w - \lambda Ev. \tag{5.101}$$

These classical equations are designated as atomic Bloch equations and can be solved with the help of matrix representation, which is expected to make the corresponding operator equations tractable. This approach has not only been used in solving the equation, but also aided in illustrating the physical nature of the density matrix.[7] The geometrical interpretation can be inferred from the Bloch equations by making use of variable replacements $w \rightarrow z$, $v \rightarrow y$ and $u \rightarrow x$. One may note that these equations resemble the Bloch equations that have been introduced for spin-$\frac{1}{2}$ system; and hence, the name.

It is also thought to be appealing applying these variable changes and defining $r^2 = u^2 + v^2 + w^2$ in writing

$$\rho_{aa} = \frac{1}{2}(1 + r\cos\theta), \qquad \rho_{bb} = \frac{1}{2}(1 - r\sin\theta), \qquad \rho_{ab} = \frac{1}{2}r\sin\theta e^{i\phi},$$

$$(5.102)$$

where

$$\theta = \cos^{-1}\left[(\rho_{aa} - \rho_{bb})/r\right], \qquad \phi = \tan^{-1}\left[\mathrm{Im}(\rho_{ab})/\mathrm{Re}(\rho_{ab})\right]. \qquad (5.103)$$

With this characterization, the superimposed two-level atomic state can be put forward as

$$|\Psi\rangle = \sin\left(\theta/2\right)e^{i\frac{\phi}{2}}|b\rangle + \cos\left(\theta/2\right)e^{-i\frac{\phi}{2}}|a\rangle, \qquad (5.104)$$

which can be related to the idea of atomic qubits (see Sect. 9.2).

5.5 Properties of the Emitted Radiation

The nature of atomic dynamics and the statistical properties of the radiation can be inferred from the pertinent interaction of the atom with radiation where the way how the atom responds to the external driving depends on the properties of the accompanying light. We now seek to explore properties of the emitted radiation that can be addressed by looking into the correlation between the radiation emitted at different times.

To this effect, the power spectrum of the radiation can be described in terms of the electric field operator as

$$S(\omega) = 2\mathrm{Re}\int_0^\infty d\tau e^{i\omega\tau}\langle\hat{E}^{(-)}(t)\hat{E}^{(+)}(t + \tau)\rangle_{ss} \qquad (5.105)$$

(see Sect. 4.2.3). Markedly, the positive part of the electric field operator for the radiation emitted by a two-level atom can be expressed in terms of the atomic

---

[7]The density operator can be expressed in terms of the Bloch vector as $\hat{\rho} = (\hat{I} + \mathbf{u}.\hat{\sigma})/2$ and visualized with the help of the Bloch sphere spanned by a vector $\mathbf{u}$ whose direction is related to the direction of the expectation value of the spin and its length is restricted to $0 \leq u \leq 1$. Note that the maximum value corresponds to the pure state of the spin lying on the surface of the Bloch sphere with unit radius and the minimum value to the completely mixed state.

operator as

$$\hat{E}^{(+)}(r, t) = \frac{\omega_0^2 \mu_{ab} \sin \theta}{4\pi \epsilon_0 c^2 r} \hat{\sigma}_-(t)\hat{\mathbf{e}}, \tag{5.106}$$

where r is the distance between the atom and the point at which the field is observed (see Appendix 3 of Chap. 6). So the power spectrum can be written as

$$S(\omega) = 2 \left(\frac{\omega_0^2}{4\pi \epsilon_0 c^2 r} \right)^2 f(r) d\tau e^{i\omega\tau} \langle \hat{\sigma}_+(t)\hat{\sigma}_-(t + \tau) \rangle_{ss}, \tag{5.107}$$

where $f(r)$ represents geometric configuration.

The photon statistics of the emitted radiation can also be studied with the aid of the pertinent second-order two-time correlation function. It might be good to confirm that the two-time correlation function for a two-level atom can be put in the form

$$G^{(2)}(\tau) = \langle \hat{\sigma}_+(t)\hat{\sigma}_+(t + \tau)\hat{\sigma}_-(t + \tau)\hat{\sigma}_-(t) \rangle. \tag{5.108}$$

On the basis of the discussion presented in Sect. 5.1.2, one can also see that

$$\langle \hat{\sigma}_+(t + \tau)\hat{\sigma}_-(t + \tau) \rangle = \frac{1}{2} \big[\langle \sigma_z(t + \tau) \rangle + 1 \big], \tag{5.109}$$

from which follows

$$G^{(2)}(\tau) = \frac{1}{2} \big[\langle \hat{\sigma}_+(t)\hat{\sigma}_-(t) \rangle + \langle \hat{\sigma}_+(t)\hat{\sigma}_z(t + \tau)\hat{\sigma}_-(t) \rangle \big]. \tag{5.110}$$

In the same spirit, the squeezing properties of the cavity radiation or atomic squeezing can be studied by applying the variances of atomic dipole operators in the normal order. To do so, one may need to ascertain that if the two quadrature components of the electric field satisfy the commutation relation

$$\big[\hat{E}_\theta, \hat{E}_{\theta-\frac{\pi}{2}} \big] = i2c\hat{I}, \tag{5.111}$$

the uncertainty relation

$$\langle (\Delta \hat{E}_\theta)^2 \rangle \langle (\Delta \hat{E}_{\theta-\frac{\pi}{2}})^2 \rangle \geq c^2 \tag{5.112}$$

holds (see also Sect. 7.1.2). The radiation that can be represented by such electric field would be in squeezed state provided that either $\langle (\Delta \hat{E}_\theta)^2 \rangle$ or $\langle (\Delta \hat{E}_{\theta-\frac{\pi}{2}})^2 \rangle$ would be below the vacuum limit c (see Sect. 2.2).

In light of this, one of the variances can be put in the normal order as

$$\langle : (\Delta \hat{E}_\theta)^2 : \rangle = \langle (\Delta \hat{E}_\theta)^2 \rangle - c. \tag{5.113}$$

The squeezing can be related to the requirement that either $\langle : (\Delta \hat{E}_\theta)^2 : \rangle$ or $\langle : (\Delta \hat{E}_{\theta - \frac{\pi}{2}})^2 : \rangle$ is negative.

Upon taking the relation between the electric field and atomic operators (see Eq. (5.106)), the variance of the field can be defined in terms of the atomic-dipole operators. It has been shown that the variances of the atomic dipole operator can be expressed in the normal order as

$$\langle : (\Delta \hat{\sigma}_i)^2 : \rangle = \langle (\Delta \hat{\sigma}_i)^2 \rangle + \frac{1}{2} \langle \hat{\sigma}_z \rangle, \tag{5.114}$$

where $i = x, y$ [5] and

$$\hat{\sigma}_x = \frac{1}{\sqrt{2}} (\hat{\sigma}_+ + \hat{\sigma}_-), \qquad \hat{\sigma}_y = \frac{i}{\sqrt{2}} (\hat{\sigma}_- - \hat{\sigma}_+). \tag{5.115}$$

In view of the fact that at steady state $\langle \hat{\sigma}_-(t) \rangle_{ss} = \langle \hat{\sigma}_+(t) \rangle_{ss}$, one can then find

$$\langle : (\Delta \hat{\sigma}_x)^2 : \rangle = \frac{1}{2} \big(1 + \langle \hat{\sigma}_z \rangle \big) - 2 \langle \hat{\sigma}_+ \rangle_{ss}^2, \tag{5.116}$$

$$\langle : (\Delta \hat{\sigma}_y)^2 : \rangle = \frac{1}{2} \big(1 + \langle \hat{\sigma}_z \rangle \big). \tag{5.117}$$

One can see from Eqs. (5.43) and (5.117) that $\langle : (\Delta \hat{\sigma}_y)^2 : \rangle$ is the same as the probability of the occupancy of the upper energy level, and so never be negative. The radiation emitted by a two-level atom is hence said to be in squeezed state provided that $\langle \hat{\sigma}_+ \rangle_{ss}^2 > \langle \hat{\sigma}_+ \hat{\sigma}_- \rangle_{ss}/2$. As one can see from Eqs. (5.42) and (5.45), the atomic squeezing would be witnessed when the atomic coherence turns out to be strong [6].

5.6   Optical Response of Dense Atomic Medium

The idea of dense atomic medium transpires from the situation when there are many atoms placed in a small space. When an external electromagnetic radiation is shined onto dense medium, its interaction with the collection of atoms can be perceived as defining property of the propagation of the light in the dense atomic medium such as the response of the medium. Pertaining to the possibility of the underlying cooperative interaction of atoms with external radiation, the propagation of the radiation in the dense medium or the nature of the radiation coming out of the medium is expected to be modified.

To look into such a modification, one may begin with Maxwell's equations in a material medium:

$$\nabla \times E + \frac{\partial B}{\partial t} = 0, \qquad \nabla \times H - \frac{\partial D}{\partial t} = J \qquad \nabla \cdot H = 0 \qquad \nabla \cdot B = 0,$$

(5.118)

where $\mathbf{J}$ and ρ are the current and free charge densities, $\mathbf{B} = \mu \mathbf{H}$ is the magnetic field and $\mathbf{D} = \varepsilon \mathbf{E} + \mathbf{P}$ is the displacement of the electric field in which $\mathbf{P}$ is the macroscopic polarization, μ and ε are the permeability and permittivity of the atomic medium.

To describe the intended interaction, it might be essential to explore how the polarization $\mathbf{P}(t)$ of the material medium depends on the strength of the applied electric field $\mathbf{E}(t)$. For instance, in the linear optical phenomena, the polarization depends on the field linearly when the direction is neglected as

$$P(t) = \chi^{(1)} E(t),$$

(5.119)

where $\chi^{(1)}$ is linear susceptibility of the medium.

The optical nonlinear response of the medium can in general be accounted for by expressing Eq. (5.119) as a power series of the electric field strength;

$$P(t) = \chi^{(1)} E(t) + \chi^{(2)} E^2(t) + \chi^{(3)} E^3(t) + \ldots,$$

(5.120)

where $\chi^{(i)}$ is the ith order nonlinear optical susceptibility. In this case, the electric polarization (5.120) can be written as

$$P^{(2)}(t) = 2\chi^{(2)} E E^* + \chi^{(2)} \left[E^2 e^{-i2\omega t} + E^{*2} e^{i2\omega t} \right].$$

(5.121)

One may note that the second and third terms in the right side of Eq. (5.121) correspond to the sum-frequency generator and stand for second-harmonic generation (see Sect. 8.1.2).

For now, imagine that a linear beam with electric field strength

$$E(t) = E e^{-i\omega t} + E^* e^{i\omega t}$$

(5.122)

is made to impinge onto a system for which the second-order susceptibility $\chi^{(2)}$ is dominant. If one then assumes that the medium is lossless, the electric field within the medium can be expressed as

$$\tilde{E}(z, t) = \tilde{E}_1(z, t) + \tilde{E}_2(z, t),$$

(5.123)

where each mode is defined in terms of a complex amplitude $E_j(z)$ as

$$\tilde{E}_j(z, t) = E_j(z)e^{-i\omega_j t} + E_j^*(z)e^{i\omega_j t} \tag{5.124}$$

with $E_j(z) = A_j e^{ik_j z}$, where A_j is slowly varying amplitude, and $k_j = \eta_j(\omega_j)/c$ in which $\eta_j(\omega_j) = \sqrt{\varepsilon^{(1)}(\omega_j)}$ and $\eta_j(\omega_j)$ is the frequency dependent refractive index of the medium.

After that, based on general wave equation, each frequency component of the electric field is expected to obey the driven wave equation:

$$\frac{\partial^2 \tilde{E}_j(z, t)}{\partial z^2} - \frac{\varepsilon^{(1)}(\omega_j)}{c^2}\frac{\partial^2 \tilde{E}_j(z, t)}{\partial t^2} = \frac{1}{\varepsilon_0 c^2}\frac{\partial^2 \tilde{P}_j(z, t)}{\partial t^2}, \tag{5.125}$$

where the last term represents the contribution of the deceleration of the charges while traveling in the medium.

Suppose one chooses to stick to the situation in which the response of the atomic medium determines the polarization and also assumes that $\varepsilon = \varepsilon_0$, where ε_0 is the permittivity of free space. Hence it may not be difficult to verify upon taking the electric field to vary slowly in the plane perpendicular to the propagation direction— that amounts to setting $\nabla.\mathbf{E} = 0$—and assuming the radiation field to propagate along z-direction that

$$\frac{\partial^2 E}{\partial z^2} - \frac{1}{c^2}\frac{\partial^2 E}{\partial t^2} = \mu_0\frac{\partial^2 P}{\partial t^2}. \tag{5.126}$$

To solve Eq. (5.126), the radiation should be taken as a classical electric field with carrier frequency ω and wave number $\omega = kc$, that is,

$$E(z, t) = \tilde{E}(z, t)e^{i(kz-\omega t)} + \tilde{E}^*(z, t)e^{-i(kz-\omega t)}, \tag{5.127}$$

where $\tilde{E}(z, t)$ is the slowly varying field envelop. The electric field is perceived in this consideration to be complex, that is, $\tilde{E} = |\tilde{E}_r|e^{i\varphi}$, where $\tilde{E}_r$ is the real amplitude and φ is the phase.

It may not be hard to foresee that the polarization can also be expressed in the same way. Then upon taking a slowly varying envelop approximation, it is possible to see that

$$\frac{\partial \tilde{E}}{\partial z} + \frac{1}{c}\frac{\partial \tilde{E}}{\partial t} = i\frac{k}{2\varepsilon_0}\tilde{P}, \tag{5.128}$$

which can also be rewritten in terms of the real and imaginary parts as

$$\left(\frac{\partial}{\partial z} + \frac{1}{c}\frac{\partial}{\partial t}\right)|\tilde{E}_r| = -\frac{k}{2\varepsilon_0}\,\mathrm{Im}\tilde{P}, \tag{5.129}$$

$$|\tilde{E}_r|\left(\frac{\partial}{\partial z} + \frac{1}{c}\frac{\partial}{\partial t}\right)\varphi = \frac{k}{2\varepsilon_0}\mathrm{Re}\tilde{P}. \tag{5.130}$$

Note that the induced polarization can be very complicated nonlinear function of the electric field for actual material medium. Even then, for a weak field that propagates in isotropic medium, the polarization can be approximated as a linear function of the electric field,

$$P(z, t) = \varepsilon_0 \int_{-\infty}^{\infty} \chi(t')E(z, t - t')dt', \tag{5.131}$$

where $\chi(t)$ is the susceptibility.

Upon making use of the value of the electric field, and then comparing it with the results already obtained, it is straightforward to see that

$$\tilde{P} = \varepsilon\chi(\omega)\tilde{E}, \tag{5.132}$$

where $\chi(\omega)$ is the Fourier transform of $\chi(t) = \int \chi(t')e^{i\omega t}dt'$. It then follows that

$$\frac{\partial\tilde{E}}{\partial z} = \frac{ik}{2}\chi(\omega)\tilde{E} \tag{5.133}$$

whose solution can be written as

$$\tilde{E}(z) = \tilde{E}(0)e^{i\varphi(z)}e^{-az}, \tag{5.134}$$

where $\varphi(z) = k\mathrm{Re}\chi(\omega)z/2$ is the phase shift and $a = k\mathrm{Im}\chi(\omega)/2$ is the linear amplitude attenuation coefficient when $a \geq 0$. One may note that the imaginary part of this susceptibility determines the dissipation of the field by the atomic medium, which is usually designated as absorption of the light while the real part signifies the refractive index.

On the other hand, to evaluate the velocity of the field profile, one may imagine the electric field with a carrier frequency ω in which the envelope function can be written as

$$\tilde{E}(t) = \int \tilde{E}(\omega + v)e^{-ivt}dv, \tag{5.135}$$

where $\tilde{E}(\omega + \nu)$ is the amplitude of the frequency component of the probe field and the corresponding expression for the polarization would be

$$\tilde{P}(t) = \varepsilon_0 \int \chi(\omega + \nu)\tilde{E}(\omega + \nu)e^{-i\nu t}d\nu. \tag{5.136}$$

In case the susceptibility is a smooth function, it can be expanded as

$$\chi(\omega + \nu) \simeq \chi(\omega) + \frac{\partial \chi(\omega)}{\partial \omega}\nu + O(\nu^2) + \dots. \tag{5.137}$$

Upon taking this consideration into account, one can then see that

$$\tilde{P}(t) = \varepsilon_0 \chi(\omega)\tilde{E}(t) + i\varepsilon_0 \frac{\partial \chi(\omega)}{\partial \omega}\frac{\partial \tilde{E}(t)}{\partial t} + \dots, \tag{5.138}$$

from which follows

$$\frac{\partial \tilde{E}}{\partial z} + \frac{1}{v_g}\frac{\partial \tilde{E}}{\partial t} = \frac{1}{2}ik\chi(\omega)\tilde{E}, \tag{5.139}$$

where the group velocity is defined as

$$v_g = \frac{c}{1 + \frac{\omega}{2}\frac{\partial \chi}{\partial \omega}}. \tag{5.140}$$

The group velocity can generally be interpreted as the velocity with which the peak of the probe pulse propagates in the medium, and heralds the modification of the speed of the profile of the wave pattern due to the pertinent interaction. This conception may be helpful when one seeks to explore how the profile of the light changes while it propagates in a particular medium, and so infer the property of the medium from the profile of the field as in the phenomena of electromagnetically induced transparency and slow light.

One may also seek to perform similar operation with the intension of studying the propagation of a weak probe field through a near resonant two-level atomic medium. To do so, one can proceed by rewriting the polarization of the medium in terms of atomic variable as

$$P(z,t) = NTr(\mu\hat{\rho}(z,t)), \tag{5.141}$$

where μ is the projection of the atomic dipole moment onto the field polarization direction, N is the number of atoms that constitute the medium and $\hat{\rho}(z,t)$ is the atomic density operator. For two-level atoms, one can see that

$$P(z,t) = N\big[\mu_{ba}\rho_{ab}(z,t) + \mu_{ab}\rho_{ba}(z,t)\big]. \tag{5.142}$$

Since the dependence of the interaction of the atoms with the propagating radiation field can be complicated, one may take the matrix elements to be slowly varying functions.

In this context, for a near resonant interaction, the spatiotemporal evolution is found to be governed by two coupled differential equations

$$\left(\frac{\partial}{\partial z} + \frac{1}{c}\frac{\partial}{\partial t}\right)\tilde{E}(z,t) = i\frac{N\omega\mu_{ba}}{2\varepsilon_0 c}\rho_{ab}(z,t), \tag{5.143}$$

$$\frac{\partial\rho_{ab}(z,t)}{\partial t} = (i\Delta - \gamma_{ab})\rho_{ab}(z,t) - i(\rho_{aa} - \rho_{bb})p_{ab}\tilde{E}(z,t) \tag{5.144}$$

(see also Sect. 5.4). Despite the elegance and simplicity of these equations, only numerical solutions are currently available except for certain special cases. Since we are interested in the weak field regime, Eq. (5.144) can be approximately solved at steady state;

$$\rho_{ab} = -\frac{\mu_{ab}(\rho_{bb} - \rho_{aa})}{\Delta + i\gamma_{ab}}\tilde{E}. \tag{5.145}$$

In the same way, dropping the time derivative in Eq. (5.143) implies that

$$\frac{\partial\tilde{E}}{\partial z} \stackrel{\sim}{=} -\alpha\tilde{E}, \tag{5.146}$$

where

$$\alpha = \frac{N\omega|\mu_{ab}|^2(\rho_{bb} - \rho_{aa})}{2\varepsilon_0 c(\gamma_{ab} - i\Delta)} \tag{5.147}$$

is the linear absorption coefficient in which $\text{Re}(\alpha)$ is the absorption and $\text{Im}(\alpha)$ is the dispersion of the medium.

In addition, in comparison with earlier discussion, the linear susceptibility for a two-level atomic medium turns out to be

$$\chi(\omega) = \frac{N\omega|\mu_{ab}|^2(\rho_{bb} - \rho_{aa})}{\varepsilon_0(\Delta + \gamma_{ab})}. \tag{5.148}$$

Note that the interference between alternative excitation pathways of the atomic states leads to modified optical behavior when compared to a coherent evolution exemplified by an oscillatory population transfer that can be taken as a foundation for forthcoming discussions.

One can infer from Fig. 5.9 that the absorption coefficient of the medium increases with the atomic dephasing when $\Delta > \gamma_{ab}$ but decreases when $\Delta < \gamma_{ab}$

Fig. 5.9 Plots of the real and imaginary parts of the linear absorption coefficient of a two-level atomic medium in the parameterized units

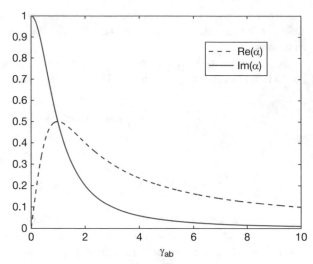

(one can see from Eq. (5.147) that $\mathrm{Re}(\alpha) = \mathrm{Im}(\alpha)$ when $\gamma_{ab} = \Delta$). The absorption would also be maximum when atomic dephasing and detuning are nearly equal, and no absorption for a case when there is no dephasing. In light of this, one might be tempted to associate the absorption of the medium with the dephasing. However, comparison with the result depicted in Fig. 5.8 may reveal that in case of open atomic transition, there could be a meaningful absorption even when dephasing is zero.

References

1. H.A. Kramers, W. Heisenberg, Z. Phys. **31**, 681 (1925); E.T. Jaynes, F.W. Cummings, Proc. IEEE **51**, 89 (1963)
2. G. Leuchsa, M. Sondermannb, J. Mod. Opt. **60**, 36 (2013)
3. V. Vedral, *Modern Foundation of Quantum Optics* (Imperial College Press, London, 2005)
4. P. Lambropoulos, D. Petrosyan, *Fundamentals of Quantum Optics and Quantum Information* (Springer, Berlin, 2007)
5. Z. Ficek, R. Tanas, Phys. Rep. **372**, 369 (2002)
6. S. Tesfa, J. Mod. Opt. **54**, 1759 (2007); S. Tesfa, J. Mod. Opt. **55**, 1587 (2008)

Cavity Mediated Interaction

6

The scope of the discussion in Chap. 5 is limited to a semiclassical regime in which the radiation is treated as classical laser light that contributes through its strength and the atom as quantum object. The atom for the most part is assumed to be placed in a free space and interact with the radiation that engulfs it. Even though such consideration can help in finding out how the atom behaves when subjected to external pumping via the evolution of the density matrix and probability amplitudes, it does not take into account the quantum nature of light; and so should not be taken as a complete description. It cannot also represent much of the analysis of the property of the radiation that comes out virtually in every direction since it is not practically possible to place detectors all over the surrounding space as demanded by the notion of free space. The other profound challenge in a free space scenario is the inability to induce a strong correlation between the atom and radiation since there is no sufficient time for the atom to reabsorb the radiation it emits. To overcome the limitation imposed by the classical treatment, the rational thing to do is to extend the discussion to the quantum realm; the approach that gives birth to quantum optics whose procedures and formulations would be utilized to expose the subtleties in quantum theory and application. Besides, as a way out from the practical challenges related to free space scenario, it is perceived that putting the atom and radiation in a cavity could be helpful.

Once the atom is placed in a cavity and the radiation is treated as quantum system, the information about the interaction of the radiation with atom can be inferred from the bare atomic state or state of the emitted radiation or combined state. The atomic dressed state particularly contains information on the atomic dynamics and state of the radiation in terms of the number of photons in the radiation, and so deemed to fairly describe the available emission-absorption events and probabilities (see Sect. 6.1). The main utility of this kind of thinking relies on the perception that the demon is hidden in the way one handles the various possible pumping mechanisms or transition probabilities. The focus of this chapter is hence geared towards highlighting the basics of the interaction of quantized light with atom

© The Author(s), under exclusive license to Springer Nature Switzerland AG 2020
S. Tesfa, *Quantum Optical Processes*, Lecture Notes in Physics 976,
https://doi.org/10.1007/978-3-030-62348-7_6

confined in a cavity. One can specifically begin with a single two-level atom placed in a cavity and assumed to interact with the cavity radiation—the scheme that can be mathematically epitomized by Jaynes-Cummings model [1]—and leads to Rabi oscillations and splitting of emission spectrum into Mallow triplet [2].

In case the resonant cavity has an absorbing wall that allows the photons to escape, the emerging atomic emission may not be significantly different from the spontaneous emission in free space (see Sect. 6.2). Nonetheless, if the walls of the cavity are assumed to be very good reflectors, unique properties of the radiation, atomic population and coherence are expected to occur. Due to the fact that the ensued enclosure of the combined system can aid in maintaining quantum coherence of atom-radiation interaction over a considerable time scale, instead of emitting a photon that goes on its way, an excited atom in a resonant cavity oscillates between its excited and unexcited energy states since the emitted photon remains in the box in the vicinity of the atom and can be reabsorbed [3].

It is the possibility of the realization of a strong coupling that makes the mechanism of cavity quantum electrodynamics relevant to the conversation in the area of quantum information processing. For instance, an elementary process in which a single quanta of excitation can be transferred back and forth between the atom and photon number state of the radiation field is very vital to institute the idea of light induced quantum bit (see Sect. 9.2). It should also be possible to transfer excitations from the atom in one cavity to different atom in another cavity, and in the process induce entanglement between the pair of atoms separated by long distance. It is the rate by which each atomic energy level is decaying and the emerging cross terms that in practice determine the correlation between various emission roots. The knowledge of the intricate relationships among various decaying processes thus could be very important to grasp the idea related to the origin of the quantum features expected to emerge from the intended interaction.

The nature of the population statistics of a certain energy level also can be exploited as a viable information to generate radiation with prescribed properties via externally manipulating the involved energy levels (see Sect. 6.3). This kind of endeavor may have a potential to incite an exchange of quantum features between the radiation and atom that can be applied in conceiving a feasible device that can be used as a carrier of the message in quantum communication and storage of the memory in quantum computation [4]. Markedly, in case the atom is placed in the cavity for quite sometime, the emission-absorption events can take place repeatedly for so many times, and at some particular time one may lose the knowledge of the energy level the atom occupies. And yet, based on the strength of the external radiation and other relevant parameters, it is possible to predict the probability of getting the atom in the required energy level. It may not be hard to construe that the circumstance of losing the knowledge of which energy level the atom occupies leads to the idea of atomic coherent superposition claimed to be the cause for the nonclassical features of the emitted radiation.

The other interesting attribute of cavity quantum electrodynamics can be related to the possibility of transferring coherence from the driving laser field to the atomic energy levels and vice versa. Available works indicate that when many atoms are placed in the cavity and allowed to interact with radiation, not all of them absorb the radiation and jump to the upper energy level at the same time (see Sect. 6.4 and also Fig. 6.18). In this situation, since there is a real chance for the radiation emitted by one of the atoms to be absorbed by the other, the absorption-emission process could be more complicated. The exchange of the radiation among the involved atoms in the long run induces correlation among the corresponding energy levels. This correlation does not only depend on the interaction of the atom with radiation, but also on the energy level in which the neighboring atoms are found. Earlier studies in this direction indicate that a bunch of atoms tend to emit more frequently than when they are alone, that is, they exhibit superradiance [5]. To illustrate the basics and application of atomic coherent superposition, the same discussion can be extended to a multi-level atom scenario (see Sects. 6.5 and 6.6).

6.1 Two-Level Atom in Lossless Cavity

Contrary to the observation that an atom placed in a free space do not have a chance to reabsorb the light it emitted, sooner or later, one would be confronted with the situation in which the radiation generated by the atomic system is required to be amplified as in the lasing process. For the amplification to happen, there should be a mechanism by which the emitted radiation can be collected so that the subsequent re-emission by the atom would be facilitated. As a result, the atom needs to be confined within restricted geometrical structure so that the emitted radiation can be successively reflected. It should also be evident that the cavity mechanism can interact with the surrounding environment, and in the process exchange energy or radiation with it, that means, the cavity can give off energy to the environment accompanied by a loss of energy depending on the strength of the radiation mode. The notion of a lossless cavity hence can be taken as the idealization, and amounts to setting the reflectivity (transitivity) of the surface of the cavity to 100% (0%)—the situation in which no radiation is presumed to escape (enter) the cavity.

We now seek to extend the discussion that leads to interaction Hamiltonian (5.26) to a case when the radiation is assumed to be quantized. To this effect, following the same reasoning as for free atom and based on atomic completeness relation, it is straightforward to propose that

$$\hat{H}_I = \sum_{i,j} \langle i|\hat{H}_I|j\rangle |i\rangle\langle j| \qquad (6.1)$$

(compare with Eq. (5.6)). With the aid of the operator of the electric field (see Eq. (6.272), the quantized interaction Hamiltonian[1] can be designated as

$$\hat{H}_I(t) = -ie\sqrt{\frac{1}{2\omega\epsilon_0 V}}[\hat{a}(t) + \hat{a}^\dagger(t)][(\mathbf{r}_{aa}E_a - E_a\mathbf{r}_{aa})\hat{\sigma}_{aa} + (\mathbf{r}_{ab}E_b - E_a\mathbf{r}_{ab})\hat{\sigma}_{ab}$$

$$+ (\mathbf{r}_{ba}E_a - E_b\mathbf{r}_{ba})\hat{\sigma}_{ba} + (\mathbf{r}_{bb}E_b - E_b\mathbf{r}_{bb})\hat{\sigma}_{bb}].\hat{\mathbf{u}}, \qquad (6.2)$$

where $\mathbf{r}_{ij} = \langle i|\mathbf{r}|j\rangle$, V is the quantization volume and $\hat{\mathbf{u}}$ is the unit vector in the direction of the electric field.

On account of the proposal that the minimum coupling approximation is not suitable for the description of the atom that possesses permanent dipole moment ($\mathbf{r}_{aa} = \mathbf{r}_{bb} = 0$), one gets

$$\hat{H}_I(t) = -ie\omega_0\sqrt{\frac{1}{2\epsilon_0\omega V}}[\hat{a}(t) + \hat{a}^\dagger(t)][\mathbf{r}_{ba}\hat{\sigma}_{ba}(t) - \mathbf{r}_{ab}\hat{\sigma}_{ab}(t)].\hat{\mathbf{u}}, \qquad (6.3)$$

where $\omega_0 = E_a - E_b$. Then upon taking $\mathbf{r}_{ab} = \mathbf{r}_{ba}$, one can see that

$$\hat{H}_I(t) = ig[\hat{a}(t) + \hat{a}^\dagger(t)][\hat{\sigma}_+(t) - \hat{\sigma}_-(t)], \qquad (6.4)$$

where

$$g = \sqrt{\frac{\omega_0^2}{2\epsilon_0\omega V}}\mu_{ab}.\hat{\mathbf{u}} \qquad (6.5)$$

is the atomic coupling constant that stands for the measure of the strength of the interaction.

On the basis of the Heisenberg equation in the absence of interaction, one can see that

$$\hat{a}(t) = \hat{a}(0)e^{-i\omega t}, \qquad (6.6)$$

$$\sigma_\pm(t) = \sigma_\pm(0)e^{\pm i\omega_0 t}, \qquad (6.7)$$

and substitution of Eqs. (6.6) and (6.7) into Eq. (6.4) yields

$$\hat{H}_I(t) = ig\left[\hat{a}\hat{\sigma}_+e^{-i(\omega-\omega_0)t} - \hat{a}^\dagger\hat{\sigma}_-e^{i(\omega-\omega_0)t} - \hat{a}\hat{\sigma}_-e^{-i(\omega+\omega_0)t} + \hat{a}^\dagger\hat{\sigma}_+e^{i(\omega+\omega_0)t}\right].$$

$$(6.8)$$

---

[1]It can also be derived starting with Eq. (5.25) and noting the general relations such as $\hat{\mathbf{P}} = -im[\hat{\mathbf{r}}, \hat{H}_0]$ and $\hat{\mathbf{A}} = (\hat{a}^\dagger + \hat{a})\hat{\mathbf{e}}/\sqrt{2\omega\varepsilon_0 V}$, where $\hat{\mathbf{e}}$ is the unit vector in the direction of the vector potential, m is the mass of the charge and V is the quantization volume.

It might not be difficult to see that, near resonance ($\omega_0 \approx \omega$), terms with twice the resonance frequency oscillate quite rapidly when compared to the others. So in view of the rotating wave approximation, in which the rapidly oscillating terms are neglected, Eq. (6.8) reduces to

$$\hat{H}_I(t) = ig\left[\hat{a}\hat{\sigma}_+ e^{-i(\omega-\omega_0)t} - \hat{a}^\dagger\hat{\sigma}_- e^{i(\omega-\omega_0)t}\right]. \tag{6.9}$$

Equation (6.9) represents the simplest form of the atom-radiation interaction in a fully quantum treatment, and commonly dubbed as Jaynes-Cummings model. One may note that the rotating wave approximation amounts to neglecting $\hat{a}\hat{\sigma}_-$ ($\hat{a}^\dagger\hat{\sigma}_+$) term that corresponds to the de-excitation (excitation) of the atom while a single radiation is annihilated (created)—a least probable process to be physically realized under normal condition; and hence, the justification for the approximation. In spite of its elegance and consistency with many experimental results, rotating wave approximation has stringent drawbacks: for one, it is restricted to the first-order approximation; and for the other, it cannot be employed to study atom-atom interaction.

6.1.1 Population Dynamics

Once the quantized interaction Hamiltonian is known, the question that one may need to answer is how significantly the radiation modifies the occupancy of the atomic energy levels. In connection to this, it might be sufficient to look for the probability amplitude of the atom to be in the upper energy level since the atom naturally occupies the lower energy level in the absence of the radiation. The evolution of the probability amplitude for the atom to be in the upper energy level is designated as atomic population dynamics since the occupancy of the energy level is directly related to the number of photons in the cavity, and the atom and the cavity exchange the available photon. With this in mind, we seek to explore the population dynamics when a two-level atom is assumed to interact with a single-mode radiation in a lossless cavity. To begin with, Eq. (6.9) can be expressed in terms of the atomic and radiation operators where the corresponding eigenvector is expected to be describable in terms of the atomic and radiation state vectors in which the combined state of the atomic and radiation system—that should also be the eigenvector of the interaction Hamiltonian—is referred to as atomic dressed state.

To identify the atomic dressed states, one may need to make use of the Hamiltonian that describes the interaction of a two-level atom with a single-mode radiation in the interaction picture (see Appendix 1):

$$\hat{H}_I = ig(\hat{a}\hat{\sigma}_+ - \hat{\sigma}_-\hat{a}^\dagger). \tag{6.10}$$

Critical scrutiny of Eq. (6.10) may envisage that it would be logically appealing to represent the lower and upper energy states of the combined system as

$$|\ell\rangle = |b, n+1\rangle, \qquad |u\rangle = |a, n\rangle. \tag{6.11}$$

The state $|\ell\rangle$ stands for the case when there is $n+1$ photons in the cavity and the atom is in the lower energy level, and $|u\rangle$ when there is one less photon in the cavity and the atom is excited to the upper energy level.

The action of Hamiltonian (6.10) on these states thus leads to

$$\hat{H}_I |\ell\rangle = ig\sqrt{n+1}|a, n\rangle, \qquad \hat{H}_I |u\rangle = -ig\sqrt{n+1}|b, n+1\rangle. \tag{6.12}$$

One may interpret this outcome as the action of the Hamiltonian on the upper (lower) energy level gives rise to the emission (absorption) of a photon and de-excitation (excitation) of the atom.

The perception that the annihilation of the radiation would be accompanied by the excitation of the atom emanates from the assumption that there are only two physically accessible transformations in the interaction process. Pertaining to the fact that the free Hamiltonian commutes with the interaction Hamiltonian, which suggests that the eigenstates of the total Hamiltonian may be written as a linear combinations of the degenerate eigenstates of the free Hamiltonian, it might be advantageous introducing a linear combination of the form

$$|n, \pm\rangle = \frac{1}{\sqrt{2}}\left[|a, n\rangle \pm i|b, n+1\rangle\right]. \tag{6.13}$$

On account of the presumption that the atomic energy levels are masked by the radiation field, a coherent combination of the atomic and radiation eigenstates is alluded to as atomic dressed state while the state that represents the atomic system alone is known as atomic bare state.

In view of Eq. (6.13), it may not be difficult to verify that

$$|a, n\rangle = \frac{1}{\sqrt{2}}\left[|n, +\rangle + |n, -\rangle\right], \qquad |b, n+1\rangle = \frac{i}{\sqrt{2}}\left[|n, -\rangle - |n, +\rangle\right], \tag{6.14}$$

where $|a, n\rangle$ and $|b, n+1\rangle$ can also be expressed as coherent superposition of $|n, \pm\rangle$ (for application; see Chaps. 7 and 11). With the aid of such consideration, it might be possible to ascertain the assertion that the dressed state is the eigenstate of interaction Hamiltonian:

$$\hat{H}_I |n, \pm\rangle = \mp\frac{\Omega_n}{2}|n, \pm\rangle, \tag{6.15}$$

where

$$\Omega_n = 2g\sqrt{n+1} \tag{6.16}$$

is defined as Rabi frequency (relate with Eq. (5.57)). By making use of Eq. (6.11), one can also see that the possible dressed atomic states are orthonormal, and so complete;

$$\hat{I} = \sum_{n=0}^{\infty} \left[|n, +\rangle\langle +, n| + |n, -\rangle\langle -, n| \right]. \tag{6.17}$$

In addition, upon expressing the total quantized unperturbed Hamiltonian as

$$\hat{H}_0 = \frac{1}{2}\omega_0\hat{\sigma}_z + \omega\hat{a}^\dagger\hat{a}, \tag{6.18}$$

and following a straightforward reasoning, one can see that the basis states are also the eigenstates of the unperturbed system:

$$\hat{H}_0|a, n\rangle = \left(\frac{\omega_0}{2} + n\omega \right) |a, n\rangle, \tag{6.19}$$

$$\hat{H}_0|b, n+1\rangle = \left[(n+1)\omega - \frac{\omega_0}{2} \right] |b, n+1\rangle. \tag{6.20}$$

One can observe by taking the action of the interaction Hamiltonian on the atomic bare states that the resulting eigenvalue equations can be put when there is detuning in the form

$$\hat{H}_n = \omega\left(n + \frac{1}{2} \right)\begin{pmatrix} 1 & 0 \\ 0 & 1 \end{pmatrix} + \begin{pmatrix} \Delta & -ig\sqrt{n+1} \\ ig\sqrt{n+1} & -\Delta \end{pmatrix}, \tag{6.21}$$

where the energy eigenvalues are found to be

$$E_{n\pm} = \omega\left(n + \frac{1}{2} \right) \pm \frac{\Omega_n}{2} \tag{6.22}$$

in which

$$\Omega_n = \sqrt{\Delta^2 + 4g^2(n+1)} \tag{6.23}$$

is the generalized Rabi frequency (relate with Eq. (5.70)). One may note that the atomic dressed state can be expressed as

$$|n, -\rangle = \frac{1}{\sqrt{(\Omega_n - \Delta)^2 + 4g^2(n+1)}}[(\Omega_n - \Delta)|a, n\rangle - 2g\sqrt{n+1}|b, n+1\rangle], \tag{6.24}$$

$$|n, +\rangle = \frac{1}{\sqrt{(\Omega_n - \Delta)^2 + 4g^2(n+1)}}[2g\sqrt{n+1}|a, n\rangle + (\Omega_n - \Delta)|b, n+1\rangle]. \tag{6.25}$$

On the basis that the dressed atomic states are the eigenstates of atom-radiation interaction, one can use them to obtain the evolution of the state vector of the combined system that can be written applying

$$\hat{U}(t) = \sum_{n=0}^{\infty} \left[e^{-i\hat{H}_I t} |n, +\rangle\langle +, n| + e^{-i\hat{H}_I t} |n, -\rangle\langle -, n| \right] \tag{6.26}$$

and Eqs. (6.13) and (6.15) as

$$\hat{U}(t) = \sum_{n=0}^{\infty} \left[\cos\left(\frac{\Omega_n t}{2}\right) \left[|a, n\rangle\langle n, a| + |b, n+1\rangle\langle n+1, b| \right] \right.$$
$$\left. - \sin\left(\frac{\Omega_n t}{2}\right) \left[|a, n\rangle\langle n+1, b| - |b, n+1\rangle\langle n, a| \right] \right]. \tag{6.27}$$

The matrix elements of the density operator can then be put upon introducing the identity operator twice in the form

$$\rho_{an,\beta m}(t) = \sum_{\alpha'\beta'n'm'} \langle \alpha, n|\hat{U}(t)|\alpha', n'\rangle\langle n', \alpha'|\hat{\rho}|\beta', m'\rangle\langle m', \beta'|\hat{U}^{\dagger}(t)|\beta, m\rangle. \tag{6.28}$$

It may not be difficult to notice from the discussion in Sect. 3.2.3 that $\langle a, n|\hat{U}(t)|\alpha', n'\rangle$ is the propagator, where the atomic aspects can be retrieved by summing over the radiation terms (see Eq. (3.186)). This suggestion is concerned with treating the radiation as reservoir in which the reduced density matrix that describes the atom alone can be expressed as $\rho_{\alpha\beta}(t) = \sum_{nm} \rho_{an,\beta m}(t)$;

$$\rho_{\alpha\beta}(t) = \sum_{nm} \sum_{\alpha'n'} \sum_{\beta'm'} K(\alpha, n, t|\alpha', n', 0) K^*(\beta, m, t|\beta', m', 0) \, \rho_{\alpha'n',\beta'm'}(0), \tag{6.29}$$

where

$$K(\alpha, n, t|\alpha', n', 0) = \langle \alpha, n|\hat{U}(t)|\alpha', n'\rangle, \qquad \rho_{\alpha'n',\beta'm'}(0) = \langle n', \alpha' \mid \hat{\rho} \mid \beta', m'\rangle. \tag{6.30}$$

Since the interaction is yet to be realized at initial time, it should be possible to decouple the combined density operator at initial time to the one that corresponds to the atom and radiation as

$$\rho_{\alpha'n',\beta'm'}(0) = \rho_{\alpha'\beta'}(0)\rho_{n'm'}(0). \tag{6.31}$$

For the case when the atom is taken to be initially in the upper (lower) energy level, it is straightforward to see that $\rho_{\alpha\beta}(0) = \delta_{\alpha a(b)}\delta_{\beta a(b)}$. One may then seek to explore

various physical phenomena when the atom is initially taken to be in the upper
energy level, in which case the density matrix elements can be expressed as

$$\rho_{\alpha\beta}(t) = \sum_{nm}\sum_{n'm'} K'(u, n, t|u, n', 0) K''^*(\beta, m, t|u, m', 0)\rho_{n'm'}(0). \qquad (6.32)$$

Consequently, the probability of the occupancy of the upper energy level at any later
time would be

$$\rho_{aa}(t) = \sum_{nm}\sum_{n'm'} K(a, n, t|a, n', 0) K^*(a, m, t|a, m', 0)\rho_{n'm'}(0), \qquad (6.33)$$

where $\rho_{n'm'}(0)$ is the probability density for finding the radiation to be in the states
describable by n' and m' at initial time.

One can also verify with the aid of Eq. (6.27) and orthonormality condition of the
atomic and radiation states that the probability of the occupancy of the upper energy
level to be

$$\rho_{aa}(t) = \sum_{nm} \cos\left(\frac{\Omega_n t}{2}\right) \cos\left(\frac{\Omega_m t}{2}\right) \rho_{nm}(0). \qquad (6.34)$$

Based on the fact that out of the matrix elements of the density operator only the
diagonal matrix elements survive, it might be interesting relating Eq. (6.34) with
photon number distribution since the photon number distribution is the trace of the
same matrix: $\rho_{nm}(0) = P_m(0)\delta_{nm}$; and one can then obtain

$$\rho_{aa}(t) = \sum_{m} P_m(0) \cos^2\left(\frac{\Omega_m t}{2}\right). \qquad (6.35)$$

Equation (6.35) designates the probability for a two-level atom initially in the
upper energy state to be still in the same state while it interacts with different types
of driving radiation fields. One may then observe that the population in the upper
energy level oscillates with time irrespective of the initial property of the radiation
and the rate of the oscillation depends on the Rabi frequency.

Moreover, to expose the issue of spontaneous emission, one may require to stick
to the quantum treatment of the interaction of excited atom with a continuum of
radiation modes. In this context, the total Hamiltonian that describes the interaction
of a two-level atom with a multi-mode radiation can be expressed as

$$\hat{H} = \sum_{k} \omega_k \hat{a}_k^\dagger \hat{a}_k + \omega_a \hat{\sigma}_a + \omega_b \hat{\sigma}_b + \sum_{k} \left(g_k \hat{a}_k \hat{\sigma}_+ + H.c\right) \qquad (6.36)$$

in which the corresponding state vector would have the form

$$|\Psi(\mathbf{r}, t)\rangle = C_{a0}(t)e^{-(i\omega_a)t}|a, 0\rangle + \sum_k C_{b1_k}(t)e^{-i(\omega_b + \varphi_k)t}|b, 1_k\rangle, \tag{6.37}$$

where φ_k is the phase factor.

In case the atom is assumed to be initially in the excited state and the radiation in the vacuum state, $C_{a0}(0) = 1$, and all other probability amplitudes vanish, the interaction part of this Hamiltonian connects the state $|a, 0\rangle$ to $|b, 1_k\rangle$. The combined state $|b, 1_k\rangle$ stands for the situation in which the atom in the lower state with one photon in the kth mode and no photons in any other mode (see Chap. 5). This can be taken as a viable transition of the atom from the upper energy level to the lower without requiring the assistance of an external agent—the physical phenomenon usually dubbed as spontaneous emission [6].

Upon making use of Eq. (3.1), one can find

$$\frac{d}{dt}C_{a0}(t) = -i \sum_k g_k e^{-i(\omega_k - \omega_0)t} C_{b1_k}(t), \tag{6.38}$$

$$\frac{d}{dt}C_{b1_k}(t) = -ig_k^* e^{-i(\omega_k - \omega_0)t} C_{a0}(t), \tag{6.39}$$

from which follows

$$\frac{d}{dt}C_{a0}(t) = -\sum_k |g_k|^2 \int_{t_0}^{t} dt' e^{-i(\omega_k - \omega_0)(t - t')} C_{a0}(t'). \tag{6.40}$$

One may note that the solution of Eq. (6.40) would not be trivial since $C_{a0}(t')$ happens to be a function of earlier time. So to solve this equation, the concept of a coarse graining approximation in which $C_{a0}(t')$ is assumed to vary little in time interval $t - t_0$ over which the remaining part of the integrand has nonzero value is utilized.

Dropping the second term of the integrand following from Eq. (6.40), which stands for the frequency shift, one obtains[2] with the help of Eq. (5.57)

---

[2]The involved sum can be converted to integration over spherical region by applying

$$\sum_k |g_k|^2 \rightarrow \frac{1}{(2\pi c)^3 V} \int r^2 dr \int_0^{\pi} \sin\theta d\theta \int_0^{2\pi} |e\mathbf{r}_{ab} \cdot \mathbf{E}|^2 \sin^2\theta d\phi,$$

and the integral equation can also be rewritten using

$$\lim_{t \to \infty} \int_{t_0}^{t} dt' e^{-i(\omega_k - \omega)(t - t')} = \pi\delta(\omega_k - \omega) - \oint d\left[\frac{i}{\omega_k - \omega}\right].$$

$$\frac{d}{dt}C_{a0}(t) = -\frac{\Gamma}{2}C_{a0}(t), \tag{6.41}$$

where Γ is the spontaneous emission decay rate

$$\Gamma = \frac{\Omega^2 \omega_0^3}{3\pi \epsilon_0 c^3}, \tag{6.42}$$

which indicates that the decay rate increases with the cube of the atomic natural frequency, and one can write

$$C_a(t) = C_a(0)e^{-\frac{\Gamma}{2}t}. \tag{6.43}$$

As one can see from Eq. (6.43), this theory—usually dubbed as Weisskopf-Wigner spontaneous emission theory—predicts irreversible exponential decay of the population of the upper energy level.

6.1.2 Radiative Properties

Since the number of photons that engage in the interaction is large, it would not be easy to know in what way the number of photons in the cavity changes as a result of the emission or absorption. By the way, looking into the evolution of the density of the occupancy of the upper energy level may help to figure out whether the number of photons in the cavity would increase. It is such a way of exploring the mechanism of emission-absorption in terms of the dynamics of the population that is designated as atomic radiative property, where the main issue is to explain the nature of the emission of the atom based on the underlying population dynamics. Incidentally, the program is to look into the means of capturing the anticipated modification in the population of the atom in the upper energy level and population inversion.

To do so, one can express the photon number distribution associated with the radiation in the number state as $P_m(0) = \delta_{nm}$ (see Eq. (2.112)). The probability of the occupancy of the upper and lower energy levels when a two-level atom is placed in a lossless cavity initially in the number state can then be put in the form

$$\rho_{aa}(t) = \cos^2\left(\frac{\Omega_n t}{2}\right), \qquad \rho_{bb}(t) = \sin^2\left(\frac{\Omega_n t}{2}\right) \tag{6.44}$$

(see Eq. (6.35)).

As portrayed in Fig. 6.1, the atom oscillates back and forth between the upper and lower energy levels with frequency Ω_n; the phenomenon usually referred to as Rabi oscillation. It is also possible to infer in view of the definition $\Omega_n = 2g\sqrt{n+1}$ that the bigger the number of the photons in the cavity, the faster the oscillation would be, and the oscillation persists even when the cavity is initially in the vacuum state. However, an excited free atom placed in a region void of external radiation

Fig. 6.1 Plots of the population dynamics of the upper and lower atomic energy levels when the atom is placed in a lossless cavity that initially contains a number state

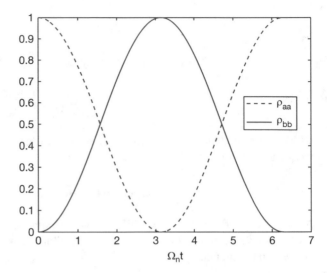

spontaneously emits radiation, and goes over to the lower energy level but with no chance to return to the upper energy level. It hence seems reasonable to think of the cavity mechanism as responsible for Rabi oscillation. It might be equally important noting that the oscillation disappears when the cavity is maintained in the vacuum state and the atom is initially taken to be in the lower energy level.

To get insight, imagine the case in which the radiation in the cavity is initially in the vacuum state ($\bar{n} = 0$) that can be achieved by keeping the bounding cavity at zero temperature and also by making sure that there is no external radiation in the cavity while the atom resides there. In this scenario, the probabilities for getting the atom to be in the upper and lower energy levels turn out to be

$$\rho_{aa}(t) = \cos^2(gt), \qquad \rho_{bb}(t) = \sin^2(gt). \tag{6.45}$$

Equation (6.45) indicates that the atom and the cavity periodically exchange the single available photon to the system (see also Fig. 6.1). This process may illustrate the simplest picture of the phenomenon of spontaneous emission in which the spontaneously emitted photon constitutes the available single-mode radiation.

Let us now turn the attention to the situation in which the cavity is assumed to contain initially a coherent light. To study the evolution of the population in the upper energy level, Eq. (6.44) needs to be rewritten with the aid of the trigonometric relation $\cos^2\theta = \frac{1}{2}(1 + \cos 2\theta)$ and the fact that $\sum_m P_m(0) = 1$ as

$$\rho_{aa}(t) = \frac{1}{2}\left[1 + \sum_m P_m(0)\cos(\Omega_m t)\right]. \tag{6.46}$$

Fig. 6.2 Plot of the
population dynamics of the
upper atomic energy level
when the lossless cavity
contains initially a coherent
light with mean photon
number of $\bar{m} = 10$

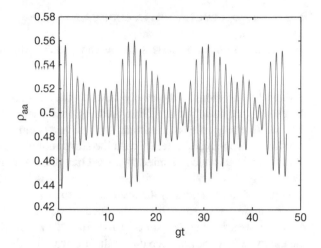

With the assertion that the photon number distribution for a coherent light is $P_m = \frac{\bar{m}^m e^{-\bar{m}}}{m!}$ (see Eq. (2.125)), the probability for getting the atom in the upper energy level at any later time and the atomic population inversion take the form

$$\rho_{aa}(t) = \frac{1}{2}\left[1 + \sum_m \frac{\bar{m}^m e^{-\bar{m}}}{m!}\cos(\Omega_m t)\right], \qquad (6.47)$$

$$W(t) = \sum_m \frac{\bar{m}^m e^{-\bar{m}}}{m!}\cos(\Omega_m t). \qquad (6.48)$$

Due to the exponential and cosine terms in Eq. (6.47), the Rabi oscillations exhibit collapse at first and partial revival later (see Fig. 6.2)—the phenomenon that would be repeated indefinitely and heralds the manifestation of the quantum features of the emitted radiation—the process commonly dubbed as atomic collapse and revival. Critical scrutiny may reveal that the existence of such periodic revivals can be linked to the discreteness of the sum over the number of the photon which ensures that after sometime the oscillating terms comeback in phase with each other, and so restore the coherent oscillation. The same physical aspect can also be interpreted in a microscopic level as coherence property of the radiation can be transferred to the dressed atomic states due to the involved successive exchange of the available radiation. Once the notion of the induced coherence is conceived, the subsequent emission and absorption process can be thought of as highly correlated—the presumption that might be anchored to the foundation of the emerging quantum features.[3]

---

[3]Taking the population inversion (6.48) into consideration leads to the same understanding. The case for which the initial radiation is in squeezed state can be considered, but no striking new feature is expected.

6.2   Two-Level Atom in Leaky Cavity

Placing the atom in a lossless cavity can prolong the interaction time; and consequently, enhances the chance for the atom and cavity to exchange the available
photon. Likewise, the situation in which there is a radiation other than the one
emitted by the atom in the cavity leads to the conception of masking the atomic
states with the radiation or atomic dressing that leads to the collapse and revival of
the radiative properties of the atom. A physical system enclosed in a cavity however
would be compelled to interact with the surrounding environment (Sect. 3.1). Such
interaction is also accompanied by the exchange of radiation between the cavity and
reservoir that instigates the radiation in the cavity to leave the cavity, and at the same
time the radiation modes of the reservoir to enter the cavity (see Fig. 6.3).

The main cause for losing coherence or observing decoherence in the cavity
quantum electrodynamics can be linked to the way in which the radiation escapes
the cavity, which may happen due to the absorption of the cavity walls or scattering
or transmission of radiation into electromagnetic modes outside the cavity [3].
The tendency of the photons escaping into a specific electromagnetic mode can
dominate other forms of external coupling such as atomic spontaneous emission,
and constitutes a well defined output channel for the open quantum system. It is
the idea of making such an output channel accessible to the system outside the
cavity that envisages the notion of the entanglement of the quantum system in the
cavity with the surrounding environment (see Sect. 4.1.4). The radiative properties
of the atom and the statistical properties of the emitted radiation are thus expected
to be significantly modified from the lossless cavity scenario presented in Sect. 6.1.
The main theme of this section is hence geared towards exploring the effects of
the external environment on the dynamics of an atom placed in a cavity that leaks
radiation to the surrounding environment.

In case the cavity experiences a measurable damping, the interaction of a two-
level atom with the modes of the reservoir can be described in the interaction picture

Fig. 6.3 Schematic
representation of an atom in a
lossy cavity when the atom
interacts with environment
through exchange of radiation
via the wall of the cavity with
the help of input and output
mechanism in terms of noise
fluctuations (relate with
Fig. 3.1)

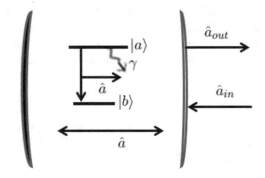

by Hamiltonian of the form

$$\hat{H}_{SR}(t') = i \sum_{k} g_k \left(\hat{\sigma}_+ \hat{a}_k e^{i(\omega_0 - \omega_k)t'} - \hat{\sigma}_- \hat{a}_k^\dagger e^{-i(\omega_0 - \omega_k)t'} \right), \tag{6.49}$$

where $\hat{a}_k$ is the annihilation operator for the reservoir modes embodied by the wave vector $\mathbf{k}$ and g_k is the coupling constant for each mode of the reservoir. The reduced density operator that describes the two-level atom is thus taken to evolve in time according to

$$\frac{d\hat{\rho}}{dt} = -i[\hat{H}_S, \ \hat{\rho}(t)]$$

$$- \int_0^t Tr_R(\hat{R}\hat{H}_{SR}(t)\hat{H}_{SR}(t'))\hat{\rho}(t)dt' - \int_0^t \hat{\rho}(t)Tr_R(\hat{R}\hat{H}_{SR}(t')\hat{H}_{SR}(t))dt'$$

$$+ \int_0^t Tr_R(\hat{H}_{SR}(t)\hat{\rho}(t)\hat{R}\hat{H}_{SR}(t'))dt' + \int_0^t Tr_R(\hat{H}_{SR}(t')\hat{\rho}(t)\hat{R}\hat{H}_{SR}(t))dt'$$

in the Born-Markov approximation (see Eq. (3.31)).

Upon sticking to the outlined approach and making use of the Hamiltonian (6.49), one can rewrite terms in this expression in view of the system and reservoir mode variables (see Sect. 3.1.1). For instance, on account of the proposal that the system and cavity variables commute at equal time and picture, one can see that

$$\hat{H}_{SR}(t)\hat{H}_{SR}(t') = -\sum_{j,k} g_k g_j \left[\hat{\sigma}_+^2 \hat{a}_k \hat{a}_j e^{i(\omega_0 - \omega_k)t' + i(\omega_0 - \omega_j)t} \right.$$

$$+ \hat{\sigma}_-^2 \hat{a}_k^\dagger \hat{a}_j^\dagger e^{-i(\omega_0 - \omega_j)t - i(\omega_0 - \omega_k)t'}$$

$$- \hat{\sigma}_- \hat{\sigma}_+ \hat{a}_k^\dagger \hat{a}_j e^{i(\omega_0 - \omega_k)t - i(\omega_0 - \omega_j)t'}$$

$$\left. - \hat{\sigma}_+ \hat{\sigma}_- \hat{a}_k \hat{a}_j^\dagger e^{i(\omega_0 - \omega_j)t - i(\omega_0 - \omega_k)t'} \right], \tag{6.50}$$

from which follows

$$Tr_R(\hat{R}\hat{H}_{SR}(t')\hat{H}_{SR}(t)) = J_1\hat{\sigma}_-\hat{\sigma}_+ + J_2\hat{\sigma}_+\hat{\sigma}_- + J_3\hat{\sigma}_-^2 + J_4\hat{\sigma}_+^2, \tag{6.51}$$

where the values of J_i's can be obtained with the aid of the correlations among the operators of a single-mode squeezed vacuum reservoir.[4]

---

[4]Recall that for a single-mode squeezed vacuum reservoir:

$$\langle \hat{a}_k^\dagger \hat{a}_j \rangle_R = N\delta_{jk}, \qquad \langle \hat{a}_k \hat{a}_j^\dagger \rangle_R = (N+1)\delta_{jk}, \qquad \langle \hat{a}_k \hat{a}_j \rangle_R = \langle \hat{a}_k^\dagger \hat{a}_j^\dagger \rangle_R = M\delta_{j,2k_0-k},$$

where $N = \sinh^2 r$ is the mean photon number and $M = \sinh r \cosh r = \sqrt{N(N+1)}$ is the measure of the correlation among reservoir modes (see Appendix 2 of Chap. 3).

The master equation that describes a two-level atom placed in a cavity coupled to a single-mode squeezed vacuum reservoir is found following the same approach to be

$$\frac{d\hat{\rho}}{dt} = -i\left[\hat{H}_S(t),\ \hat{\rho}(t)\right] + \frac{\gamma(N+1)}{2}\left[2\hat{\sigma}_-\hat{\rho}\hat{\sigma}_+ - \hat{\sigma}_+\hat{\sigma}_-\hat{\rho} - \hat{\rho}\hat{\sigma}_+\hat{\sigma}_-\right]$$
$$+ \frac{\gamma N}{2}\left[2\hat{\sigma}_+\hat{\rho}\hat{\sigma}_- - \hat{\sigma}_-\hat{\sigma}_+\hat{\rho} - \hat{\rho}\hat{\sigma}_-\hat{\sigma}_+\right] - \gamma M\left[\hat{\sigma}_-\hat{\rho}\hat{\sigma}_- + \hat{\sigma}_+\hat{\rho}\hat{\sigma}_+\right],$$

$$(6.52)$$

where

$$\gamma = \frac{\omega_0^3 \mu_{ab}^2}{3\pi \epsilon_0 c^3} \tag{6.53}$$

is atomic decay rate (see Appendix 2 and compare also with Eq. (6.42)).

In depth scrutiny may reveal that this damping constant can be taken as par with the rate at which the radiation emitted by the atom leaks via the wall of the cavity, and may have the same significance as the cavity damping constant provided earlier for the radiation in the cavity (compare with Eq. (3.41)). Even then, one should note that atomic damping constant depends on the atomic parameters that include the dipole moment and atomic transition frequency since it is the atom that is presumed to interact with the reservoir modes directly contrary to earlier discussion when the cavity radiation is taken to interact with the reservoir modes.

6.2.1   Coupled to Thermal Reservoir

We now seek to study the effects of the reservoir mode fluctuations on the radiative properties of the atom and statistical properties of the emitted radiation when the cavity that contains a two-level atom is placed in thermal reservoir.[5] Recall that the time evolution of the population of the atom that occupies one of the atomic energy levels can be derived by using the pertinent Hamiltonian (see Eq. (3.2)). With the same intention in mind, we opt to derive the dynamical equation that can be utilized to expose the atomic radiative property and statistical properties of the emitted radiation. To this end, for thermal reservoir, one can notice that $N = \bar{n}$ and $M = 0$ for which Eq. (6.52) reduces to

$$\frac{d\hat{\rho}}{dt} = \frac{\gamma(\bar{n}+1)}{2}[2\hat{\sigma}_-\hat{\rho}\hat{\sigma}_+ - \hat{\sigma}_+\hat{\sigma}_-\hat{\rho} - \hat{\rho}\hat{\sigma}_+\hat{\sigma}_-]$$
$$+ \frac{\gamma\bar{n}}{2}[2\hat{\sigma}_+\hat{\rho}\hat{\sigma}_- - \hat{\sigma}_-\hat{\sigma}_+\hat{\rho} - \hat{\rho}\hat{\sigma}_-\hat{\sigma}_+]. \tag{6.54}$$

---

[5]The discussion on the contribution of the radiation available in the cavity other than the reservoir modes is differed to later sections (see also [7]).

Once the master equation is known, various expectation values can be obtained by employing the definition,

$$\frac{d}{dt}\langle\sigma_i(t)\rangle = Tr\left(\frac{d\hat{\rho}(t)}{dt}\sigma_i\right),\tag{6.55}$$

where $i = +, -, z$. In view of the information provided in Sect. 5.1.2, one can get

$$\frac{d}{dt}\langle\hat{\sigma}_+(t)\rangle = -\frac{\gamma(2\bar{n}+1)}{2}\langle\hat{\sigma}_+(t)\rangle\tag{6.56}$$

whose solution can be written as

$$\langle\hat{\sigma}_+(t)\rangle = \langle\hat{\sigma}_+(0)\rangle e^{-\frac{\gamma}{2}(2\bar{n}+1)t}.\tag{6.57}$$

One can infer from Eq. (6.57) that the atomic coherence decays exponentially, where one of the contributions of thermal reservoir is to enhance this decaying process. Such a result can be taken as the basis for introducing the idea of decoherence in this kind of discussion. In the same way, one can see that

$$\frac{d}{dt}\langle\hat{\sigma}_z(t)\rangle = -\gamma(2\bar{n}+1)\langle\hat{\sigma}_z(t)\rangle - \gamma\tag{6.58}$$

the solution of which can be proposed as

$$\langle\hat{\sigma}_z(t)\rangle = \left[\langle\hat{\sigma}_z(0)\rangle + \frac{1}{2\bar{n}+1}\right]e^{-\gamma(2\bar{n}+1)t} - \frac{1}{2\bar{n}+1}.\tag{6.59}$$

Once $\langle\hat{\sigma}_z(t)\rangle$ is known, it would be straightforward to analyze population dynamics and atomic population inversion (see Sect. 5.1.2). In this regard, upon using Eqs. (5.43) and (6.59), one can write

$$\hat{\rho}_{aa}(t) = \frac{\bar{n}}{2\bar{n}+1} + \frac{1}{2}\left[\langle\hat{\sigma}_z(0)\rangle + \frac{1}{2\bar{n}+1}\right]e^{-\gamma(2\bar{n}+1)t},\tag{6.60}$$

which represents the probability for the two-level atom to occupy the upper energy level at time t. One can assert at steady state and for $\bar{n} \gg 1$ that the atom would have nearly equal probability to occupy either the upper or lower atomic energy level ($\rho_{aa} \simeq 0.5$). But when the reservoir is in the vacuum state ($\bar{n} = 0$), the probability for the atom to be in the upper energy level reduces to

$$\rho_{aa}(t) = \rho_{aa}(0)e^{-\gamma t}.\tag{6.61}$$

This outcome entails that the probability of the occupancy of the upper atomic energy level would be zero at steady state in case there is no radiation in the cavity other than the one emitted by the atom. It might be important hence identifying two interesting aspects of assuming the cavity to allow the radiation to pass partially through it: for one, the revival phenomenon disappears when the radiation has a chance to leak through the walls of the cavity to empty environment (compare with the result in Sect. 6.1.2 specially with Figs. 6.1 and 6.2); and for the other, even though the thermal fluctuation that enters the cavity can destroy the coherence, it has a potential to pump the atom in the same way as an external driving radiation does.

In the same token, with the aid of Eqs. (5.43), (5.44) and (6.59), the atomic population inversion is found to be

$$W(t) = 2\left(\rho_{aa}^{(0)} - \frac{\bar{n}}{2\bar{n} + 1}\right)e^{-\gamma(2\bar{n}+1)t} - \frac{1}{2\bar{n} + 1}, \tag{6.62}$$

which reduces when the atom is assumed to be initially in the upper energy level to

$$W(t) = \frac{1}{2\bar{n} + 1}\left[2(\bar{n} + 1)e^{-\gamma(2\bar{n}+1)t} - 1\right]. \tag{6.63}$$

One can see from Fig. 6.4 that the rate at which the atom decays could be so sharp and fast at early stages of the evolution in case the reservoir is strong. Nonetheless, at steady state, as usually expected, the atom tends to stay in the lower energy level when the reservoir is very weak. It may worth underlining that although the probability for the atom to occupy the upper energy level at steady state increases with the strength of the reservoir, it turns out to be below 50% regardless.

We next seek to obtain the normalized absorption spectrum of the radiation emitted by a two-level atom placed in thermal reservoir by making use of the

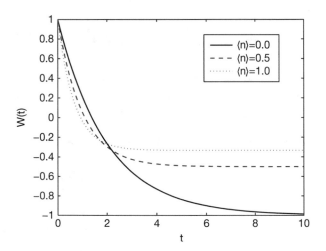

Fig. 6.4 Plots of the time evolution of the atomic inversion for $\gamma = 0.5$ and different values of $\bar{n}$

procedure already outlined (see Sects. 4.2 and 5.5). To begin with, applying Eqs. (6.57) and (6.60), one can write

$$\langle \hat{\sigma}_+(t)\hat{\sigma}_-(t+\tau)\rangle = \left[\frac{n}{2\bar{n}+1} + \frac{1}{2}\left(\langle \hat{\sigma}_z(0)\rangle + \frac{1}{2\bar{n}+1} \right) e^{-\gamma(2\bar{n}+1)t} \right] e^{\frac{\gamma}{2}(2\bar{n}+1)\tau},$$

(6.64)

where at steady state

$$\langle \hat{\sigma}_+(t)\hat{\sigma}_-(t+\tau)\rangle_{ss} = \frac{\bar{n}}{2\bar{n}+1} e^{-\frac{\gamma}{2}(2\bar{n}+1)\tau}.$$

(6.65)

Upon using Eqs. (5.107) and (6.65), one can see that

$$S(\omega) = \frac{4\gamma\bar{n}f(r)}{4\omega^2 + \left[\gamma(2\bar{n}+1)\right]^2}.$$

(6.66)

Obtaining the exact value of $f(r)$ requires the detailed knowledge of the atom involved in the interaction such as the orientation of the dipole moment.

To understand the main features of the statistical properties of the emitted radiation that can be captured by Eq. (6.66), the normalized version of the absorption spectrum is more often found to be sufficient. In light of this, the normalization process is found to be helpful in getting rid of the contribution of $f(r)$ so that one should not be bothered to look into the details of the interaction to interpret the pertinent photon statistics. To expedite the normalization process, one can utilize the integral identity of the form $\int_{-\infty}^{\infty} \frac{d\omega}{\omega^2+a^2} = \frac{\pi}{a}$. The normalized absorption spectrum can hence be put straightaway in the form

$$S(\omega) = \frac{2\gamma(2\bar{n}+1)/\pi}{4\omega^2 + \left[\gamma(2\bar{n}+1)\right]^2};$$

(6.67)

and for $\bar{n}=0$

$$S(\omega) = \frac{2\gamma/\pi}{4\omega^2 + \gamma^2}.$$

(6.68)

Equations (6.67) and (6.68) indicate that the normalized absorption spectrum of the radiation emitted by a two-level atom placed in a lossy cavity is Lorentzian. It is also possible to affirm that the broadening in the absorption spectrum is induced by damping in which the spectrum becomes broader and shorter for stronger damping. The fluctuation in the modes of the thermal input is found in this sense to broaden the spread of the absorption spectrum.

In addition, recall that the time delayed scenario can display the correlation between the radiation emitted by a two-level atom at different times and can be determined by calculating the two-time second-order correlation (see Sect. 4.2.2).

In this respect, first and foremost, it is good to note that Eq. (5.110) encompasses two terms where the first denotes the population in the upper energy level which is already rigorously calculated. One may thus need to rewrite the second term in a more appealing form starting with Eq. (6.60) and in view of the property of the atomic operator $\hat{\sigma}_z \hat{\sigma}_- = -\hat{\sigma}_-$ as

$$\langle \hat{\sigma}_+(t) \hat{\sigma}_z(t + \tau) \hat{\sigma}_-(t) \rangle = \left[-\frac{2\bar{n}}{2\bar{n} + 1} e^{-\gamma(2\bar{n}+1)\tau} - \frac{1}{2\bar{n} + 1} \right] \langle \hat{\sigma}_+(t) \hat{\sigma}_-(t) \rangle.$$
(6.69)

Upon applying Eqs. (5.110), (6.59), (6.60) and (6.69), it is possible afterward to find

$$G^{(2)}(\tau) = \frac{\bar{n}}{2\bar{n} + 1} \langle \hat{\sigma}_+(t) \hat{\sigma}_-(t) \rangle \left[1 - e^{-\gamma(2\bar{n}+1)\tau} \right].$$
(6.70)

One may also need to look for the normalized two-time second-order correlation function to directly infer the photon statistics (see Sect. 4.2). In line with this, the normalized correlation function at steady state is found to be

$$g^{(2)}(\tau) = 1 - e^{-\gamma(2\bar{n}+1)\tau}.$$
(6.71)

It is evident from Fig. 6.5 that $g^{(2)}(0) = 0$ regardless of the strength of the reservoir, and $g^{(2)}(\tau) > g^{(2)}(0)$ for $\tau \neq 0$. This outcome can serve as justification for a two-level atom to tend to emit radiation with more correlation at different times, which demonstrates that the emitted photon exhibits a phenomenon of photon anti-bunching irrespective of the strength of thermal reservoir. Besides, since $g^{(2)}(\tau) < 1$, one can witness that the emitted radiation exhibits sub-Poissonian photon statistics. One can hence realize that the main effect of thermal reservoir

Fig. 6.5 Plots of the two-time correlation function for $\gamma = 0.5$ and different values of $\bar{n}$

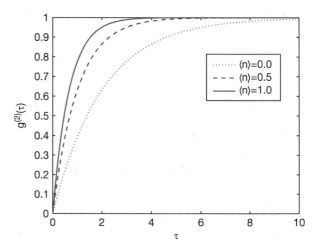

on the radiative properties of the atom is to preserve the probability of finding the atom in the upper energy level and somehow enhance the rate at which the atomic coherence would be lost. The thermal modes that get a chance to enter the cavity are found to enhance the broadening of the absorption spectrum in the same way as the external pumping but with no splitting (compare with the result in Sect. 6.3 specially Fig. 6.9). It should also be emphasized that even though both anti-bunching and sub-Poissonian photon features depend on the strength of the reservoir modes, thermal reservoir is not directly responsible to observe the nonclassical features of the radiation.

6.2.2 Coupled to Squeezed Vacuum Reservoir

We now intend to expose the extent of the influence of the correlation between the quadratures of the reservoir modes on the radiative and statistical properties of a two-level atom placed in a cavity coupled to a single-mode squeezed vacuum reservoir. Aiming at attesting to the proposal that the quantum traits of the emitted radiation would be enhanced since the quantum features associated with the squeezed vacuum reservoir modes can enter the cavity, we opt to concentrate on the atomic dynamics and absorption spectrum[6] of the generated radiation.

To this effect, upon applying the outlined procedure along with the property of atomic operators (see Sects. 5.1.2 and 6.2.1), one can verify that

$$\frac{d}{dt}\langle\hat{\sigma}_-(t)\rangle = -\frac{\gamma(2N+1)}{2}\langle\hat{\sigma}_-(t)\rangle - \gamma M\langle\hat{\sigma}_+(t)\rangle. \tag{6.72}$$

Since this differential equation is nonlinear, it may not be trivial to look for $\langle\hat{\sigma}_-(t)\rangle$ directly. To overcome the emerging challenge, in view of the fact that $\hat{\sigma}_+^\dagger = \hat{\sigma}_-$, it appears appealing introducing

$$\frac{d}{dt}\beta_\pm(t) = -\frac{\gamma(2N+1\pm2M)}{2}\beta_\pm(t), \tag{6.73}$$

where $\beta_\pm(t) = \langle\hat{\sigma}_-(t)\rangle \pm \langle\hat{\sigma}_+(t)\rangle$.

With this variable change, it is straightforward to show that

$$\langle\hat{\sigma}_\mp(t)\rangle = \frac{1}{2}\big[\langle\hat{\sigma}_-(0)\rangle + \langle\hat{\sigma}_+(0)\rangle\big]e^{-\frac{\gamma}{2}(2N+1+2M)t}$$

$$\pm \frac{1}{2}\big[\langle\hat{\sigma}_-(0)\rangle - \langle\hat{\sigma}_+(0)\rangle\big]e^{-\frac{\gamma}{2}(2N+1-2M)t}. \tag{6.74}$$

---

[6]Detailed calculation shows that the time-delayed correlation has the same form as what we have already presented, and so we are not going to include the two-time second-order correlation in present discussion.

One may infer from Eqs. (5.45) and (6.74) that placing the atom in the cavity coupled to the squeezed vacuum reservoir is not directly responsible for inducing atomic coherence since $2N + 1 > 2M$. Even then, one may learn from the second term that the squeezed input can help to somehow preserve initially introduced atomic coherence under certain conditions contrary to the thermal input that significantly inhibits coherence regardless.

We next strive to obtain the probability of the occupancy of the upper atomic energy level. To do so, one may first require to verify that

$$\frac{d}{dt}\langle \hat{\sigma}_z(t) \rangle = -\gamma(2N + 1)\langle \hat{\sigma}_z(t) \rangle - \gamma \tag{6.75}$$

whose solution can be proposed as

$$\langle \hat{\sigma}_z(t) \rangle = \left[\langle \hat{\sigma}_z(0) \rangle + \frac{1}{2N + 1} \right] e^{-\gamma(2N+1)t} - \frac{1}{2N + 1}. \tag{6.76}$$

After that, the probability of the occupancy of the upper energy level is found upon using Eq. (5.43) to be

$$\rho_{aa}(t) = \frac{1}{2}\left[\langle \hat{\sigma}_z(0) \rangle + \frac{1}{2N + 1} \right] e^{-\gamma(2N+1)t} + \frac{N}{2N + 1}. \tag{6.77}$$

With the aid of Eqs. (5.43), (5.44) and (6.77), the atomic population inversion then turns to be

$$W(t) = 2\left[\rho_{aa}^{(0)} - \frac{N}{2N + 1} \right] e^{-\gamma(2N+1)t} - \frac{1}{2N + 1}, \tag{6.78}$$

which reduces for a two-level atom initially in the upper energy level to

$$W(t) = \frac{1}{2N + 1}\left[2(N + 1)e^{-\gamma(2N+1)t} - 1 \right]. \tag{6.79}$$

One can deduce from Eqs. (6.77) and (6.79) that the atom would prefer to stay in the lower energy level at steady state specially for a weak reservoir ($N \to 0$). It is also possible to envisage that the stronger the reservoir input, the more the atom tends to be excited; and yet, the probability for the atom to occupy the lower energy level is always greater than that of the upper energy level in the same way as for thermal reservoir. One can also notice that the dynamics of the population when the cavity is coupled to squeezed reservoir is similar to when it is coupled to thermal environment. As a result, it might be appropriate stressing that the population dynamics would be directly related to the intensity of the external radiation but not to the correlation among the reservoir modes.

We also try to obtain the absorption spectrum following the procedure provided in Sect. 6.2.1. In view of this proposal and Eq. (6.74), it is possible to find

$$\langle \hat{\sigma}_+(t)\hat{\sigma}_-(t+\tau)\rangle_{ss} = \frac{1}{2}\left[\langle \hat{\sigma}_+(t)\hat{\sigma}_-(t)\rangle_{ss} + \langle \hat{\sigma}_+(t)\hat{\sigma}_+(t)\rangle_{ss}\right]e^{-\frac{\gamma}{2}(2N+1+2M)\tau}$$

$$+ \frac{1}{2}\left[\langle \hat{\sigma}_+(t)\hat{\sigma}_-(t)\rangle_{ss} - \langle \hat{\sigma}_+(t)\hat{\sigma}_+(t)\rangle_{ss}\right]e^{-\frac{\gamma}{2}(2N+1-2M)\tau},$$

(6.80)

which can be rewritten using the properties of the atomic operators as

$$\langle \hat{\sigma}_+(t)\hat{\sigma}_-(t+\tau)\rangle_{ss} = \langle \hat{\sigma}_+(t)\hat{\sigma}_-(t)\rangle_{ss}\cosh(\gamma M\tau)e^{-\frac{\gamma}{2}(2N+1)\tau}.$$

(6.81)

Upon applying Eqs. (5.107) and (6.77), the normalized absorption spectrum turns out to be

$$S(\omega) = \frac{2\gamma \cosh 2r}{\pi}\left[\frac{\gamma^2 + 4\omega^2}{\gamma^4 + 8\gamma^2\omega^2\cosh 4r + 16\omega^4}\right].$$

(6.82)

Figure 6.6 indicates that, when compared to vacuum and thermal reservoirs, the squeezed input results in a narrower spread and a larger absorption spectrum. This modification can be directly linked to the correlation among the reservoir modes of the squeezed input, which is found to impede the spectral spread of the successively emitted photons, and is more or less consistent with the outcome that can be deduced from the atomic coherence.

Fig. 6.6 Plots of the absorption spectrum for $\gamma = 5$ and different values of the squeeze parameter

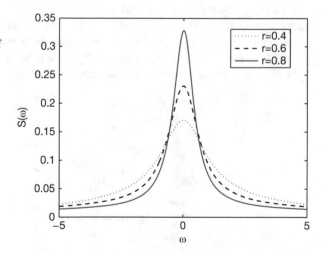

6.3   Driven Two-Level Atom

In addition to placing the atom in free space and cavity, there should be other valid
setups such as externally driving the cavity containing the atom—the system usually
dubbed as driven atomic scheme. Driving the atom in practice amounts to pumping
it continuously by a coherent light whose frequency matches with the atomic
transition frequency. Such interaction can be dealt with by treating the external
driving radiation as classical or strong light beam, that is, by treating the operator
related to the radiation in the interaction Hamiltonian as a c-number variable in
which Eq. (6.10) can be rewritten as

$$\hat{H}_I = ig\varepsilon(\hat{\sigma}_+ - \hat{\sigma}_-), \tag{6.83}$$

where ε is the amplitude of the driving field.

The overall program demands putting the atom in the cavity externally driven by
a coherent light from one side and coupled to various reservoirs from the other (see
Fig. 6.7), and then study the resulting radiative properties of the atom and statistical
properties of the emitted radiation. In this case, the atom interacts with three variants
of the radiation in the cavity: the coherent light, noise fluctuations and radiation
emitted by the atom. The result of such a study is expected to depend significantly
on the properties of the external sources and the coupling constants, that is, the
properties of the radiation emitted by the atom or the absorption-emission events
are expected to be substantially modified.

6.3.1   Coupled to Vacuum Reservoir

The atom-cavity combined system is presumed to be placed in a free space or
coupled to a vacuum-reservoir. It might be helpful relating the present scheme with
the phenomenon of resonance fluorescence in which the atom is assumed to be
placed in a free space; and yet, externally driven by a strong external coherent
light. The system we intend to explore can be studied by rewriting Eq. (6.52) for

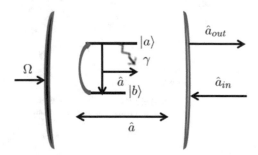

Fig. 6.7 Schematic representation of a driven two-level atom when the cavity containing a two-
level atom is pumped with a coherent light resonant with atomic transition and the atom can
exchange the radiation across the other side of the wall with the environment (relate with Fig. 6.3)

$N = M = 0$ and Hamiltonian (6.83) as

$$\frac{d\hat{\rho}}{dt} = \frac{\Omega}{i}\left(\hat{\sigma}_+\hat{\rho} - \hat{\sigma}_-\hat{\rho} + \hat{\rho}\hat{\sigma}_- - \hat{\rho}\hat{\sigma}_+\right) + \frac{\gamma}{2}\left[2\hat{\sigma}_-\hat{\rho}\hat{\sigma}_+ - \hat{\sigma}_+\hat{\sigma}_-\hat{\rho} - \hat{\rho}\hat{\sigma}_+\hat{\sigma}_-\right],$$

(6.84)

where $\Omega = 2\epsilon g$ is a constant proportional to the amplitude of the classical radiation and perceived as a measure of the strength of the driving radiation.

With the understanding that the system under consideration to be examined with the aid of the time evolution of the atomic operators, and following the procedure outlined in Sects. 6.1 and 6.2, one can verify that

$$\frac{d}{dt}\langle\hat{\sigma}_\pm(t)\rangle = -\frac{\gamma}{2}\langle\hat{\sigma}_\pm(t)\rangle - \frac{\Omega}{2}\langle\hat{\sigma}_z(t)\rangle,$$

(6.85)

$$\frac{d}{dt}\langle\hat{\sigma}_z(t)\rangle = -\gamma\langle\hat{\sigma}_z(t)\rangle + \Omega\left[\langle\hat{\sigma}_-(t)\rangle + \langle\hat{\sigma}_+(t)\rangle\right] - \gamma,$$

(6.86)

from which follows

$$\frac{d}{dt}\left[\langle\hat{\sigma}_-(t)\rangle + \langle\hat{\sigma}_+(t)\rangle\right] = -\frac{\gamma}{2}\left[\langle\hat{\sigma}_-(t)\rangle + \langle\hat{\sigma}_+(t)\rangle\right] - \Omega\langle\hat{\sigma}_z(t)\rangle.$$

(6.87)

Aiming at making the process of solving these equations more manageable, we opt to follow the factorization technique. So differentiation of Eq. (6.86) with respect to time once again, and then making use of Eq. (6.87) lead to

$$\frac{d^2}{dt^2}\langle\hat{\sigma}_z(t)\rangle + \frac{3\gamma}{2}\frac{d}{dt}\langle\hat{\sigma}_z(t)\rangle + \left(\Omega^2 + \frac{\gamma^2}{2}\right)\langle\hat{\sigma}_z(t)\rangle + \frac{\gamma^2}{2} = 0,$$

(6.88)

where applying the factorization technique results

$$\left(\frac{d}{dt} + \eta_+\right)\left(\frac{d}{dt} + \eta_-\right)\langle\hat{\sigma}_z(t)\rangle = -\frac{\gamma^2}{2},$$

(6.89)

in which

$$\eta_\pm = \frac{1}{4}\left[3\gamma \pm \sqrt{\gamma^2 - 16\Omega^2}\right].$$

(6.90)

To solve Eq. (6.89), it is found appealing introducing a new variable such that

$$\frac{d}{dt}\langle\hat{\sigma}_z(t)\rangle + \eta_-\langle\hat{\sigma}_z(t)\rangle = q(t),$$

(6.91)

which on the basis of Eq. (6.89) can be written as

$$\frac{dq(t)}{dt} + \eta_+ q(t) = -\frac{\gamma^2}{2}.$$

(6.92)

With this technique, we manage to obtain a linear differential equation whose solution can be put forward as

$$q(t) = q(0)e^{-\eta_+ t} - \frac{\gamma^2}{2\eta_+}(1 - e^{-\eta_+ t}). \tag{6.93}$$

Nonetheless, to determine the time evolution of the atomic operators, one may need to return to Eq. (6.91) in which the related solution can be written as

$$\langle \hat{\sigma}_z(t) \rangle = \langle \hat{\sigma}_z(0) \rangle e^{-\eta_- t} + \int_0^t e^{-\eta_-(t-t')} q(t') dt'. \tag{6.94}$$

After that, upon substituting Eq. (6.93) into (6.94), and then carrying out the resulting integration, one can find

$$\langle \hat{\sigma}_z(t) \rangle = \langle \hat{\sigma}_z(0) \rangle e^{-\eta_- t} + \frac{q(0)}{\eta_- - \eta_+} \left(e^{-\eta_+ t} - e^{-\eta_- t} \right) - \frac{\gamma^2}{2\eta_+ \eta_-} \left(1 - e^{-\eta_- t} \right)$$

$$+ \frac{\gamma^2}{2\eta_+(\eta_- - \eta_+)} \left(e^{-\eta_+ t} - e^{-\eta_- t} \right). \tag{6.95}$$

To obtain the time development of $\langle \hat{\sigma}_z(t) \rangle$, it is desirable evaluating the explicit value of $q(0)$ in which inserting Eq. (6.86) into (6.91), and so setting $t = 0$ in the resulting expression, yield

$$q(0) = \langle \hat{\sigma}_z(0) \rangle (\eta_- - \gamma) + \Omega \left[\langle \hat{\sigma}_-(0) \rangle + \langle \hat{\sigma}_+(0) \rangle \right] + \gamma, \tag{6.96}$$

from which follows

$$\langle \hat{\sigma}_z(t) \rangle = \left(\langle \hat{\sigma}_z(0) \rangle + \frac{\gamma^2}{2\eta_+ \eta_-} \right) e^{-\eta_- t} - \frac{\gamma^2}{2\eta_+ \eta_-} + \frac{1}{\eta_- - \eta_+} \left[\langle \hat{\sigma}_z(0) \rangle (\eta_- - \gamma) \right.$$

$$+ \Omega \left(\langle \hat{\sigma}_-(0) \rangle + \langle \hat{\sigma}_+(0) \rangle \right) + \frac{2\gamma \eta_+ + \gamma^2}{2\eta_+} \left] \left(e^{-\eta_+ t} - e^{-\eta_- t} \right). \tag{6.97}$$

It is possible to infer from Eq. (6.97) that the dynamics of the atomic population inversion depends on the inversion and energy level occupied by the atom at initial time in somewhat complicated manner. Since optical experiments fortunately require some time to perform, one can limit the discussion to the steady state regime;

$$W = -\frac{\gamma^2}{\gamma^2 + 2\Omega^2} \tag{6.98}$$

(see also Eq. (5.44)).

Fig. 6.8 Plots of the population in the upper atomic energy level and atomic coherence at steady state

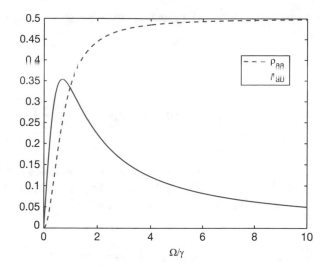

Besides, in view of Eqs. (5.36) and (6.97), the probability of the occupancy of the upper energy level is found to have the form

$$\rho_{aa}(t) = \frac{2\eta_+\eta_- - \gamma^2}{4\eta_+\eta_-} + \frac{1}{2}\left(\langle\hat{\sigma}_z(0)\rangle + \frac{\gamma^2}{2\eta_+\eta_-}\right)e^{-\eta_- t}$$

$$+ \frac{1}{2(\eta_- - \eta_+)}\Bigg[\langle\hat{\sigma}_z(0)\rangle(\eta_- - \gamma) + \Omega\big(\langle\hat{\sigma}_-(0)\rangle + \langle\hat{\sigma}_+(0)\rangle\big)$$

$$+ \frac{2\eta_+\gamma + \gamma^2}{2\eta_+}\Bigg]\left(e^{-\eta_+ t} - e^{-\eta_- t}\right), \tag{6.99}$$

where at steady state one gets

$$\rho_{aa} = \frac{\Omega^2}{\gamma^2 + 2\Omega^2}. \tag{6.100}$$

When there is no external pumping radiation, the atom prefers to occupy the lower energy level at steady state since $\rho_{aa} \to 0$ for $\Omega \to 0$ (see Fig. 6.8). But if the strength of the external radiation would be increased, the atom would have a chance to occupy the upper energy level; and yet, the probability for the atom to occupy the upper energy level would not be greater than half regardless. This outcome may suggest that the pumping mechanism alone cannot grant atomic population inversion.

It is also possible to explore the atomic coherence as defined in Eq. (5.45) starting with writing the solution of Eq. (6.85) as

$$\langle\hat{\sigma}_+(t+\tau)\rangle = \langle\hat{\sigma}_+(t)\rangle e^{-\frac{\gamma}{2}\tau} - \frac{\Omega}{2}e^{-\frac{\gamma}{2}\tau}\int_0^\infty e^{\frac{\gamma}{2}\tau'}\langle\hat{\sigma}_z(t+\tau')\rangle d\tau', \tag{6.101}$$

where $\langle \hat{\sigma}_z(t+\tau) \rangle$ can be generated from Eq. (6.97) by replacing t by $t+\tau$ and 0 by t as

$$
\langle \hat{\sigma}_z(t+\tau) \rangle = \left(\langle \hat{\sigma}_z(t) \rangle + \frac{\gamma^2}{2\eta_+\eta_-} \right) e^{-\eta_- \tau} - \frac{\gamma^2}{2\eta_+\eta_-} + \frac{1}{\eta_- - \eta_+} \left[\langle \hat{\sigma}_z(t) \rangle (\eta_- - \gamma) \right.
$$
$$
\left. + \Omega \left(\langle \hat{\sigma}_-(t) \rangle + \langle \hat{\sigma}_+(t) \rangle \right) + \frac{2\gamma\eta_+ + \gamma^2}{2\eta_+} \right] \left(e^{-\eta_+\tau} - e^{-\eta_-\tau} \right).
$$

$$(6.102)$$

If required, the time dependent atomic coherence can be obtained by inserting Eq. (6.102) into (6.101), and after that carrying out the resulting integration. But we still stick to the atomic coherence at steady state that can be directly obtained from Eq. (6.85);

$$
\langle \hat{\sigma}_+ \rangle_{ss} = -\frac{\Omega}{\gamma} \langle \hat{\sigma}_z \rangle_{ss},
$$

$$(6.103)$$

which in view of Eqs. (5.45) and (6.97) takes the form

$$
\rho_{ab} = \frac{\gamma \Omega}{\gamma^2 + 2\Omega^2}.
$$

$$(6.104)$$

One can see from Eq. (6.104) and Fig. 6.8 that the atomic coherence at steady state is attributed to the pumping mechanism, but found to be appreciably small for a strong pumping. To establish a meaningful coherence, one needs the external pumping that is not so strong—a delicate feat to achieve in practice. It might be insightful emphasizing that pumping the atom to the upper energy level and establishing the coherence between the atomic energy levels should not be taken as the same process since atomic damping needs to be taken into account to induce the coherence as opposed to ordinary pumping.

In addition, to obtain the absorption spectrum of the emitted radiation (5.107), one can verify using the information provided in Sect. 5.1.2 that

$$
\langle \hat{\sigma}_+(t) \hat{\sigma}_-(t+\tau) \rangle_{ss} = \frac{\Omega^2}{\gamma^2 + 2\Omega^2} \left\{ \frac{\gamma^2}{\gamma^2 + 2\Omega^2} + \frac{1}{2} e^{-\frac{\gamma\tau}{2}} + \frac{1}{4(\gamma^2 + 2\Omega^2)} \right.
$$
$$
\times \left[\left(2\Omega^2 - \gamma^2 - \frac{\gamma}{\sqrt{\gamma^2 - 16\Omega^2}} (10\Omega^2 - \gamma^2) \right) \right.
$$
$$
\times e^{\left(\frac{\sqrt{\gamma^2 - 16\Omega^2} - 3\gamma}{4} \right)\tau}
$$
$$
+ \left(2\Omega^2 - \gamma^2 - \frac{\gamma}{\sqrt{\gamma^2 - 16\Omega^2}} (10\Omega^2 - \gamma^2) \right)
$$
$$
\left. \left. \times e^{-\left(\frac{\sqrt{\gamma^2 - 16\Omega^2} + 3\gamma}{4} \right)\tau} \right] \right\}.
$$

$$(6.105)$$

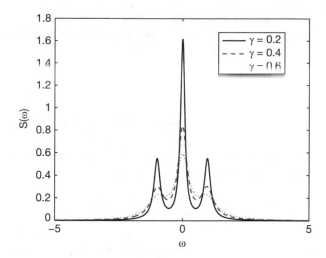

Fig. 6.9 Plots of the absorption spectrum for different values of atomic damping constants when $\Omega = 1$

The normalized absorption spectrum would hence turn out to be

$$S(\omega) = \delta(\omega) + \frac{\gamma/\pi}{\gamma^2 + 4\omega^2}$$

$$+ \text{Re}\left[\frac{\left(3\gamma - \sqrt{\gamma^2 - 16\Omega^2}\right)/\pi}{\left(3\gamma - \sqrt{\gamma^2 - 16\Omega^2}\right)^2 + 16\omega^2} + \frac{\left(3\gamma + \sqrt{\gamma^2 - 16\Omega^2}\right)/\pi}{\left(3\gamma + \sqrt{\gamma^2 - 16\Omega^2}\right)^2 + 16\omega^2} \right].$$

$$(6.106)$$

The delta function in Eq. (6.106) stands for the spectrum of the coherent light and the remaining terms indicate that the external pumping broadness the absorption spectrum of the atom. When there is a strong external pumping radiation, the spectrum specifically acquires sidebands at equal distance from the central line (see Fig. 6.9)—Mollow triplet. One can also see that the height of the spectrum decreases but the width increases with increasing atomic damping constant for a constant pumping strength. Deeper scrutiny may also reveal that the sidebands disappear for weak pumping.

It is also possible to write the two-time second-order normalized correlation function in terms of the atomic operators;

$$g^{(2)}(\tau) = \frac{\langle \hat{\sigma}_+(t)\hat{\sigma}_+(t+\tau)\hat{\sigma}_-(t+\tau)\hat{\sigma}_-(t)\rangle}{\langle \hat{\sigma}_+(t)\hat{\sigma}_-(t)\rangle^2}$$

$$(6.107)$$

(see Sects. 4.2 and 5.5). It should be possible to obtain $\langle \hat{\sigma}_+(t)\hat{\sigma}_+(t+\tau)\hat{\sigma}_-$ $(t+\tau)\hat{\sigma}_-(t)\rangle$ from $\langle \hat{\sigma}_+(t+\tau)\hat{\sigma}_-(t+\tau)\rangle$ by multiplying with $\hat{\sigma}_+(t)$ from left and by $\hat{\sigma}_-(t)$ from right (see Eq. (4.86)).

To expedite this task, it may suffice to obtain the probability of the occupancy of the upper energy level at time $t+\tau$ as denoted by Eq. (5.43);

$$\langle \hat{\sigma}_+(t+\tau)\hat{\sigma}_-(t+\tau)\rangle = \frac{1}{2}\big[\hat{\sigma}_z(t+\tau)+1\big]. \tag{6.108}$$

It then turns out to be straightaway to obtain by making use of the properties of the atomic operators (see Sect. 5.1.2) at steady state that

$$g^{(2)}(\tau)_{ss} = 1 - \cosh\left(\tau\sqrt{\gamma^2 - 16\Omega^2}/4\right)e^{-\frac{3\gamma\tau}{4}}$$

$$- \frac{3\gamma(\gamma^2 + 2\Omega^2)}{2\Omega\sqrt{\gamma^2 - 16\Omega^2}} \sinh\left(\tau\sqrt{\gamma^2 - 16\Omega^2}/4\right)e^{-\frac{3\gamma\tau}{4}}. \tag{6.109}$$

As can be seen from Fig. 6.10, the emitted light is found to exhibit photon anti-bunching with oscillation of the photon statistics between sub- and super-Poissonian character. The photon anti-bunching phenomenon can be understood as the atom that goes over to the lower energy level, after it emits a photon, takes some time before it undergoes the transition to the upper energy level, and then re-emits another photon. It is also important to distinguish the source of the two distinct phenomena: the oscillation and damping of $g^{(2)}(\tau)$ in which the oscillation is related to the strength of the pumping mechanism while the damping to the atomic decaying process as can be deduced from Eq. (6.109).

One may come to the conclusion that pumping the atom placed in the cavity externally with a resonant coherent light significantly modifies the radiative prop-

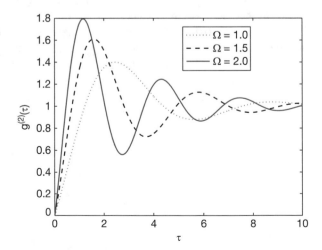

Fig. 6.10 Plots of the two-time correlation function when the atom is externally pumped by a strong coherent light for $\gamma = 0.5$ and different values of Ω

erties of the atom and the photon statistics of the emitted radiation. The external radiation particularly drives the atomic population to the upper energy level, and as a result can induce a strong coherence between the eigenstates of the atom, and that may also lead to entanglement. The pumping mechanism on the other hand makes the successive emission-absorption process possible in the sense that the successively emitted photons turn out to be correlated; the phenomenon believed to be the source of nonclassical features of the emitted radiation as in correlated emission laser.

6.3.2 Coupled to Broadband Reservoir

When a cavity containing a two-level atom and placed in a free space is externally driven, the external radiation would be involved in the pumping process, and so atomic coherence can be induced (see Eq. (6.104)). It may be reasonable thus to inquire how far the atomic dynamics, radiative properties of the atom, and statistical and quantum properties of the emitted radiation would be modified in case the cavity is coupled to a broadband squeezed vacuum fluctuation modes (see also [8]); which is the main theme of present discussion.

To begin with, a two-level atom in a cavity placed in a squeezed vacuum reservoir and externally driven by a strong coherent light can be described by

$$
\frac{d\hat{\rho}}{dt} = \frac{\Omega}{2}\left(\hat{\sigma}_+\hat{\rho} - \hat{\sigma}_-\hat{\rho} + \hat{\rho}\hat{\sigma}_- - \hat{\rho}\hat{\sigma}_+\right)
$$
$$
+ \frac{\gamma(N+1)}{2}\left[2\hat{\sigma}_-\hat{\rho}\hat{\sigma}_+ - \hat{\sigma}_+\hat{\sigma}_-\hat{\rho} - \hat{\rho}\hat{\sigma}_+\hat{\sigma}_-\right]
$$
$$
+ \frac{\gamma N}{2}\left[2\hat{\sigma}_+\hat{\rho}\hat{\sigma}_- - \hat{\sigma}_-\hat{\sigma}_+\hat{\rho} - \hat{\rho}\hat{\sigma}_-\hat{\sigma}_+\right] - \gamma M\left[\hat{\sigma}_-\hat{\rho}\hat{\sigma}_- + \hat{\sigma}_+\hat{\rho}\hat{\sigma}_+\right],
$$

$$(6.110)$$

from which follows

$$
\frac{d}{dt}\langle\hat{\sigma}_\pm(t)\rangle = -\frac{\gamma(2N+1)}{2}\langle\hat{\sigma}_\pm(t)\rangle - \gamma M\langle\hat{\sigma}_\mp(t)\rangle - \frac{\Omega}{2}\langle\hat{\sigma}_z(t)\rangle, \quad (6.111)
$$

$$
\frac{d}{dt}\langle\hat{\sigma}_z(t)\rangle = -\gamma(2N+1)\langle\hat{\sigma}_z(t)\rangle + \Omega\left[\langle\hat{\sigma}_-(t)\rangle + \langle\hat{\sigma}_+(t)\rangle\right] - \gamma. \quad (6.112)
$$

Aiming at utilizing the factorization technique, it is admissible to put forward

$$
\frac{d}{dt}\left[\langle\hat{\sigma}_+(t)\rangle - \langle\hat{\sigma}_-(t)\rangle\right] = -\frac{\gamma e^{-2r}}{2}\left[\langle\hat{\sigma}_+(t)\rangle - \langle\hat{\sigma}_-(t)\rangle\right], \quad (6.113)
$$

$$
\frac{d}{dt}\left[\langle\hat{\sigma}_+(t)\rangle + \langle\hat{\sigma}_-(t)\rangle\right] = -\Omega\langle\hat{\sigma}_z(t)\rangle - \frac{\gamma e^{2r}}{2}\left[\langle\hat{\sigma}_+(t)\rangle + \langle\hat{\sigma}_-(t)\rangle\right], \quad (6.114)
$$

$$\frac{d^2}{dt^2}\langle\hat{\sigma}_z(t)\rangle = \Omega\frac{d}{dt}\big[\langle\hat{\sigma}_+(t)\rangle + \langle\hat{\sigma}_-(t)\rangle\big] - \gamma(2N+1)\frac{d}{dt}\langle\hat{\sigma}_z(t)\rangle, \tag{6.115}$$

which leads to

$$\left(\frac{d}{dt}+\alpha_+\right)\left(\frac{d}{dt}+\alpha_-\right)\langle\hat{\sigma}_z(t)\rangle = -\frac{\gamma^2 e^{2r}}{2}, \tag{6.116}$$

where

$$\alpha_{\pm} = \frac{\gamma}{4}(6N+3+2M)\left[1\mp\sqrt{1-\frac{8[2\Omega^2+\gamma^2 e^{2r}(2N+1)]}{[\gamma(6N+3+2M)]^2}}\right]. \tag{6.117}$$

Since Eq. (6.116) is expressed in a similar mathematical form as Eq. (6.89), one can follow the same approach as in Sect. 6.3.1 to explore properties of the atom and the emitted radiation. To this effect, one can verify that

$$\langle\hat{\sigma}_z(t)\rangle = \left(\langle\hat{\sigma}_z(0)\rangle + \frac{e^{2r}\gamma^2}{2\alpha_+\alpha_-}\right)e^{-\alpha_- t} - \frac{e^{2r}\gamma^2}{2\alpha_+\alpha_-}$$

$$+ \frac{1}{\alpha_- - \alpha_+}\Big[\langle\hat{\sigma}_z(0)\rangle\big[\alpha_- - \gamma(2N+1)\big]$$

$$+ \Omega\big(\langle\hat{\sigma}_-(0)\rangle + \langle\hat{\sigma}_+(0)\rangle\big) + \frac{e^{2r}\gamma^2 - 2\alpha_+\gamma}{2\alpha_+}\Big]\big(e^{-\alpha_+ t} - e^{-\alpha_- t}\big), \tag{6.118}$$

from which follows at steady state

$$W = -\frac{\gamma^2 e^{2r}}{2\Omega^2 + \gamma^2(2N+1)e^{2r}}, \tag{6.119}$$

$$\rho_{aa} = \frac{\Omega^2 + \gamma^2 N e^{2r}}{2\Omega^2 + \gamma^2(2N+1)e^{2r}}, \tag{6.120}$$

$$\rho_{ab} = \frac{\Omega\gamma}{2\Omega^2 + \gamma^2(2N+1)e^{2r}}. \tag{6.121}$$

One can see from Fig. 6.11 that the probability for the atom to occupy the upper energy level increases with the strength of the external radiation and degree of squeezing of the reservoir modes for a general case, but decreases with the squeeze parameter for a strong coherent radiation due to the emerging phase competition. This observation can be perceived as if the squeezed input somehow inhibits absorption. Even then, irrespective of the strength of the external radiations, the atom prefers to occupy the lower energy level at steady state. The effect of the

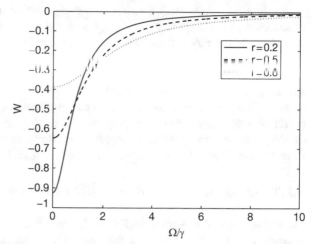

Fig. 6.11 Plots of the atomic population inversion at steady state when the atom is externally driven for different values of r

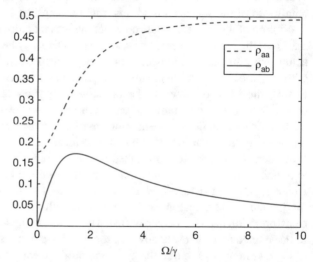

Fig. 6.12 Plots of the occupancy of the upper atomic energy level and atomic coherence for $r = 0.8$ (compare with Fig. 6.8)

squeezed input is markedly found to be minimal when the atom is strongly driven externally but show a tangible potential in pumping the atom in case the external radiation is weak.

The result captured by Fig. 6.12 entails that the atomic coherence can be induced by the external pumping, and can increase with the atomic damping rate for a strong driving case, which can also be inferred from Eq. (6.121) (see also [8]). This realization can be linked to the fact that the atomic coherence is induced via successive absorption-emission events that could be enhanced for a larger damping rate, and the strength of the atomic coherence decreases with the strength of the squeezed input (compare with the interpretation following from Eq. (6.104)). Note that this result could be reversed by manipulating phase relationship between the squeezed input and cavity radiation.

It might be worthy underscoring as a recap that although the atom is capable of absorbing a radiation from the cavity, and making transition to the upper energy level, it prefers to stay in the lower energy state more often irrespective of the amplitude of the coherent radiation and the strength of the squeeze parameter. One may also come to the understanding that although the squeezing is found to broaden the spectrum and also to reduce the height significantly, it does not directly contribute to the splitting [8]. The emitted radiation can also be shown to exhibit a nonclassical correlation such as coherent superposition, photon anti-bunching (independent of the squeeze parameter and amplitude of the coherent radiation) and super-Poissonian photon character (happened to be enhanced by the squeezed input).

6.3.3 Coupled to Finite Bandwidth Reservoir

Most of the available studies dealing with a two-level atom coupled to a squeezed vacuum reservoir suppose that the reservoir input is broadband, that is, the bandwidth of the squeezed modes is much larger than atomic line width. Practical realization of the squeezed state however evinces that the bandwidth of the squeezed light is in the order of atomic line width. Such disparity between the assumption and actual experimental fact may call for a deeper inspection, which is the rationale for taking up the issue of finite bandwidth squeezed vacuum reservoir [9]. In connection to this, the issue of how far the radiative properties of the atom and statistical properties of the emitted radiation can be modified by the finiteness of the bandwidth of the noise fluctuations that enter the cavity might be a relevant concern.

One may note that the system can be conceived as the cavity containing a two-level atom is coupled to a squeezed vacuum reservoir with finite bandwidth, and externally driven by a strong single-mode laser field with frequency ω_L, Rabi frequency Ω and detuned by $\Delta = \omega_L - \omega_0$ (relate with Fig. 6.7). To include the interaction of the atom with the driving field, first and foremost, the dressing transformation needs to be performed. The resulting dressed system after that needs to be coupled to a narrow bandwidth squeezed vacuum reservoir in the same way as a single system does so that the approach outlined earlier could be followed.

In view of the fact that the interaction picture is the intermediate alternative that lies between Schrödinder and Heisenberg picture, we strive to obtain the interaction Hamiltonian in interaction picture. To begin with, the Hamiltonian in the rotating wave and electric dipole approximations can be expressed as

$$\hat{H} = \hat{H}_R + \hat{H}_{AL} + \hat{H}_A + \hat{H}_{AR}, \tag{6.122}$$

where $\hat{H}_A = -\Delta\hat{\sigma}_z/2 + \omega_L\hat{\sigma}_z/2$ is the free atom Hamiltonian (see Eqs. (5.14) and (5.63)), $\hat{H}_R = \int_0^\infty \omega\hat{b}^\dagger(\omega)\hat{b}(\omega)d\omega$ is the Hamiltonian that describes the reservoir (see Eq. (3.3)),

$$\hat{H}_{AL} = \frac{i\Omega}{2}\left[\hat{\sigma}_+ e^{-i(\omega_L t + \phi)} - \hat{\sigma}_- e^{i(\omega_L t + \phi)}\right] \tag{6.123}$$

represents the interaction between the atom and classical radiation field (see Eq. (6.9) and

$$\hat{H}_{AR} = i \int_0^\infty g(\omega) \left[\hat{\sigma}_+ \hat{b}(\omega) - \hat{b}^\dagger(\omega)\hat{\sigma}_- \right] d\omega \qquad (6.124)$$

designates the interaction of the atom with the reservoir (see Eq. (6.49)). In the same way, $g(\omega)$ is the frequency dependent measure of the coupling of the atom with the reservoir modes, ϕ is the phase and $\hat{b}(\omega)$ is the annihilation operator of the reservoir modes that satisfies the boson commutation relation $[\hat{b}(\omega), \hat{b}^\dagger(\omega')] = \hat{I}\delta(\omega - \omega')$.

Once the interaction Hamiltonian is identified, it might be suitable to perform a two-step unitary transformations in which the atomic Hamiltonian and free field Hamiltonian are utilized to transform the operators to the frame rotating with the laser frequency, and then transform the resulting Hamiltonian to the interaction picture with respect to the reservoir modes. In doing so, the rotating frame is also shifted in phase with ϕ that amounts to setting new raising and lowering operators that absorb the phase factor according to $\hat{\sigma}_\pm e^{\mp i(\phi + \omega_0 t)} \rightarrow \hat{\sigma}_\pm$ [10]. Once these transformations are done, the system under consideration can be designated by the Hamiltonian

$$\hat{H} = \hat{H}_0 + \hat{H}_I(t), \qquad (6.125)$$

where

$$\hat{H}_0 = -\frac{\Delta}{2}\hat{\sigma}_z + \frac{i\Omega}{2}(\hat{\sigma}_+ - \hat{\sigma}_-), \qquad (6.126)$$

$$\hat{H}_I(t) = i \int_0^\infty g(\omega)\left[\hat{\sigma}_+\hat{b}(\omega)e^{i(\phi+(\omega_L-\omega)t)} - \hat{b}^\dagger(\omega)\hat{\sigma}_- e^{-i(\phi+(\omega_L-\omega)t)}\right]d\omega. \qquad (6.127)$$

The second step is the unitary dressing transformation performed with the aid of the Hamiltonian $\hat{H}_0$ that can be done by employing $\hat{\sigma}_\pm(t) = e^{-i\hat{H}_0 t}\hat{\sigma}_\pm e^{i\hat{H}_0 t}$, which leads following the discussion provided earlier

$$\hat{\sigma}_\pm(t) = \mp\frac{1}{2\sqrt{\Omega^2 + \Delta^2}}\left[\left(\sqrt{\Omega^2 + \Delta^2} \pm \Delta\right)\hat{\Sigma}_- e^{-it\sqrt{\Omega^2+\Delta^2}}\right.$$
$$\left. + \left(\sqrt{\Omega^2 + \Delta^2} \mp \Delta\right)\hat{\Sigma}_+ e^{it\sqrt{\Omega^2+\Delta^2}} \mp \Omega\,\hat{\Sigma}_z\right], \qquad (6.128)$$

where

$$\hat{\Sigma}_\pm = \mp\frac{1}{2\sqrt{\Omega^2+\Delta^2}}\left[\left(\sqrt{\Omega^2+\Delta^2} \pm \Delta\right)\hat{\sigma}_- - \left(\sqrt{\Omega^2+\Delta^2} \mp \Delta\right)\hat{\sigma}_+ \pm \Omega\hat{\sigma}_z\right], \qquad (6.129)$$

$$\hat{\Sigma}_z = \frac{\Omega}{\sqrt{\Omega^2 + \Delta^2}}(\hat{\sigma}_- + \hat{\sigma}_+) - \frac{\Delta}{\sqrt{\Omega^2 + \Delta^2}}\hat{\sigma}_z, \qquad (6.130)$$

in which $\hat{\Sigma}_{\pm}$ and $\hat{\Sigma}_z$ are the dressed operators that oscillate at frequencies $\pm\sqrt{\Omega^2 + \Delta^2}$ and 0. On the basis of these transformations (for $\Omega \to 0$ and a negative detuning, $\hat{\Sigma}_{\pm} \to \hat{\sigma}_{\pm}$ and $\hat{\Sigma}_z \to \hat{\sigma}_z$, and for a positive detuning, $\hat{\Sigma}_{\pm} \to -\hat{\sigma}_{\pm}$ and $\hat{\Sigma}_z \to -\hat{\sigma}_z$), the interaction Hamiltonian in the interaction picture is found to take the form

$$\hat{H}_{SR}(t) = i \int_0^\infty g(\omega) \left[\hat{\sigma}_+(t)\hat{b}(\omega)e^{i(\phi+(\omega_L-\omega)t)} - \hat{b}^\dagger(\omega)\hat{\sigma}_-(t)e^{-i(\phi+(\omega_L-\omega)t)} \right] d\omega.$$

(6.131)

The master equation for the reduced density operator of this system can be obtained by applying the standard methods (see Sects. 3.1 and 6.2). In light of Born-Markov approximation, the time evolution of the reduced density operator can then be expressed as

$$\frac{\partial\hat{\rho}}{\partial t} = -\int_0^t Tr \left\{ \left[\hat{H}_{SR}(t), \left[\hat{H}_{SR}(t-\tau), \hat{R}\hat{\rho}(t-\tau) \right] \right] \right\} d\tau.$$

(6.132)

It turns out to be suitable proposing that

$$Tr_R\left(\hat{b}^\dagger(\omega)\hat{b}(\omega')\hat{R}\right) = N(\omega)\delta(\omega - \omega'),$$

(6.133)

$$Tr_R\left(\hat{b}(\omega)\hat{b}^\dagger(\omega')\hat{R}\right) = [N(\omega) + 1]\delta(\omega - \omega'),$$

(6.134)

$$Tr_R\left(\hat{b}(\omega)\hat{b}(\omega')\hat{R}\right) = M(\omega)e^{i\varphi_s}\delta(2\omega_L - \omega - \omega'),$$

(6.135)

where $N(\omega)$ is the mean photon number at frequency ω, $M(\omega)$ is the strength of the correlation among reservoir modes and φ_s is the associated phase (see also Eqs. (3.36) and (3.37)).

Since the master equation can be derived following the procedures already outlined, in case the contribution of the principal term is neglected,[7] one can verify at resonance that

$$\frac{\partial\hat{\rho}}{\partial t} = \frac{\Omega}{2} \left[\hat{\sigma}_+ - \hat{\sigma}_-, \hat{\rho} \right] - \frac{A}{4} \left[\hat{\sigma}_+ + \hat{\sigma}_-, \left[\hat{\sigma}_z, \hat{\rho} \right] \right]$$

---

[7]Since for a degenerate parametric oscillator [11],

$$N(\omega) = \kappa\varepsilon \left[\frac{1}{(\omega - \omega_L)^2 + (\kappa - \varepsilon)^2} - \frac{1}{(\omega - \omega_L)^2 + (\kappa + \varepsilon)^2} \right],$$

$$M(\omega) = \kappa\varepsilon \left[\frac{1}{(\omega - \omega_L)^2 + (\kappa - \varepsilon)^2} + \frac{1}{(\omega - \omega_L)^2 + (\kappa + \varepsilon)^2} \right],$$

it is possible to extend the upper limit of the involved integration over τ to infinity in which the integrations can be carried out citing $\int_0^\infty \exp(\pm i\xi\tau)d\tau = \pi\delta(\xi) \pm iP\frac{1}{\xi}$, where P denotes the Cauchy principal value and ξ takes the values $\omega_L - \omega$ and $\omega_L - \omega \pm \sqrt{\Omega^2 + \Delta^2}$.

$$+ \frac{\gamma N'}{2} (2\hat{\sigma}_+ \hat{\rho}\hat{\sigma}_- - \hat{\sigma}_- \hat{\sigma}_+ \hat{\rho} - \hat{\rho}\hat{\sigma}_- \hat{\sigma}_+)$$

$$+ \frac{\gamma (N' + 1)}{2} (2\hat{\sigma}_- \hat{\rho}\hat{\sigma}_+ - \hat{\sigma}_+ \hat{\sigma}_- \hat{\rho} - \hat{\rho}\hat{\sigma}_+ \hat{\sigma}_-) - \gamma M'(\hat{\sigma}_+ \hat{\rho}\hat{\sigma}_+ + \hat{\sigma}_- \hat{\rho}\hat{\sigma}_-),$$

(6.136)

where

$$N' = \frac{N(\omega_0 + \Omega) + N(\omega_0)}{2} + \frac{M(\omega_0 + \Omega) - M(\omega_0)}{2}, \qquad (6.137)$$

$$M' = \frac{M(\omega_0 + \Omega) + M(\omega_0)}{2} + \frac{N(\omega_0 + \Omega) - N(\omega_0)}{2}, \qquad (6.138)$$

$$A = \frac{\gamma^2 \Omega^2}{2} \left(\frac{\lambda_+^2 - \lambda_-^2}{\lambda_-(\Omega^2 + \lambda_-^2)} \right) \qquad (6.139)$$

in which $\lambda_\pm$ are related to the cavity damping rate and real amplification constant of the parametric oscillator according to $\lambda_\pm = (\kappa \pm 2\varepsilon)/2$ and γ is the usual atomic damping rate.

Pertaining to the perception that the frequency dependence of N' and M' can be related to the physical properties of the source of the squeezed vacuum, in case the source of the squeezed input is chosen to be a degenerate parametric oscillator, one may write

$$N(x) = \frac{\lambda_+^2 - \lambda_-^2}{4} \left[\frac{1}{x^2 + \lambda_-^2} - \frac{1}{x^2 + \lambda_+^2} \right], \qquad (6.140)$$

$$M(x) = \frac{\lambda_+^2 - \lambda_-^2}{4} \left[\frac{1}{x^2 + \lambda_-^2} + \frac{1}{x^2 + \lambda_+^2} \right], \qquad (6.141)$$

where x stands for various choice of parameters.

In view of Eqs. (6.55) and (6.136), it is straightforward to see that

$$\frac{d}{dt} \langle \hat{\sigma}_\pm(t) \rangle = -\frac{\gamma (2N' + 1)}{2} \langle \hat{\sigma}_\pm(t) \rangle - \gamma M' \langle \hat{\sigma}_\mp(t) \rangle - \frac{\Omega}{2} \langle \hat{\sigma}_z(t) \rangle, \qquad (6.142)$$

$$\frac{d}{dt} \langle \hat{\sigma}_z(t) \rangle = -\gamma (2N' + 1) \langle \hat{\sigma}_z(t) \rangle + \Omega' (\langle \hat{\sigma}_-(t) \rangle + \langle \hat{\sigma}_+(t) \rangle) - \gamma, \qquad (6.143)$$

where

$$\Omega' = \Omega \left[1 + \frac{\gamma^2 \Omega}{2} \left(\frac{\lambda_+^2 - \lambda_-^2}{\lambda_-(\Omega^2 + \lambda_-^2)} \right) \right]. \qquad (6.144)$$

Then following the factorization technique, one can obtain

$$
\langle \hat{\sigma}_z(t) \rangle = -\frac{\gamma^2(1 + 2N' + 2M')}{2\alpha\beta} + \left(\langle \hat{\sigma}_z(0) \rangle + \frac{\gamma^2(1 + 2N' + 2M')}{2\alpha\beta} \right) e^{-\beta t}
$$

$$
+ \left[\frac{\beta - \gamma(2N' + 1)}{\beta - \alpha} \langle \hat{\sigma}_z(0) \rangle + \frac{\gamma^2(1 + 2N' + 2M')}{2\alpha(\beta - \alpha)} \right.
$$

$$
\left. + \frac{\Omega'}{\beta - \alpha} \left(\langle \hat{\sigma}_-(0) \rangle + \langle \hat{\sigma}_+(0) \rangle \right) - \frac{\gamma}{\beta - \alpha} \right] \left(e^{-\alpha t} - e^{-\beta t} \right), \qquad (6.145)
$$

where

$$
\alpha = \frac{\gamma}{4}\left(6N' + 3 + 2M'\right) - \xi, \qquad \beta = \frac{\gamma}{4}\left(6N' + 3 + 2M'\right) + \xi, \qquad (6.146)
$$

in which

$$
\xi = \sqrt{\frac{\gamma^2}{16}\left(6N' + 3 + 2M'\right)^2 - \Omega\Omega' - \frac{\gamma^2}{2}\left(1 + 2N' + 2M'\right)(2N' + 1)}.
$$
$$(6.147)$$

Once the evolution of the energy operator is obtained, it should be straightforward to explore the effects of the finiteness of the bandwidth on the properties of the atomic dynamics and statistical properties of the emitted radiation (see previous sections). It might be insightful emphasizing that when a finite bandwidth biased noise fluctuations enter the cavity, the atomic population would acquire a characteristic dip for certain values of the intensity of the external driving radiation, where the dipping phenomenon—that can be associated with the phase of the squeezed input—is found to increase with the strength of the squeezed input and witnessed for smaller values of the amplitude of the coherent radiation [12]. When compared to other options, the rate of the absorption of the atom would also be strongly inhibited. The inhibition of the absorption may bear a significant implication on practical utilization of the interaction of a two-level atom with a squeezed radiation since the modes of the reservoir have some degree of finiteness in the bandwidth. As can be expected, the frequency for which the absorption spectrum is peaked and the associated spectral line width are also considerably modified in which the narrower the widths of the modes of the reservoir, the narrower the width of the emitted radiation turns out to be.

6.4   Many Two-Level Atoms

It is a well established fact that interaction of single two-level with vacuum fluctuation leads to relaxation due to emission of photons at atomic transition frequency. However, in many atoms scenario, the presence of another atom leads to the absorption of the photon emitted by the neighboring atom. Just as individual photons can perturb atomic dynamics in the strong coupling regime, even a single atom can influence the state of the cavity field. One may then foresee that the effect the atom casts on the properties of the photon can be modulated by the presence of the second photon—the interaction process that brought us to the quantum regime of nonlinear optics. Since properties of individual systems can be affected by the evolution of the state of the combined system, it appears reasonable to explore how far the dynamics of many two-level atoms compelled to interact with each other would be modified. Particularly, since the population can be coherently transferred back and forth from one atom to the other, different collective atomic phenomena are expected to appear. The most notable example is the phenomenon of superradiance in which the interaction among the atoms leads to a cooperative emission [13].

To begin with, imagine N large but finite number of two-level atoms to be placed close enough in a relatively small region so that a strong interaction can be initiated among them. Assume also the individual two-level atom to be characterized by $|a_i\rangle$ upper energy state, $|b_i\rangle$ lower energy state with ω_i transition frequency, μ_i transition dipole moment and $\mathbf{r}_i$ position (see Fig. 6.13). In line with the perception that the atoms placed at different positions may have different transition frequencies ($\omega_1 \neq \omega_2 \neq \omega_3, \ldots, \neq \omega_N$) and dipole moments ($\mu_1 \neq \mu_2 \neq \mu_3, \ldots, \neq \mu_N$), the total Hamiltonian can be expressed in the electric dipole approximation as

$$\hat{H} = \frac{1}{2}\sum_{i=1}^{N}\omega_i\,\hat{S}_i^z + \sum_{\mathbf{k},s}\omega_k\left(\hat{a}_{\mathbf{k},s}^\dagger\hat{a}_{\mathbf{k},s} + \frac{1}{2}\right)$$

$$- i\sum_{\mathbf{k},s,i=1}^{N}\left[\boldsymbol{\mu}_i\cdot\mathbf{g}_{\mathbf{k},s}(\mathbf{r}_i)(\hat{S}_i^+ + \hat{S}_i^-)\hat{a}_{\mathbf{k},s} - H.c.\right], \qquad (6.148)$$

Fig. 6.13 Schematic representation of many two-level atoms when each atom is placed at position $\mathbf{r}_i$ from the origin and possesses arbitrarily directed dipole moment μ_i in which the interaction among the atoms is considered as the dipole-dipole interaction between neighboring atoms

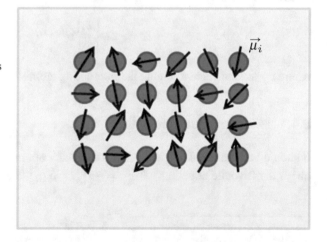

where

$$\hat{S}_i^+ = |a_i\rangle\langle b_i|, \qquad \hat{S}_i^- = |b_i\rangle\langle a_i|, \qquad \hat{S}_i^z = |a_i\rangle\langle a_i| - |b_i\rangle\langle b_i| \qquad (6.149)$$

are the dipole raising, dipole lowering and energy operators for the ith atom, respectively.

On the basis of the discussion in Sect. 5.1.2, it may not be difficult to verify that

$$\left[\hat{S}_i^+, \hat{S}_j^-\right] = \hat{S}_i^z \delta_{ij}, \quad \left[\hat{S}_i^z, \hat{S}_j^{\pm}\right] = \pm 2\hat{S}_i^{\pm}\delta_{ij}, \quad \hat{S}_i^+\hat{S}_j^- + \hat{S}_i^-\hat{S}_j^+ = \hat{I}\delta_{ij} \qquad (6.150)$$

and $\left(\hat{S}_i^{\pm}\right)^2 = 0$. $\hat{a}_{\mathbf{k},s}$ and $\hat{a}_{\mathbf{k},s}^{\dagger}$ on the other hand are the annihilation and creation operators of the radiation mode that has wave vector $\mathbf{k}$ (frequency ω_k), where s denotes the index of polarization. The coupling can be partly designated in this case as

$$\mathbf{g}_{\mathbf{k},s}(\mathbf{r}_i) = \sqrt{\frac{\omega_k}{2\epsilon_0 V}}\hat{e}_{\mathbf{k},s}e^{i\mathbf{k}\cdot\mathbf{r}_i}, \qquad (6.151)$$

which can be perceived as the mode function for the three-dimensional vacuum field evaluated at the position of the ith atom, V is the quantization volume and $\hat{e}_{\mathbf{k},s}$ is the unit polarization vectors of the radiation (compare with Eq. (6.5)).

In case one seeks to analyze the effects of the external driving radiation, the pertinent Rabi frequency can be defined as

$$\Omega_i = \boldsymbol{\mu}_i \cdot \mathbf{E}_L e^{i\mathbf{k}_L \cdot \mathbf{r}_i} \qquad (6.152)$$

(compare with Eq. (5.57)), where $\mathbf{k}_L$ is the wave vector of the driving laser field and Eq. (6.152) indicates that the Rabi frequency depends on the location of the atom and its interaction with the neighboring atoms. For instance, if the electric dipole moment of the atoms are parallel, the Rabi frequencies of two arbitrarily separated atoms are related by

$$\Omega_j = \Omega_i \frac{|\boldsymbol{\mu}_j|}{|\boldsymbol{\mu}_i|} e^{i\mathbf{k}_L \cdot \mathbf{r}_{ij}}, \qquad (6.153)$$

where $\mathbf{r}_{ij}$ are the vectors joining the interacting atoms.

6.4.1 Master Equation

It should be common knowledge that the density operator of the combined system obeys the Livoulle equation:[8] $\frac{\partial\hat{\rho}_{AR}}{\partial t} = -i\left[\hat{H}, \hat{\rho}_{AR}\right]$, and the density operator can

---

[8]In deriving the master equation for many two-level atoms enclosed in small space, the technique provided in [14, 15] can be generalized with the aid of the procedure outlined for a single two-level atom.

be transformed into the interaction picture by employing $\hat{\rho}_{AR}^I(t) = e^{i\hat{H}_0 t}\hat{\rho}_{AR}e^{-i\hat{H}_0 t}$
if required;

$$\frac{\partial \hat{\rho}_{An}^I}{\partial t} = -i[\hat{H}_{AR}(t), \hat{\rho}_{AR}^I(0)] - i\int_0^{t} dt'[\hat{H}_{AR}(t), [\hat{H}_{AR}(t'), \hat{\rho}_{AR}^I(t')]],$$

$$(6.154)$$

where $\hat{H}_{AR}$ stands for the interaction Hamiltonian

$$\hat{H}_{AR}(t) = -i\sum_{\mathbf{k},s}\sum_{i=1}^{N}\left[\boldsymbol{\mu}_i \cdot \mathbf{g}_{\mathbf{k},s}(\mathbf{r}_i)\left(\hat{S}_i^+\hat{a}_{\mathbf{k},s}e^{-i(\omega_k-\omega_i)t} + \hat{S}_i^-\hat{a}_{\mathbf{k},s}^\dagger e^{-i(\omega_k+\omega_i)t}\right)\right.$$

$$\left. - H.c.\right].$$

$$(6.155)$$

The time evolution of the density operator that describes the atom alone can then be obtained by tracing over the field variables,

$$\frac{\partial \hat{\rho}}{\partial t} = -iTr_R[\hat{H}_{AR}(t), \hat{\rho}_{AR}^I(0)]$$

$$- i\int_0^t dt'\, Tr_R([\hat{H}_{AR}(t), [\hat{H}_{AR}(t'), \hat{\rho}_{AR}^I(t')]]),$$

$$(6.156)$$

where employing the Born approximation leads to

$$\frac{\partial \hat{\rho}}{\partial t} = -iTr_R[\hat{H}_{AR}(t), \hat{\rho}_A(0)\hat{\rho}_R(0)]$$

$$- \int_0^t d\tau\, Tr_R([\hat{H}_{AR}(t), [\hat{H}_{AR}(t-\tau), \hat{\rho}_R(0)\hat{\rho}(t-\tau)]]).$$

$$(6.157)$$

In the process of obtaining the time evolution of the density operator,[9] it might be helpful employing the correlation of noise fluctuations for perfectly matched reservoir:

$$\langle \hat{a}_{\mathbf{k},s}^\dagger \hat{a}_{\mathbf{k'},s'}\rangle = N(\omega_k)\delta^3(\mathbf{k} - \mathbf{k'})\delta_{ss'},$$

$$(6.158)$$

---

[9]The calculation depends on the angle over which the reservoir mode is made to spread. So since the atoms occupy large space and the sum over $\mathbf{k}$ can be converted into integration with the transformation

$$\sum_{\mathbf{k},s} \rightarrow \frac{V}{(2\pi c)^3}\sum_{s=1}^{2}\int_0^\infty d\omega_k \omega_k^2 \int d\Theta_k,$$

where Θ_k is the solid angle (relate with the derivation in Appendix 2). In this case, while carrying out the polarization sums and integrals over $d\Theta_k$, the dipole moments of the atoms are taken to be

$$\langle \hat{a}_{\mathbf{k},s} \hat{a}_{\mathbf{k}',s'}^{\dagger} \rangle = \left[N(\omega_k) + 1 \right] \delta^3 (\mathbf{k} - \mathbf{k}') \delta_{ss'}, \tag{6.159}$$

$$\langle \hat{a}_{\mathbf{k},s} \hat{a}_{\mathbf{k}',s'} \rangle = M(\omega_k) \delta^3 (2\mathbf{k}_s - \mathbf{k} - \mathbf{k}') \delta_{ss'}, \tag{6.160}$$

where $N(\omega_k)$ and $M(\omega_k)$ characterize the reservoir modes in which $N(\omega_k)$ is related to the number of photons in mode $\mathbf{k}$ and $M(\omega_k) = |M(\omega_k)|e^{i\phi_s}$ to the magnitude of the two-photon correlation between the reservoir modes, whereas ϕ_s is the phase (relate with Eqs. (3.36) and (3.37)) and $\langle \hat{a}_{\mathbf{k},s} \rangle = \langle \hat{a}_{\mathbf{k},s}^{\dagger} \rangle = 0$.

It is then possible to verify in the rotating wave approximation that the master equation of the system of N two-level atoms placed in a broadband reservoir that propagates in the opposite direction to the emitted radiation and when the contribution of the principal part is neglected takes the form [15]

$$\frac{\partial \hat{\rho}}{\partial t} = -i \sum_{i=1}^{N} \omega_i \left[\hat{S}_i^z, \ \hat{\rho} \right] - i \sum_{i \neq j}^{N} \Omega_{ij} \left[\hat{S}_i^+ \hat{S}_j^-, \ \hat{\rho} \right]$$

$$- \frac{1}{2} \sum_{i,j=1}^{N} \Gamma_{ij} N(\omega_s) \left[\hat{\rho} \hat{S}_i^- \hat{S}_j^+ + \hat{S}_i^- \hat{S}_j^+ \hat{\rho} - 2\hat{S}_j^+ \hat{\rho} \hat{S}_i^- \right]$$

$$+ \frac{1}{2} \sum_{i,j=1}^{N} \Gamma_{ij} (1 + N(\omega_s)) \left[\hat{\rho} \hat{S}_i^+ \hat{S}_j^- + \hat{S}_i^+ \hat{S}_j^- \hat{\rho} - 2\hat{S}_j^- \hat{\rho} \hat{S}_i^+ \right]$$

$$+ \frac{1}{2} \sum_{i,j=1}^{N} \Gamma_{ij} M(\omega_s) \left[\hat{\rho} \hat{S}_i^+ \hat{S}_j^+ + \hat{S}_i^+ \hat{S}_j^+ \hat{\rho} - 2\hat{S}_j^+ \hat{\rho} \hat{S}_i^+ \right.$$

$$\left. + \hat{\rho} \hat{S}_i^- \hat{S}_j^- + \hat{S}_i^- \hat{S}_j^- \hat{\rho} - 2\hat{S}_j^- \hat{\rho} \hat{S}_i^- \right], \tag{6.161}$$

where Ω_{ij} represents the vacuum induced dipole-dipole interaction and $\Gamma_{ij} = \sqrt{\Gamma_i \Gamma_j}$ stands for pertinent spontaneous emission rate (compare with Eq. (6.110)).

As one may recall, the spontaneous emission rate for a single two-level atom is given by $\Gamma_i = \Gamma_{ii} = \omega_i^3 \mu_i^2 / (3\pi \epsilon_0 c^3)$ (see also Eq. (6.53)). Γ_{ij}'s portray coupling between the atoms through the vacuum fluctuations in which the spontaneous emission from one of the atoms influences the spontaneous emission of the other when the dipole-dipole interaction terms introduce a coherent coupling between the

---

parallel and the spherical coordinate system for the propagation vector $\mathbf{k}$ in which

$$\mathbf{k} = k \left[\sin \theta \cos \phi \hat{x} + \sin \theta \sin \phi \hat{y} + \cos \theta \hat{z} \right],$$

$$\hat{e}_{\mathbf{k}1} = -\cos \theta \cos \phi \hat{x} - \cos \theta \sin \phi \hat{y} + \sin \theta \hat{z}, \qquad \hat{e}_{\mathbf{k}2} = \sin \phi \hat{x} - \cos \phi \hat{y}$$

are used and the orientation of the atomic dipole moments can be taken in the x-direction [15].

states of the atoms. Following a similar approach, the Rabi frequency related to the collective spontaneous emission rate that contributes to the shift of atomic energy level is then found in the small sampling regime to be

$$\Omega_{ij} = \frac{\Im I_{ij}}{4}\left(1 - \frac{\Im(\mu_i.r_{ij})'}{(k_0 r_{ij})^2}\right), \tag{6.162}$$

where $\hat{\mu}_i$'s and $\hat{r}_{ij}$'s are unit vectors along the transmission dipole moments, $\mathbf{r}_{ij} = \mathbf{r}_j - \mathbf{r}_i$ and $k_0 = (\omega_i + \omega_j)/2c$. The presence of collective parameters Γ_{ij} and Ω_{ij} changes the structure of the master equation since the second term is entirely absent in case of a single atom.

Nonetheless, if the whole system is coupled to vacuum reservoir and pumped externally with a coherent light, the master equation would be

$$\frac{\partial\hat{\rho}}{\partial t} = -i\left[\sum_{i=1}^{N}\omega_i\hat{S}_i^z + \sum_{i,j=1}^{N}\Omega_{ij}\hat{S}_i^+\hat{S}_j^- - \frac{1}{2}\sum_{i=1}^{N}\left(\Omega_i\hat{S}_i^+ e^{i(\omega_L t+\phi_L)} + H.c\right), \hat{\rho}\right]$$
$$+ \frac{1}{2}\sum_{i,j=1}^{N}\Gamma_{ij}\left[\hat{\rho}\hat{S}_i^+\hat{S}_j^- + \hat{S}_i^+\hat{S}_j^-\hat{\rho} - 2\hat{S}_j^-\hat{\rho}\hat{S}_i^+\right] \tag{6.163}$$

in which ω_L is the Rabi frequency and ϕ_L is the phase of the classical laser. One may note that the derived master equation is intended for exploring the dynamics of a large (but finite) number of two-level atoms placed in a confined space and when the interaction among the involved atoms is presumed to be quite strong.

6.4.2 Ensemble of Two-Level Atoms

Since the number of atoms would not be finite except in a controlled experimental setting, the atomic system can be considered as a sea of uniformly distributed gas whose dynamics is statistically determined—the presumption that invokes the notion of atomic ensemble. Pertaining to the fact that one cannot address the radiative properties of individual atoms in the ensemble scenario, the intention is to concentrate on the most plausible characteristics of the emitted radiation. In other words, the aim is geared towards directing the attention to the coupling of external coherent light with the sea of atoms engaged in a spontaneous dipole-dipole interaction perceived to affect the radiative properties of the combined system.

In light of the analogy of optical parametric oscillation to the process that excites pairs of atoms from the lower energy level to a well defined upper energy level, one may construe the occurrence of the quantum correlations among the energy states of the ensemble of atoms (see Chap. 8). Available works in this direction show that it is technically feasible to map the quantum states of the light onto the quantum states of the atoms by using Schweniger's representation of the angular momentum in terms of the boson operators [16]. Since the operators related to the radiation and atoms should be put in equal footing in this consideration, the coherent coupling among atoms in the ensemble is expected to induce a nonclassical correlation epitomized by the reduced fluctuation in one of the quadratures in the expense of canonical conjugate quadrature just like entanglement and squeezing in the radiation—which leads to the idea of spin squeezing [17].

To attest to such proposal, assume that each atom in the ensemble responds to the external radiation in the same way irrespective of where it is situated in the configuration. The Hamiltonian that describes the interaction of N two-level atoms with radiation then can be given by

$$\hat{H}_I = ig(\hat{a}^\dagger \hat{J}_- - \hat{a}\hat{J}_+),   \tag{6.164}$$

where g is the coupling constant taken to be the same for all atoms, $\hat{a}$ is the cavity radiation and $\hat{J}_\pm = \sum_{j=1}^{N} \hat{S}_j^\pm$ are collective atomic operators that can be expressed in the Schweniger's representation of angular momentum operators in terms of the boson operators as

$$\hat{J}_- = \hat{b}^\dagger \hat{c}, \qquad \hat{J}_+ = \hat{c}^\dagger \hat{b}   \tag{6.165}$$

(relate with the discussion in Sect. 7.1.4 and see also [18]).

One may infer that $\hat{b}$ and $\hat{c}$ are related to the atoms that occupy the lower and upper energy levels. With this denotation,

$$\hat{H}_I = ig\left[\hat{a}^\dagger \hat{b}^\dagger \hat{c} - \hat{a}\hat{b}\hat{c}^\dagger\right],   \tag{6.166}$$

where the order of the operators suggests that the annihilation of the atomic state in the upper energy level leads to a simultaneous creation of a photon and observation of an atomic state in the lower energy level and vice versa.

One should note that the differential equations following from a trilinear Hamiltonian are nonlinear, and so are not solvable analytically. One may as a result compelled to consider certain special conditions. To overcome this challenge, in the first place, the circumstance in which almost all the atoms are prepared to be initially in the upper energy level is considered. It might be justifiable in this case to treat the operator $\hat{c}$ as a c-number γ_0 chosen to be a real positive constant;

$$\hat{H}_I = i\lambda\left[\hat{a}^\dagger \hat{b}^\dagger - \hat{a}\hat{b}\right],   \tag{6.167}$$

where $\lambda = g\gamma_0$. It may not be difficult to note that one can study the nonclassical properties and photon statistics of the radiation resulting from such interaction by solving the emerging differential equations (see Chap. 3).

With this in mind, the correlation that can be induced between the ensemble of atoms and the radiation via exchange of spontaneously emitted photons when the two-level atoms are initially prepared in the upper energy level, and then trapped in a lossless cavity containing a source of a squeezed radiation would be explored (see also [19]). To do so, one can recall that the Q-function for a two-mode cavity radiation (2.80) can be expressed as

$$Q(\alpha, \beta, t) = \frac{\langle \alpha, \beta | \hat{U}(t) | r, 0 \rangle \langle 0, r | \hat{U}^{\dagger}(t) | \alpha, \beta \rangle}{\pi^2} \tag{6.168}$$

in which $\hat{U}(t) = \exp(-i\hat{H}_I t)$ is the evolution operator, $|r, 0\rangle$ is the initial state of the radiation and r is the squeeze parameter. One may note that the bosonic state associated with the atoms is taken to be in vacuum state until the onset of the interaction.

With this information and employing the fact (relate with Eq. 2.91) that

$$\langle \gamma | r \rangle = \sqrt{\frac{1}{\cosh r}} \exp\left[-\frac{\gamma^* \gamma}{2} - \frac{\gamma^{*2}}{2} \tanh r \right], \tag{6.169}$$

one can obtain

$$Q(\alpha, \beta, t) = \frac{1}{\pi^2 \cosh r} \int \frac{d^2\gamma}{\pi} \frac{d^2\eta}{\pi} \frac{d^2\mu}{\pi} \frac{d^2\nu}{\pi} K(\alpha, \beta, t|\gamma, \eta, 0) K^*(\alpha, \beta, t|\nu, \mu, 0)$$

$$\times \exp\left[-\frac{1}{2}\left(\gamma^*\gamma + \eta^*\eta + \nu^*\nu + \mu^*\mu + \left(\gamma^{*2} + \mu^2 \right) \tanh r \right) \right], \tag{6.170}$$

where $K(\alpha, \beta, t|\gamma, \eta, 0) = \langle \alpha, \beta | \hat{U}(t) | \gamma, \eta \rangle$ (see also Eq. (3.185)) is found following the procedure outlined in Sect. 3.2.3 to be

$$K(\alpha, \beta, t|\gamma, \eta, 0) = \frac{1}{\cosh(\lambda t)} \exp\left[-\frac{\alpha^*\alpha}{2} - \frac{\beta^*\beta}{2} - \frac{\gamma^*\gamma}{2} - \frac{\eta^*\eta}{2} \right.$$

$$\left. + \left(\alpha^*\beta^* - \gamma\eta \right) \tanh(\lambda t) + \frac{1}{\cosh(\lambda t)} \left(\alpha^*\gamma + \beta^*\eta \right) \right], \tag{6.171}$$

as a result follows

$$Q(\alpha, \beta, t) = \frac{1}{\pi^2 \cosh r \cosh^2(\lambda t)} \exp\left[-\alpha^*\alpha - \beta^*\beta \right.$$

$$\left. + (\alpha^*\beta^* + \alpha\beta) \tanh(\lambda t) - \frac{\tanh r}{2 \cosh^2(\lambda t)} \left(\alpha^{*2} + \alpha^2 \right) \right]. \tag{6.172}$$

Notice that α and β are c-number variables that stand for the radiation and atomic properties.

Obviously, to explore properties of the available radiation, one needs to integrate over atomic variables;

$$Q(\alpha, t) = \frac{1}{\pi \cosh r \cosh^2(\lambda t)} \exp\left[-\frac{1}{\cosh^2(\lambda t)}\left[\alpha^*\alpha + \frac{\tanh r}{2}\left(\alpha^{*2} + \alpha^2\right)\right]\right],$$

(6.173)

which represents the cavity radiation since the contribution of the atoms is incorporated to the normalization constant. In the absence of the coupling of the atoms with the radiation ($\lambda = 0$), one can see that

$$Q(\alpha, t) = \frac{1}{\pi \cosh r} \exp\left[-\alpha^*\alpha - \frac{\tanh r}{2}\left(\alpha^{*2} + \alpha^2\right)\right],$$

which designates the squeezed radiation, and when there is no squeezed light ($r = 0$);

$$Q(\alpha, t) = \frac{1}{\pi(1 + \sinh^2(\lambda t))} \exp\left[-\frac{\alpha^*\alpha}{(1 + \sinh^2(\lambda t))}\right],$$

(6.174)

which exemplifies the radiation that would have been emitted spontaneously from the two-level atoms placed in the cavity, and describes the radiation in the chaotic state with mean photon number of $\sinh^2(\lambda t)$ (compare with Eq. (2.154)).

Carrying out the integration over radiation variables on the other hand indicates that

$$Q(\beta, t) = \frac{1}{\pi \cosh r} \sqrt{\frac{1}{\cosh^4(\lambda t) - \tanh^2 r}} \exp\left[-\frac{1}{\cosh^4(\lambda t) - \tanh^2 r}\right.$$

$$\times \left.\left(\beta^*\beta\left(\cosh^2(\lambda t) - \tanh^2 r\right) + \left(\beta^{*2} + \beta^2\right)\frac{\tanh r \sinh^2(\lambda t)}{2}\right)\right],$$

(6.175)

which represents the atomic properties associated with the population in the lower energy level, and reduces for $r = 0$ to

$$Q(\beta, t) = \frac{1}{\pi(1 + \sinh^2(\lambda t))} \exp\left[-\frac{\beta^*\beta}{(1 + \sinh^2(\lambda t))}\right].$$

(6.176)

Equation (6.176) also designates a quantum system in a chaotic state with $\sinh^2(\lambda t)$ mean photon number. Likewise, comparison of Eqs. (6.174) and (6.176) may reveal that the radiation in the cavity and atomic properties in terms of the lower energy level in the absence of the squeezed radiation are governed by similar Q-function since the source of the cavity radiation is the spontaneous emission process.

Fig. 6.14 Plots of the positive quadrature variance of an ensemble of two-level atoms when initially almost all the atoms are presumed to occupy the upper energy level for different values of r

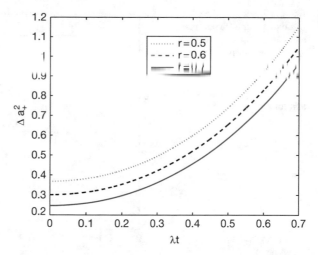

Once the Q-function is obtained, the expectation values of the pertinent moments can be calculated:

$$\langle \hat{a} \rangle = \langle \hat{a}^\dagger \rangle = 0, \tag{6.177}$$

$$\langle \hat{a}\hat{a}^\dagger \rangle = \cosh^2 r \cosh^2(\lambda t), \tag{6.178}$$

$$\langle \hat{a}^{\dagger 2} \rangle = \langle \hat{a}^2 \rangle = -\sinh r \cosh r \cosh^2(\lambda t). \tag{6.179}$$

In light of these results and Eq. (2.153), it may not be hard to arrive at

$$\Delta a_\pm^2 = 2\cosh^2(\lambda t)\cosh r\left(\cosh r \mp \sinh r\right) - 1. \tag{6.180}$$

One can see from Eq. (6.180) (also Fig. 6.14) that the squeezing does not exist for $r = 0$ but happens to be maximum just after the commencement of the interaction. This outcome may entail that the squeezing in the cavity radiation is attributed to the squeezed radiation generated by the source placed in the cavity, whereas the observed degradation of the degree of squeezing with time to the increased chance for each atom to undergo spontaneous emission in time (see also [20]).

The mean photon number can also be obtained from Eq. (6.178), that is,

$$\langle \hat{a}^\dagger \hat{a} \rangle = \cosh^2 r \cosh^2(\lambda t) - 1. \tag{6.181}$$

One can see from Eq. (6.181) that the mean photon number of the radiation increases with the squeeze parameter and interaction time due to the fact that as time progresses much of the atoms initially in the upper energy level would have a chance to decay to the lower energy level; and subsequently, emit radiation. One can as a result claim that a significantly intense light can be generated if the atoms are allowed to stay in the cavity for sufficiently long time. As one can see from

Fig. 6.15 Plots of the
positive quadrature variance
and mean photon number of
an ensemble of two-level
atoms when initially almost
all of them are presumed to
occupy the upper energy level
and for $r = 0.5$

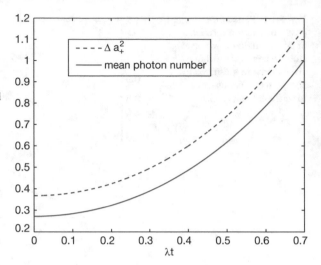

Fig. 6.15, the degree of squeezing however decreases with the chance of getting an
intense light—the circumstance that might compromise the feasibility of this system
for application.

One may need to examine at some point the statistical properties of the radiation
employing the photon number distribution. To this end, upon making use of
Eqs. (2.116) and (6.173) along with the fact that factorials are not well defined for
negative integers, one can verify that

$$P(n, t) = \frac{n!}{\cosh^2(\lambda t)\cosh r} \sum_{j=0}^{n} \frac{\tanh^{2j}(\lambda t)\left(\frac{\tanh r}{\cosh^2(\lambda t)}\right)^{n-j}}{2^{n-j}j!\left[\left(\frac{n-j}{2}\right)!\right]^2}. \tag{6.182}$$

In the absence of the squeezed light, one can see that

$$P(n, t) = \frac{(\sinh^2(\lambda t))^n}{(1 + \sinh^2(\lambda t))^{n+1}}, \tag{6.183}$$

which designates the radiation spontaneously emitted from the ensemble of two-
level atoms, and shows that the emitted radiation is in a chaotic state as it should be
(compare with Eq. (2.124)).

On the other hand, to overcome the challenge posed by a nonlinearity associated
with a trilinear Hamiltonian, one may consider the case in which the two-level atoms
are initially prepared with nearly equal number of atoms in each energy level. The
interaction Hamiltonian in this case can take the from

$$\hat{H}_I = i\varsigma(\hat{a}^\dagger - \hat{a}), \tag{6.184}$$

where ς is proportional to coupling constant. For this case, we opt to solve the pertinent Fokker-Planck equation for the P-function when the combined system is presumed to be coupled to a single-mode thermal reservoir in which the master equation can be adopted from Eq. (3.42) as

$$\frac{\partial \hat{\rho}}{\partial t} = \varsigma(\hat{a}^\dagger \hat{\rho} - \hat{a}\hat{\rho} - \hat{\rho}\hat{a}^\dagger + \hat{\rho}\hat{a}) + \frac{\kappa}{2}(\bar{n}+1)(2\hat{a}\hat{\rho}\hat{a}^\dagger - \hat{a}^\dagger \hat{a}\hat{\rho} - \hat{\rho}\hat{a}\hat{a}^\dagger)$$

$$+ \frac{\kappa}{2}\bar{n}(2\hat{a}^\dagger \hat{\rho}\hat{a} - \hat{a}\hat{a}^\dagger \hat{\rho} - \hat{\rho}\hat{a}\hat{a}^\dagger). \tag{6.185}$$

It is not possible to directly obtain the Fokker-Planck equation by applying the definition of the P-function in terms of the density operator since there are some mixed terms (see Sect. 3.2.3). It is therefore necessary first putting every operator in the anti-normal ordering into normal ordering with the aid of the commutation relation (3.170). One can then see that the Fokker-Planck equation for the P-function takes in the phase space representation the form

$$\frac{\partial P(x, y, t)}{\partial t} = \frac{\kappa}{2}\left[\left(\frac{\partial}{\partial x}((x-\varsigma)P) + \frac{\partial}{\partial y}(yP)\right) + \frac{\kappa\bar{n}}{2}\left(\frac{\partial^2 P}{\partial x^2} + \frac{\partial^2 P}{\partial y^2}\right)\right], \tag{6.186}$$

which reduces at steady state to

$$\frac{\partial}{\partial x_i}(A(x_i)P) + \frac{\kappa\bar{n}}{2}\frac{\partial^2 P}{\partial x_i^2} = 0, \tag{6.187}$$

where $A(x_1) = x - \varsigma$ and $A(x_2) = y$. It is also straightforward to verify that

$$A(x_i)P + \frac{\kappa\bar{n}}{2}\frac{\partial P}{\partial x_i} = 0 \tag{6.188}$$

while the constant of the integration is fixed at zero.

Note that

$$P(x, y) = C\exp\left[-\frac{1}{\kappa\bar{n}}(x^2 + y^2 - 2\varsigma x + \varsigma^2)\right], \tag{6.189}$$

where C is the normalization constant related to initial condition. With the help of the variable conversions in Sect. 3.2.3 once again, the P-function can be expressed as

$$P(\alpha, \alpha^*) = \frac{1}{\pi\kappa\bar{n}}\exp\left[-\frac{\alpha\alpha^*}{\kappa\bar{n}} + \frac{\varsigma(\alpha^* + \alpha)}{\kappa\bar{n}} - \frac{\varsigma^2}{\kappa\bar{n}}\right]. \tag{6.190}$$

Once the quasi-statistical distribution function is known, it should be straightforward to calculate the variance of the photon number that can emulate the statistical properties of the radiation. To this end, it might worth noting that the variance of the photon number,

$$\Delta n^2 = \langle \hat{n}^2 \rangle - \langle \hat{n} \rangle^2, \tag{6.191}$$

can be expressed in the normal order in terms of the boson operators as

$$\Delta n^2 = \langle \hat{a}^\dagger \hat{a} \rangle + \langle \hat{a}^\dagger \hat{a}^\dagger \hat{a} \hat{a} \rangle - \langle \hat{a}^\dagger \hat{a} \rangle^2. \tag{6.192}$$

It is required to evaluate only two of these terms since the last term is the square of the first term. In this respect, one can see that

$$\langle \hat{a}^\dagger \hat{a} \rangle = \frac{1}{\kappa \bar{n}} \int \frac{d^2\alpha}{\pi} \alpha^* \alpha \exp\left[-\frac{\alpha^* \alpha}{\kappa \bar{n}} + \frac{\varsigma(\alpha^* + \alpha)}{\kappa \bar{n}} - \frac{\varsigma^2}{\kappa \bar{n}} \right], \tag{6.193}$$

from which follows

$$\langle \hat{a}^\dagger \hat{a} \rangle = \kappa \bar{n} + \varsigma^2. \tag{6.194}$$

Equation (6.194) indicates that the mean photon number is contributed by the thermal fluctuations entering the cavity and the photons emitted by the atoms. This result also supports the assertion that although the atoms are initially prepared with equal occupancy of the lower and upper energy levels, after some time, more of them prefer to spontaneously emit radiation and go over to the lower energy level. This on the other hand heralds that the initially induced or prepared coherence can be lost in case the system is left alone (relate with the discussion in Sect. 10.3).

In a similar manner, it is possible to see that

$$\langle \hat{a}^{\dagger^2} \hat{a}^2 \rangle = 2(\kappa \bar{n})^2 + 4\varsigma^2 \kappa \bar{n} + \varsigma^4, \tag{6.195}$$

and as a result follows

$$\Delta n^2 = \kappa \bar{n}(1 + \kappa \bar{n}) + \varsigma^2(1 + 2\kappa \bar{n}). \tag{6.196}$$

It may not be difficult to infer from Eq. (6.196) that the cavity radiation exhibits super-Poissonian photon statistics ($\Delta n^2 > \kappa \bar{n} + \varsigma^2$) for $\kappa \neq 0$, that means, photon number correlation exhibits classical property despite the initial coherent preparation.[10]

---

[10]Owing to the problem associated with the P-function in quantifying nonclassical properties of radiation (see the remark in Sect. 2.3), one may expose the squeezing properties of the cavity radiation with the help of the quantum Langevin equation or solving the Fokker-Planck equation for Q-function.

6.5 Multi-Level Atom

As already noted, a two-level approximation can adequately explain the interaction of atom with a single mode radiation whose frequency is nearly resonant with the atomic transition between the involved energy levels. But when the atom interacts with the radiation field that encompasses many frequencies (multi-mode), several energy levels of the atom need to be incorporated in the process since different atomic transitions are expected to undergo near resonant interaction with separate components of the field; the idea that constitutes the notion of multi-level atom. The presumption that the engaged many modes of the radiation can be coherently superimposed may envisage that the interaction of a multi-mode radiation with different energy levels of the atom can lead to correlated emission [21]. The coherent interaction between a broadband electromagnetic radiation and multi-level atoms for example can induce atomic coherence among the pertinent populations of the energy states that leads to some quantum interference effects such as lasing without population inversion and electromagnetically induced transparency.

6.5.1 General Consideration

One may recall that the study of the evolution of the atom-radiation combined system requires characterization of the parameters in terms of the amplitude or density matrix (see Sect. 5.2). Notably, it is possible to retain the phase information related to the evolution of the amplitude of atomic state—it is in this sense that one refers to atomic coherence and coherent preparation [22]. This description should not be understood in the same way as the rate equation treatment of the populations that turns out to be appropriate when damping is large or coupling is weak, and so the coherence can be ignored. Even then, it happens to be difficult to solve coupled differential equations that follow from Schrödinger or master equation when many energy levels of the atom participate in the interaction. To overcome this obstacle, certain approximation techniques such as taking the effect of the transition between the two levels on the other pair as minimal would be utilized.

To begin with, the eigenstate of a multi-level atom can be expressed as

$$|\Psi\rangle = \sum_j C_j(t)e^{-i\omega_j t}|j\rangle, \tag{6.197}$$

where j stands for each energy level (see also Eq. (5.2)) and the probability amplitudes satisfy the normalization condition $\sum_j |C_j(t)|^2 = 1$. One can then write with the aid of Schrödinger equation that

$$\frac{d}{dt}C_l(t) = -i\sum_j C_j(t)\langle l|\hat{V}|j\rangle e^{i\omega_{jl}t}, \tag{6.198}$$

where $\hat{V} = -\hat{\mu}.\hat{E}$ is the interaction Hamiltonian (see also Eq. (5.29)) and $\omega_{lj} = E_l - E_j$ is atomic transition frequency. Equation (6.198) can be solved by assuming that initially the system is in arbitrary energy level, let us say, $C_j(0) = 1$ and $C_{l \neq j}(0) = 0$. Even though such a set of equations can be solvable for a simple case like two-level scheme, directly solving the resulting coupled differential equations would become formidable task when many energy levels are involved.

The solution of the emerging coupled equations however can be obtained by employing perturbation technique in which the driving field is taken to be so weak, that is, when the atomic population changes very little. Consistent with the general approach in perturbation, one can propose the Hamiltonian and the probability amplitudes of the form

$$\hat{H} = \hat{H}_0 + \lambda \hat{H}', \tag{6.199}$$

$$C_l(t) = C_l^{(0)}(t) + \lambda C_l^{(1)}(t) + \lambda^2 C_l^{(2)} + \dots, \tag{6.200}$$

where $\hat{H}'$ stands for the perturbation and λ is taken to be a small constant. Insertion of Eqs. (6.199) and (6.200) into Eq. (3.1) results

$$\frac{d}{dt} C_l^{(0)}(t) = 0, \tag{6.201}$$

$$\frac{d}{dt} C_l^{(1)}(t) = -i \sum_j C_j^{(0)}(t) \langle l | \hat{H}' | j \rangle e^{i\omega_{lj}t}, \tag{6.202}$$

$$\frac{d}{dt} C_l^{(2)}(t) = -i \sum_j C_j^{(1)}(t) \langle l | \hat{H}' | j \rangle e^{i\omega_{lj}t}, \tag{6.203}$$

$$\vdots$$

$$\frac{d}{dt} C_l^{(n)}(t) = -i \sum_j C_j^{(n-1)}(t) \langle l | \hat{H}' | j \rangle e^{i\omega_{lj}t}. \tag{6.204}$$

With this information, the discussion on two-level atom might be required to be extended to three-level atom; the simplest and easiest scheme that can be used to exemplify a multi-level scenario. One may note that the three-level atom scheme embodies the situation in which three energy levels of the atom interact with nearly resonant two-mode electromagnetic radiation. Due to the possibility of coherence superposition among participating three energy levels of the atom that can be induced either by initial preparation or external pumping, a correlated emission is expected. The possibility of preparing the three-level atom in a coherent superposition of the involved energy levels can thus lead to entangled atomic states and generation of nonclassical light (see Sect. 9.2). A three-level atom can be described by three atomic energy levels generally denoted by the lower $|l\rangle$, intermediate $|e\rangle$ and upper $|u\rangle$. Since only two from the designated three transitions

Fig. 6.16 Schematic representation of a three-level atom where the upper, intermediate and lower energy levels are denoted by states $|u\rangle$, $|e\rangle$ and $|l\rangle$

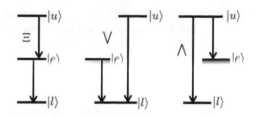

are dipole allowed, a three-level atom approximation should be consistent with a two-photon process such as parametric oscillation and correlated emission. Based on various possible energy transition permutations, one can then think of three valid schemes (see Fig. 6.16 and also [23]).

The first and most straightforward arrangement is the ladder (Ξ)-configuration for which the lower and upper atomic energy levels have the same parity but the intermediate state has the opposite, that is, $|e\rangle \rightarrow |l\rangle$ and $|u\rangle \rightarrow |e\rangle$ are dipole allowed transitions but $|u\rangle \rightarrow |l\rangle$ is dipole forbidden. When the atom in this configuration is excited to the upper energy level, it can spontaneously decay to the lower energy level via successive emission (or cascading) of two photons with frequency resonant to respective transitions; as a result, this system is usually dubbed as three-level cascade atom. The other possibility is the $\vee$-configuration in which the atom is assumed to have a stable lower energy state where the two excited states are coupled to the lower energy level via dipole allowed transitions but not to each other. If the atom occupies either of these states, it decays to the lower state, and emits a single photon whose frequency is close to the resonant frequency of the respective transition. Besides, in the $\wedge$-configuration, the atom would have two lower energy levels—one stable $|l\rangle$ and the other meta-stable $|e\rangle$—and one excited state. In this configuration, the upper excited state can decay to one of the lower states but the transition between the lower states is dipole forbidden. In these schemes, the atomic coherent superposition can also be induced by pumping the dipole forbidden transition with external coherent radiation.

In this scenario, two radiations designated by electric fields $\mathbf{E}_1$ and $\mathbf{E}_2$ are required to couple the dipole allowed transitions. Upon using the denotation as in Sect. 5.1, the atomic operator can be expressed as $\hat{\sigma}_{mj} = |m\rangle\langle j|$, where m and j stand for l, e, u. So the Hamiltonian that represents a three-level atom that interacts with a two-mode radiation can be expressed as $\hat{H} = \hat{H}_A + \hat{V}$, where $\hat{V} = -\hat{\mu}.\left[\hat{\mathbf{E}}_1 + \hat{\mathbf{E}}_2\right]$ is the interaction Hamiltonian. If the electric field $\hat{\mathbf{E}}_1$ is assumed to interact with the atomic transition $|m\rangle \leftrightarrow |j\rangle$ and $\hat{\mathbf{E}}_2$ with $|j\rangle \leftrightarrow |n\rangle$, the interaction Hamiltonian can be written as

$$\hat{V} = -\hat{\mu}_{mj}\hat{E}_1 e^{-i\omega_1 t} - \hat{\mu}_{nj}\hat{E}_2 e^{-i\omega_2 t} + c.c, \tag{6.205}$$

where $\hat{\mu}_{mj} = \langle m|e\mathbf{r}.\hat{e}_k|j\rangle$ is the dipole matrix elements of atomic transition and m, j, n stand for the accessible dipole transitions that can be modified depending on the configuration one chooses (relate with Eq. (5.30)).

The state vector that describes a three-level atom on the other hand can be expressed as

$$|\psi(t)\rangle = C_l(t)|l\rangle + C_e(t)|e\rangle + C_u(t)|u\rangle. \tag{6.206}$$

Note that the time dependence is included in the probability amplitudes (compare with Eq. (5.2)). With this information, one can write

$$\frac{dC_l}{dt} = -i\omega_l C_l + iC_u \Omega_1 e^{i\omega_1 t}, \tag{6.207}$$

$$\frac{dC_e}{dt} = -i\omega_e C_e + iC_l \Omega_1 e^{i\omega_1 t} + iC_u \Omega_2 e^{-i\omega_2 t}, \tag{6.208}$$

$$\frac{dC_u}{dt} = -i\omega_u C_u + iC_e \Omega_2 e^{-i\omega_2 t}, \tag{6.209}$$

where $\Omega_{1,2}$ are the Rabi frequencies of the pertinent transitions (see Eq. (5.57)). One may note that Eqs. (6.207)–(6.209) can also be rewritten in the interaction picture by applying appropriate transformation.

The time evolution of the populations and atomic coherence can be evaluated with the help of the master equation that can be derived following the approach outlined for the two-level case. With this in mind, the interaction Hamiltonian in the frame rotating with the frequencies of the optical fields can be imagined:

$$\hat{H}_\wedge = -\Delta_1 \hat{\sigma}_{uu} + (\Delta_2 - \Delta_1)\hat{\sigma}_{ee} - [\Omega_1 \hat{\sigma}_{ul} + \Omega_2 \hat{\sigma}_{ue} + H.c], \tag{6.210}$$

$$\hat{H}_\vee = -\Delta_1 \hat{\sigma}_{uu} - \Delta_2 \hat{\sigma}_{ee} - [\Omega_1 \hat{\sigma}_{ul} + \Omega_2 \hat{\sigma}_{el} + H.c], \tag{6.211}$$

$$\hat{H}_\Xi = -\Delta_1 \hat{\sigma}_{ee} - (\Delta_1 + \Delta_2)\hat{\sigma}_{uu} - [\Omega_1 \hat{\sigma}_{el} + \Omega_2 \hat{\sigma}_{ue} + H.c], \tag{6.212}$$

where $\Delta_{1,2}$ are respective detunings (relate with Eq. (5.63)) and $H.c$ stands for Hermitian conjugate. Note that the decay process and external pumping can be included in the same manner, and equation of the elements of the density matrix can be obtained by making use of the master equation such as (5.51). One may also need to extend this kind of discussion to the case when more than three atomic energy levels are engaged in the interaction.

In this respect, one may need to begin with the assumption that the dipole moment that corresponds to different transitions oscillate with different frequencies. In doing so, the Hamiltonian that describes the interaction can be expressed as

$$\hat{H} = \hat{H}_A + \hat{H}_R + \hat{H}_{AR}, \tag{6.213}$$

where $\hat{H}_A = \sum_{j=0}^{N} \omega_j |j\rangle\langle j|$ is the Hamiltonian of the free atom with energy levels $|0\rangle$ that corresponds to the lower energy level, $|1\rangle, |2\rangle, \ldots, |N-1\rangle$ the intermediate excited states and $|N\rangle$ the upper energy level, $\hat{H}_R$ is the free radiation Hamiltonian: $\hat{H}_R = \sum_{k,s} \omega \hat{a}_{k,s}^\dagger \hat{a}_{k,s}$, where $\hat{a}_{k,s}$ is the boson annihilation of the radiation with mode k and s is the index of polarization. Whereas $\hat{H}_{AR}$ is the interaction Hamiltonian: $\hat{H}_{AR} = -\hat{\mu}.\hat{\mathbf{E}}(\mathbf{r})$ with

$$\hat{\mu} = \sum_{j=0}^{N} \sum_{l>j}^{N-1} \left(\boldsymbol{\mu}_{lj}\hat{\sigma}_{lj} + \boldsymbol{\mu}_{jl}\hat{\sigma}_{jl} \right) \tag{6.214}$$

in which $\boldsymbol{\mu}_{lj} = \boldsymbol{\mu}_{jl}^* = \langle l|\boldsymbol{\mu}|j\rangle$ is the dipole matrix elements and $\hat{\sigma}_{lj} = |l\rangle\langle j|$ is the atomic transition dipole operator.

One can see that the atomic dipole operators satisfy the commutation relation

$$\left[\hat{\sigma}_{jl}, \hat{\sigma}_{pq} \right] = \hat{\sigma}_{jp}\delta_{lq} - \hat{\sigma}_{pl}\delta_{qj} \tag{6.215}$$

and the completeness relation would be

$$\hat{I} = \sum_{j=0}^{N} |j\rangle\langle j| = \sum_{j=0}^{N} \hat{\sigma}_{jj}. \tag{6.216}$$

With this consideration, the interaction Hamiltonian would be expressed as

$$\hat{H}_{AR} = -i \sum_{k,s} \sum_{j=0}^{N} \sum_{l>j}^{N-1} \left[\hat{\boldsymbol{\mu}}_{lj}.\mathbf{u}_{k,s}\left(\hat{\sigma}_{lj}\hat{a}_{k,s} + \hat{\sigma}_{jl}\hat{a}_{k,s} \right) - H.c \right], \tag{6.217}$$

where

$$\mathbf{u}_{k,s} = \sqrt{\frac{\omega_{ks}}{2\varepsilon_0 V}} \hat{e}_{k,s} \tag{6.218}$$

is the mode function of a multi-mode resonant radiation field in free space (compare with Eq. (6.151)), V is the quantization volume and $\hat{e}_{k,s}$ is the unit polarization vector.

The atomic transitions are taken to be realized by the dipole transition, on the basis of which the time evolution of a multi-dipole system interacting with a multi-mode dissipative electromagnetic radiation is described in terms of the density operator that embodies the combined system, and the Liouvelle-Neuman equation: $\frac{\partial \hat{\rho}_{AR}^I}{\partial t} = -i \left[\hat{H}_{AR}, \hat{\rho}_{AR}^I(t) \right]$, where $\hat{\rho}_{AR}^I(t) = e^{i\hat{H}_0 t} \hat{\rho}_{AR} e^{-i\hat{H}_0 t}$ and the interaction

Hamiltonian is

$$\hat{H}_{AR}(t) = -i \sum_{k,s} \sum_{j=0}^{N} \sum_{l>j}^{N-1}$$

$$\times \left[\hat{\boldsymbol{\mu}}_{lj}.\mathbf{u}_{k,s} \left(\hat{\sigma}_{lj} \hat{a}_{k,s} e^{-i(\omega_{ks}-\omega_{lj})t} + \hat{\sigma}_{jl} \hat{a}_{k,s} e^{-i(\omega_{ks}+\omega_{lj})t} \right) - H.c \right].$$

(6.219)

One may note that many of the $\hat{\sigma}_{lj}$'s correspond to dipole forbidden transitions ($\hat{\boldsymbol{\mu}}_{lj} = 0$).

Suppose the transition we need to keep is denoted by l, which stands for the numeration of allowed transition in which $\omega_{lj} = \omega_m$, $\hat{\sigma}_{lj} = \hat{s}_m^+$ and $\hat{\sigma}_{jl} = \hat{s}_m$. With this variable change, the interaction Hamiltonian becomes

$$\hat{H}_{AR}(t) = -i \sum_{k,s} \sum_{m} \left[\hat{\boldsymbol{\mu}}_m.\mathbf{u}_{k,s} \left(\hat{s}_m^+ \hat{a}_{k,s} e^{-i(\omega_{k,s}-\omega_m)t} + \hat{s}_m^- \hat{a}_{k,s} e^{-i(\omega_{ks}+\omega_m)t} \right) - H.c \right].$$

(6.220)

Once the interaction Hamiltonian is set, the time evolution of the reduced density operator can be obtained by tracing over radiation variable:

$$\frac{\partial \hat{\rho}}{\partial t} = -i \left[\hat{H}_{AR}(t), \hat{\rho}(0) \right] - \int_0^t \left[\hat{H}_{AR}(t), \left[\hat{H}_{AR}(t'), \hat{\rho}(t') \right] \right] dt'. \qquad (6.221)$$

The master equation for arbitrary configuration is thus found (compare with Eq. (6.54)) to be

$$\frac{\partial \hat{\rho}}{\partial t} = \frac{(1+\bar{n})}{2} \sum_{l,j=1}^{N} \Gamma_{lj} \left(2\hat{s}_j^- \hat{\rho} \hat{s}_l^+ - \hat{\rho} \hat{s}_l^+ \hat{s}_j^- - \hat{s}_l^+ \hat{s}_j^- \hat{\rho} \right) e^{i(\omega_l-\omega_j)t}$$

$$+ \frac{\bar{n}}{2} \sum_{l,j=1}^{N} \Gamma_{lj} \left(2\hat{s}_j^+ \hat{\rho} \hat{s}_l^- - \hat{\rho} \hat{s}_l^- \hat{s}_j^+ - \hat{s}_l^- \hat{s}_j^+ \hat{\rho} \right) e^{-i(\omega_l-\omega_j)t}, \qquad (6.222)$$

where $\Gamma_{ll} = \gamma_l = \omega_l^3 \mu_l^2 / (3\pi\varepsilon_0 c^3)$ is the atomic damping rates and $\Gamma_{lj(l\neq j)} = \sqrt{\gamma_l \gamma_j}$ are cross damping rates or dephasing (see Sect. 6.2 and also [22]). $\Gamma_{lj(l\neq j)} \neq 0$ in case the spontaneous emission from one level alters the dipole moment of the other transition, that is, when the dipole moments have a nonzero parallel component.

For a large gap in the atomic transition frequencies ($\omega_l - \omega_j \gg 0$), it might be possible to neglect the contribution of the off-diagonal terms. Equation (6.222) can also be modified if the cavity is coupled to an external pumping of strength Ω as

$$\frac{\partial \hat{\rho}}{dt} = \sum_{l=1}^{N} \wp_l [\hat{s}_l^+ - \hat{s}_l^- \quad \hat{\rho}] + \frac{(1+\bar{n})}{\gamma} \sum_{l,j=1}^{N} \Gamma_{lj} (2\hat{s}_j^- \hat{\rho}\hat{s}_l^+ - \hat{\rho}\hat{s}_l^+ \hat{s}_l^- - \hat{s}_l^+ \hat{s}_l^- \hat{\rho})$$

$$+ \frac{\bar{n}}{2} \sum_{l,j=1}^{N} \Gamma_{lj} (2\hat{s}_j^+ \hat{\rho}\hat{s}_l^- - \hat{\rho}\hat{s}_l^- \hat{s}_j^+ - \hat{s}_l^- \hat{s}_l^+ \hat{\rho}) \tag{6.223}$$

when only the resonance case is retained.

6.5.2 Origin and Consequences of Atomic Coherence

A large collection of atoms can be efficiently coupled to the quantized light when the collective superposition among various states of many atoms is utilized (see Sect. 5.6). If the modes of light impinged onto a collection of atoms in the ground state exhibit quantum correlations or entanglement, and absorbed by the atomic medium, one can argue that the quantum correlations can be mapped onto the collective superposition of the final states of the atoms [24]. With the help of the atom-radiation interface sanctioned by the interaction of an ensemble of multi-level atoms with radiation, it should be possible to achieve the quantum state transfer from light to atoms. The other vital features of the notion of the ensemble of multi-level atoms is the generation of entanglement between remote atomic ensembles that can be used in the realization of quantum teleportation and entanglement swapping protocols. Since the resonances can be broadened due to a relatively rapid decaying process caused by the interaction between dipole moments of the excited electrons with the continuum of the radiation modes, one can propose that, in case the atoms go over to the lower energy level via emitting radiation, the final continuum state can be reached either via direct excitation or dipole interaction that leads to the decay that provides the second channel to the final state.

It should be stressed that the addition of the second driving field to the arrangement does not only cause qualitative changes in the shape of the emission and absorption spectra, but also modifies the line width in a way that depends on the atomic parameters and the relative strength of the fields. The interference between these channels can be constructive or destructive, and so constitutes frequency dependent suppression or enhancement of photoionization cross-section. It might be as result acceptable to foresee that the interference between two closely spaced resonances decaying to the same continuum does lead to lasing without population inversion as in correlated emission laser (see Sect. 8.1.4). Such a coherent interaction also institutes the idea of electromagnetically induced transparency[11]— the phenomenon that renders the medium transparent over a narrow spectral range within the absorption line [25, 26].

---

[11]Even through there are several options, we opt to use the concept of electromagnetically induced transparency to illustrate the consequences of atomic coherent superposition.

The notion of electromagnetically induced transparency can also be related to the phenomenon of coherent population trapping in which the atom prefers to occupy one of the energy levels. In this case, electromagnetically induced transparency relies on the destructive interference of the transition probability amplitudes between atomic states where the main cause of the modification in the optical response of the atomic medium is the coherence induced by external pumping in the sense that the coherence leads to quantum interference between the excitation pathways. Incidentally, due to destructive interference, the absorption capability of the medium might be inhibited, and so the refraction or linear susceptibility at the resonant frequency of the transition can be modified.

To explain the origin of electromagnetically induced transparency, one can begin with the conception that the atomic dressed states can be viewed as comprising two closely spaced resonances that effectively decay to the same continuum. If the probe field is tuned exactly to the zero point or midway resonance frequency, the contribution of the linear susceptibility due to the two resonances—which are equally spaced but with opposite signs of detuning—would be equal and opposite, and thus leads to the cancelation of the response at this frequency. In the view that the involved transitions can be seen as following different pathways between the bare states, the effect of the laser lights is perceived to transfer a small but finite amplitude to one of the energy levels of the excited states. Upon taking the coupling field to be much more intense than the probe, the indirect pathway is expected to have a probability amplitude that equals in magnitude to the direct pathway but for resonant fields it is of opposite sign, and so leads to cancelation.

Imagine that the probe field is tuned near resonance between the two states, and a much stronger coupling field is also tuned near resonance at different transition. If the states are selected properly, the presence of the coupling field is expected to create spectral window of transparency that can be detected by the probe. To understand the phenomenon of electromagnetically induced transparency in a more qualitative manner, let us begin with the assumption that, in the $\wedge$-configuration, the two excited atomic energy levels are coupled to external laser of frequency ω_c and the upper energy level is coupled to the lower energy level to the probe laser of frequency ω_p with detunings:

$$\Delta_1 = \omega_{ul} - \omega_p, \qquad \Delta_2 = \omega_{ul} - \omega_c, \tag{6.224}$$

and damping rates Γ_{ue} and Γ_{ul}.

The interaction Hamiltonian would be expressed in the form

$$\hat{H}_I = -\frac{\Omega_p}{2}|l\rangle + \left(\Delta_1 - \Delta_2 - \frac{\Omega_c}{2}\right)|e\rangle + \left(2\Delta_1 - \frac{\Omega_c}{2} - \frac{\Omega_p}{2}\right)|u\rangle. \tag{6.225}$$

For a two-photon resonance ($\Delta_1 = \Delta_2$), one can introduce new parameters:

$$\theta = \tan^{-1}\left(\frac{\Omega_p}{\Omega_c}\right), \qquad 2\psi = \tan^{-1}\left(\frac{\sqrt{\Omega_p^2 + \Omega_c^2}}{\Delta}\right), \qquad (6.226)$$

New bare atomic states are then introduced following the outline in Sect. 6.1.1;

$$|a\rangle = \cos\theta|l\rangle - \sin\theta|e\rangle, \qquad (6.227)$$

$$|a\rangle_+ = \sin\theta\sin\phi|l\rangle + \cos\phi|u\rangle + \cos\theta\sin\phi|e\rangle, \qquad (6.228)$$

$$|a\rangle_- = \sin\theta\cos\phi|l\rangle - \sin\phi|u\rangle + \cos\theta\cos\phi|e\rangle. \qquad (6.229)$$

One can see that $|a\rangle$ has no contribution from the state $|u\rangle$: for the atom that occupies this state, there is no chance to excite to $|u\rangle$; and hence, no subsequent spontaneous emission. Such a state is usually referred to as dark state. The evolution into the dark state via optical pumping is one way of trapping the population into this state. Besides, in case of a weak probe ($\Omega_p \ll \Omega_c$), one can see that $\sin\theta \to 0$ and $\cos\theta \to 1$;

$$|a\rangle = |l\rangle, \quad |a\rangle_+ = \cos\phi|u\rangle + \sin\phi|e\rangle, \quad |a\rangle_- = -\sin\phi|u\rangle + \cos\phi|e\rangle. \qquad (6.230)$$

The lower state becomes the dark state from which no excitation can occur, and when the probe is assumed to be in resonance ($\Delta = 0$), one can see that $\tan\phi \to 1$. So one can write

$$|a\rangle = |l\rangle, \qquad |a\rangle_\pm = \frac{1}{\sqrt{2}}\big[|e\rangle \pm |u\rangle\big]. \qquad (6.231)$$

Note that it is the coherent superposition between $|e\rangle$ and $|u\rangle$ states which is responsible for a peculiar response of the medium. These are the states relevant in the analysis of electromagnetically induced transparency that can be done in the same way as the discussion on the ensemble of two-level atoms (see Sect. 5.6 and also [25]).

It should also be admissible to consider a normal three-level ladder scheme with most of its population in the lower energy level. In this case, suppose the probe beam in resonance with the $|e\rangle \to |l\rangle$ transition is strongly absorbed. But when a strong coupling beam resonant with the $|u\rangle \to |e\rangle$ transition is added, the absorption of the probe beam is expected to be highly reduced although most of the population is still in the ground state. In this situation, the phenomenon of electromagnetically induced transparency can be inferred [27]. To look into this proposal, take for example ω_{el} to be the frequency of the $|e\rangle \to |l\rangle$ transitions and ω_p the frequency of the probe field that couples these states with detuning $\Delta_1 = \omega_p - \omega_{el}$, and similarly ω_c is the frequency of the laser that couples transition $|u\rangle \to |e\rangle$ and having a detuning

$\Delta_2 = \omega_c - \omega_{ue}$. In view of the approach introduced in Sect. 5.2, one can see that

$$\frac{d\rho_{ue}}{dt} = -\left(\gamma_{ue} - i\Delta_2\right)\rho_{ue} + ig\Omega_c\left(\rho_{uu} - \rho_{ee}\right) + ig\Omega_p\rho_{ue}, \qquad (6.232)$$

$$\frac{d\rho_{el}}{dt} = -\left(\gamma_{el} - i\Delta_1\right)\rho_{el} + ig\Omega_p\left(\rho_{ee} - \rho_{ll}\right) - ig\Omega_c\rho_{el}, \qquad (6.233)$$

$$\frac{d\rho_{ul}}{dt} = -\left[\gamma_{ul} - i(\Delta_1 + \Delta_2)\right]\rho_{ul} - ig\Omega_c\rho_{el} + ig\Omega_p\rho_{ue}. \qquad (6.234)$$

As long as one limits the solution of this coupled differential equations to a lowest order in the weak probe field, the time dependence of the coherence may not be required. It may not be hard to see that the emerging results would be independent of the channels through which the different states may get populated or depleted. The density matrix required to determine the response of the medium to the external probing at steady state is thus found to be

$$\rho_{el} = -\frac{ig\left[\gamma_{el} - i(\Delta_1 + \Delta_2)\right]}{(\gamma_{el} - i\Delta)[\gamma_{ul} - i(\Delta_1 + \Delta_2)] + g^2\Omega_c^2}. \qquad (6.235)$$

Once the atomic coherence is known, the electromagnetically induced transparency can be generally explored by calculating the index of refraction and group velocity using the procedure outlined in Sect. 5.6.

6.6 Highly Excited Atom

With the advent of efficient laser and sophisticated experimental procedure, it becomes possible to excite the atom to high-lying energy levels—the scheme often referred to as Rydberg atom. Due to the occupancy of the electrons at a high-lying orbit, let us say, over thousands of Bohr's radii, a Rydberg atom is expected to have a very large effective size, and so interact over a long distance. One may note that the atom can generally be sent to a highly-excited energy level with the help of charge exchange mechanism in which a positive ion is made to collide with a ground state atom, where in the process, the ion can capture one electron from the ground state atom and the induced electron impact can send the atom to a very high-lying state. It is also feasible to realize this feast by utilizing optical excitation of the atom to a high energy level via two or many-step pumping, which is found to be more advantageous when compared to charge exchange and electron impact scenarios since individual high-lying states can be accurately addressed by gauging the photon energy used to drive the atoms [28].

Once high-lying energy levels are created, one can exploit the prevailing strong interaction between the atoms to explore the emerging many body quantum dynamics. When the atom is excited to the high-lying energy level, virtually all of its properties would become exaggerated. Since the size of the Rydberg atom scales quadratically with the principal quantum number, and the motion of the valence electron slows down by the same fraction, the time for the atom to radiatively decay to the lower lying energy levels drastically increases. The strength of the interaction between Rydberg atoms can also be tuned over several orders of magnitude upon changing the density of the energy levels to which they are excited: the Rydberg atom is thus anticipated to feel strong interaction from its neighbor, that means, the potential energy between two Rydberg atoms increases by a considerable order where such a strong interaction can shift the energy levels of the atoms, and is hence sufficient to modify the Rydberg excitation over a comparatively longer distance [29].

6.6.1 Characterization

When the Rydberg atom spontaneously decays from a high-lying energy level, it goes over to the ground state or first to another low-lying energy level, and then cascades down till it reaches the ground state. Aiming at reducing the emerging complications and assuming that the low-lying energy levels have relatively short lifetimes, the contribution of the intermediate transitions is often ignored, and so each atom can be approximated as a two-level system and the spontaneous emission is accounted for by the pertinent line width. Even then, each atom can emit to different radiation modes due to the accompanying myriads of possible emission roots.

With this information, the emerging dynamics of the atom and radiation can be explored by applying the master equation (see Eq. (6.163) and also [30])

$$\frac{d\hat{\rho}}{dt} = -i[\hat{H}_I, \hat{\rho}] - \gamma \sum_j \left[\langle e_j|\hat{\rho}|e_j\rangle|g\rangle\langle g| - \frac{1}{2}|e_j\rangle\langle e_j|\hat{\rho} - \frac{1}{2}\hat{\rho}|e_j\rangle\langle e_j| \right],$$

(6.236)

where γ is the atomic line width, $|g\rangle$ is the manifold of the ground state, $|e_j\rangle$ is one of the excited states and $\hat{H}_I$ is the interaction Hamiltonian

$$\hat{H}_I = \sum_j \left[-\Delta_j|e_j\rangle\langle e_j| + \frac{\Omega_j}{2}\left(|e_j\rangle|\langle g + |g\rangle\langle e_j|\right) \right.$$

$$\left. + \frac{V}{N-1} \sum_{j<k} |e_j\rangle\langle e_j| \otimes |e_k\rangle\langle e_k| \right]$$

(6.237)

in which Δ_j is the detuning, Ω_j is the Rabi frequency and V is the potential that describes the interaction between pairs of Rydberg atoms expected to be large due to resonant dipole-dipole interaction.

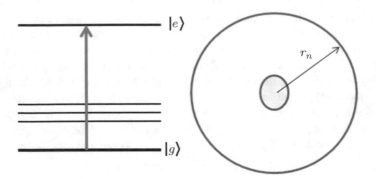

Fig. 6.17 Schematic representation of a highly excited atom where $|e\rangle$ stands for a high-lying Rydberg state, $|g\rangle$ for ground state and r_n is the radius of nth excited energy level

It is worth noting that an excited atom with a very large principal quantum number demonstrates an exaggerated response to electric and magnetic fields, and acquires long decay periods and wave functions that mimic classical orbit. Owing to the large orbit, the core electrons are expected to shield the outer electron from the electric field of the nucleus such that from a distance the electric potential looks like that of hydrogen atom. Bohr's model is thus thought to be adequate to explain some of the properties exhibited by the Rydberg atom (see Fig. 6.17). One may recall that the electron revolving in a circular orbit of radius r about a nucleus of charge $+e$ can be described by Newton's second law: $\frac{e^2}{4\pi\varepsilon_0 r^2} = \frac{mv^2}{r}$, where ε_0 is the permittivity of free space, m is the mass of electron and v is the magnitude of the velocity with which the electron revolves about the nucleus. Nonetheless, in quantum realm, the orbital momentum of the electron revolving about the nucleus should be quantized; $mvr = n$, where n is the principal quantum number.

Combination of these ideas then leads to

$$r_n = \frac{4\pi\varepsilon_0}{e^2 m} n^2 = cn^2, \tag{6.238}$$

where c is just a constant and r_n is the radius of nth excited energy level. One can see from Eq. (6.238) that the radius of the orbit scales as n^2 and the geometric cross section as n^4, which entails that the size of the Rydberg atom would be big with a loosely bound valence electrons, and so can easily be perturbed or ionized by collision or external field since the binding energy of the Rydberg electron is proportional to $1/r$; and hence, falls like $1/n^2$ while the energy level spacing falls like $1/n^3$, which leads to a more closely spaced energy levels.

Pertaining to these unique features, the outermost electron is expected to feel almost hydrogenic type Coulomb potential from the compact ion core that consists of the nucleus with Z number of protons and the lower electron shells filled with $Z - 1$ electrons: $U_C = -\frac{e^2}{4\pi\varepsilon_0 r}$. The similarity of the effective potential that can be seen by the outer electron to the hydrogenic potential is one of the defining attributes of the Rydberg states, which explains why the electron wave function resembles the classical orbit, and so adopting a two-level approximation might be sufficient. Even then, there are other notable exceptions that can be exemplified by the term added to the potential energy. For example, in case the atom has two or more electrons in a highly excited states with comparable orbital radii, the electron-electron interaction gives rise to a significant deviation from the hydrogenic potential, which includes $U_{ee} = \frac{e^2}{4\pi\varepsilon_0} \sum_{i<j} \frac{1}{|r_i - r_j|}$. In addition, in case the valence electron has a very low angular momentum, it acquires a potential capable of polarizing the ion core that gives rise to a $1/r^4$ core polarization term in the potential: $U_{pol} = -\frac{e^2\alpha_d}{(4\pi\varepsilon_0)^2} r^4$, where α_d is the dipole polarizability.

Whatever the case may be, it turns out to be appropriate and appealing to obtain a much simplified version of the energy. To do so, if one imagines the outer electron to see the nucleus with one proton and behaves much like the electron of the Hydrogen atom, the high energy levels in the Hydrogenic atom can be modeled by the Rydberg equation:

$$E_n = T - \frac{Ry}{n^2}, \tag{6.239}$$

where $Ry = 1.097 \times 10^7 m^{-1}$ is Rydberg constant, E_n is the energy above the ground state and T is the ionization limit. In line with the observation that the energy spectrum of the Rydberge atom can deviate from Eq. (6.239) in case one lifts the circular orbit approximation, it is customary introducing a correction term referred to as quantum defect.

In this regard, the Rydberg equation (6.239) needs to be modified so that it can account for the quantum defect;

$$E_n = T - \frac{Ry}{(n - \delta)^2}, \tag{6.240}$$

where δ is the quantum defect. Note that the quantum defect should be different for atomic states with different angular momenta. For example, an electron with no angular momentum passes directly through the core of the shielding electrons, and so has a high quantum defect. On the other hand, the energy level with a high principal quantum number describes the atom whose valence electrons are excited into formerly unpopulated electron orbital with a lower binding energy. The existence of a low binding energy at high values of the principal quantum number n explains why the Rydberg states are susceptible to ionization.

Moreover, a viable means to obtain a coherently superposed system is to use a collection of atoms where the collective excitation degree of freedom of the atomic ensemble is taken to be approximately equivalent to harmonic oscillator (see Sects. 6.4 and 6.5). The atom-cavity coupled system can then be described by a simple quadratic Hamiltonian in the raising and lowering operators for the atomic and radiation excitations (see Eq. (6.167)). In connection to this, the atomic ensemble is assumed to be placed in a high finesse cavity and the effect of the nonlinearity is induced by the interaction among the atoms resonantly coupled to a high-lying Rydberg state. Particularly, in a relatively small ensemble or Dicke's model, the dipole-dipole interaction can shift the energy of the states with two or more Rydberg excitations in which the atom-cavity system can be described by Jaynes-Cummings model, and the emerging collective behavior may lead to a strong coupling.

Upon omitting the interatomic interactions, one can represent the interaction of the ensemble of Rydberg atoms with radiation as

$$
\hat{H} \approx -\sum_{j=1}^{N} \Delta_j |i_j\rangle\langle i_j| + \left[g_0 \hat{a}^\dagger \sum_{j=1}^{N} |g_j\rangle\langle i_j| + H.c. \right] + \left[\sum_{j=1}^{N} \Omega_j |e_j\rangle\langle i_j| + H.c. \right],
$$
$$(6.241)$$

where g_0 is the atomic coupling constant and $|g_i\rangle$ stands for the ith lower-lying energy levels. In the limit of many atoms, one can define collective operator: weighed sum of the individual atomic operators that obeys the harmonic oscillator like commutator relations to good approximation (see Sect. 6.4);

$$
\hat{S}^\dagger = \frac{1}{\sqrt{N}} \sum_{j=1}^{N} |e_j\rangle\langle g_j|.
$$
$$(6.242)$$

As long as the atoms do not move appreciably and change the coupling strength, this weighed collective atomic mode plays the same role as the ideal symmetric mode and its coupling would be enhanced by the same factor.

6.6.2   Atomic Dipole Blockade

Assuming that the so far omitted dipole-dipole interaction would be strong between neighboring atoms and in inducing the transition $|g\rangle \rightarrow |e\rangle$, one expects to resonantly couple the collective ground state manifold $|G\rangle \equiv |g_1, \ldots, g_N\rangle$ to the collective symmetric state with a single Rydberg excitation: $|E\rangle \equiv \hat{S}^\dagger |G\rangle$. The other higher excited energy levels are presumed to be detuned too far to contribute to the interaction; the idea that constitutes the notion of the Rydberg blockade phenomenon (see also Fig. 6.18). The excitation of atoms to higher Rydberg level generally results in a very strong interaction having a resonant dipole-dipole character scaling as $1/R^3$ at short distances and van der Waals character scaling as $1/R^6$ at long distances. The two-atom dipole-dipole interaction is thus found to be turned on and off with a contrast of high orders of magnitude over a wide range, which might be a unique property to the Rydberg scheme [31].

When a low-lying atoms are excited to high-lying Rydberg states, the resulting interaction gives rise to energies that exceed the translational energy by many orders of magnitude, and so strongly modifies the excitation dynamics. In a small ensemble

Fig. 6.18 Schematic representation of dipole blockade where $|G\rangle$ stands for the ground state manifold and $|R\rangle$ for the Rydberg state in which only one atom can excite to $|R\rangle$ state out of the atoms in πr^2 area

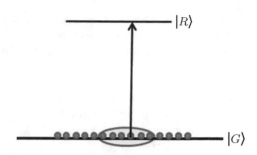

confined to small volume, a single Rydberg atom is found to block excitations in its sphere of influence, that means, in case there is an excitation hopping via a resonant dipole-dipole interaction between Rydberg atoms, and if the ensemble is confined within a finite volume, the manifold of doubly excited states has energy gap much larger than the line width of the Rydberg state. That is, a resonant excitation from singly to doubly excited states would be strongly suppressed, which can be taken as the underlying physical process that leads to the observation of dipole blockade. It may not be practically evident to achieve a perfect blockade since pairs of atoms can couple to noninteracting states, and so can evade the blockade phenomenon unless a meticulous care is taken in the choice of atomic states and relative orientation.

In the blockade scenario, the high-lying excitation can reduce to a two-level system characterized by atomic ground state $|g\rangle$, excited state $|e\rangle$ and the photon number state is $|N\rangle$. The interaction in this case can be perceived as exchange of energy between the atomic excitation and light in the cavity. So the atom-radiation coupled system can be represented by the atomic dressed states

$$|\pm, N\rangle = \frac{1}{\sqrt{2}}\left[c_g^{\pm}|g, N+1\rangle \pm c_e^{\pm}|e, N\rangle\right], \qquad (6.243)$$

where $c_{g,e}^{\pm}$ are the probability amplitudes (compare with Eq. (6.13)). One may note that the dressed state represents the subsequent stimulated absorption and emission process where the spontaneous emission accounts for the photons resulting from the cascading of the atom down the ladder. It is such possibility of manipulating the emerging entangled dressed state that encourages expecting this system as a viable resource for application (see Sects. 9.2 and 12.2, and also [32]).

In addition to the imagined benefits of exploiting the exaggerated properties of Rydberg blockade, one can still extend the present discussion to the mapping of the outcome of the interaction between the Rydberg excitations in the ensemble onto photon via electromagnetically induced transparency, which heralds optical nonlinearity. Since the interaction between highly excited Rydberg states is long-range, unlike the conventional nonlinear optics such as Kerr effect, the Rydberg mediated nonlinearity is expected to be long range. The main idea in Rydberg nonlinear optics is thus to take the long range dipolar interaction between highly excited Rydberg atoms, and map it onto photons. The main challenge perhaps would be to localize the photon to the characteristic length scale of the dipole-dipole interaction. The origin of the nonlinearity in the Rydberg environment can be

explained in the same way as in Sect. 5.6, that is, the nonlinearity can occur either
when the medium cannot absorb or scatter the second photon at the same time, in
case there is one emitter or when the resonance condition for the second photon is
different.

In case the optical nonlinearity arises due to a field dependent shift in the atomic
resonance, the first-order nonlinear response would be proportional to the product
of the shift and the gradient of a frequency dependence of the refractive index. It
might be convenient to parameterize the gradient in the refractive index in terms of
the group refractive index:

$$\eta_g = 1 + \omega \frac{\partial \eta}{\partial \omega}, \tag{6.244}$$

where ω is the angular frequency of the light with refractive index η. In a dilute
medium where the refractive index is close to unity, one can approximately write
$\eta = 1 + \frac{1}{2}\mathrm{Re}\chi$ in which χ is the susceptibility.

For a large group index, so one can approximately write

$$\eta_g = \frac{\omega}{2} \frac{\partial \mathrm{Re}\chi}{\partial \omega}. \tag{6.245}$$

It might also be reasonable to consider the shift of the Rydberg level that can be
defined for external electric field as

$$\Delta_R = -\frac{\alpha}{2}E^2, \tag{6.246}$$

where α is atomic polarizability at the frequency of the external field. This general
consideration may highlight the origin of large nonlinearities in the Rydberg
ensembles.

Now upon writing the nonlinear optical response as a slope, it is possible to get
a term that scales quadratically with the field [33]; which looks like Kerr effect,

$$\chi^{(3)}E^2 = -\frac{\eta_g \alpha}{\omega}E^2. \tag{6.247}$$

Equation (6.247) shows that the Kerr nonlinearity $\chi^{(3)}$ is proportional to the
product of the group index and polarizability. Such consideration however has the
disadvantage that the imaginary part of the susceptibility would also be large; giving
rise to off-axis scattering; and hence, leads to loss. The remedy for such a setback
may include introducing electromagnetically induced transparency into the setup,
which can be conceived by assuming that, in the ladder configuration, there is a
two-step pumping mechanism by external laser light.

The external light that couples the lower atomic energy level with the intermediate is taken as probe, and assumed to have strength Ω_p and detuning Δ_p, and the
control light of strength Ω_c sends the atom to the Rydberg state $|r\rangle$ with detuning of

$\Delta_p + \Delta_c$. Then following the approach in Sect. 6.5.2, the steady state solution for a weak-probe limit ($\rho_{ll} = 1$ and $\rho_{ee} = \rho_{rr} = 0$) turns out to be

$$\rho_{ol} = \frac{i\Omega_p(\gamma_r \qquad i(\Delta_p \mid \Delta_c))}{2[(\gamma - i\Delta_p)(\gamma_r - i(\Delta_p + \Delta_c)) + \Omega_c^2]} \qquad (6.248)$$

where γ_r is the decay rate of the two-photon transition and γ is the spontaneous atomic decay rate (compare with Eq. (6.235)). Once this solution is obtained, it is expected that the refractive index and group velocity can readily be calculated.

Appendix 1: Interaction Hamiltonian in Interaction Picture

It might be required sometimes to express the interaction Hamiltonian in the interaction picture. The Hamiltonian in the interaction picture then can be put in terms of the pertinent free Hamiltonian as

$$\hat{H}_I = e^{i\hat{H}_0 t} \hat{H}_I e^{-i\hat{H}_0 t}. \qquad (6.249)$$

With the aid of a general operator identity,

$$e^{\alpha\hat{A}} \hat{B} e^{-\alpha\hat{A}} = \hat{B} + \alpha[\hat{A}, \ \hat{B}] + \frac{\alpha^2}{2!}[\hat{A}, \ [\hat{A}, \ \hat{B}]] + \dots, \qquad (6.250)$$

and upon making use of the commutation relations of the boson and atomic operators (5.37), one may verify that

$$e^{i\omega\hat{a}^\dagger\hat{a}t} \hat{a} e^{-i\omega\hat{a}^\dagger\hat{a}t} = \hat{a} e^{i\omega t}, \qquad e^{i\frac{\omega_0\hat{\sigma}_z}{2}t}\hat{\sigma}_+ e^{-i\frac{\omega_0\hat{\sigma}_z}{2}t} = \hat{\sigma}_+ e^{i\omega_0 t}. \qquad (6.251)$$

So it may not be hard to show that the atom-radiation interaction Hamiltonian (6.9) can be rewritten in the interaction picture in the form

$$\hat{H}_I = ig\left[\hat{a}\hat{\sigma}_+ - \hat{a}^\dagger\hat{\sigma}_-\right]. \qquad (6.252)$$

Appendix 2: Atomic Damping Constant

To obtain the master equation with the aid of the methods as outlined in Sect. 3.1.1, it might be necessary to go for the spherical coordinate system in k-space. Imagine the atomic dipole matrix elements $\boldsymbol{\mu}_{ab}$'s to point along z-axis, where the unit vector $\hat{\mathbf{u}}_k$ is assumed to lie on the plane formed by the vectors $\boldsymbol{\mu}_{ab}$ and $\mathbf{k}$. Upon taking the transversality condition for each radiation modes ($\hat{\mathbf{u}}_k.\mathbf{k} = 0$), one finds that $\boldsymbol{\mu}_{ab}.\hat{\mathbf{u}}_k = \mu_{ab}\sin\theta$, where θ is the angle between $\boldsymbol{\mu}_{ab}$ and $\mathbf{k}$.

Then upon applying the definition of the coupling constant (6.5), one can write

$$\sum_k g_k^2 = \frac{\omega_0^2}{2\epsilon_0 V} \sum_k \frac{\mu_{ab}^2}{\omega_k} \sin^2 \theta_k. \tag{6.253}$$

In case the reservoir modes are taken to be closely spaced, the summation over k can be changed to the integration upon utilizing the transformation

$$\sum_k \rightarrow \frac{V}{(2\pi)^3} \int d^3k; \tag{6.254}$$

so one can see that

$$\sum_k g_k^2 e^{\pm i(\omega_0 - \omega_k)(t - t')} = \frac{\omega_0^2}{8\pi^3 \epsilon_0} \int_k \frac{\mu_{ab}^2}{\omega_k} \sin^2 \theta e^{\pm i(\omega_k - \omega_0)(t - t')} d^3k, \tag{6.255}$$

where the right side of Eq. (6.255) is multiplied by 2 so that the contribution of the polarization is properly taken care of.

One may also note that the volume in k-space can be denoted as $d^3k = k^2 \sin \theta d\theta d\varphi dk$, that is,

$$\sum_k g_k^2 e^{\pm i(\omega_0 - \omega_k)(t - t')} = \frac{\omega_0^2 \mu_{ab}^2}{8\pi^3 \epsilon_0 c^3} \int_{-\infty}^{\infty} \omega d\omega \int_0^{\pi} \int_0^{2\pi} \sin^3 \theta e^{\pm i(\omega - \omega_0)(t - t')} d\theta d\varphi. \tag{6.256}$$

Carrying out the integration over angular variables results in

$$\sum_k g_k^2 e^{\pm i(\omega_0 - \omega_k)(t - t')} = \frac{\omega_0^2 \mu_{ab}^2}{3\pi^2 \epsilon_0 c^3} \int_{-\infty}^{\infty} e^{\pm i(\omega - \omega_0)(t - t')} \omega d\omega. \tag{6.257}$$

Note that the atomic damping rate can be obtained upon carrying out this integration based on the assumption that ω varies very little around ω_0 and taking the noise as Markovian (see Eq. (6.53) and also Sect. 3.1.1).

Appendix 3: Electric Field Operator

According to the discussion in Sect. 2.1, the positive part of the electric field operator (2.13) can be expressed as

$$\hat{\mathbf{E}}^{(+)}(\mathbf{r}, t) = i \sum_{k,s} \sqrt{\frac{\omega_{ks}}{2\epsilon_0 V}} \hat{a}_{k,s}(t) e^{i\mathbf{k}\cdot\mathbf{r}} \hat{\mathbf{u}}, \tag{6.258}$$

where the summation over s is meant to take care of the contributions of the polarization. It might also be good to observe that the time evolution of the boson operator can be expressed using Eq. (6.9) as

$$\frac{d}{dt}\tilde{a}_{k,s}(t) = g_{k,s}\tilde{\sigma}_-(t)e^{-i(\omega_0-\omega_k)t}, \tag{6.259}$$

where the pertinent slowly varying terms can be defined as

$$\tilde{a}_{k,s}(t) = \hat{a}_{k,s}(t)e^{i\omega_k t}, \qquad \tilde{\sigma}_-(t) = \hat{\sigma}_-(t)e^{i\omega_0 t}. \tag{6.260}$$

So the formal solution of Eq. (6.259) can be written as

$$\tilde{a}_{k,s}(t) = \tilde{a}_{k,s}(0) + g_{k,s}e^{-i(\omega_0-\omega_k)t}\int_0^t \tilde{\sigma}_-(t')e^{i(\omega_0-\omega_k)(t-t')}dt'. \tag{6.261}$$

When the situation prior to the interaction is disregarded, one can see with the aid of Eq. (6.258) that

$$\hat{\mathbf{E}}^{(+)}(\mathbf{r},t) = i\sum_{k,s} g_{k,s}\sqrt{\frac{\omega_{ks}}{2\epsilon_0 V}}e^{i\mathbf{k}\cdot\mathbf{r}}e^{-i\omega_0 t}\int_0^t \tilde{\sigma}_-(t')e^{i(\omega_0-\omega_k)(t-t')}dt'\hat{\mathbf{u}}, \tag{6.262}$$

which can also be written upon applying Eq. (6.5) as

$$\hat{\mathbf{E}}^{(+)}(\mathbf{r},t) = i\sum_{k,s} \frac{\omega_{ks}}{2\epsilon_0 V}\hat{\mathbf{u}}(\boldsymbol{\mu}_{ab}\cdot\hat{\mathbf{u}})e^{i\mathbf{k}\cdot\mathbf{r}}e^{-i\omega_0 t}\int_0^t \tilde{\sigma}_-(t')e^{i(\omega_0-\omega_k)(t-t')}dt'. \tag{6.263}$$

Since $\hat{\mathbf{u}}$ is arbitrarily directed unit vector that should be perpendicular to $\mathbf{k}$, one can propose that

$$\hat{\mathbf{u}}(\boldsymbol{\mu}_{ab}\cdot\hat{\mathbf{u}}) = \boldsymbol{\mu}_{ab} - \frac{\hat{\mathbf{k}}(\boldsymbol{\mu}_{ab}\cdot\hat{\mathbf{k}})}{k^2}; \tag{6.264}$$

$$\hat{\mathbf{E}}^{(+)}(\mathbf{r},t) = i\sum_{k,s} \frac{\omega_{ks}}{2\epsilon_0 V}\left[\boldsymbol{\mu}_{ab} - \frac{\hat{\mathbf{k}}(\boldsymbol{\mu}_{ab}\cdot\hat{\mathbf{k}})}{k^2}\right]e^{i\mathbf{k}\cdot\mathbf{r}}e^{-i\omega_0 t}\int_0^t \tilde{\sigma}_-(t')e^{i(\omega_0-\omega_k)(t-t')}dt'. \tag{6.265}$$

For closely spaced radiation modes, the summation over k can be changed into the integration over k;

$$\hat{\mathbf{E}}^{(+)}(\mathbf{r}, t) = i \frac{V}{(2\pi)^3} \int_{-\infty}^{\infty} \int_0^{\pi} \int_0^{2\pi} \frac{\omega}{2\epsilon_0 V} e^{-i\omega_0 t} e^{i\mathbf{k}.\mathbf{r}} \left[\boldsymbol{\mu}_{ab} - \frac{\hat{\mathbf{k}}(\boldsymbol{\mu}_{ab}.\hat{\mathbf{k}})}{k^2} \right]$$

$$\times \, k^2 \sin\theta d\theta d\varphi dk \int_0^t \tilde{\sigma}_-(t') e^{i(\omega_0 - \omega_k)(t - t')} dt'. \qquad (6.266)$$

One may also assume that the electric dipole moment lies in xz plane and makes angle ϑ with the z-axis. It is then possible to express $\mathbf{k}$ and $\boldsymbol{\mu}_{ab}$ in the Cartesian coordinate as

$$\mathbf{k} = k(\sin\theta\cos\varphi\hat{\mathbf{x}} + \sin\theta\sin\varphi\hat{\mathbf{y}} + \cos\theta\hat{\mathbf{z}}), \qquad \boldsymbol{\mu}_{ab} = \mu_{ab}(\sin\vartheta\hat{\mathbf{x}} + \cos\vartheta\hat{\mathbf{z}}).$$
$$(6.267)$$

Hoping to ease the rigor, one may integrate term by term in which

$$\int_0^{2\pi} \left[\boldsymbol{\mu}_{ab} - \frac{\hat{\mathbf{k}}(\boldsymbol{\mu}_{ab}.\hat{\mathbf{k}})}{k^2} \right] d\varphi = 2\pi \mu_{ab} \left[\sin\vartheta \left(1 - \frac{\sin^2\theta}{2} \right) \hat{\mathbf{x}} + \cos\vartheta \sin^2\theta\hat{\mathbf{z}} \right].$$
$$(6.268)$$

Besides, employing the fact that $\mathbf{k}.\mathbf{r} = kr\cos\theta$,

$$\int_0^{\pi} \sin\theta \left[\sin\vartheta \left(1 - \frac{\sin^2\theta}{2} \right) \hat{\mathbf{x}} + \cos\vartheta \sin^2\theta\hat{\mathbf{z}} \right] e^{ikr\cos\theta} d\theta = \sin\vartheta \left[\frac{e^{ikr} - e^{-ikr}}{ikr} \right] \hat{\mathbf{x}}.$$
$$(6.269)$$

To do this integration, one may need to go over to the ω space following the procedure adopted earlier. In this space, it is possible to see that

$$\int_k k \left[e^{ikr} - e^{-ikr} \right] e^{-i\omega_0 t} \int_0^t \tilde{\sigma}_-(t') e^{i(\omega_0 - \omega)(t - t')} dt' dk$$

$$= \frac{1}{c} \int_0^t \int_{-\infty}^{\infty} \hat{\sigma}_-(t')\omega \left[e^{i\omega(\frac{r}{c} - t + t')} - e^{-i\omega(\frac{r}{c} + t - t')} \right] dt' \, d\omega. \qquad (6.270)$$

Assuming that $\omega \approx \omega_0$ (near resonance), it is possible to replace ω with ω_0; and as result

$$\int_k k \left[e^{ikr} - e^{-ikr} \right] e^{-i\omega_0 t} \int_0^t \tilde{\sigma}_-(t') e^{i(\omega_0 - \omega)(t - t')} dt' dk$$

$$= \frac{2\pi\omega_0}{c} \left[\hat{\sigma}_- \left(\frac{r}{c} - t \right) - \hat{\sigma}_- \left(\frac{r}{c} + t \right) \right]. \qquad (6.271)$$

The second term in Eq. (6.271) represents the incoming radiation field. Then upon ignoring the external influence, the positive part of the electric field turns out to be

$$\hat{E}^{(+)}(\mathbf{r}, t) \quad \frac{\omega_0^2 \mu_{ge} \sin \vartheta}{4\pi r \epsilon_0 c^2} \hat{\sigma} (t)_{-}. \qquad (6.272)$$

One should note that Eq. (6.272) relates the atomic operator with the corresponding electric field operator.

References

1. E.T. Jaynes, F.W. Cummings, Proc. IEEE **51**, 89 (1963)
2. B.R. Mollow, Phys. Rev. **188**, 1969 (1969)
3. H. Mabuchi, A.C. Doherty, Science **298**, 1372 (2002)
4. M. Lewenstein, T.W. Mossberg, R.J. Glauber, Phys. Rev. Lett. **59**, 775 (1987); H.J. Carmichael, A.S. Lane, D.F. Walls, J. Mod. Opt. **34**, 821 (1987); O. Kocharovskaya et al., Phys. Rev. A **49**, 4928 (1994); A.K. Kudlaszyk, R. Tanas, J. Mod. Opt. **48**, 347 (2001)
5. R.H. Dicke, Phys. Rev. **93**, 99 (1954); X. Maitre et al., Phys. Rev. Lett. **79**, 769 (1997); M. Hennrich et al., ibid. **85**, 4872 (2000); A. Kuhn, M. Hennrich, G. Rempe, ibid. **89**, 067901 (2002)
6. P. Meystre, M. Sargent III, *Elements of Quantum Optics*, 4th edn. (Springer, Berlin, 2007)
7. S. Tesfa, J. Mod. Opt. **55**, 1587 (2008)
8. S. Tesfa, J. Mod. Opt. **54**, 1759 (2007)
9. J.I. Cirac, L.L.S. Soto, Phys. Rev. A **44**, 1948 (1991)
10. G. Yeoman, S.M. Barnett, J. Mod. Opt. **43**, 2037 (1996); R. Tanas et al., ibid. **45**, 1859 (1998), J. Opt. B **4**, 142 (2002)
11. C.W. Gardiner, A.S. Parkins, M.J. Collett, J. Opt. Soc. Am. B **4**, 1683 (1987); M.J. Collett, R. Loudon, C.W. Gardiner, J. Mod. Opt. **34**, 881 (1987)
12. S. Tesfa, J. Mod. Opt. **56**, 105 (2009)
13. M. Gross, S. Haroche, Phys. Rep. **93**, 301 (1982)
14. R.H. Lehmberg, Phys. Rev. A **2**, 883, 889 (1970)
15. Z. Ficek, R. Tanas, Phys. Rep. **372**, 369 (2002)
16. S.L. Mielke, G.T. Foster, L.A. Orozco, Phys. Rev. Lett. **80**, 3948 (1998)
17. D.J. Wineland et al., Phys. Rev. A **46**, 6797 (1992); M. Kitagawa, M. Ueda, ibid. **47**, 5138 (1993); J. Ma et al., Phys. Rep. **509**, 89 (2011)
18. T.A.B. Kennedy, D.F. Walls, Phys. Rev. A **42**, 3015 (1990)
19. S. Tesfa, N two-level atoms coupled to a squeezed vacuum reservoir. M.Sc. thesis, Addis Ababa University, 1997; A. Andre, M.D. Lukin, Phys. Rev. A **65**, 053819 (2002)
20. M.M. Cola, M.G.A. Paris, N. Poivella, Phys. Rev. A **70**, 043809 (2004); S. Tesfa, J. Mod. Opt. **55**, 1683 (2008)
21. M.O. Scully, Phys. Rev. Lett. **55**, 2802 (1985); S.E. Harris, ibid. **62**, 1033 (1989); A. Imamoglu, Phys. Rev. A **40**, 2835 (1989); T. Pellizzari et al., Phys. Rev. Lett. **75**, 3788 (1995); S.J. van Enk, J.I. Cirac, P. Zoller, ibid. **78**, 4293 (1997)
22. Z. Ficek, S. Swain, *Quantum Interference and Coherence: Theory and Experiments* (Springer Science+Business Media Inc., New York, 2005)
23. L.M. Narducci et al., Phys. Rev. A **42**, 1630 (1990); M.O. Scully, M.S. Zubairy, *Quantum Optics* (Cambridge University Press, Cambridge, 1997); P. Lambropoulos, D. Petrosyan, *Fundamentals of Quantum Optics and Quantum Information* (Springer, Berlin, 2007)
24. J. Hald et al., Phys. Rev. Lett. **83**, 1319 (1999)
25. M. Fleischhauer, A. Imamoglu, J.P. Marangos, Rev. Mod. Phys. **77**, 633 (2005)

26. S.E. Harris, J.E. Field, A. Imamoglu, Phys. Rev. Lett. **64**, 1107 (1990)
27. J.G. Banacloche et al., Phys. Rev. A **51**, 576 (1995); N. Thaicharoen et al., ibid. **100**, 063427 (2019)
28. K.A. Safinya et al., Phys. Rev. Lett. **47**, 405 (1981); Y. Miroshnychenko et al., Phys. Rev. A **82**, 013405 (2010)
29. M.P.A. Jones, L.G. Marcassa, J.P. Shaffer, J. Phys. B **50**, 060202 (2017)
30. T.E. Lee, H. Haffner, M.C. Cross, Phys. Rev. Lett. **108**, 023602 (2012)
31. M. Saffman, T.G. Walker, K. Molmer, Rev. Mod. Phys. **82**, 2313 (2010)
32. D. Jaksch et al., Phys. Rev. Lett. **85**, 2208 (2000); M.D. Lukin et al., ibid. **87**, 037901 (2001); R.G. Unanyan, M. Fleischhauer, Phys. Rev. A **66**, 032109 (2002); M. Saffman, T.G. Walker, ibid. **72**, 042302 (2005); L.H. Pedersen, K. Molmer, ibid. **79**, 012320 (2009); L. Isenhower et al., Phys. Rev. Lett. **104**, 010503 (2010); T. Wilk et al., ibid., 010502 (2010); C.S. Adams, J.D. Pritchard, J.P. Shaffer, J. Phys. B **53**, 012002 (2020)
33. O. Firstenberg, C.S. Adams, S. Hofferberth, J. Phys. B **49**, 152003 (2016)

Part III

Quantum Aspects of Light

Quantum Features of Light

7

Studies in the field of optics galvanize that properties of light that can be visualized by the wave front profile are classical by nature. Even in this regard, the outcome of experiments such as Young's double-slit experiment led to reevaluation of a large class of properties of light related to the interference phenomenon. With the advent of laser, and subsequent understanding of the interaction of atom with radiation, a pressing demand to reassess the previously gained knowledge has also arisen. Owing to the far-reaching and in-depth research in the area of nonlinear optics [1], it becomes evident that there are certain aspects of light that cannot be explained within the scope of classical theory alone. As a special case, properties of light that can be explained only within the regime of quantum theory are alluded to as quantum features of light. One of these extraordinary traits of light is the squeezing in which the noise in one of the quadratures goes below the classical limit at the expense of the conjugate quadrature. Motivated by the possibility of experimenting with a more complex nonclassical behaviors such as higher-order and multi-mode squeezing, we intend to extend the available criteria that meant to characterize the lowest order nonclassicality to a higher-order.

It might be necessary, first and foremost, to underscore that a quantum state displays, let us say, nth order nonclassical behavior with respect to arbitrary quantum operator in case the nth order variance of the operator in that state is below the corresponding value of the nth order variance in the coherent state, that is, when

$$\Delta A^n_{|\psi\rangle} < \Delta A^n_{|\alpha\rangle}. \tag{7.1}$$

Taking this into account, the mathematical inequalities that can be used to capture the nth order, amplitude square, three-mode, sum and difference squeezing would be provided in Sect. 7.1.

Based on the inherent inconsistencies in the foundation of quantum theory associated with measurement, various criteria that are applicable to quantify the degree of quantumness or nonclassicality can also be formulated. For example, based on the fact that $Tr(\hat{\rho}\hat{A}^{\dagger}\hat{A}) \geq 0$, which implies the nth order correlation function to satisfy $G^{(n)}(\mathbf{r}_i; \tau_i) \geq 0$ and so the classical correlation matrix to be a positive definite in the quadratic form $(\det[G^{(1)}(x_i; x_j)] \geq 0)$, Schwartz's inequality (4.78) can be proposed to be

$$G^{(n)}(x_1, \ldots, x_n; x_n, \ldots, x_1)G^{(n)}(x_{n+1}, \ldots, x_{2n}; x_{2n}, \ldots, x_{n+1})$$

$$\geq |G^{(n)}(x_1, \ldots, x_n; x_{n+1}, \ldots, x_{2n})|^2; \tag{7.2}$$

the violation of which can be taken as the manifestation of nonclassical behavior.

In another context, while the observer performs the experiment that can be described by variables x_i's, and if a_i's values are observed, the probability of the outcome can be written as $P(a_1, \ldots, a_n | x_1, \ldots, x_n)$. Such a conditional probability can be interpreted as correlation, and believed to carry a lot of information about how various measurement values are related, which can lead to the notion of quantum discord and entanglement steering. Apart from being nonnegative and normalized, the local magnitudes in a given measurement are taken to be independent of experiment:

$$\sum_a P(a, b | x, y) = P(b | y). \tag{7.3}$$

To extend this consideration to a multi-party scenario, one may need to begin with the assumption that every two parties, or pairwise groupings, should perform similar experiment but at two space like separated locations so that a classical like communication between them is precluded, that is, the local probabilities are presumed to be defined on the local state of affairs that may not be known. Imagine also that almost all correlations between the independent observers known are local except some predicted by quantum physics.

According to quantum theory, in case two systems that initially interact with each other are separated; and in the process, if the position or momentum of one of the systems is measured, the observer can be aware of the value of the second particle even without directly measuring due to the initial interconnectedness[1] (see Sect. 4.1). Such realization is the predominant thinking that goes into setting up the current conversation; and yet, seems to violate the uncertainty principle as both the position and momentum of a single particle are presumed to be known with certainty (see Sect. 7.2). In line with the understanding that quantum states of light are useful resource for implementing quantum information processing within the scope of continuous variable entanglement, we also seek to examine the quantum features of light pertaining to the relative simplicity and high efficiency in the generation,

---

[1]This can be taken as the essence of the well celebrated quantum nonlocality in relation to the EPR contention. Even though there is no as such concrete experimental evidence that fully justify quantum nonlocality, no serious physicist doubts the assertion that nature exhibits quantum nonlocality [2].

manipulation and detection of optical entangled states, and implementation of the corresponding gate operation (see Chap. 9). We particularly aspire to look into details of nonlocality, entanglement, quantum discord and steering (see Sects. 7.3–7.5).

7.1 Higher-Order Squeezing

The reduction of noise fluctuation related to the second-order moment or standard deviation of a given observable below coherent state level can be referred to as the lowest-order nonclassical feature [3]. In light of this delineation, a radiation field is presumed to exhibit squeezing property if the noise fluctuation in one of the quadratures reduces below the coherent state level: $\Delta Q_i^2 < 1$, and photon anti-bunching phenomenon if the fluctuation in the photon number is below the Poisson level: $\Delta N^2 < \langle N \rangle$. The squeezing and anti-bunching phenomena discussed in Sect. 2.2.1.4 belong to the lowest-order nonclassical properties of light. Likewise, the situation in which the noise in one of the combined quadrature operators would be below the corresponding classical limit for a higher-order uncertainty at the expense of the conjugate quadrature can be designated as higher-order squeezing.

7.1.1 Three-Mode Squeezing

On the basis of the proposal that three-mode squeezing can be used in the implementation of quantum networks, it appears imperative exploring the possible ways of its quantification in terms of the pertinent position and momentum operators [4] (see also Chap. 2). To this effect, one can begin with the unitary operator

$$\hat{U} = \exp\left[ir\left(\hat{Q}_1\hat{P}_2 + \hat{Q}_2\hat{P}_3 + \hat{Q}_3\hat{P}_1\right)\right]. \tag{7.4}$$

Since this unitary operator has the form of the squeeze operator that embodies three different generalized positions and momenta operators, it is expected to describe the three-mode squeezed radiation.

One can see that the operator pairs $\hat{Q}_1$ and $\hat{P}_2$, $\hat{Q}_2$ and $\hat{P}_3$, and $\hat{Q}_3$ and $\hat{P}_1$ commute, and so difficult to factorize $\hat{U}$ but possible to express it as $\hat{U} = \exp\left(ir\,Q_i A_{ij} P_j\right)$, where Q_i is the row matrix constructed from $\hat{Q}_1$, $\hat{Q}_2$, $\hat{Q}_3$, and P_j is the column matrix constructed from $\hat{P}_1$, $\hat{P}_2$, $\hat{P}_3$ while A_{ij} is a matrix that can be defined as

$$A_{ij} = \begin{pmatrix} 0 & 1 & 0 \\ 0 & 0 & 1 \\ 1 & 0 & 0 \end{pmatrix}. \tag{7.5}$$

Observe that quantifying a three-mode squeezing is more subtle than the two-mode case since in the three-mode squeezing the correlations among the three modes should be inferred.

Once the unitary operator is properly described upon taking the three-mode coordinate eigenstate as $|\mathbf{q}\rangle$, the three-mode squeeze operator can be designated as

$$\hat{S}(r) = \frac{1}{\sqrt{|\Lambda|}} \int d^3 q \, |\Lambda \mathbf{q}\rangle \langle \mathbf{q}|, \tag{7.6}$$

where $\hat{U}|\mathbf{q}\rangle = \sqrt{|\Lambda|}|\Lambda \mathbf{q}\rangle$ in which $\Lambda = e^{-r\tilde{A}}$, and with the help of which required correlations can be calculated.

To quantify the available degree of squeezing, it might be vital introducing the quadrature operators:

$$\hat{d}_+ = \frac{\hat{Q}_1 + \hat{Q}_2 + \hat{Q}_3}{\sqrt{6}}, \qquad \hat{d}_- = \frac{\hat{P}_1 + \hat{P}_2 + \hat{P}_3}{\sqrt{6}}, \tag{7.7}$$

with $[\hat{d}_+, \hat{d}_-] = i\hat{I}/2$. With this in mind, a three-mode light is said to be in squeezed state if either of

$$\Delta d_\pm^2 < \frac{1}{4} \tag{7.8}$$

and $\Delta d_+^2 \Delta d_-^2 \geq 1/4$.

Imagine a three-wave mixing mechanism in which the radiation denoted by mode $\hat{a}$ interacts nonlinearly with modes $\hat{b}$ and $\hat{c}$, and the pertinent three-mode radiation is expressed in terms of the interaction Hamiltonian as

$$\hat{H}_I = \exp i \left[g_{ab}(\hat{a}\hat{b} - \hat{a}^\dagger \hat{b}^\dagger) + g_{ac}(\hat{a}\hat{c} - \hat{a}^\dagger \hat{c}^\dagger) \right], \tag{7.9}$$

where g_{ab} and g_{ac} are the measure of the coupling between the radiation in mode $\hat{a}$ with radiations in modes $\hat{b}$ and $\hat{c}$ [5]. Equation (7.9) can also be rewritten in terms of the quadratures as

$$\hat{H}_I = \exp \left[i g_{ab}(\hat{Q}_b \hat{P}_a + \hat{Q}_a \hat{P}_b) + i g_{ac}(\hat{Q}_c \hat{P}_a + \hat{Q}_a \hat{P}_c) \right], \tag{7.10}$$

which can also be written as

$$\hat{S} = \exp \left[i \hat{Q}_i \Lambda_{ij} \hat{P}_j \right], \tag{7.11}$$

where $i, j = a, b, c$ and[2]

$$
A = \begin{pmatrix} 0 & g_{ab} & g_{ac} \\ g_{ab} & 0 & 0 \\ g_{ac} & 0 & 0 \end{pmatrix},
$$

Following the approach in Chap. 2, one can verify the transformations:

$$
\hat{S}^{-1}\hat{a}\hat{S} = \hat{a}\cosh r - \hat{b}^\dagger \cos\theta \sinh r - \hat{c}^\dagger \sin\theta \sinh r, \tag{7.12}
$$

$$
\hat{S}^{-1}\hat{b}\hat{S} = -\hat{a}^\dagger \cosh r \cos\theta + \hat{b}(\sin^2\theta + \cos^2\theta \cosh r) + \frac{\hat{c}}{2}(\cosh r - 1)\sin 2\theta, \tag{7.13}
$$

$$
\hat{S}^{-1}\hat{c}\hat{S} = -\hat{a}^\dagger \sinh r \sin\theta + \frac{\hat{b}}{2}(\cosh r - 1)\sin 2\theta + \hat{c}(\cos^2\theta + \sin^2\theta \cosh r), \tag{7.14}
$$

where

$$
r = \sqrt{g_{ab}^2 + g_{ac}^2}, \qquad \cos\theta = \frac{g_{ab}}{r}, \qquad \sin\theta = \frac{g_{ac}}{r}. \tag{7.15}
$$

Once again, following the same procedure as for the two-mode case, the quadrature variances turn out to be

$$
\Delta d_\pm^2 = \frac{1}{12}\left[1 + 2\cosh 2r + \sin 2\theta(\cos 2r - 1) \pm 2(\cos\theta + \sin\theta)\sinh 2r\right]. \tag{7.16}
$$

It may not be difficult to notice from Fig. 7.1 that the three-mode radiation that can be generated by the mechanism of three-wave mixing can exhibit three-mode squeezing under certain restricted condition. As one may expect on physical ground, the degree of squeezing is found to be large when the couplings between the radiations are taken to be relatively large. Deeper inspection however reveals that mode a radiation cannot be strongly correlated with modes b and c under similar condition, and so it may not be admissible to increase the involved couplings arbitrarily.

---

[2]To appreciate the difference between the representations of the corresponding matrices, it might be advisable looking into the way the radiation is envisioned to be different from the one given by Eq. (7.4).

Fig. 7.1 Plots of the three-mode minus quadrature variance for various coupling constants

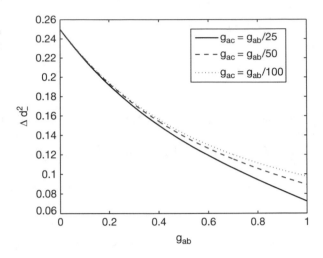

7.1.2 nth Order Squeezing

With the advent of viable techniques that can capture higher-order correlations, the feasibility for studying higher-order nonclassical features such as nth order squeezing appears to be more promising. In support of this suggestion, let us begin with the presumption that the two quadrature components of the electric field can be described by

$$\hat{E}_x = \hat{E}^{(+)}e^{-i\phi} + \hat{E}^{(-)}e^{i\phi}, \qquad \hat{E}_p = \hat{E}^{(+)}e^{-i(\phi+\frac{\pi}{2})} + \hat{E}^{(-)}e^{i(\phi+\frac{\pi}{2})}, \qquad (7.17)$$

where $\hat{E}^{(+)}$ and $\hat{E}^{(-)}$ are the positive and negative frequency parts of the electric field operator, and ϕ is the phase to be chosen based on the condition that the two quadrature components should be out of phase by $\frac{\pi}{2}$ while studying the squeezing. It is also possible to propose that $[\hat{E}^{(+)}, \hat{E}^{(-)}] = c\hat{I}$, where c is taken to be a finite real positive number. With this consideration, the two quadrature components would satisfy $[\hat{E}_x, \hat{E}_p] = i2c\hat{I}$ from which follows

$$\sqrt{\langle(\Delta\hat{E}_x)^2\rangle\langle(\Delta\hat{E}_p)^2\rangle} \geq c. \qquad (7.18)$$

The pertinent quantum state is designated as nonclassical or squeezed if either of

$$\langle(\Delta\hat{E}_{x,p})^2\rangle < c \qquad (7.19)$$

and when the uncertainty relation holds. In addition, with the help of the commutation relation of the quadrature components, the normally ordered variance for one of the quadrature components, let us say, x can be expressed as

$$\langle: (\Delta\hat{E}_x)^2 :\rangle = \langle(\Delta\hat{E}_x)^2\rangle - c. \qquad (7.20)$$

So for the squeezing to manifest, $\langle : (\Delta \hat{E}_x)^2 : \rangle$ should be negative since the quantum fluctuation of $\hat{E}_x$ for the squeezed state is expected to be smaller than the coherent state: $\langle (\Delta \hat{E}_x)^2 \rangle < c$

One may next seek to generalize the same consideration for a higher-order squeezing. To do so, it might be advantageous stressing that the variance of the fluctuation exhibits a nonclassical behavior for even moments since the expectation values of odd powers of electric field are zero. One can also designate a given state as $(2n)th$ order squeezed in x quadrature if there exists a phase ϕ for which $\langle (\Delta \hat{E}_x)^{2n} \rangle$ is smaller than what it should be for the coherent radiation as invoked in deriving atomic squeezing (see Sect. 5.5 and also [6]). One may also found useful to express these higher-order moments in their normally ordered form with the aid of mathematical identity

$$\langle \exp(\Delta \hat{E}_x x) \rangle = \langle : \exp(\Delta \hat{E}_x x) : \rangle \exp\left(\frac{x^2 c}{2}\right). \tag{7.21}$$

One can then see that expanding both sides of Eq. (7.21) in a power series of x, and comparing the coefficients lead to

$$\langle (\Delta \hat{E}_x)^n \rangle = \langle : (\Delta \hat{E}_x)^n : \rangle + \frac{n(n-1)}{1!}\left(\frac{c}{2}\right)\langle : (\Delta \hat{E}_x)^{n-2} : \rangle$$

$$+ \frac{n(n-1)(n-2)(n-3)}{2!}\left(\frac{c}{2}\right)^2 \langle : (\Delta \hat{E}_x)^{n-4} : \rangle + \dots$$

$$+ \begin{cases} (n-1)!!c^{n/2}, & \text{if n is even;} \\ \frac{n!c^{(n-3)/2}}{3!2^{(n-3)/2}[(n-3)/2]!}\langle : (\Delta \hat{E}_x)^3 : \rangle, & \text{if n is odd.} \end{cases} \tag{7.22}$$

Since the normally ordered moments of the deviation of all orders for the coherent state vanish, the field is said to be squeezed in x quadrature to $(2n)th$ order provided that[3]

$$\langle (\Delta \hat{E}_x)^{2n} \rangle < (2n-1)!!c^n, \tag{7.23}$$

where

$$n!! = \begin{cases} n(n-2)\dots 5.3.1, & \text{if n is odd positive integer;} \\ n(n-2)\dots 6.4.2, & \text{if n is even positive integer;} \\ 1, & \text{if n=0.} \end{cases} \tag{7.24}$$

---

[3]The double factorial (!!) can also be expressed in terms of the Gamma function and ordinary factorial as

$$(2n-1)!! = \frac{2^n \Gamma(n+1/2)}{\sqrt{\pi}}, \qquad (2n)!! = 2^n n!, \qquad (2n-1)!! = \frac{(2n)!}{2^n n!}.$$

For example, there would be squeezing to fourth order when

$$\langle : (\varDelta \hat{E}_x)^4 : \rangle + 6c\langle : (\varDelta \hat{E}_x)^2 : \rangle < 0, \tag{7.25}$$

and to six order if

$$\langle : (\varDelta \hat{E}_x)^6 : \rangle + 15c\langle : (\varDelta \hat{E}_x)^4 : \rangle + 45c^2\langle : (\varDelta \hat{E}_x)^2 : \rangle < 0. \tag{7.26}$$

One may see that there is a possibility for Eqs. (7.25) and (7.26) to be satisfied without necessarily having $\langle : (\varDelta \hat{E}_x)^{2n} : \rangle < 0$ for $n \geq 2$ when $\langle : (\varDelta \hat{E}_x)^2 : \rangle$ dominates the series. It may not be appropriate in such a case claiming that there is a higher-order squeezing based on Eq. (7.23) alone, that is, the most convenient way of capturing $(2n)th$ order squeezing may require additional condition such as

$$0 \leq \langle (\varDelta \hat{E}_x)^{2n} \rangle < (2n - 1)!!c^n. \tag{7.27}$$

One should note that various physical systems such as degenerate parametric down conversion, harmonic generation, resonance fluorescence, higher-order harmonic generation and anharmonic oscillation exhibit strong squeezing at least to the sixth order [7].

7.1.3   Amplitude Square Squeezing

Amplitude square squeezing that designates a single-mode radiation is linked to the square of the annihilation and creation operators, and can be described in terms of the quadrature operators given by

$$\hat{X}_1 = \frac{1}{2}\left(\hat{a}^2 + \hat{a}^{\dagger^2}\right), \qquad \hat{X}_2 = \frac{1}{2i}\left(\hat{a}^2 - \hat{a}^{\dagger^2}\right) \tag{7.28}$$

in which

$$\left[\hat{X}_1, \ \hat{X}_2\right] = i\left(2\hat{a}^{\dagger}\hat{a} + 1\right), \qquad \varDelta X_1 \varDelta X_2 \geq \langle \hat{a}^{\dagger}\hat{a} \rangle + \frac{1}{2}. \tag{7.29}$$

With this designation, the amplitude square squeezing is said to be manifested if either of

$$\langle \hat{a}^{\dagger^2}\hat{a}^2 \rangle + \langle \hat{a}^2\hat{a}^{\dagger^2} \rangle \pm \langle \hat{a}^{\dagger^4} \rangle \pm \langle \hat{a}^{\dagger^4} \rangle \mp \left[\langle \hat{a}^2 \rangle^2 - \langle \hat{a}^{\dagger^2} \rangle^2 + 2\langle \hat{a}^2 \rangle\langle \hat{a}^{\dagger^2} \rangle\right] - 4\langle \hat{a}^{\dagger}\hat{a} \rangle < 2, \tag{7.30}$$

and when the uncertainty relation is satisfied [8].

One may note that it is possible to extend this discussion to a higher-order scenario by applying

$$\hat{V}_1 = \frac{1}{\sqrt{2}}(\hat{a}^n + \hat{a}^{\dagger n}), \qquad \hat{V}_2 = \frac{i}{\sqrt{2}}(\hat{a}^{\dagger n} - \hat{a}^n), \qquad (7.31)$$

which represent the quadrature operators with $n = 1, 2, 3, \ldots$ that denotes the respective order. It might thus be instructive writing the generalized uncertainty condition as

$$\Delta V_1^2 \Delta V_2^2 \geq \frac{1}{4} \langle [\hat{V}_1, \hat{V}_2] \rangle^2, \qquad (7.32)$$

where ΔV_i^2 is the variance. So the quantum state is said to be amplitude square squeezed in one of the quadratures when

$$\Delta V_i^2 < \frac{1}{2} \langle [\hat{V}_1, \hat{V}_2] \rangle \qquad (7.33)$$

and the uncertainty relation is satisfied.

To express the above results in a more appealing manner, it is found suitable invoking the commutation relation $[\hat{a}^n, \hat{a}^{\dagger n}] = \hat{F}^n$, where $\hat{F}^n$ is a polynomial of order $(n-1)$ and defining the quantity

$$D^n = \frac{(\Delta V_i)^2 - \frac{1}{2}\langle \hat{F}^n \rangle}{\frac{1}{2}\langle \hat{F}^n \rangle}, \qquad (7.34)$$

which can be rewritten in terms of $\hat{a}^n$ and $\hat{a}^{\dagger n}$ as

$$D^n = \frac{2[Re(\langle \hat{a}^{2n} \rangle) - 2(Re\langle \hat{a}^n \rangle)^2 + \langle \hat{a}^{\dagger n} \hat{a}^n \rangle]}{\langle \hat{F}^n \rangle}. \qquad (7.35)$$

A given state is hence said to exhibit nth power amplitude squeezing in $\hat{V}_i$ when

$$-1 \leq D^n < 0 \qquad (7.36)$$

and the pertinent uncertainty relation is satisfied.

7.1.4 Sum and Difference Squeezing

In the same way as the amplitude squaring, the higher-order squeezing in terms of the sum and difference of two separate modes, let us say, $\hat{a}$ and $\hat{b}$ can be foreseen [9]. With this information, the sum squeezing can be defined in terms of the quadrature operators

$$\hat{V}_{s1} = \frac{1}{2}(\hat{a}\hat{b} + \hat{a}^\dagger\hat{b}^\dagger), \qquad \hat{V}_{s2} = \frac{i}{2}(\hat{a}^\dagger\hat{b}^\dagger - \hat{a}\hat{b}), \tag{7.37}$$

where the commutation and uncertainty relations are found to be

$$[\hat{V}_{s1}, \hat{V}_{s2}] = \frac{i}{2}(2 + \hat{a}^\dagger\hat{a} + \hat{b}^\dagger\hat{b}), \qquad \Delta V_{s1}\Delta V_{s2} \geq \frac{1}{4}(\langle\hat{a}^\dagger\hat{a}\rangle + \langle\hat{b}^\dagger\hat{b}\rangle) + \frac{1}{2}.$$
$$\tag{7.38}$$

So the sum squeezing is said to be manifested if either of

$$2\langle\hat{a}^\dagger\hat{a}\hat{b}^\dagger\hat{b}\rangle \pm \langle\hat{a}^2\hat{b}^2\rangle \pm \langle\hat{a}^{\dagger^2}\hat{b}^{\dagger^2}\rangle \mp \langle\hat{a}\hat{b}\rangle^2 \mp \langle\hat{a}^\dagger\hat{b}^\dagger\rangle^2 - 2\langle\hat{a}\hat{b}\rangle\langle\hat{a}^\dagger\hat{b}^\dagger\rangle < 1 \tag{7.39}$$

and the quadrature operators satisfy the uncertainty relation.

The difference squeezing can also be quantified by making use of the quadrature operators

$$\hat{V}_{d1} = \frac{1}{2}(\hat{a}\hat{b}^\dagger + \hat{a}^\dagger\hat{b}), \qquad \hat{V}_{d2} = \frac{i}{2}(\hat{a}\hat{b}^\dagger - \hat{a}^\dagger\hat{b}), \tag{7.40}$$

with the aid of which one can verify

$$[\hat{V}_{d1}, \hat{V}_{d2}] = \frac{i}{2}(\hat{a}^\dagger\hat{a} - \hat{b}^\dagger\hat{b}), \qquad \Delta V_{d1}\Delta V_{d2} \geq \frac{1}{4}(\langle\hat{a}^\dagger\hat{a}\rangle - \langle\hat{b}^\dagger\hat{b}\rangle). \tag{7.41}$$

The difference squeezing is thus said to be manifested if either of

$$2\langle\hat{b}^\dagger\hat{b}\rangle + 2\langle\hat{a}^\dagger\hat{a}\hat{b}^\dagger\hat{b}\rangle \pm \langle\hat{a}^2\hat{b}^2\rangle \pm \langle\hat{a}^{\dagger^2}\hat{b}^{\dagger^2}\rangle \mp \langle\hat{a}\hat{b}\rangle^2 \mp \langle\hat{a}^\dagger\hat{b}^\dagger\rangle^2 - 2\langle\hat{a}\hat{b}\rangle\langle\hat{a}^\dagger\hat{b}^\dagger\rangle < 0$$
$$\tag{7.42}$$

and the uncertainty relation is satisfied.

7.2 EPR Conjecture

Even though there is no single experiment that defies its prediction to date, quantum theory has been in the limelight for so long due to the inconsistencies in its theoretical foundation, which emanate from the interpretation that the wave function can be linked to the coherent superposition of the various distinguishable outcomes: $|\phi\rangle = \sum_j \langle\phi_j|\phi\rangle|\phi_j\rangle$. In the process of measurement, the collapse of the wave function in this context is presumed to take place in the sense that a single definite state $|\phi_j\rangle$ of the system would be chosen out of the possible continuum (see Chap. 4). The conceptual drawbacks often stem from the interpretation of the manner in which this definite state is chosen from all possible outcomes. One of such inconsistencies is exemplified by the contention dubbed as EPR paradox: named after Einstein, Podolsky and Rosen[4] (compare with the discussion in Sect. 4.1). The EPR conjecture—as we prefer to call it—is alluded to the dichotomy that either the measurement of a physical quantity in one system should affect the measurement of the physical quantity in another spatially separated system or the description of reality given by the wave function should be incomplete.

Imagine two particles that interact until sufficiently strong interconnectedness can be established between them, and then set off in opposite directions. According to Heisenberg's uncertainty principle, it is impossible to measure both the momentum and position of one of the particles exactly. It is however admissible to measure the exact position of one of the particles and the exact momentum of the other. A straightforward calculation may then reveal that with the exact position of one of the particles, let us say, A is known, the exact position of the other particle B can be known. If the exact momentum of particle B is known, the exact momentum of particle A can also be worked out. On the basis of such argument, EPR claimed that a simultaneous exact values of position and momentum of particle B can be known contrary to the prediction of uncertainty principle; it is this kind of discrepancy that typifies the conception of EPR contention. One may as a result argue as EPR conjecture has laid a foundation for a deepened understanding of quantum theory via exposing the nonclassical characteristics of the measurement process.

It might be worthwhile noting that EPR also argued that in a complete theory there should be an element that corresponds to each element of the reality, that is, the sufficient condition for the reality of the physical quantity is linked by them to the possibility of predicting it with certainty without necessarily disturbing the system in anyway. Nonetheless, for two physical quantities described by non-commuting dynamical operators, the exact knowledge of one of them precludes the slightest understanding of the other. Hoping to reconcile these lines of thought, they start out with defining the completeness of the theory as every element of the physical reality must have a counterpart in the physical theory, and setting a criterion that if without in anyway disturbing a system one can predict with certainty the value of the physical quantity, they proposed that there exists an element of the physical reality corresponding to this quantity. Based on the criteria they have introduced, they sought out to test the completeness of the quantum theory.

Even though the dynamics of a given system can be described by the pertinent wave function, it is not possible to know the state in which each system belongs unless one carries out the measurement such as the reduction of the wave packet. Let $a_1, a_2, a_3, \ldots$ be the eigenvalues of some physical quantity A pertaining to system 1, and $u_1(x_1), u_2(x_1), u_3(x_1), \ldots$ be the associated eigenfunctions, where x_1 stands

---

[4]There is no paradox as far as current understanding goes; rather, there are differences in the way one perceives the physical reality in everyday life against the probabilistic nature of quantum theory. Even then, EPR conjecture is widely treated in quantum physics and philosophy of science concerned with the measurement and description of microscopic systems (see also the seminal work of Einstein et al. [10]).

for the variable used to describe the first system. The combined wave function can then be expressed as

$$\Phi(x_1, x_2) = \sum_{n=1}^{\infty} \phi_n(x_2) u_n(x_1), \tag{7.43}$$

where x_2 is the variable that describes the second system and $\phi_n(x_2)$'s are the coefficients of the expansion of $\Phi(x_1, x_2)$ into series of orthogonal functions $u_n(x_1)$.

If a quantity A is measured and the outcome turns out to be a_k, one can conclude that after measurement the first system would be left in the state given by the wave function $u_k(x_1)$ and the second system in the state given by the wave function $\phi_k(x_2)$. Instead of choosing A, one can take the quantity B having eigenvalues b_1, $b_2, b_3, \ldots$ and eigenfunctions $v_1(x_1)$, $v_2(x_1)$, $v_3(x_1)$, ... One can now propose that

$$\Phi(x_1, x_2) = \sum_{s=1}^{\infty} \varphi_s(x_2) v_s(x_1), \tag{7.44}$$

where $\varphi_s(x_2)$'s are the new coefficients. If the measurement of B yields b_r, it can be understood as the second system is left in the state $\varphi_r(x_2)$. As a consequence of the two different measurements of the first system, the second system might be left in states with two different wave functions.

Pertaining to the inability of the two systems to interact during measurement, a real change is not expected to take place in the second system due to anything that might have been done to the first system. It can also be argued further that if the wave function is assumed to completely describe the physical reality, these two wave functions should correspond to two non-commuting dynamical operators, that is, by measuring either A or B it appears possible to predict the other with certainty without in anyway disturbing it. This proposal insinuates the realization of an outcome that defies the long standing principle of quantum theory.

Since the assumption that the wave function completely describes the physical reality seems to lead to the conclusion that negates the quantum theory, EPR resorted to claim that the physical reality given by the wave function is not complete. Even though it might not be straightforward to see outright, the central message of EPR contention appears to allege that quantum theory should not be taken as a complete theory of nature; and also imply that some other theory—such as a hidden variable formulation—should have to be invoked in order to fully describe the physical nature.

Despite the elegance in its exposition, quantifying the EPR conjecture may require detailed discussion that encompasses formidable rigor. One may consider as demonstration the criterion set forward based on the inferred variance of the quadrature operators $\hat{X}_{a,b}$ and $\hat{P}_{a,b}$, where

$$\hat{X}_a = \hat{a} + \hat{a}^\dagger, \qquad \hat{P}_a = i(\hat{a}^\dagger - \hat{a}) \tag{7.45}$$

are taken in place of the positions and momenta envisioned in the original treatment.[5] Since these operators have the same commutation properties as canonical positions and momenta, it is acceptable to define the inferred variances for position and momentum variables as

$$V^{inf}(\hat{X}_a) = V(\hat{X}_a) - \frac{\left[V(\hat{X}_a, \hat{X}_b)\right]^2}{V(\hat{X}_b)}, \tag{7.46}$$

where

$$V(\hat{A}, \hat{B}) = \langle \hat{A}\hat{B} \rangle - \langle \hat{A} \rangle \langle \hat{B} \rangle. \tag{7.47}$$

Owing to the fact that $\hat{X}$ and $\hat{P}$ for the same radiation field do not commute, the product of the variances obeys Heisenberg uncertainty condition: $V(\hat{X})V(\hat{P}) \geq 1$. A quantum system hence can be said to demonstrate EPR conjecture whenever

$$V^{inf}(\hat{X})V^{inf}(\hat{P}) < 1 : \tag{7.48}$$

the condition that can be applied to quantify the degree by which the EPR conjecture can be verified or the degree by which the nonclassical feature, specially EPR steering, is exhibited [11].

7.2.1 Quantum Nonlocality

The other conspicuous drawback related to quantum formalism is the inability of the pertinent state to capture the local description of events. Imagine two systems whose wave functions are known and allowed to interact for some time. It is a well established fact that one can determine the wave function resulting from the interaction with the aid of Schrödinger formulation. One may wish at some point to know what happens when the momentum of system A is measured in case A and the other system B move so far apart that they no more able to interact in any fashion. Based on conservation of momentum and characteristics of the system before interaction, it should be possible to infer the momentum of system B via measurement of system A. The perceived separation in space like distance implies that B can have the inferred values of the momentum not only in the instant after one makes the measurement on A at its site, but also in the few moments before the measurement was made. If the measurement at site A somehow compels B

---

[5]Suppose there are two spatially separated subsystems with observable such as $\hat{X}_a$ and $\hat{P}_a$ that satisfy $[\hat{X}_a, \hat{P}_a] = 2i\hat{I}$ and $\Delta X_a^2 \Delta P_a^2 \geq 1$. One can predict the result of measurement of $\hat{X}_a$ based on the result of a causally separated measurement of $\hat{X}_b$ performed at site of B when the two subsystems are partially correlated. But the prediction is imperfect due to inference error, and so the basis for using inference as a measure of nonclassical features.

to enter into certain momentum state, it is expected that system A has a way of signaling B and to inform him that a measurement has taken place. Nonetheless, due to the separation between them, it is already presumed that the two systems do not communicate in any way.

In case one examines the wave function at the moment just before the measurement at the site of A has taken place, one would find that there is no certainty as to the momentum of B since the combined system would be in the superposition of a multiple momentum eigenstates of A and B. Even though system B must be in a definite state before the measurement at A takes place, the wave function description of this system cannot tell us what that momentum is. The verdict is: since system B is conned to have a definite momentum although quantum theory do not predict definite momentum, the perceived conclusion would be quantum theory as described by the wave function would not allow the local description of events, that is, quantum state exhibits nonlocal character [12]. For two spatially separated entangled state, the idea that the measurement on one subsystem would allow to predict definitely the outcome of a similar measurement on the other subsystem is thus taken as the basis for attesting to the claim that quantum theory is nonlocal. Such understanding had posed a serious conceptual challenge to physicists by then since faster than light transmission of information is prohibited by the theory of special relativity. But it is now understood that although these nonlocal correlations do occur, they cannot be used to transmit information, and so do not violate causality.

The emerging difficulties in verifying nonlocality are linked to the assumption employed in deriving Bell's inequalities that can be categorized as: realism—the measurement results are determined by properties that the particles carry prior to and are independent of the measurement; locality—results obtained at one location are independent of the actions performed at space like separation; and free will—the setting of local apparatus is independent of the hidden variables that determine the local results. Bell evidently construed that these assumptions impose constraints in the form of certain inequalities on the statistical correlations in experiment involving bipartite systems [13]. He then showed that the probabilities for the outcomes obtained when some entangled state is suitably measured violate the inequality presently named after him. He came in this way to the understanding that entanglement is a peculiar feature of the quantum formalism, which forges the impossibility of simulating quantum correlations within classical formalism.

Bell's inequality, in the nutshell, refers to a class of correlation inequalities that hold under local realism: if a given quantum system failed to satisfy one of these inequalities, it is designated as nonlocal. Bell's theorem is thus believed to offer perspective in quantifying some concepts and issues related to EPR conjecture, and eventually provide experimental basis to test the contention of entanglement versus local realism [13]. Markedly, a given state should not have to violate all possible Bell's inequalities to be considered as nonlocal: a state is said to be nonlocal if it happens to violate any of Bell's inequalities. It is hence imperative inferring that the degree of nonlocality that can be revealed depends not only on the quantum state, but also on the selected Bell's operator. Even so, nonlocality is believed to be one of the most profound features of quantum theory that enables current developments in the area of quantum information processing (see Chaps. 11 and 12, and also [14]).

7.2.2 Bell Chain Inequality

To grasp the basics of Bell's experiment, it may suffice to take an arrangement in which two observers perform independent measurement on a given system prepared in some fixed state upon choosing between various detector settings on each trial. In one version of the setup, suppose Alice can choose between two detector settings to measure output X_A or Y_A while Bob can choose between X_B and Y_B where each measurement has one of the two possible outcomes: ± 1. After repeated trials, let Alice and Bob collect statistics on their measurement, and then correlate the obtained results. Imagine that the pertinent composite system consists of two particles prepared in special state believed to be in the superposition of the two states, and one of which is sent to Alice and the other to Bob.

7.2.2.1 General Consideration

With the assumption that each measurement reveals objective physical property of the system, and measurement taken by one observer has no effect on the measurement taken by the other, one can argue that successive results of the experiment can allow to test the failure of local realism. To do so, one needs to define the correlation for repeated independent trials as

$$C(X, Y) = \lim_{N \to \infty} \frac{1}{N} [X_1 Y_1 + X_2 Y_2 + \ldots + X_N Y_N], \tag{7.49}$$

which is true provided that the limit exits and it is robust, that is, the limit exists for enough subspaces and takes up the same value. It is thus presumed that each expression is equal to those obtained by taking the limit of the average on the entire run including those values that are not related to the measurement.

For discrete variables, one may need to initiate the locality assumption, that is, for each trial at least one of $X_B^n + Y_B^n$ and $X_B^n - Y_B^n$ should be zero regardless of whether X_A or Y_A is measured by Alice. One can thus show by employing the fact that the possible outcome of each measurement is either 1 or -1 that

$$\sum_{n=1}^{N} [X_A^n X_B^n + X_A^n Y_B^n + Y_A^n X_B^n - Y_A^n Y_B^n] \leq 2N \tag{7.50}$$

since each term can be grouped as

$$[X_A^n X_B^n + X_A^n Y_B^n + Y_A^n X_B^n - Y_A^n Y_B^n] = X_A^n (X_B^n + Y_B^n) + Y_A^n (X_B^n - Y_B^n). \tag{7.51}$$

In light of this, it is possible to write

$$C(X_A, X_B) + C(X_A, Y_B) + C(Y_A, X_B) - C(Y_A, Y_B)$$

$$= \lim_{N \to \infty} \frac{1}{N} [X_A^n X_B^n + X_A^n Y_B^n + Y_A^n X_B^n - Y_A^n Y_B^n], \qquad (7.52)$$

from which follows

$$C(X_A, X_B) + C(X_A, Y_B) + C(Y_A, X_B) - C(Y_A, Y_B) = 2. \qquad (7.53)$$

As a result, under local realism;

$$C(X_A, X_B) + C(X_A, Y_B) + C(Y_A, X_B) - C(Y_A, Y_B) \le 2, \qquad (7.54)$$

which is usually referred to as Bell-CHSH (Clauser, Horne, Shimony and Holt) inequality, and the violation of which can be interpreted as manifestation of quantum feature [15].

The same inequality can also be obtained by applying a general approach starting with the fact that $a + a' = 0$ and $a - a' = \pm 2$ or $a + a' = \pm 2$ and $a - a' = 0$ since the particles are assumed to move in opposite directions or has opposite spins, and a, a', b and b' to take either 1 or -1. It may not be difficult to see in this respect that

$$(a + a')b + (a - a')b' = \pm 2, \qquad (7.55)$$

where expanding Eq. (7.55) leads to

$$ab + a'b + ab' - a'b' = \pm 2. \qquad (7.56)$$

The ensemble average should also follow the same suit:

$$\langle ab \rangle + \langle a'b \rangle + \langle ab' \rangle - \langle a'b' \rangle \le \pm 2, \qquad (7.57)$$

which implies that

$$|\langle ab \rangle + \langle a'b \rangle + \langle ab' \rangle - \langle a'b' \rangle| \le 2, \qquad (7.58)$$

where $\langle ab \rangle = C(a, b)$ is the correlation.

7.2.2.2 In Relation to Light

It is a well established fact that in the nonlinear optical processes such as nondegenerate optical parameter oscillation, the signal and idler modes can differ in polarization and/or frequency that allows the spatial separation of the two beams, and so helps in the construction of continuous variable correlations of the type originally envisioned by Einstein et al. [10, 16]. The manifestation of

EPR correlation during nondegenerate parametric oscillation can be attributed to a two-photon process that ensures a strong correlation between the radiations (see Sect. 8.1.1). To modify Bell's inequalities for a continuous variable, one thus can begin with a seemingly obvious assumption that the measurements at separated places are independent, which enables to obtain the joint probabilities of the pairs of outcomes upon multiplying the separate probabilities for any selected values of the hidden variable.

Upon assuming that the hidden variable ϵ is drawn from a fixed distribution of possible states of the source, the probability of the source in the state ϵ for any particular trial can then be described by the density function $\rho(\epsilon)$, where the integral of this density function over the complete hidden variable space equals to 1. Once the probability density is known, it is possible to express the correlation as

$$E(a, b) = \int \bar{A}(a, \epsilon)\bar{B}(b, \epsilon)\rho(\epsilon)d\epsilon, \qquad (7.59)$$

where $\bar{A}(a, \epsilon)$ and $\bar{B}(b, \epsilon)$ are the average values that change with the hidden variables. Since the possible values of $\bar{A}$ and $\bar{B}$ are taken to be -1, 0 and $+1$, one may see that $|\bar{A}(a, \epsilon)| \leq 1$ and $|\bar{B}(b, \epsilon)| \leq 1$.

It should also be possible to design a, a', b and b' as alternative settings for the detectors, which leads to

$$E(a, b) - E(a, b') = \int \bar{A}(a, \epsilon)\bar{B}(b, \epsilon)\big[1 \pm \bar{A}(a', \epsilon)\bar{B}(b', \epsilon)\big]\rho(\epsilon)d\epsilon$$

$$- \int \bar{A}(a, \epsilon)\bar{B}(b', \epsilon)\big[1 \pm \bar{A}(a', \epsilon)\bar{B}(b, \epsilon)\big]\rho(\epsilon)d\epsilon, \qquad (7.60)$$

where applying the fact that $[1 \pm \bar{A}(a', \epsilon)\bar{B}(b', \epsilon)]\rho(\epsilon)$ and $[1 \pm \bar{A}(a', \epsilon)\bar{B}(b, \epsilon)]\rho(\epsilon)$ are nonnegative implies that

$$|E(a, b) - E(a, b')| \leq \int [1 \pm \bar{A}(a', \epsilon)\bar{B}(b', \epsilon)]\rho(\epsilon)d\epsilon$$

$$+ \int [1 \pm \bar{A}(a', \epsilon)\bar{B}(b, \epsilon)]\rho(\epsilon)d\epsilon. \qquad (7.61)$$

Upon using the fact that the integral of $\rho(\epsilon)$ is 1, one can then verify that

$$|E(a, b) - E(a, b')| \leq 2 \pm [E(a', b') + E(a', b)]. \qquad (7.62)$$

We now opt to explore as illustration the way of studying the dynamical behavior of quantum nonlocality based on parity measurement of a two-mode squeezed light [17]. To embody the continuous nature of the variables that denote light within discrete variable description, the odd and even parities of the light, that are, projection operators that measure the probabilities of the field having even and odd numbers of photons,

$$\hat{e} = \sum_{n=0}^{\infty} |2n\rangle\langle 2n|, \qquad \hat{o} = \sum_{n=0}^{\infty} |2n+1\rangle\langle 2n+1|, \tag{7.63}$$

can be applied as measurement variables (see also Sect. 9.3).

Recall that the Wigner function can be expressed at the origin of the phase space for the state with density operator $\hat{\rho}$ as

$$W(0) = \frac{2}{\pi} Tr[(\hat{e} - \hat{o})\hat{\rho}] \tag{7.64}$$

and the mean parity for the displaced state as

$$W(\alpha) = \frac{2}{\pi} Tr[(\hat{e} - \hat{o})\hat{D}(\alpha)\hat{\rho}\hat{D}^{\dagger}(\alpha)], \tag{7.65}$$

where $\hat{D}(\alpha)$ is displacement operator defined by Eq. (2.24).

Such representation would be applicable for a single-mode radiation although nonlocality is related to two-mode field. To reconcile this disparity, the quantum correlation operator related to the joint parity measurement needs to be defined as

$$\hat{\Pi}^{ab}(\alpha, \beta) = \hat{\Pi}_e^a(\alpha)\hat{\Pi}_e^b(\beta) - \hat{\Pi}_e^a(\alpha)\hat{\Pi}_o^b(\beta) - \hat{\Pi}_o^a(\alpha)\hat{\Pi}_e^b(\beta) + \hat{\Pi}_o^a(\alpha)\hat{\Pi}_o^b(\beta), \tag{7.66}$$

where the superscripts a and b denote the modes and

$$\hat{\Pi}_{e(o)}(\alpha) = \hat{D}(\alpha)\hat{O}_{e(o)}\hat{D}^{\dagger}(\alpha) \tag{7.67}$$

in which the displaced parity operator acts like a rotated spin projection operator. It can be derived for the local hidden variable theory on account of Eq. (7.66), that is,

$$|B(\alpha, \beta)| - \left|\hat{\Pi}^{ab}(\alpha, \beta) + \hat{\Pi}^{ab}(\alpha, \beta') + \hat{\Pi}^{ab}(\alpha', \beta) - \hat{\Pi}^{ab}(\alpha', \beta')\right| \leq 2, \tag{7.68}$$

where $B(\alpha, \beta)$ is the Bell function.

It is common knowledge that the mean of the quadrature operators can be determined with the aid of two-mode Wigner function

$$W(\alpha, \beta) = 4\pi^2 Tr\left[\hat{\rho}_{ab}\hat{\Pi}^{ab}(\alpha, \beta)\right]. \tag{7.69}$$

In view of Eq. (7.69), the Bell function can be described in terms of the Wigner function at different phase space points as

$$\beta(\alpha, \beta) = \frac{\pi^\perp}{4}\left[W(0, 0) + W(\alpha, 0) \mid W(0, \beta) - W(\alpha, \beta)\right],$$ (7.70)

where the violation of Bell-CHSH inequality can be expressed as

$$W(0, 0) + W(\alpha, 0) + W(0, \beta) - W(\alpha, \beta) > \frac{8}{\pi^2}.$$ (7.71)

So upon relating the even and odd parities to the number of photons generated in a given physical process, it is believed to be admissible to study the nonlocal feature of the radiation.

To be more specific, one can extend the discussion on Bell's inequalities to the case of light, and check whether the original EPR states maximally violate Bell-CHSH inequalities. For a single-mode light, one can thus introduce pseudo-spin operators

$$\hat{s}_z = \sum_{n=0}^{\infty}\left[|2n + 1\rangle\langle 2n + 1| - |2n\rangle\langle 2n|\right], \qquad \hat{s}_- = \sum_{n=0}^{\infty}|2n\rangle\langle 2n + 1| = (\hat{s}_+)^+,$$ (7.72)

where $|n\rangle$ is the number state (see Sect. 9.3 and also [18]). It is also possible to define $\hat{s}_z = -(-1)^{\hat{N}}$ as the photon parity operator where $\hat{N}$ is the number operator, and $\hat{s}_+$ and $\hat{s}_-$ are the parity flip operators.

In terms of boson creation and annihilation operators, one can also express

$$\hat{s}_- = \left[\hat{I} + (-1)^{\hat{N}}\right]e^{i\hat{\theta}},$$ (7.73)

where $\hat{I}$ is the identity operator and $\frac{1}{\sqrt{\hat{N}+1}}\hat{a} = e^{i\hat{\theta}}$ in which $\hat{\theta}$ is the phase operator. With this designation, one can verify that

$$[\hat{s}_z, \hat{s}_\pm] = \pm 2\hat{s}_\pm, \qquad [\hat{s}_+, \hat{s}_-] = \hat{s}_z.$$ (7.74)

Since these commutation relations turn out to be identical to that of the spin operators (see Sect. 5.1.2), they are often referred to as parity spin of photon due to their resemblance with the parity space of the photon.

Then upon choosing arbitrary vector lying on the surface of a unit sphere,

$$\mathbf{h} = \sin\theta_h\cos\phi_h\mathbf{i} + \sin\theta_h\sin\phi_h\mathbf{j} + \cos\theta_h\mathbf{k},$$ (7.75)

where θ_h and ϕ_h are the polar and azimuthal angles, $\mathbf{h}$ is the vector along the direction of the measurement, and the fact that $\hat{s} = \hat{s}_x \mathbf{i} + \hat{s}_y \mathbf{j} + \hat{s}_z \mathbf{k}$ with $\hat{s}_x = \hat{s}_+ + \hat{s}_-$ and $\hat{s}_y = i(\hat{s}_- - \hat{s}_+)$, it may not be difficult to verify that

$$\mathbf{h} \cdot \hat{s} = \hat{s}_z \cos\theta_h + \sin\theta_h \left(e^{i\phi_h}\hat{s}_- + e^{-i\phi_h}\hat{s}_+\right), \tag{7.76}$$

which leads to $(\mathbf{h} \cdot \hat{s})^2 = \hat{I}$. This discussion may indicate that there can exist an analogy between continuous variable and spin-$\frac{1}{2}$ systems.

On the basis of the formalism developed for a two-mode light, the Bell operator can be defined as

$$\hat{B}_{\text{CHSH}} = (\mathbf{h} \cdot \hat{s}_1) \bigotimes (\mathbf{v} \cdot \hat{s}_2) + (\mathbf{h} \cdot \hat{s}_1) \bigotimes (\mathbf{v}' \cdot \hat{s}_2)$$
$$+ (\mathbf{h}' \cdot \hat{s}_1) \bigotimes (\mathbf{v} \cdot \hat{s}_2) - (\mathbf{h}' \cdot \hat{s}_1) \bigotimes (\mathbf{v}' \cdot \hat{s}_2), \tag{7.77}$$

where $\mathbf{v}$ is assumed to be the vector that is not along the measurement line of the spin. It should be clear by now that a quantum state exhibits a nonlocal character provided that

$$|\langle \hat{B}_{\text{CHSH}} \rangle| > 2, \tag{7.78}$$

where $|\langle \hat{B}_{\text{CHSH}} \rangle|$ is the expectation value of $\hat{B}_{\text{CHSH}}$ with respect to the quantum state of continuous variable. In view of the fact that

$$(\mathbf{h} \cdot \hat{s}_1)^2 = (\mathbf{v} \cdot \hat{s}_2)^2 = (\mathbf{h}' \cdot \hat{s}_1)^2 = (\mathbf{v}' \cdot \hat{s}_2)^2 = \hat{I} \tag{7.79}$$

and commutation relations of spin operators, one can see that

$$B_{\text{CHSH}}^2 = 4I + 4(\mathbf{h} \times \mathbf{h}' \cdot \hat{s}_1) \bigotimes (\mathbf{v} \times \mathbf{v}' \cdot \hat{s}_2), \tag{7.80}$$

which entails that $\langle B_{\text{CHSH}}^2 \rangle \leq 8$. One can thus say that a given state maximally violates Bell-CHSH inequalities provided that

$$|\langle \hat{B}_{\text{CHSH}} \rangle| = 2\sqrt{2}. \tag{7.81}$$

Let us now consider a practical example for a maximal violation of Bell-CHSH inequality for continuous variable scenario. To do so, one can begin with the fact that the state of a nondegenerate parametric amplifier can be defined as

$$|r, r\rangle = e^{r(\hat{a}^\dagger \hat{b}^\dagger - \hat{a}\hat{b})}|00\rangle, \tag{7.82}$$

which can also be put in the form

$$|n, n\rangle = \sum_{n=0}^{\infty} \frac{(\tanh(r))^n}{n! \cosh(n)} (\hat{a}^\dagger)^n (\hat{b}^\dagger)^n |00\rangle, \tag{7.83}$$

The states that represent a nondegenerate parametric amplifier are suggested to be the optical analogy of the EPR type entangled states in the limit of infinite squeezing, and can be utilized in the teleportation of continuous variables (see Chap. 11).

On account of previous discussion, one may see that

$$\langle \hat{B}_{CHSH} \rangle = E(\theta_h, \theta_v) + E(\theta_h, \theta_{v'}) + E(\theta_{h'}, \theta_v) - E(\theta_{h'}, \theta_{v'}), \tag{7.84}$$

where the correlation function can be defined as

$$E(\theta_h, \theta_v) = \langle r, r | \hat{s}_{\theta h}^{(a)} \bigotimes \hat{s}_{\theta v}^{(b)} | r, r \rangle \tag{7.85}$$

with $\hat{s}_{\theta h}^{(j)} = \hat{s}_{jz} \cos \theta_h + \hat{s}_{jx} \sin \theta_h$ and j denotes the modes in which all the azimuthal angles are set to zero.

It might be good to note that the quadrature operator,

$$\hat{X}_\theta = \hat{a}^\dagger e^{i\theta} + \hat{a} e^{-i\theta}, \tag{7.86}$$

reduces to $\hat{X}_\theta = 2(\hat{x} \cos \theta + \hat{p}_x \sin \theta)$, where $\hat{a} = \hat{x} + i \hat{p}_x$ which holds also for $\hat{b}$ mode. So with the aid of Eq. (7.83), one gets

$$E(\theta_h, \theta_v) = \cos \theta_h \cos \theta_v + \tanh(2r) \sin \theta_h \sin \theta_v, \tag{7.87}$$

which would be lesser or equal to 1 and regarded as a qualitative measure of the nonlocality. As demonstration, upon choosing $\theta_h = 0$, $\theta_{h'} = \frac{\pi}{2}$ and $\theta_v = -\theta_{v'}$, one can obtain

$$\langle \hat{B}_{CHSH} \rangle = 2[\cos \theta_v + \tanh(2r) \sin \theta_v]. \tag{7.88}$$

One may see that the maximal violation of Bell-CHSH inequality is observed when

$$\langle \hat{B}_{CHSH} \rangle_{max} = 2\sqrt{1 + \tanh^2(2r)}. \tag{7.89}$$

Fig. 7.2 Plot of the
maximized expectation value
of the Bell-CHSH operator

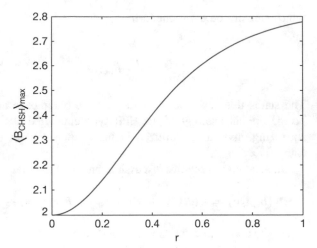

One can see from Fig. 7.2 that the light generated by a nondegenerate parametric optical amplifier violate the Bell-CHSH inequality provided that $r \neq 0$, that is, as long as parametric down conversion process is physically possible.[6] Critical scrutiny also reveals that $\langle \hat{B}_{CHSH} \rangle_{max} \leq 2\sqrt{2}$ (relate with Eq. (7.81)) where the equality can be realized for a considerably large squeeze parameter. But pumping a radiation with infinite squeeze parameter is not as such an easy task.

7.3   Entanglement

Entanglement can be perceived as the property of the states of the quantum system that contains distinct parts in which the information that describes the combined system is inextricably linked to each other in such a way that the act of measurement on one party immediately alters the properties of the other even when they are placed at a distance. Entanglement is thus the most enthralling trait of quantum theory that has been studied extensively in connection to Bell's inequalities. Even though the notion of entanglement was thought to be the consequence of the mathematical formulation of quantum theory, it manages eventually to inspire the research ranging from the EPR conjecture to quantum information processing due to the expectation

---

[6]The pseudo-spin and parity are not strictly speaking a genuine continuous variable considerations since ultimately one sticks to the discrete variable formalism similar to the spin-$\frac{1}{2}$ particle. For a general case, the definition of the correlation in Eq. (7.49) can be modified as

$$C(X_A, X_B) = \frac{\langle \hat{X}_A \hat{X}_B \rangle}{\sqrt{\langle \hat{X}_A^2 \rangle \langle \hat{X}_B^2 \rangle}}$$

so that the Bell's inequality (7.54) can be inferred.

that the emerging interconnectedness can enhance the pertinent efficiency. Owing to the fact that the transposition of a given quantum state would lead to a quantum state, the negativity of the transposition of the density operator is in most part taken as a sufficient condition for inseparability that can be taken as an indicator of the manifestation of quantum entanglement.

In light of this, a bipartite state is said to be separable if it can be written as a convex combination of factorable states:

$$\hat{\rho} = \sum_{n=0}^{\infty} p_n \hat{\rho}_1^{(n)} \bigotimes \hat{\rho}_2^{(n)}, \tag{7.90}$$

where $\hat{\rho}_{1,2}^{(n)}$ are the density of states that constituent the parts of the composition and p_n's are nonnegative numbers that satisfy $\sum_{n=0}^{\infty} p_n = 1$ for normalized state. In case the density of these states cannot be put in the form of Eq. (7.90), the composite quantum state is designated as inseparable. Since the entangled state contains information on each composite separately and on the correlation between the measurements on composite systems, the degree of entanglement can also be related to the amount of information going to be lost when one of the combined systems is interrogated.

One simple way to visualize entanglement for the nonspecialist is perhaps through imagining the constituents of some composite system such as a pair of photons as able to provide the same random answer whenever they are asked the same question. It should be emphasized that the answer or measurement result is random, but it is precisely the same randomness that appears at two distant locations provided that the observers perform the same experiment.

Despite the validity of such illustration, characterizing whether a given composite state is entangled is usually far from being simple and trustworthy. Even then, although they are not always straightforward to employ, measures such as positive partial transpose, entanglement witness and hierarchies of entanglement have been proposed [19]. After serious consideration, various simpler and more applicable inseparability criteria for a continuous variable Gaussian states have also been established based on the realization that certain variances can be bounded due to the uncertainty relation or other quantum theoretical principles; the emerging classical bound obviously would be violated for entangled state where the available criteria are in the form of some sort of inequality and in most instances provide only sufficient conditions.

Regardless of the rigor required in using the emerging criteria to quantify it, quantum entanglement has three disagreeable but interesting features: it has very complex structure; it is fragile with respect to environment; and it cannot be increased on average when the systems are not in direct contact but distributed in spatially separated regions. The discussion on entanglement then should try to answer questions such as how to optimally detect entanglement; how to reverse the inevitable process of degradation of entanglement; how to characterize, control and quantify entanglement. On account of the observation that different entanglement criteria do not necessarily lead to the same conclusion, we mainly seek to outline various criteria for quantifying discrete as well as continuous variable entanglement hoping that the comparison among the results obtained following the suggested criteria may give further insight to the overall nature of the quantumness.

7.3.1   Introduction

To outline some approaches for quantifying entanglement based on concurrence (related to the variances or quantum uncertainties of appropriate variable [20]), it is possible to begin with the proposal that the uncertainty of the observable can be expressed in terms of the correlation

$$C(x, \psi) = \langle \psi | \hat{A}^2 | \psi \rangle - \langle \psi | \hat{A} | \psi \rangle^2 \tag{7.91}$$

in which the total variance can be put in the form

$$V(\psi) = \sum_{\alpha} \left[\langle \psi | \hat{A}_{\alpha}^2 | \psi \rangle - \langle \psi | \hat{A}_{\alpha} | \psi \rangle^2 \right], \tag{7.92}$$

where the sum can be extended over orthonormal bases of the traceless local operators.

The total variance can naturally be taken as the trace of the quadratic form

$$V(\hat{A}) = \langle \psi | \hat{A}^2 | \psi \rangle - \langle \psi | \hat{A} | \psi \rangle^2, \tag{7.93}$$

which is independent of the bases, and believed to measure the quantum fluctuations of the system. One can see that when $\langle \psi | \hat{A} | \psi \rangle = 0$, the states are completely disordered, that means, there exists some local basis in which the reduced state is given by a diagonal matrix related to a uniform probability distribution. This assertion embodies the idea of maximally entangled states where Eq. (7.93) designates entanglement quantifier and the state is taken to be completely entangled when $\langle \psi | \hat{A} | \psi \rangle = 0$. One can hence propose that a completely entangled states can be portrayed by the maximality of the total variance, that is, one can perceive entanglement as a manifestation of high quantum fluctuation.

The concurrence for a two-party system can also be defined in terms of the variances as

$$C^2(\psi) = \frac{V(\psi) - V_{\text{coh}}}{V_{\text{ent}} - V_{\text{coh}}}, \tag{7.94}$$

where V_{ent} and V_{coh} are levels of quantum fluctuations in a completely entangled and coherent state (state with a minimal total level of quantum fluctuation), and the total variance $V(\psi)$ so takes the value between and including 0 and 1. Such representation therefore may clarify the meaning of the concurrence as a measure of an overall quantum fluctuation in a system; and at the same time, may lead to the natural measure of entanglement of pure state:

$$C(\psi) = \sqrt{\frac{V(\psi)}{V_{\text{ent}} - V_{\text{coh}}}}$$ (7.95)

in which $0 \leq C(\psi) \leq 1$: one can infer that a pure bipartite state would be entangled when $C(\psi)$ happens to be nonzero.

A pure composite state of a pair of quantum systems is also designated as entangled if it cannot be factorized, and a mixed state when it cannot be represented as a mixture of pure states that can be factored. With this delineation, the density matrix of a pair of quantum systems can be expressed in terms of the ensembles of states and probabilities as $\hat{\rho} = \sum_i p_i |\psi_i\rangle\langle\psi_i|$. Then for a pure state, the entanglement of formation can be defined as the von Neumann entropy of either of the two subsystems:

$$E(\psi) = -Tr(\hat{\rho}\log_2\hat{\rho}),$$ (7.96)

where $\hat{\rho}$ is the partial trace of $|\psi_i\rangle\langle\psi_i|$ over one of the subsystems; it does not matter which.

The entanglement of formation for a mixed state can also be expressed as the average entanglement of the pure states of the decomposition minimized over the density operator, that is,

$$E(\rho) = \min \sum_i p_i E(\psi_i).$$ (7.97)

Markedly, the entanglement of formation has a very distinct property that it would be zero if the state can be expressed as a mixture of product states. The entanglement of formation can also be expressed in a more elegant manner as

$$E(\psi) = E(C(\psi_i)),$$ (7.98)

where the concurrence can be designated as the overlap between the states at different settings: $C(\psi) = |\langle\psi|\psi'\rangle|$ in which $|\psi'\rangle$ is the transformed state. Entanglement of formation can then be describe in terms of the concurrence C as [21]

$$E(C) = -\left(\frac{1 + \sqrt{1 - C^2}}{2}\right) \log_2 \left(\frac{1 + \sqrt{1 - C^2}}{2}\right)$$
$$- \left(\frac{1 - \sqrt{1 - C^2}}{2}\right) \log_2 \left(\frac{1 - \sqrt{1 - C^2}}{2}\right).$$ (7.99)

One of the physically motivated quantitative measures of entanglement is the entanglement of formation, which is intended to quantify the resources required to create the entangled state. There are also some peculiar properties of the entanglement of formation that makes it natural entanglement measure for a pure state: the entanglement of independent systems is additive; it is conserved under local unitary operations; the expectation value of E cannot be increased by a local nonunitary operations. These features may entail that the entanglement of a pure bipartite state can be completely captured by the entanglement of formation.

The entanglement of formation for a mixed state can also be proposed to be

$$E(\rho) = E(C(\rho)), \tag{7.100}$$

where

$$C(\rho) = \max(0, \lambda_1 - \lambda_2 - \lambda_3 - \lambda_4) \tag{7.101}$$

in which λ_i's are nonnegative eigenvalues in the decreasing order of the Hermitian matrix: $\sqrt{\sqrt{\hat{\rho}}\hat{\rho}^\dagger\sqrt{\hat{\rho}}}$, where $\hat{\rho}^\dagger$ is the complex conjugate of $\hat{\rho}$.

7.3.2 Continuous Variable Bipartite Entanglement

Intuitive consideration may reveal that with increasing dimension of the Hilbert space, the test for a separability would become more difficult to implement. In the limit of infinite dimension that practically corresponds to continuous variable, there are fortunately certain criteria that lead to the separability tests that are manageable to implement. In support of the proposal that it is the quantum feature related to the interconnectedness among infinite dimensional states of the radiation that is taken as continuous variable entanglement, the operation of the partial transposition in the continuous variable scenario can be taken as a mirror reflection in the Wigner phase space. It is also common knowledge nowadays that one of the valuable attributes of continuous variable quantum entanglement is the high efficiency in practical realization. For example, continuous variable entangled state has been successfully utilized in unconditional quantum teleportation, quantum dense coding, quantum error correction, entanglement swapping and universal quantum computation (see Chaps. 9, 11 and 12). With the fact that separability imposes far more restriction on the second moments or uncertainties than the traditional uncertainty principle, a necessary and sufficient separability condition for a bipartite Gaussian state could be proposed.

Imagine for instance a bipartite system composed of two modes that can be represented by annihilation operator

$$\hat{a}_i = \frac{1}{\sqrt{2}}[\hat{q}_i + i\hat{p}_i], \tag{7.102}$$

where $i = 1, 2$ (see also Sect. 2.4). For illustration, let mode 1 is supposed to be in possession of an experimenter A and mode 2 in possession of B. By definition, a given quantum state that can be described by the density operator of a bipartite system is inseparable if it cannot be expressed in the form $\hat{\rho} = \sum_i r_i \hat{\rho}_{i1} \otimes \hat{\rho}_{iL}$ where ρ_{i1}'s and ρ_{i2}'s are the density operators of the modes of the systems that belong to the experimenters A and B.

One may argue that a partial transpose operation takes a separable density operator necessarily into a nonnegative operator, which may imply that a bipartite density operator can be transcribed with respect to the Wigner distribution function into the transformation

$$W(q_1, p_1, q_2, p_2)^{PT} \rightarrow W(q_1, p_1, q_2, -p_2). \tag{7.103}$$

One may note that transformation (7.103) is related to the aforementioned mirror reflection or a local time reversal that inverts only the p_2 coordinate, that is, if the density operator is separable, its Wigner distribution function necessarily goes over to the Wigner distribution function under the phase space mirror reflection.

In the context of Gaussian state, one can define the variance as $\Delta\hat{\xi}_\alpha = \hat{\xi}_\alpha - \langle\hat{\xi}_\alpha\rangle$, where $\hat{\xi}_\alpha = \hat{q}_1, \ \hat{p}_1, \ \hat{q}_2, \ \hat{p}_2$ and $[\hat{\xi}_\alpha, \hat{\xi}_\beta] = i\Omega_{\alpha\beta}$ while $\Omega_{\alpha\beta}$ is a constant. To obtain the required measure, one may begin with arranging the uncertainties or variances into $4X4$ real matrix as

$$V_{\alpha\beta} = \frac{1}{2}\left[\langle\Delta\hat{\xi}_\alpha \Delta\hat{\xi}_\beta\rangle + \langle\Delta\hat{\xi}_\beta \Delta\hat{\xi}_\alpha\rangle\right] \tag{7.104}$$

from which the compact form of the uncertainty principle follows

$$V + \frac{i}{2}\Omega \geq 0, \tag{7.105}$$

where Ω is the matrix constructed from $\Omega_{\alpha\beta}$ for different values of α and β (relate with Eq. (2.176)). Since the uncertainty relation (7.105) is the consequence of the commutation relation and nonnegativity of the density operator, it might be reasonable to consider the uncertainty principle as a basis for inseparability criteria.

Entangled continuous variable state on the other hand can be designated as co-eigenstate of a pair of EPR type quadrature operators such as $\hat{x}_1 + \hat{x}_2$ and $\hat{p}_1 - \hat{p}_2$ (see Sect. 7.2). One can then assert that the total variances of these operators reduce to zero for maximally entangled states; and there also exists a lower bound of the total variance for separable states. Without loss of generality, one can thus introduce two EPR type quadrature operators:

$$\hat{u} = \mu\hat{x}_1 + \frac{1}{\mu}\hat{x}_2, \qquad \hat{v} = \mu\hat{p}_1 - \frac{1}{\mu}\hat{p}_2, \tag{7.106}$$

where μ is a nonzero real number. The total variance of such a pair of EPR type quadrature operators satisfies for any separable state[7]

$$\langle(\Delta\hat{u})^2\rangle + \langle(\Delta\hat{v})^2\rangle \geq \mu^2 + \frac{1}{\mu^2}; \tag{7.107}$$

and hence, a two-party quantum system is said to be entangled if this inequality is violated.

Pertaining to the possibility of obtaining entanglement measures by fixing μ, a simpler inseparability criterion for a continuous variable Gaussian states would be proposed. Note that the states of a bipartite system are entangled if the sum of the variables of a pair of EPR type operators for $\mu = 1$

$$\hat{u} = \hat{x}_a + \hat{x}_b, \qquad \hat{v} = \hat{p}_a - \hat{p}_b \tag{7.108}$$

with

$$\hat{x}_a = \frac{1}{\sqrt{2}}[\hat{a} + \hat{a}^\dagger], \qquad \hat{x}_b = \frac{1}{\sqrt{2}}[\hat{b} + \hat{b}^\dagger], \tag{7.109}$$

$$\hat{p}_a = \frac{i}{\sqrt{2}}[\hat{a}^\dagger - \hat{a}], \qquad \hat{p}_b = \frac{i}{\sqrt{2}}[\hat{b}^\dagger - \hat{b}], \tag{7.110}$$

are the quadrature operators for mode a and b, and satisfy

$$\Delta u^2 + \Delta v^2 < 2. \tag{7.111}$$

This condition is designated for purpose of clarity as Duan-Giedke-Cirac-Zoller (DGCZ) entanglement criterion.[8]

One may recall that these quadrature operators have the same properties as the canonical positions and momenta for the harmonic oscillator, although they correspond to the real and imaginary parts of the electromagnetic mode;

$$\hat{u} = \frac{1}{\sqrt{2}}[\hat{a}^\dagger + \hat{a} + \hat{b} + \hat{b}^\dagger], \qquad \hat{v} = \frac{i}{\sqrt{2}}[\hat{a}^\dagger - \hat{a} + \hat{b} - \hat{b}^\dagger]. \tag{7.112}$$

---

[7]Even though other ways of deriving the entanglement measures for continuous variable are available, the discussion presented here is due to Simon [22].

[8]It is also possible to take $\hat{u} = \hat{x}_a - \hat{x}_b$ and $\hat{v} = \hat{p}_a + \hat{p}_b$ depending on the way the Hamiltonian of the system is expressed [24]. It is also good to note that—as applied in [25]—the operators in this book might be suitable when one seeks to include the contribution of the phase.

Note that with the help of boson commutation relation, the variances of these operators can be put in the form

$$\Delta u^2 = 1 + \langle \hat{a}^\dagger \hat{a} \rangle + \langle \hat{b}^\dagger \hat{b} \rangle + \langle \hat{a}\hat{b} \rangle + \langle \hat{a}^\dagger \hat{b}^\dagger \rangle + \langle \hat{a}\hat{b}^\dagger \rangle + \langle \hat{a}^\dagger \hat{b} \rangle$$
$$- \langle \hat{a}^\dagger \rangle \langle \hat{a} \rangle - \langle \hat{b}^\dagger \rangle \langle \hat{b} \rangle - \langle \hat{a} \rangle \langle \hat{b} \rangle - \langle \hat{a}^\dagger \rangle \langle \hat{b}^\dagger \rangle - \langle \hat{a} \rangle \langle \hat{b}^\dagger \rangle - \langle \hat{a}^\dagger \rangle \langle \hat{b} \rangle$$
$$+ \frac{1}{2}[\langle \hat{a}^{\dagger 2} \rangle + \langle \hat{a}^2 \rangle + \langle \hat{b}^{\dagger 2} \rangle + \langle \hat{b}^2 \rangle - \langle \hat{a}^\dagger \rangle^2 - \langle \hat{a} \rangle^2 - \langle \hat{b}^\dagger \rangle^2 - \langle \hat{b} \rangle^2], \tag{7.113}$$

$$\Delta v^2 = 1 + \langle \hat{a}^\dagger \hat{a} \rangle + \langle \hat{b}^\dagger \hat{b} \rangle + \langle \hat{a}\hat{b} \rangle + \langle \hat{a}^\dagger \hat{b}^\dagger \rangle - \langle \hat{a}\hat{b}^\dagger \rangle - \langle \hat{a}^\dagger \hat{b} \rangle$$
$$+ \langle \hat{a}^\dagger \rangle \langle \hat{a} \rangle + \langle \hat{b}^\dagger \rangle \langle \hat{b} \rangle + \langle \hat{a} \rangle \langle \hat{b} \rangle - \langle \hat{a}^\dagger \rangle \langle \hat{b}^\dagger \rangle - \langle \hat{a} \rangle \langle \hat{b}^\dagger \rangle - \langle \hat{a}^\dagger \rangle \langle \hat{b} \rangle$$
$$- \frac{1}{2}[\langle \hat{a}^{\dagger 2} \rangle + \langle \hat{a}^2 \rangle + \langle \hat{b}^{\dagger 2} \rangle + \langle \hat{a}^2 \rangle + \langle \hat{a}^\dagger \rangle^2 + \langle \hat{a} \rangle^2 + \langle \hat{b}^\dagger \rangle^2 + \langle \hat{b} \rangle^2]. \tag{7.114}$$

It is also possible to argue that several sufficient inseparability criteria of a composite state can be proposed for a continuous variable system. One such possibility is the logarithmic negativity that can be captured by

$$E_n = \max[0, -\log_2 V_s], \tag{7.115}$$

where V_s is the smallest eigenvalue of the symplectic matrix (see Sect. 2.4 and also [23]). One can infer from Eq. (7.115) that details can be explored by solving the eigenvalue equations for the symplectic spectrum of the covariance matrix of the partially transposed density operator since a two-party Gaussian state can be fully specified up to a local displacement by its covariance matrix.

One may recall that the smallest eigenvalue for a bipartite system can be expressed in the form

$$V_s = \left[\frac{\zeta - \sqrt{\zeta^2 - 4\det\Xi}}{2} \right]^{1/2}, \tag{7.116}$$

with $\zeta = \det A + \det B - 2\det C_{AB}$ in which A and B are the covariance matrices that describe each modes separately while C_{AB} represents the intermodal correlations (relate with Eq. (2.183)).

For a two-mode light, the 2X2 block form of a covariance matrix can be expressed as

$$\Xi = \begin{pmatrix} A & C_{AB} \\ C_{AB}^T & B \end{pmatrix}, \tag{7.117}$$

where

$$\varXi_{ij} = \frac{1}{2}\left(\langle \hat{X}_i \hat{X}_j \rangle + \langle \hat{X}_j \hat{X}_i \rangle\right) - \langle \hat{X}_i \rangle \langle \hat{X}_j \rangle \tag{7.118}$$

and the pertinent quadrature operators are defined as

$$\hat{X}_1 = \hat{a} + \hat{a}^\dagger, \qquad \hat{X}_2 = i(\hat{a}^\dagger - \hat{a}), \qquad \hat{X}_3 = \hat{b} + \hat{b}^\dagger, \qquad \hat{X}_4 = i(\hat{b}^\dagger - \hat{b}) \tag{7.119}$$

(compare with Eq. (2.176)). Note that the quadrature operators can be expressed in the same way as before, and so the entanglement criterion that follows can be adjusted.

With this information, Eq. (7.118) can be expanded as

$$\varXi = \begin{pmatrix} m & 0 & c & 0 \\ 0 & m & 0 & -c \\ c & 0 & n & 0 \\ 0 & -c & 0 & n \end{pmatrix} \tag{7.120}$$

in which

$$m = 1 + 2\langle \hat{a}^\dagger \hat{a} \rangle + \langle \hat{a}^2 \rangle + \langle \hat{a}^{\dagger 2} \rangle - [\langle \hat{a} \rangle^2 + \langle \hat{a}^\dagger \rangle^2 + 2\langle \hat{a} \rangle \langle \hat{a}^\dagger \rangle], \tag{7.121}$$

$$n = 1 + 2\langle \hat{b}^\dagger \hat{b} \rangle + \langle \hat{b}^2 \rangle + \langle \hat{b}^{\dagger 2} \rangle - [\langle \hat{b} \rangle^2 + \langle \hat{b}^\dagger \rangle^2 + 2\langle \hat{b} \rangle \langle \hat{b}^\dagger \rangle], \tag{7.122}$$

$$c = \langle \hat{a}\hat{b} \rangle + \langle \hat{a}^\dagger \hat{b}^\dagger \rangle + \langle \hat{a}\hat{b}^\dagger \rangle + \langle \hat{a}^\dagger \hat{b} \rangle - [\langle \hat{a} \rangle \langle \hat{b} \rangle + \langle \hat{a}^\dagger \rangle \langle \hat{b}^\dagger \rangle + \langle \hat{a} \rangle \langle \hat{b}^\dagger \rangle + \langle \hat{a}^\dagger \rangle \langle \hat{b} \rangle]. \tag{7.123}$$

It is also insightful underlining that a composite state is said to be entangled provided that $E_n > 0$, that is, a two-mode Gaussian composite state is designated as entangled if

$$E_n = -\log_2 V_s, \tag{7.124}$$

which entails that $\log_2 V_s$ to be negative, that is,

$$V_s < 1. \tag{7.125}$$

Equation (7.125) epitomizes a necessary and sufficient entanglement condition for a two-mode Gaussian state (compare this interpretation with what follows from Eq. (2.185)). In light of this assertion, one may realize that a bipartite continuous variable entanglement can be quantified by applying either Eq. (7.124) or (7.125), which is usually designated as criteria of logarithmic negativity, and the degree of the quantumness can be equated to how much V_s is closer to zero [26].

One may recall that quantities that are quadratic in the mode creation and annihilation operators can be used to study the sum and difference squeezing (see Sect. 7.1). With the observation that the quadratic operators and their uncertainties can be measured more comfortably than the operators, the corresponding entanglement can be practically manipulable by employing such representations.

With this in mind, it should be possible to construct various operators from the creation and annihilation operators such as

$$\hat{L}_1 = \hat{a}\hat{b}^\dagger + \hat{a}^\dagger\hat{b}, \qquad \hat{L}_2 = i(\hat{a}\hat{b}^\dagger - \hat{a}^\dagger\hat{b}), \qquad \hat{L}_3 = \hat{a}^\dagger\hat{a} + \hat{b}^\dagger\hat{b}. \tag{7.126}$$

Comparison reveals that $\hat{L}_1$ and $\hat{L}_2$ are related to the frequency difference quadratures $\hat{V}_{d1}$ and $\hat{V}_{d2}$, and $\hat{L}_3$ to the total photon number (compare also with Schwenger's angular momentum representation provided in Sect. 6.4.2). Following a straightforward algebra and applying Eq. (7.2), it may not be difficult to show that

$$\Delta L_1^2 + \Delta L_2^2 \geq 2[\langle \hat{a}^\dagger \hat{a} \rangle + \langle \hat{b}^\dagger \hat{b} \rangle]. \tag{7.127}$$

One may hence notice that the associated state would be entangled provided that

$$|\langle \hat{a}\hat{b}^\dagger \rangle| > \sqrt{\langle \hat{a}^\dagger \hat{a} \hat{b}^\dagger \hat{b} \rangle}. \tag{7.128}$$

One can also construct operators such as

$$\hat{L}_4 = \hat{a}^\dagger\hat{b}^\dagger + \hat{a}\hat{b}, \qquad \hat{L}_5 = i(\hat{a}^\dagger\hat{b}^\dagger - \hat{a}\hat{b}) \tag{7.129}$$

(see also Eq. (7.37)). Following the same line of argument leads to the observation that a composite state is entangled if

$$|\langle \hat{a}\hat{b} \rangle| > \sqrt{\langle \hat{a}^\dagger \hat{a} \rangle \langle \hat{b}^\dagger \hat{b} \rangle}. \tag{7.130}$$

This entanglement measure can be designated as Hillery-Zubairy (HZ) criterion [27]. Since the operators are already put in the normal order, Eq. (7.130) can also be expressed in terms of c-number variables associated with the normal ordering as

$$|\langle \alpha(t)\beta(t) \rangle| > \sqrt{\langle \alpha^*(t)\alpha(t) \rangle \langle \beta^*(t)\beta(t) \rangle}. \tag{7.131}$$

7.3.3 Continuous Variable Multipartite Entanglement

Entanglement among more than two quantum states can potentially be important resource in the quantum information processing since quantum teleportation, dense coding, telecloning, key distribution etc. schemes that encompass two sites can be extendible to arbitrary number of parties sharing multipartite entanglement. Since different classes of entanglement needs to be defined depending on how the system's density matrix could be partitioned, it might be legitimate to emphasize that the characterization of, let us say, tripartite entanglement can be more subtle than that of bipartite entanglement [28]. The classification ranges from fully inseparable, the density matrix is not separable for any grouping of the constituents, to fully separable. With the progress in continuous variable entanglement research still going on, the generation of more than two-party entanglement has garnered significant attention lately. Despite the anticipated challenge, quantification procedures for certain inequalities that can be applicable to continuous variable scenario are introduced based on various ways of manipulating the composite system's symmetry profiles. For example, a continuous variable tripartite entangled state has been constructed via combining three independent squeezed vacuum states that can be generated upon using a cascaded nonlinear interaction.

For three light modes that can be described by annihilation operators $\hat{a}_j$'s, the quadrature operators can be defined for each mode as

$$\hat{X}_j = \hat{a}_j + \hat{a}_j^\dagger, \qquad \hat{P}_j = -i(\hat{a}_j - \hat{a}_j^\dagger) \tag{7.132}$$

in which the Heisenberg uncertainty relation signifies

$$V(\hat{X}_j)V(\hat{P}_j) \geq 1, \tag{7.133}$$

where $V(.)$ is the variance of the pertinent parameters and $j = 1, 2, 3$. Based on the proposal that interference measurement has a potential to reveal a nonclassical feature, one can imagine a linear estimation of the quadrature $\hat{X}_i$ from the properties of the combined mode $j + k$ by making use of the parameters that can be optimized.

One may note that minimizing the root mean square error in the estimation leads to an optimal inferred variance:

$$V^{\text{inf}}(\hat{X}_i) = V(\hat{X}_i) - \frac{[V(\hat{X}_i, \hat{X}_j \pm \hat{X}_k)]^2}{V(\hat{X}_j \pm \hat{X}_k)}, \tag{7.134}$$

$$V^{\text{inf}}(\hat{P}_i) = V(\hat{P}_i) - \frac{[V(\hat{P}_i, \hat{P}_j \pm \hat{P}_k)]^2}{V(\hat{P}_j \pm \hat{P}_k)}. \tag{7.135}$$

(see Eq. (7.46) and also [29]). As highlighted in Sect. 7.2, the affirmation to the EPR conjecture can be acclaimed whenever

$$V^{\text{inf}}(\hat{X}_i)V^{\text{inf}}(\hat{P}_i) < 1. \tag{7.136}$$

Equation (7.136) is expected to demonstrate a sufficient tripartite entanglement condition without necessarily invoking assumption such as whether the states are Gaussian.

Since it is possible to set the expectation value of a Gaussian variable to zero, Eq. (7.134) can be rewritten for a Gaussian system as

$$V^{\text{inf}}(\hat{X}_i) = \langle \hat{X}_i^2 \rangle - \frac{[\langle \hat{N}_i \hat{N}_i \rangle + \langle \hat{N}_i \hat{N}_k \rangle]^2}{\langle \hat{X}_j^2 \rangle + \langle \hat{X}_k^2 \rangle + 2\langle \hat{X}_j \hat{X}_k \rangle}, \tag{7.137}$$

where a similar expression holds also for the momentum counterpart (see also Sect. 2.4). In case one utilizes mode i to infer properties of the combined mode $j+k$, one finds that there is a possibility of demonstrating the other three-mode form of the EPR conjecture (usually referred to as Olsen-Bradley-Reid (OBR) entanglement criteria) whenever

$$V^{\text{inf}}(\hat{X}_j \pm \hat{X}_k) V^{\text{inf}}(\hat{P}_j \pm \hat{P}_k) < 4, \tag{7.138}$$

where

$$V^{\text{inf}}(\hat{X}_j \pm \hat{X}_k) = V(\hat{X}_j \pm \hat{X}_k) - \frac{[V(\hat{X}_i, \hat{X}_j \pm \hat{X}_k)]^2}{V(\hat{X}_i)} \tag{7.139}$$

for the position quadrature and true also for momentum quadrature. One perhaps able to show that

$$V^{\text{inf}}(\hat{X}_j + \hat{X}_k) = \langle \hat{X}_j^2 \rangle + \langle \hat{X}_k^2 \rangle + 2\langle \hat{X}_j \hat{X}_k \rangle - \frac{[\langle \hat{X}_i \hat{X}_j \rangle + \langle \hat{X}_i \hat{X}_k \rangle]^2}{\langle \hat{X}_i^2 \rangle}. \tag{7.140}$$

Over the years, a set of other conditions that are sufficient to quantify a tripartite entanglement for a quantum state have been proposed. As illustration, by applying the quadrature operators already defined, the conditions usually dubbed as van Loock-Furusawa entanglement criteria are provided by inequalities:

$$V_{12} = V(\hat{X}_1 - \hat{X}_2) + V(\hat{P}_1 + \hat{P}_2 + g_3 \hat{P}_3) \geq 4, \tag{7.141}$$

$$V_{13} = V(\hat{X}_1 - \hat{X}_3) + V(\hat{P}_1 + g_2 \hat{P}_2 + \hat{P}_3) \geq 4, \tag{7.142}$$

$$V_{23} = V(\hat{X}_2 - \hat{X}_3) + V(g_1 \hat{P}_1 + \hat{P}_2 + \hat{P}_3) \geq 4, \tag{7.143}$$

where g_i's are arbitrary real numbers [30]. Note that the mean of a multi-mode state can be set to zero without loss of generality since the entanglement properties of a multiparty state are invariant under local phase space displacements (see Sect. 2.4). Equations (7.141)–(7.143) can be fairly interpreted as the violation of the first inequality ($V_{12} < 4$) leaves the possibility that mode 3 could be separated from modes 1 and 2 which is negated by violation of the second inequality and so on. So the system is said to be fully inseparable in case any two of these inequalities are violated: a genuine tripartite entanglement is guaranteed.

After that, upon rewriting the required variance for example as

$$V(\hat{X}_1 - \hat{X}_2) = \langle \hat{X}_1^2 \rangle + \langle \hat{X}_2^2 \rangle - 2\langle \hat{X}_1 \hat{X}_2 \rangle, \tag{7.144}$$

$$V(\hat{P}_1 + \hat{P}_2 + g_3 \hat{P}_3) = \langle \hat{P}_1^2 \rangle + \langle \hat{P}_2^2 \rangle + g_3^2 \langle \hat{P}_3^2 \rangle$$
$$+ 2[\langle \hat{P}_1 \hat{P}_2 \rangle + g_3(\langle \hat{P}_1 \hat{P}_3 \rangle + \langle \hat{P}_2 \hat{P}_3 \rangle)], \tag{7.145}$$

it is possible to explore the optimization of these criteria by applying the freedom allowed by the choice of g_i. Thus minimization with respect to g_i may suggest that

$$g_1 = -\frac{\langle \hat{P}_1 \hat{P}_2 \rangle + \langle \hat{P}_1 \hat{P}_3 \rangle}{\langle \hat{P}_1^2 \rangle}, \quad g_2 = -\frac{\langle \hat{P}_1 \hat{P}_2 \rangle + \langle \hat{P}_2 \hat{P}_3 \rangle}{\langle \hat{P}_2^2 \rangle}, \quad g_3 = -\frac{\langle \hat{P}_1 \hat{P}_3 \rangle + \langle \hat{P}_2 \hat{P}_3 \rangle}{\langle \hat{P}_3^2 \rangle}. \tag{7.146}$$

Once such optimization process has been performed, one can see that

$$V(\hat{P}_1 + \hat{P}_2 + g_3 \hat{P}_3) = V(\hat{P}_1 + \hat{P}_2) - \frac{[V(\hat{P}_3, \hat{P}_1 + \hat{P}_2)]^2}{V(\hat{P}_3)}, \tag{7.147}$$

where the right side expression is the inferred variance.

On account of earlier discussion, the optimized correlations can then be put in the form

$$V_{12} = V(\hat{X}_1 - \hat{X}_2) + V^{\text{inf}}(\hat{P}_1 + \hat{P}_2) \geq 4, \tag{7.148}$$

$$V_{13} = V(\hat{X}_1 - \hat{X}_3) + V^{\text{inf}}(\hat{P}_1 + \hat{P}_3) \geq 4, \tag{7.149}$$

$$V_{23} = V(\hat{X}_2 - \hat{X}_3) + V^{\text{inf}}(\hat{P}_2 + \hat{P}_3) \geq 4. \tag{7.150}$$

These criteria appear to have a similar form as the criteria of the bipartite entanglement, but with the actual variance $V(\hat{P}_j + \hat{P}_k)$ replaced by the inferred variance $V^{\text{inf}}(\hat{P}_j + \hat{P}_k)$ (compare Eq. (7.148) with DGCZ type entanglement criteria). Markedly, the violation of two out of three of these inequalities is sufficient to demonstrate a genuine tripartite entanglement.

It might be possible at some point to develop a single sufficient condition to detect a genuine tripartite entanglement from the combined quadrature variances. One may hence realize that to fully capture the inseparability of a three-party state, it is sufficient to satisfy

$$V\left(\hat{X}_i - \frac{\hat{X}_j + \hat{X}_k}{\sqrt{2}} \right) + V\left(\hat{P}_i + \frac{\hat{P}_j + \hat{P}_k}{\sqrt{2}} \right) < 4, \tag{7.151}$$

where the mode indices i, j, k ought to be different. It might also be good to stress that the multiparty entanglement criteria provided here do not rely on the assumption of Gaussian states, although optical states commonly produced in laboratory are Gaussian, and the theoretical classification of different types of multipartite entanglement thought to be simpler for Gaussian states.

The other possible strategy in quantifying entanglement in a multipartite system is to invoke pairwise inequalities to check for entanglement in every possible bipartite cut. The amount of the work required to perform the task however can grow enormously as the number of the subsystems increases. It is so desirable to have multipartite inequalities that allow one to check for overall entanglement in a multipartite system in straightforward and transparent manner without necessarily looking into each imaginable bipartite cut [31].

In line with this proposal, for a system that consists of n subsystems that can be described by the operator $\hat{A}_k$ on the Hilbert space of the kth subsystem, a state is said to be entangled if either of the two conditions

$$\left|\left\langle \prod_{k=1}^{n} \hat{A}_k \right\rangle\right| > \prod_{k=1}^{n} \left\langle (\hat{A}_k^\dagger \hat{A}_k)^{n/2} \right\rangle^{1/n} \tag{7.152}$$

or

$$\left|\left\langle \prod_{k=1}^{n} \hat{A}_k \right\rangle\right| > \frac{1}{n^{n/2}} \left\langle \left(\sum_{k=1}^{n} \hat{A}_k^\dagger \hat{A}_k \right)^{n/2} \right\rangle \tag{7.153}$$

is satisfied.

It should be possible to rewrite these equations for a tripartite ($n = 3$) case upon assuming the boson operators that describe the pertinent system to commute as

$$|\langle \hat{a}\hat{b}\hat{c}\rangle|^2 > \langle \hat{a}^\dagger \hat{a}\rangle \langle \hat{b}^\dagger \hat{b}\rangle \langle \hat{c}^\dagger \hat{c}\rangle, \tag{7.154}$$

$$|\langle \hat{a}\hat{b}\hat{c}\rangle|^{2/3} > \frac{\langle \hat{a}^\dagger \hat{a}\rangle + \langle \hat{b}^\dagger \hat{b}\rangle + \langle \hat{c}^\dagger \hat{c}\rangle}{3}. \tag{7.155}$$

The tripartite entanglement for a continuous variable system can be quantified by either Eq. (7.154) or (7.155) usually named as Hillery-Dund-Zheng (HDZ) entanglement criteria. Since the emerging inequalities do not necessarily impose upper bound, the main challenge is to obtain the correlation of the three operators and to figure out the degree of the quantumness.

One can also see for a bipartite system ($n = 2$) that the continuous variable entanglement exists if either

$$|\langle \hat{a}\hat{b}\rangle|^2 > \langle \hat{a}^\dagger \hat{a}\rangle \langle \hat{b}^\dagger \hat{b}\rangle \tag{7.156}$$

or

$$|\langle \hat{a}\hat{b}\rangle|^2 > \frac{\langle \hat{a}^\dagger \hat{a}\rangle + \langle \hat{b}^\dagger \hat{b}\rangle}{2} \tag{7.157}$$

is satisfied (compare Eq. (7.130) with (7.156)).

It might not be hard to conceive that a multipartite quantum correlation can be quantified in a variety of ways. One of these approaches is related to the concept of monogamy of entanglement—if two particles share a quantum state with a high quantum correlation, they cannot have a significant amount of quantum correlation with any third particle—of shared bipartite quantum entanglement in a multiparty quantum system [32]. A multiparty quantum correlation measure constructed via monogamy of a bipartite quantum correlation may add a benefit of providing information about the shareability of the bipartite quantum correlations. Multipartite quantum correlations can also be quantified for example with the help of a geometric measure [33]. It might also be required to establish a clear connection between the multiparty entanglement content of the multipartite quantum states with their ability to act as a basic block of circuity in quantum information protocols (see Chap. 11).

7.4   Quantum Discord

It should be clear by now that the idea of entanglement stems from the underlying principles of quantum theory, and believed to be a suitable resource for quantum implementation due to the embodied nonlocal character. Whereas separable states are often considered as purely classical since they do not violate Bell's inequalities and can be prepared using local operations. Given these facts, it might be imperative and valid to inquire whether a highly mixed state is useless for quantum information processing tasks. It might worth noting that recent advances indicate that they have a potential to demonstrate computational advantage without possessing much entanglement content. To elucidate this, one may begin with the well known understanding that two systems are designated as correlated if they happen to contain more information together than taken separately.[9] One may hence apprehend that entanglement is not the only aspect of the composite system with acclaimed quantum advantages.

One of the most appealing and significant approaches in this regard is based on the fact that certain equivalent classical correlations lead to different quantum versions due to the non-commutativity of the engaged operators. Since non-commutativity is one of the fundamental traits of quantum theory, it is envisaged

---

[9]Schrödinger's conjecture—which construed that entangled state provides more information about the total system than about subsystems—is quantifiable by a means of von Neumann entropy: the entropy of the subsystem would be greater than the entropy of the total system, that is, the subsystems of the entangled system may exhibit more disorder than the system as a whole; the phenomenon which never happens in the classical world [34].

that this difference can be exploited to characterize and quantify the quantum nature of the emerging correlations. Owing to the observation that two classically identical expressions of mutual information differ when the composite system exhibits quantum features, the non vanishing difference between quantum mutual information and the corresponding classical correlation is taken to epitomize quantum discord; which heralds the emerging nonclassicality. As a result, quantum discord that can also be linked to the internal correlations among parts of the system such as the information the observer can gain by probing part of a multipartite system controlled by another observer might be fancied to capture the quantum feature of the states of the system even more effectively than entanglement. In other words, since the separability of a density matrix does not automatically guarantee vanishing of quantum discord, one may comprehend that the absence of entanglement does not necessarily lead to classicality [35, 36].

7.4.1 Theoretical Background

The mutual information of two random variables can be interpreted as a quantity that measures the mutual dependence of the two variables. The classical mutual information of two discrete random variables can thus be expressed as

$$I(X:Y) = \sum_{y \in Y} \sum_{x \in X} p(x, y) \log \left(\frac{p(x, y)}{p(x) \, p(y)} \right), \tag{7.158}$$

where $p(x, y)$ is the joint probability distribution function of X and Y, whereas $p(x)$ and $p(y)$ are the marginal probability distribution functions of X and Y that can be defined as

$$p(x) = \sum_{y} p(x, y), \qquad p(y) = \sum_{x} p(x, y). \tag{7.159}$$

For a continuous function, Eq. (7.158) can be converted into a definite double integral of the form

$$I(X:Y) = \int_Y \int_X p(x, y) \log \left(\frac{p(x, y)}{p(x) \, p(y)} \right) dx \, dy. \tag{7.160}$$

Markedly, a mutual information measures the information that X and Y share: how far knowing one of these variables reduces the uncertainty about the other. For example, if X and Y are independent, knowing X does not give any information about Y, that is, their mutual information would be zero. Conversely, when X and Y are identical, information conveyed by X is shared with Y: knowing X would be tantamount to knowing the value of Y. It would hence be possible to concur as the mutual information quantifies the interdependence between the joint distribution of X and Y. The classical mutual information can thus be expressed in a more appealing manner as

$$I(X : Y) = H(p(x)) + H(p(y)) - H(p(x, y)), \tag{7.161}$$

where $H(.)$ denotes the Shannon entropy that can be interpreted as the ignorance about the random variable, and mathematically designated by

$$H(X) = -\sum_x p_{X=x} \log p_{X=x}. \tag{7.162}$$

The correlation between two random variables X and Y can also be measured by the mutual information expressed somehow in different forms:

$$J(X : Y) = H(X) - H(X|Y), \tag{7.163}$$

where

$$H(X|Y) = \sum_y p_{Y=y} H(X|Y = y) \tag{7.164}$$

is the conditional entropy of X given Y. Whereas the corresponding probability distribution can be derived from the joint probability distribution as

$$p_X = \sum_y p_{X,Y=y}, \qquad p_Y = \sum_x p_{X=x,Y}, \qquad p_{X|Y=y} = \frac{p_{X,Y=y}}{p_{Y=y}}, \tag{7.165}$$

where the right side expression is dubbed as Baye's rule.

One may note that the mutual information measures the average decrease of entropy via X when Y is found out. With the help of Baye's rule, one can see that

$$H(X|Y) = H(X, Y) - H(Y). \tag{7.166}$$

Inserting Eq. (7.166) into (7.163) leads to another classically equivalent expression for the mutual information:

$$I(X : Y) = H(X) + H(Y) - H(X, Y). \tag{7.167}$$

It might not be difficult to observe that $H(X)$ and $H(Y)$ are the marginal entropies, $H(X|Y)$ and $H(Y|X)$ are the conditional entropies and $H(X, Y)$ is the joint entropy of X and Y. Owing to the nonnegativity of the entropy, it is possible to notice that $H(X) \geq H(X|Y)$.

Intuitive consideration may also reveal that if entropy $H(X)$ is regarded as the measure of uncertainty about random variable, $H(X|Y)$ can then be interpreted as a measure of what Y does not inform about X or the amount of uncertainty remaining about X after Y is known, and so the right side of Eq. (7.167) can be read as the amount of uncertainty in X minus the amount of uncertainty in X that remains after Y is known: the amount of uncertainty in X removed by knowing Y. This might corroborate with intuitive definition of mutual information as the amount of information that knowing either variable tells us about the other. Note that $H(X|X) = 0$ for discrete case; and therefore, $H(X) = I(X : X)$, which leads to

$$I(X : X) \geq I(X : Y). \tag{7.168}$$

One can hence assert that a variable contains as much information about itself as any other variable can provide.

To extend this discussion to quantum realm, it might be appealing to define I and J for a pair of quantum systems delineated as system and apparatus for the time being [37]. The ingredients involved in the definition of I can be generated to deal with arbitrary quantum system by replacing the classical probability distribution by appropriate density matrix and the Shannon entropy by the von Neumann entropy:

$$H(S) = -Tr_S \hat{\rho}_S \log \hat{\rho}_S, \tag{7.169}$$

which leads to

$$I(S : A) = H(S) + H(A) - H(S, A), \tag{7.170}$$

where $H(S) + H(A)$ stands for the uncertainty of the system and apparatus treated separately, and $H(S, A)$ for the uncertainty of the composite system. The generalization for J however is not as automatic as for I since the conditional entropy $H(S|A)$ requires the state of the system to be known given the state of the apparatus. Such a statement would be ambiguous in the quantum theory until a set of states to be measured are selected.

It might be thus good to focus on the measurements of, let us say, parameter A defined by a set of one dimensional projectors $\{\Pi_j^A\}$, where the label j distinguishes various outcomes of the measurement. Note that the state of the system after the outcome pertinent to $\{\Pi_j^A\}$ has been detected is

$$\rho_{S|\Pi_j^A} = \frac{\hat{\Pi}_j^A \hat{\rho}_{S,A} \hat{\Pi}_j^A}{Tr_{S,A} \hat{\Pi}_j^A \hat{\rho}_{S,A}}, \tag{7.171}$$

where $p_j = Tr_{S,A} \hat{\Pi}_j^A \hat{\rho}_{S,A}$ is the probability (relate with Eq. (4.17)). Making use of such definition, $H(\rho_{S|\Pi_j^A})$ can be interpreted as the missing information about the system.

One may also see that the entropies $H(\rho_{S|\Pi_j^A})$ weighed by probability p_j yield the conditional entropy of the system given the measurement Π_j^A on the apparatus;

$$H\big(S|\{\Pi_j^A\}\big) = \sum_j p_j H\big(\rho_{S|\Pi_j^A}\big),$$

which leads to quantum generalization of the form

$$J(S:A)_{\{\Pi_j^A\}} = H(S) - H\big(S|\{\Pi_j^A\}\big), \tag{7.172}$$

and stands for the information gained about the system as a result of the measurement $\{\Pi_j^A\}$ on the apparatus.

7.4.2   Quantification of Quantum Discord

Since a generic bipartite quantum state is a hybrid object with both classical and quantum characteristics, and perceived to encode classical as well as quantum correlations via superposition, entanglement and mixing, the discussion on quantum discord is geared towards distinguishing these correlations, and then devising the way of exploiting them in quantum information processing [38]. With this in mind and owing to the possibility that two classically equivalent expressions of correlations lead to different quantum analogues, the forward quantum discord can be defined as

$$D_A(S:A)_{\{\Pi_j^A\}} = I(S:A) - J(S:A)_{\{\Pi_j^A\}}, \tag{7.173}$$

from which follows

$$D_A(S:A)_{\{\Pi_j^A\}} = H(A) - H(S:A) + H\big(S|\{\Pi_j^A\}\big). \tag{7.174}$$

One may see that quantum discord as defined is asymmetric[10] under the exchange of S and A since the definition of the conditional entropy encompasses measurement on one end (in our case the apparatus) that allows the observer to infer the state of the system, and leads to enhancement of entropy:

$$H\big(S|\{\Pi_j^A\}\big) \geq H(S, A) - H(A), \tag{7.175}$$

---

[10]There is a zoo of various quantum discord definitions and interpretations including symmetric discord and higher order extensions [39].

which implies for any measurement $\{\Pi_j^A\}$ that

$$D_A(S : A)_{\{\Pi_j^A\}} \geq 0. \tag{7.176}$$

One is often concerned with the set $\{\Pi_j^A\}$ that minimizes the discord for certain composite density of states. The process of minimization of the discord over possible measurement on apparatus amounts to finding the measurement that disturbs the overall quantum state; and at the same time, allows one to extract the most available information about the system. D_A large so quantitatively represents the situation in which a lot of information is missed or destroyed by the measurement on the apparatus alone, whereas D_A small can be taken as if almost all information about the system that exists in the system-apparatus correlations can be locally recoverable from the state of the apparatus.

Quantum discord can thus be equated to a measure of the information that cannot be extracted by reading the state of the apparatus alone; and as a result, taken as a good indicator of the quantum nature of the correlations. Even then, despite the available evidence for the relevance of quantum discord in describing nonclassical resources in information processing, there is no as such a direct criterion to verify the presence of discord for a given quantum state since optimization of the conditional entropy is required due to local measurement process.

Fortunately, under restricted Gaussian conditions, a closed analytical relation that can be utilized to quantify quantum discord for a two-mode Gaussian state can be outlined. To begin with, one needs to subtract the classical correlations from the total correlation to witness the quantum correlations between the subsystems. For a two party quantum system, the total correlation can then be calculated with the aid of

$$I(A : B) = S(A) - S(A|B) \tag{7.177}$$

in which

$$S(A|B) = S(A, B) - S(B), \tag{7.178}$$

where $S(X) = TrX\log X$ is the von Neumann entropy for arbitrary density matrix X and when the variable transformations $AB = \rho_{AB}$, $A = \rho_A = Tr_B(\rho_{AB})$ and $B = \rho_B = Tr_A(\rho_{AB})$ are adopted.

The assertion that the projective measurements on the subsystem remove the nonclassical correlations between the parts implies that after the measurement on a particular subsystem, quantum correlations would be destroyed (see Sect. 4.1). The quantity that describes the mutual information after the measurement on one of the subsystems is carried out thus can be defined as

$$I(A : B)\{\Pi_j^B\} = S(A) - S(A|\{\Pi_j^B\}), \tag{7.179}$$

where $S(A|\{\Pi_j^B\})$ is the conditional entropy after the measurement, $\{\Pi_j^B\}$ represents the complete set of one dimensional projectors and j's account for the different outcomes of the measurement.

To quantify the available quantum correlations, one may need to look for the maximum of $I(A : B)_{\{\Pi_j^B\}}$ since $I(A : B)_{\{\Pi_j^B\}}$ depends on the projector basis $\{\Pi_j^B\}$. It is now insightful expressing a measure of the total classical correlations between the two subsystems as

$$J(A : B) = S(A) - \min_{\Pi_j^B} S(A|\{\Pi_j^B\}). \tag{7.180}$$

The corresponding quantum discord can hence be expressed as

$$D_A(A : B) = I(A : B) - J(A : B). \tag{7.181}$$

Note that if all information can be obtained locally from B, this system exhibits zero quantum discord: the measurement on B does not alter the state.

With the observation that quantum discord can be valuable as a resource, and motivated by the expected importance and potential of light that exhibits quantum discord in emerging technology, it turns out to be imperative extending the discussion on quantum discord that has been restricted to a finite dimensional systems to infinite dimensional systems. In the context of quantum optics, one may require to outline the method of the optimization constrained by the measurement that preserves the Gaussian character of the state. In doing so, one may require to take the challenge head on in the hope that a two-mode light can exhibit applicable quantum aspect that can be captured by quantum discord. The discussion in this section is however limited to a bipartite system that can be described by a two-mode Gaussian states.

The idea of quantum discord so grows out of the fact that the quantum versions of the mutual information of a bipartite composite state can be represented in two nonequivalent ways, that is, the straightforward quantization of,

$$I(\rho_{AB}) = S(\rho_A) + S(\rho_B) - S(\rho_{AB}) \text{ and } J_A(\rho_{AB}) = S(\rho_A) - S(\rho_{A|B}), \tag{7.182}$$

where

$$\hat{\rho}_{A|i} = \frac{Tr_B(\hat{\rho}_{AB}\hat{\Pi}_i)}{Tr_{A,B}(\hat{\rho}_{AB}\hat{\Pi}_i)} \tag{7.183}$$

is the conditional entropy in which $\sum_j \hat{\Pi}_j = \hat{I}$, would not be the same. Quantization based on the conditional entropy or the extractable information involves the conditional state of the subsystem after a measurement is performed on the other. This consideration has mainly three relevant consequences: the symmetry between the two subsystems would be broken; the conditional entropy depends on the choice of the measurement; the resulting expression would be different from the mutual information.

With this information, the quantum analog of the conditional entropy can be expressed as

$$S_{\Pi_i}(A|B) = \sum_i p_i S(\rho_{A|i}),$$ (7.184)

from which the one way classical correlation maximized overall possible measurements follows

$$J_A(\rho_{AB}) = S(\rho_A) - \inf_{\Pi_i} S_{\Pi_i}(A|B).$$ (7.185)

The quantum discord A can thus be written as

$$D_A(\rho_{AB}) = S(\rho_A) - S(\rho_{AB}) + \inf_{\Pi_i} S_{\Pi_i}(A|B).$$ (7.186)

States with zero discord represent a classical probability distribution embedded in the quantum system while a positive discord constitutes the quantumness.

To optimize the measurement, one can begin with a well established fact that a two-mode Gaussian state would be fully specified up to a local displacement by its covariance matrix of the form

$$\sigma_{AB} = \begin{pmatrix} \alpha & \gamma \\ \gamma^T & \beta \end{pmatrix},$$ (7.187)

where

$$\alpha = \text{diag}(m, m), \qquad \beta = \text{diag}(n, n), \qquad \gamma = \text{diag}(c, d)$$ (7.188)

(see also Eqs. (2.182) and (7.117)).

One can also recall that the covariance matrix designates a physically allowable state when the symplectic eigenvalues,[11]

$$v_{\pm}^2 = \frac{\Delta \pm \sqrt{\Delta^2 - 4D}}{2}$$ (7.189)

---

[11]This covariance matrix is the same as the one given by Eq. (7.117) except the change of variable in which the smallest symplectic eigenvalue V_s can be obtained from v_- by time reversal operation, that is, by replacing C with $-C$ (see Eq. (7.116)).

with $\Delta = I_A + I_B + 2C$, are greater or equal to zero in which

$$I_A = \det \alpha, \qquad I_B = \det \beta, \qquad C = \det \gamma, \qquad D = \det \sigma_{AB} \qquad (7.190)$$

(compare with Eqs. (2.183) and (7.116)).

Since J_A and D_A are invariant under local unitary transformation, a two-mode Gaussian state is envisaged to demonstrate quantum discord when the conditional entropy is restricted to generalized Gaussian positive operator valued measurements on B, which are executable in the realm of linear optics with the help of homodyne detection (see Sects. 4.1 and 4.5). With this proposal, one can write

$$D_A(\rho_{AB}) = S(\rho_A) - S(\rho_{AB}) + \inf_{\Pi_B(\eta)} \int d\eta \; p_B(\eta) S(\rho_{A\eta}). \qquad (7.191)$$

The Gaussian measurement $\{\Pi_B(\eta)\}$ on the subsystem B can also be expressed as

$$\Pi_B(\eta) = \frac{1}{\pi} \hat{W}_B(\eta) \Pi_B^0 \hat{W}_B^\dagger(\eta), \qquad (7.192)$$

where $\hat{W}_B(\eta) = \exp[\eta \hat{b}^\dagger - \eta^* \hat{b}]$ is the Weyl operator in which $\hat{b} = (\hat{x}_B + i\hat{p}_B)/\sqrt{2}$ and Π_B^0 is the density matrix of a mixed single-mode Gaussian state.

Confining to states $\{\Pi_B^0\}$ that are pure (single-mode Gaussian state whose covariance matrix is denoted as σ_0), the conditional state $\rho_{A|\eta}$ of the subsystem A after carrying out the measurement $\{\Pi_B(\eta)\}$ on B would have a covariance matrix independent of the measurement outcome and given by [40]

$$\varepsilon = \alpha - \gamma(\beta + \sigma_0)^{-1}\gamma^T. \qquad (7.193)$$

The von Neumann entropy of n-mode Gaussian state with a covariance matrix σ on the other hand can be expressed as

$$S(\sigma) = \sum_{i=1}^{N} g(v_i), \qquad (7.194)$$

where v_i's are the symplectic eigenvalues and

$$g(x) = \left(\frac{x+1}{2}\right) \log \left(\frac{x+1}{2}\right) - \left(\frac{x-1}{2}\right) \log \left(\frac{x-1}{2}\right) \qquad (7.195)$$

(see also Eq. (2.186)).

The general strategy to get a closed formula is to minimize det ε overall covariance matrices σ_0. So the minimization relying on different approaches for a general two-mode Gaussian state σ_{AB} leads to

$$\inf_{\sigma_0} \det \varepsilon$$

$$= \begin{cases} \frac{[2C^2+(I_B-1)(D-I_A)+2|C|\sqrt{C^2+(I_B-1)(D-I_A)}](D-I_AI_B)^2}{(I_B-1)^2} & \leq (1+I_B)C^2(I_A+D); \\ \frac{I_AI_B-C^2+D-\sqrt{C^4+(D-I_AI_B)^2-2C^2(I_AI_B+D)}}{2I_B}; & \text{otherwise.} \end{cases}$$

$$(7.196)$$

The first case with which we are interested corresponds to a more general measurement where the projection measurement is expected to lead to unbalanced finite variances of the quadrature operators such as $\hat{x}_B$ and $\hat{p}_B$. There are a notable class of states that satisfy this condition including the pure states characterizable by $c = -d$ for which the conditional entropy can be minimized by heterodyne measurements.

One may note that it is based on such observation that the Gaussian quantum discord is believed to be feasible in the regime of linear optics. Besides, for every entangled state, it might not be hard to conceive that the quantum discord is strictly positive since $S(\rho_B) > S(\rho_{AB})$. It then follows that two-mode Gaussian states with zero Gaussian quantum discord are product states with no correlations at all. Correlated two-mode Gaussian states can exhibit nonclassical correlations that can be certified by the presence of a nonzero quantum discord. One may consider at some point the special case for which $c = -d$ and ρ_0 is a generic zero mean Gaussian state (it is equivalent to assuming the initial state to be in the vacuum state) with matrix $(\sigma_0)_{11} = \alpha$, $(\sigma_0)_{22} = \beta$, $(\sigma_0)_{21} = (\sigma_0)_{12} = \gamma$. In this case, the quantum discord turns out to be

$$D_A(\rho) = g(\sqrt{I_B}) - g(v_-) - g(v_+) + \inf_{\sigma_0} g(\sqrt{\det \varepsilon}). \qquad (7.197)$$

To expedite the minimization, one can start with the fact that the entropy of Gaussian states depends on the covariance matrix but the covariance matrix σ_0 of the conditional state does not depend on the outcome of the measurement itself. The minimum of the mismatch $(I(\rho) - J(\rho))$ would be obtained for $\alpha = \beta = 1/2$ and $\gamma = 0$, that is, when the covariance matrix of the measurement is the identity which is equivalent to the case when initially there is a vacuum state. Taking this into account, and upon setting $\sigma_0 \longrightarrow \hat{I}/2$, the Gaussian discord for the generic bipartite system can be expressed in terms of the symplectic values and pertinent correlation [36] as

$$D_A(\rho) = g(\sqrt{I_B}) - g(v_-) - g(v_+) + g\left(\frac{\sqrt{I_A} + 2\sqrt{I_AI_B} + 2C}{1+2\sqrt{I_B}}\right). \qquad (7.198)$$

Notice that one can pass from A discord to B upon swapping I_A with I_B.

7.5   Quantum Steering

Suppose there is no knowledge of either the state of the particle one of the experimenter, let us say, Alice holds or the measurement she performs. But what one knows is that she can select to perform one measurement out of a set of choices each of which has certain number of possible outcomes. Imagine that we have a complete knowledge of the measurement her companion (Bob) performs, that is, he has performed state tomography, and so can acquire quantum description of his system. Suppose Bob can accept that he and Alice share entanglement even when he does not trust her measurement. Bob can rectify his judgment by ruling out the possibility that Alice can prepare, and send a message to him by utilizing her knowledge of announcing fabricated measurement that she expects to be correlated with his results. The intended protocol to ensure this proposal then requires Alice and Bob to compare results from rounds of local measurements on each pair of quantum states. Pertaining to the presumption that Bob does not trust Alice, there is a possibility that a dishonest Alice may try to cheat. The admissible way for the honest Alice to distinguish herself as such is by demonstrating her ability to steer Bob's state, notwithstanding the possibility that the dishonest Alice may still employ powerful cheating strategies that would appear to Bob indistinguishable from loss. To distinguish between the dishonest Alice and a natural detection loophole, Bob is required to devise a dependable strategy.

To begin with, Bob can rule out other possibilities and verify that he and Alice must be sharing entanglement when the measurement correlations across complementary observable is found to be sufficiently high. In regards to such measurement procedure, one of the important things one should do is to look into how much one of the two remotely entangled parties can alter the state of the other. Quantum (EPR) steering[12] as a result refers to the situation in which when one of the parties, let us say, Alice who shares entangled state with Bob can remotely modify the state of his subsystem in such a way that it would be impossible if their systems were classically correlated. Quantum steering can also be taken as quantum phenomenon that allows one party to change (or steer) the states of the system that belongs to the distant party via manipulating their shared entanglement without necessarily requiring a physical interaction between the subsystems.

Suppose Alice wants to convince Bob who does not trust her that she shares entangled state with him. To be convinced, assume that Bob demands from Alice to remotely prepare a collection of states of his subsystems. Then imagine that Alice performed the requested measurement unknown to Bob, and communicated the results to him. Based on the nature of quantum measurement and by looking at the conditional states prepared by Alice, it is expected that Bob can ascertain by using the notion of quantum steering whether they come from measurement on entangled

---

[12]The idea of quantum steering was introduced by Schrödinger in response to the suggestion forwarded by Einstein et al. that quantum mechanics lacks physical reality in its elements [10, 11, 41] (for corresponding experimental demonstration; see for instance [42]).

state. So EPR steering should be demonstrated by a simple test, that is, it might be sufficient to consider two measurements with two outcomes for Alice and prepare a collection of four states for Bob where the measured uncertainties violate certain inequality. Steering inequalities are useful not only because they assist in witnessing entanglement without performing a complete state tomography, but they are also capable of verifying entanglement between two parties even when the measurement of one of the parties is not trusted.

It is a well established fact that the joint probability for obtaining the measurement outcome a for Alice and b for Bob can be expressed as

$$P(a, b) = \sum_j P(j) P(a|j) P_Q(b|j), \tag{7.199}$$

where $P(j)$ is the distribution of the hidden variable j and Q denotes that the Bob's part of the system is governed by quantum principle. Even though Alice cannot affect Bob's local measurement result as presumed,

$$P(b) = \sum_a P(a, b) = \sum_j P(j) P_Q(b|j), \tag{7.200}$$

which otherwise violates the no-signaling condition, she can prepare states belonging to a particular ensemble with probability $P(a) = \sum_b P(a, b)$ by measuring her own state, and can obtain the outcome a.

But Bob might be contemplating if Alice could possibly deceive him by using her knowledge of stochastic map from j to a ($P(a|j)$) known to Bob. The probability for Bob's getting the outcome b conditioned on Alice's outcome can be expressed by

$$P(b|a) = \frac{1}{P(a)} \sum_j P(j) P(a|j) P_Q(b|j). \tag{7.201}$$

The idea is then: if Bob failed to find such a map, that is, in case his measurement result cannot be explained by the outlined model, he should admit that the shared state is truly entangled.

It might be worthy noting that the more general condition can be related to the uncertainty relation based on inferred variance. The bipartite system therefore can be proclaimed as EPR steerable when the inequality

$$V^{\text{inf}}(X_b|X_a) V^{\text{inf}}(P_b|P_a) \geq \frac{1}{2} \tag{7.202}$$

is violated [29].

One may resort at some point to discuss the steerability of a bipartite continuous variable system for which the idea of quantum steering was originally introduced, where the focus is on the mixed multi-mode bipartite Gaussian states. To this effect, consider a generic Gaussian $(n + m)$-mode state of a bipartite system composed of subsystem A of n-modes and subsystem B of m-modes. For each mode j that belongs to A (B), one may note that the quadratures can be grouped as

$$\hat{R} = (\hat{x}_1^A, \hat{p}_1^A, \ldots, \hat{x}_n^A, \hat{p}_n^A, \hat{x}_1^B, \hat{p}_1^B, \ldots, \hat{x}_m^B, \hat{p}_m^B)^T \qquad (7.203)$$

(see also Eq. (2.169)), where each pair satisfies the relations (2.170). Recall that the Gaussian state that can be designated by a density operator is fully specified up to a local displacement by its covariance matrix epitomized by Eq. (2.176). Recall also that every covariance matrix that denotes a physical quantum system has to satisfy the bona fide condition (2.178).

With such description of continuous Gaussian variable, quantum steerability can be perceived as a set of measurements on Alice's variable in such a way that a bipartite state ρ_{AB} is $A \rightarrow B$ steerable, if it is not possible to express the joint probability as

$$P(r_A, r_B | R_A, R_B, \rho_{AB}) = \sum_j \rho_j \rho(r_A | R_A, j) P(r_B | R_B, \rho_j), \qquad (7.204)$$

where R_A is the element on the measurement on A and R_B on B with respect to r_A and r_B, and $P(r_B | R_B, \rho_j)$ is the probability distribution subject to extra condition of being evaluated on the quantum state ρ_j: the complete knowledge of Bob's devices but not of Alice's which is required to quantify the steering.

To witness entanglement steerability, at least one measurement pair of R_A and R_B should violate Eq. (7.204) when ρ_j is fixed across the measurements. One may hence insist that the measure or the degree of entanglement steerability should quantify how much the correlations of a given quantum state depart from this condition. Besides, since the manifestation of these correlations is expected to be realized by the violation of a suitable EPR steering criterion, one can get a quantitative estimation of the degree of steerability by evaluating the maximum violation of the chosen criterion.

One can also argue that a Gaussian state that can be described by a density operator $\hat{\rho}_{AB}$ is $A \rightarrow B$ steerable by Alice's measurement if the Bona fide condition,

$$B - C_{AB}^T A^{-1} C_{AB} + i\Omega_A \geq 0, \qquad (7.205)$$

is violated (related with Eq. (7.117)), and the corresponding measure of a Gaussian $B \rightarrow A$ steerability can be obtained by swapping the role of A with B. The quantum steering witnessed in this form can then be related to a bipartite entanglement quantifiable via logarithmic negativity and quantum discord.

With the help of the symplectic eigenvalues (see Eqs. (7.187) and (7.189)), the measure of a steerability for a continuous variable Gaussian state can be expressed as

$$S_{A \to B(B \to A)} = \sqrt{\frac{\det \sigma_{AB}}{\det \alpha \ (\beta)}}, \qquad (7.206)$$

where $S_{A \to B \ (B \to A)} < 1$ when quantum steering $(A \to B \ (B \to A))$ is admissible [43].

References

1. R.W. Boyd, *Nonlinear Optics* (Elsevier, Amsterdam, 2008)
2. M. Giustina et al., Nature **497**, 227 (2013); B. Hensen et al., ibid. **526**, 682 (2015); L.K. Shalm et al., Phys. Rev. Lett. **115**, 250402 (2015)
3. M.C. Teich, B.E.A. Saleh, Quant. Opt. **1**, 153 (1989)
4. H. Fan, G. Yu, Phys. Rev. A **65**, 033829 (2002); H.Y. Fan, N.Q. Jiang, J. Opt. Phys. **6**, 238 (2004); F. Wen et al., Sci. Rep. **6**, 25554 (2016)
5. X.X. Xu, Int. J. Theor. Phys. **51**, 2056 (2012)
6. C.K. Hong, L. Mandel, Phys. Rev. Lett. **54**, 323 (1985); D.K. Giri, P.S. Gupta, J. Mod. Opt. **52**, 1769 (2005); S. Tesfa, ibid. **54**, 1759 (2007)
7. P. Marian, Phys. Rev. A **44**, 3325 (1991)
8. M. Hillery, Opt. Commun. **62**, 135 (1987), Phys. Rev. A **36**, 3796 (1987); D.K. Giri et al., Opt. Quant. Elec. **42**, 215 (2010)
9. M. Hellery, Phys. Rev. A **40**, 3147 (1989); D.M. Truong, H.T.X. Nguyen, A.B. Nguyen, Int. J. Theor. Phys. **53**, 899 (2014); K.K. Mishra et al., Opt. Quant. Elec. **22**, 186 (2020)
10. A. Einstein, B. Podolsky, N. Rosen, Phys. Rev. **47**, 777 (1935)
11. M.D. Reid et al., Rev. Mod. Phys. **81**, 1727 (2009)
12. A. Aspect, J. Dalibard, G. Roger, Phys. Rev. Lett. **49**, 1804 (1982); M. Genovese, Phys. Rep. **413**, 319 (2005); D.F. Walls, G.J. Milburn, *Quantum Optics*, 2nd edn. (Springer, Berlin, 2008); H. Kwon, H. Jeong, Phys. Rev. A **88**, 052127 (2013); R.B. Griffiths, Phys. Rev. A **101**, 022117 (2020)
13. J.S. Bell, Phys. Phys. Fizika **1**, 195 (1965); N. Brunner et al., Rev. Mod. Phys. **86**, 419 (2014)
14. H. Buhrman et al., Rev. Mod. Phys. **82**, 665 (2010); S. Pironio et al., Nature **464**, 1021 (2010); S. Bravyi, D. Gosset, R. König, Science **362**, 308 (2018)
15. J.F. Clauser et al., Phys. Rev. Lett. **23**, 880 (1969); P. Meystre, M. Sargent III, *Elements of Quantum Optics*, 4th edn. (Springer, Berlin, 2007)
16. Z.Y. Ou et al., Phys. Rev. Lett. **68**, 3663 (1992); C. Vitelli et al., **85**, 012104 (2012)
17. K. Banaszek, K. Wodkiewicz, Phys. Rev. A **58**, 4345 (1998); C.Y. Park, H. Jeong, Phys. Rev. A **91**, 042328 (2015)
18. Z.B. Chen et al., Phys. Rev. Lett. **88**, 040406 (2002)
19. R. Horodecki et al., Rev. Mod. Phys. **81**, 865 (2009); O. Guhne, G. Toth, Phys. Rep. **474**, 1 (2009)
20. W.K. Wootters, Quant. Inf. Comp. **1**, 24 (2001); A.A. Klyachko, B. Öztop, A.S. Shumovsky, Phys. Rev. A **75**, 032315 (2007)
21. C.H. Bennett et al., Phys. Rev. A **54**, 3824 (1996); S. Hill, W.K. Wootters, Phys. Rev. Lett. **78**, 5022 (1997); W.K. Wootters, ibid. **80**, 2245 (1998); G. Giedke et al., ibid. **91**, 107901 (2003)
22. R. Simon, Phys. Rev. Lett. **84**, 2726 (2000)
23. G. Vidal, R.F. Wener, Phys. Rev. A **65**, 032314 (2002); G. Adesso, A. Serafini, F. Illuminati, ibid. **70**, 022318 (2004); J. Fiurasek, N.J. Cerf, Phys. Rev. Lett. **93**, 063601 (2004)

24. L.M. Duan et al., Phys. Rev. Lett. **84**, 2722 (2000)
25. S. Tesfa, Phys. Rev. A **74**, 043816 (2006)
26. S. Tesfa, J. Phys. B **42**, 215506 (2009)
27. M. Hillery, M.S. Zubairy, Phys. Rev. Lett. **96**, 050503 (2006); M. Hillery, M.S. Zubairy, Phys. Rev. A **74**, 032333 (2006)
28. M.K. Oslen, A.S. Bradley, Phys. Rev. A **74**, 063809 (2006)
29. M.D. Reid, Phys. Rev. A **40**, 913 (1989); K. Dechoum et al., ibid. **70**, 053807 (2004)
30. P. van Loock, A. Furusawa, Phys. Rev. A **67**, 052315 (2003)
31. M. Hillery, H.T. Dund, H. Zheng, Phys. Rev. A **81**, 062322 (2010)
32. A.K. Ekert, Phys. Rev. Lett. **67**, 661 (1991); V. Coffman, J. Kundu, W.K. Wootters, Phys. Rev. A **61**, 052306 (2000); G. Adesso, F. Illuminati, New J. Phys. **8**, 15 (2006); T.J. Osborne, F. Verstraete, Phys. Rev. Lett. **96**, 220503 (2006); J. S. Kim, Phys. Rev. A **85**, 062302 (2012)
33. A. Sen(De), U. Sen, Phys. Rev. A **81**, 012308 (2010); R. Nepal et al., ibid. **87**, 032336 (2013)
34. R. Horodecki, P. Horodecki, Phys. Lett. A **194**, 147 (1994)
35. H. Ollivier, W.H. Zurek, Phys. Rev. Lett. **88**, 017901 (2001)
36. P. Giorda, M.G.A. Paris, Phys. Rev. Lett. **105**, 020503 (2010); G. Adesso, A. Datta, ibid. **105**, 030501 (2010); F.F. Fanchini, L.K. Castelano, A. O. Caldeira, New J. Phys. **12**, 073009 (2010); S. Tesfa, Opt. Commun. **285**, 830 (2012)
37. N.J. Cerf, C. Adami, Phys. Rev. Lett. **79**, 5194 (1997)
38. S. Lou, Phys. Rev. A **77**, 042303 (2008); B. Dakic et al., Nat. Phys. **8**, 666 (2012)
39. C.C. Rulli, M.S. Sarandy, Phys. Rev. A **84**, 042109 (2011); A. Bera et al., Rep. Prog. Phys. **81**, 024001 (2018); C. Radhakrishnan, M. Lauriere, T. Byrnes, Phys. Rev. Lett. **124**, 110401 (2020)
40. A.S. Holevo, R.F. Werner, Phys. Rev. A **63**, 032312 (2001); G. Giedke, J.I. Cirac, ibid. **66**, 032316 (2002)
41. E. Schrödinger, Proc. Camb. Phil. Soc. **31**, 555 (1935); H.M. Wiseman, S.J. Jones, A.C. Doherty, Phys. Rev. Lett. **98**, 140402 (2007); E. Cavalcanti et al., Phys. Rev. A **80**, 032112 (2009); P. Skrzypczyk, M. Navascues, D. Cavalcanti, ibid. **112**, 180404 (2014); D. Cavalcanti, P. Skrzypczyk, Rep. Prog. Phys. **80**, 024001 (2017); R. Uola et al., Rev. Mod. Phys. **92**, 015001 (2020)
42. S. Wollmann, R. Uola, A.C.S. Costa, Phys. Rev. Lett. **125**, 020404 (2020); H. Yang et al., Phys. Rev. A **101**, 022324 (2020)
43. B. Wittmann et al., New J. Phys **14**, 053030 (2012); A.J. Bennet et al., Phys. Rev. X **2**, 031003 (2012); S. Steinlechner et al., Phys. Rev. A **87**, 022104 (2013); Q.Y. He, Q.H. Gong, M.D. Reid, Phys. Rev. Lett. **114**, 060402 (2015); I. Kogias et al., ibid., 060403 (2015); I. Kogias, G. Adesso, J. Opt. Phys. B **32**, A27 (2015)

Sources and Illustrations of Nonclassical Light

With the dawning of the prospect of realizing quantum information processing in the regime of continuous variable, the need for generating a bright light with a robust quantum feature is growing with a faster rate than ever before [1]. Various mechanisms that have a potential to generate a nonclassical light are explored over the years where the main sources of light with nonclassical attributes are related either to nonlinear optical process such as a crystal pumped with an external coherent radiation or linear optical process that can be realized by a correlated emission of multi-level atoms (see Chap. 6 and also [2]). The nonclassicality in such cases can be traced to optical coherence induced while the energy levels of the source atoms are coupled. This can also be linked to the interaction of radiation with a continuum medium in a nonlinear fashion as in the Kerr medium.

At low intensities, for typical non-laser source, properties of the material medium remain independent of the intensity of illumination in which the superposition principle holds true; and so the light waves can pass through the material medium or reflected from the boundaries and interfaces without undergoing interaction. A laser source however can provide a light with high intensity that has a capacity to modify the optical properties of the material medium. Due to the emerging momentum and energy exchange, the involved light waves are expected to interact with each other, and the superposition principle would no longer valid. One perhaps needs to stress that such interaction of light waves can give rise to the generation of optical fields at new frequencies that include optical harmonics of incident radiation or sum or difference frequency signals, and lead to the realization of optical schemes such as parametric oscillation and multi-wave mixing (see Sect. 8.1).

© The Author(s), under exclusive license to Springer Nature Switzerland AG 2020
S. Tesfa, *Quantum Optical Processes*, Lecture Notes in Physics 976,
https://doi.org/10.1007/978-3-030-62348-7_8

It is common knowledge that an optical phenomenon is said to be nonlinear[1] when the response of the medium to the external electric field depends in a nonlinear manner on the strength of the electric field (see Sect. 5.6 and also [3]). To describe optical nonlinearity, one may begin with how the polarization of the material medium depends on the strength of the applied electric field:

$$P(t) = \chi^{(1)} E(t) + \chi^{(2)} E^2(t) + \chi^{(3)} E^3(t) + \ldots, \tag{8.1}$$

where $\chi^{(n)}$ is designated as the nth order nonlinear optical susceptibility (see also Eq. (5.120)). Even though the nonlinear susceptibility generally depends on the frequency of the applied external field, we opt to consider an instantaneous response in which the susceptibility is taken to be constant of frequency.

Imagine the electric field that consists of two distinct frequency components and incident upon a nonlinear optical medium;

$$E(t) = E_1 e^{-i\omega_1 t} + E_2 e^{-i\omega_2 t} + E_1^* e^{i\omega_1 t} + E_2^* e^{i\omega_2 t}. \tag{8.2}$$

While confining to the second-order contribution, the nonlinear polarization given by Eq. (8.1) can be rewritten as

$$P^{(2)}(t) = \chi^{(2)} \left[E_1^2 e^{-i2\omega_1 t} + E_2^2 e^{-i2\omega_2 t} + 2E_1 E_2 e^{-i(\omega_1 + \omega_2)t} \right.$$

$$\left. + 2E_1 E_2^* e^{-i(\omega_1 - \omega_2)t} + c.c. + 2(E_1 E_1^* + E_2 E_2^*) \right], \tag{8.3}$$

where $c.c.$ stands for the complex conjugate of the preceding terms. The electric field in a more general case can be taken as a traveling wave in which a typical electric field at position $\mathbf{r}$ takes the form

$$E(\mathbf{r}, t) = E e^{i(\mathbf{k} \cdot \mathbf{r} - \omega t)} + c.c., \tag{8.4}$$

where $k = \eta(\omega)\omega/c$ stands for the wave number, c for the velocity of light and $\eta(\omega)$ for the index of refraction of the medium at angular frequency ω.

With this characterization, the second-order polarization at angular frequency $\omega_3 = \omega_1 \pm \omega_2$ can be written as

$$P^{(2)}(\mathbf{r}, t) \propto E_1 E_2 e^{i[(\mathbf{k}_1 + \mathbf{k}_2) \cdot \mathbf{r} - \omega_3 t]} + c.c. \tag{8.5}$$

Equation (8.5) indicates that at position $\mathbf{r}$, within the nonlinear medium, the oscillating second-order polarization can radiate at angular frequency ω_3 with a

---

[1]Nonlinear optical effect belongs to a broader class of electromagnetic phenomena described within the framework of macroscopic Maxwell equations that serves to identify and classify nonlinear phenomena in terms of the relevant nonlinear optical susceptibilities or nonlinear terms in the induced polarization, and governs the nonlinear optical propagation effects.

wave number $k_3 = \eta(\omega_3)\omega_3/c$. In this situation, a constructive interference can happen, and so a field of frequency ω_3 with higher or lower intensity can be generated provided that $\mathbf{k}_3 = \mathbf{k}_1 \pm \mathbf{k}_2$; the phase matching condition. One may infer that it is the manipulation of the polarization in terms of the frequency that leads to light with quantum properties.

It is also found to be feasible to generate a light with nonclassical properties when multi-level atoms interact with a multi-mode radiation (see Sect. 6.5 and also [4]). In view of this suggestion, we seek to examine a correlated emission laser as an example of a linear source of light with quantum features. With such understanding and the intention of developing the procedure one can use while analyzing the nonclassical aspects of light, the techniques of quantifying the nonclassical features of radiation by applying the master equation, quantum Langevin equation, stochastic differential equation and the linearization procedure would be provided in Sect. 8.2. Aiming at gaining some feeling for the nature of the pertinent quantum features, some but limited number of examples are also included.

8.1   Sources of Light with Quantum Features

Since the light generated by conventional sources such as the sun and electric bulb invariably loses intensity and visibility, it would not be suitable for experiment, that is, in many interferometric settings, the coherent light generated by a laser source is preferred. Even then, the coherent light is found to be not suitable in case one aspires to demonstrate certain quantum tasks. Evidently, the need for generating a robust and bright light that possesses quantum character is steadily growing with the advance in continuous variable quantum implementations (see Chaps. 9, 11 and 12). On account of the observation that the coherent superposition is responsible for inducing quantum features, various physical schemes that have a potential to generate a nonclassical light would thus be explored.

8.1.1   Optical Parametric Oscillator

A parametric oscillator is mainly the same as conventional harmonic oscillator except for some oscillating parameters such as frequency or amplitude; in which, let us say, a child by periodically squatting the harmonic oscillator can increase the amplitude of the swing of the oscillation. Note that it is after the oscillator starts the motion that the swing can be parametrically driven by alternately squatting at key points in the swing or changing the center of mass of the child. So the process of inflicting a regular disturbance on the otherwise smoothly oscillating system changes the moment of inertia of the swing, which may lead to the possibility of quickly attaining large amplitudes. Analogous to this, if a highly reflective mirrors are placed on either sides of a nonlinear medium to form an optical resonator as shown in Fig. 3.1, the oscillation of the cavity modes occurs due to parametric amplification—a device commonly dubbed as optical parametric oscillator [5]. One

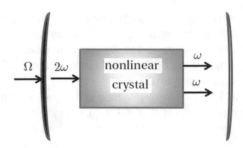

Fig. 8.1 Schematic description of a degenerate parametric oscillator that can be realized by placing a nonlinear crystal inside an optical resonator in which a beam of light with frequency 2ω falling on the crystal is down converted to two beams of light with frequency ω (see also Fig. 3.3 and 3.6)

may note that since the second-order nonlinearity responsible for the parametric oscillation is weak, the crystal is often placed in optical cavity so that the interaction time can be increased (see Sect. 3.1). Owing to the fact that the parametric oscillator can oscillate at optical frequencies, it is envisaged that it can convert the incident coherent light or pump mode into two output lights of lower frequency (idler and signal modes).

One can gather that the interaction between the emerging three modes of light may induce amplitude gain for the signal and idler modes and de-amplification for the pump mode. The resulting gain may enable the resonating modes to oscillate in the cavity; and in the process, compensate for the loss due to the out coupling of the cavity radiation with the mirrors of the resonator. It is therefore possible to foresee that the correlation among the pump, signal and idler modes in the nonlinear crystal can lead to the manifestation of a very interesting quantum optical phenomena. The leakage through the mirrors however leads to the amplification of the noise, and subsequent degradation of the available nonclassical features. This conflicting physical processes might be evinced by the fact that there is a 50% upper bound to the degree of squeezing of the light generated by the parametric oscillator (see Fig. 8.3).

In addition to the possibility of generating light with a nonclassical character, a parametric oscillator can also offer the potential for a wide wavelength tuning ranges that depends on the phase matching details related to the change in the crystal temperature or the angular orientation of the crystal or the poling period. Parametric oscillator is taken as a tunable in the sense that the idler frequency ω_2, which is less than the external pump frequency ω_1, can satisfy the condition that $\omega_3 = \omega_1 - \omega_2$ for some desired signal frequency ω_3. A degenerate parametric oscillator, particularly, can be taken as special kind of optical parametric oscillator in which the light of frequency 2ω falling on the crystal would be down converted to two beams of light of frequency ω; wherein the correlation between them would be strong enough to induce one-mode nonclassical features such as single-mode first-order squeezing;

Fig. 8.2 Schematic representation of the energy levels that engage in degenerate parametric oscillation process that can be described by a three-level cascade scheme in which the absorption of the light with frequency 2ω leads to cascade excitation

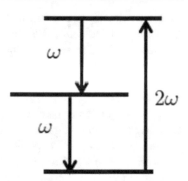

or else, the device is referred to as nondegenerate and corresponds to the generation of two-mode nonclassical light (see Fig. 8.1).

The down conversion process occurs at a single photon regime where each pump photon is annihilated inside the cavity and gives rise to a pair of photons (see Fig. 8.2), and is taken to be responsible for inciting the quantum correlation between the signal and idler fields—the perception that envisages the realization of the nonclassical features.[2] Note that the possibility of generating two light beams at the same time may invoke the conception of twin beams in which the emerging phase can exhibit a measurable correlation that makes the manifestation of the pertinent quantum entanglement conceivable [7]. Optical parametric oscillator is thus deemed to be one of the most interesting and well studied devices in the regime of nonlinear quantum optics.

To provide a meaningful and consistent description of a light generated by a degenerate optical parametric oscillator, one may resort to the explanation offered by a nonlinear optics [3]. In this regard, notice that the nonlinear polarization (8.3) encompasses various contributions, where in relation to parametric oscillator the process in which the frequency of the polarization takes the difference of the two frequencies can be designated as difference-frequency generation and expressed as

$$P(\omega_1 - \omega_2) = 2\chi^{(2)} E_1 E_2^* e^{-i(\omega_1 - \omega_2)t}. \tag{8.6}$$

The complex conjugate of Eq. (8.6) can also represent the difference-frequency generation.

As one may observe from classical electrodynamics, the pumping electromagnetic field would interact with the electric dipoles in the medium that subsequently compels them to oscillate. This induced oscillation can also serve as the source of electromagnetic radiation. As the intensity of the incident radiation increases, it might be reasonable expecting the interconnectedness between the pumping radiation and the emerging amplitude of the vibration to become nonlinear. It is

---

[2]When a nonlinear crystal is exposed to external radiation of frequency 2ω, only some part of the radiation can be down converted to a pair of photons (see Sect. 3.2.4, Fig. 3.6 and also [6]).

such a nonlinear interaction that leads to the generation of frequencies different from those of incident fields. In other words, in view of energy conservation, one can observe that for every photon created at difference frequency, $\omega_3 = \omega_1 - \omega_2$, a photon at the higher frequency ω_1 is destroyed and a photon at the lower input frequency ω_2 is created. It may not be consequently hard to conceive that when an excited three-level atomic system spontaneously decays in a cascade fashion from the upper to the lower energy level, two photons are emitted, which makes optical parametric oscillator a genuine two-photon device.

We now attempt to explore the quantum nature of the radiation emitted by degenerate parametric oscillator (for Hamiltonian, master equation, dynamics of c-number variables, various correlations and phase space distributions of a light generated by degenerate parametric oscillator; see Chap. 3 and for some interesting examples [8]). To do so, one can begin with the variances of the quadratures that can be expressed as

$$\Delta a_{\pm}^2 = 1 + 2\langle \alpha^*(t)\alpha(t)\rangle \pm \left[\langle \alpha^{*2}(t)\rangle + \langle \alpha^2(t)\rangle\right], \tag{8.7}$$

where $\alpha(t)$ is defined by Eq. (3.159). One should also note that Eq. (3.159) does not explode at steady state provided that $\lambda_{\pm} \geq 0$. That means, to get a physically realizable outcome, one should be confined to the case when $\kappa \geq 2\varepsilon$; and hence, $\varepsilon = \kappa/2$ is taken as threshold condition.

Following the procedure already established, one can determine the correlations in Eq. (8.7) by applying Eqs. (3.159)–(3.161) (see Sect. 3.2.2). To this end, one can begin with

$$\langle \alpha^*(t)\alpha(t)\rangle = A_+(t)\langle \alpha^*(0)\alpha(t)\rangle + A_-(t)\langle \alpha(0)\alpha(t)\rangle + \langle B_+^*(t)\alpha(t)\rangle - \langle B_-^*(t)\alpha(t)\rangle. \tag{8.8}$$

Upon citing the proposal that the noise force at later time t does not correlate with the system variables at earlier times, it may not be hard to see that

$$\langle \alpha^*(t)\alpha(t)\rangle = \langle B_+^*(t)B_+(t)\rangle - \langle B_-^*(t)B_+(t)\rangle - \langle B_+^*(t)B_-(t)\rangle + \langle B_-^*(t)B_-(t)\rangle. \tag{8.9}$$

One can also write with the aid of Eq. (3.161) that

$$\langle B_{\pm}^*(t)B_{\pm}(t)\rangle = \frac{1}{4}\int_0^\infty \int_0^\infty e^{-\frac{\lambda_{\pm}}{2}(2t - t' - t'')}\langle f_{\pm}^*(t')f_{\pm}(t'')\rangle dt' dt'', \tag{8.10}$$

$$\langle B_+^*(t)B_-(t)\rangle = \frac{1}{4}\int_0^\infty \int_0^\infty e^{-\frac{1}{2}[(\lambda_+ + \lambda_-)t - \lambda_+ t' - \lambda_- t'']}\langle f_+^*(t')f_-(t'')\rangle dt' dt''. \tag{8.11}$$

So in view of Eq. (3.156), it may not be difficult to verify that

$$\langle B_{\pm}^{*}(t) B_{+}(t)\rangle = \pm \frac{\varepsilon}{2\lambda_{\pm}}, \tag{8.12}$$

$$\langle B_{-}^{*}(t) B_{+}(t)\rangle = \langle B_{-}(t) B_{+}^{*}(t)\rangle = 0. \tag{8.13}$$

Consequently, the mean photon number at steady state would be

$$\langle \alpha^{*}(t)\alpha(t)\rangle_{ss} = \frac{2\varepsilon^{2}}{\kappa^{2} - 4\varepsilon^{2}}. \tag{8.14}$$

In the same token, one can write

$$\langle \alpha^{2}(t)\rangle = A_{+}(t)\langle \alpha(0)\alpha(t)\rangle + A_{-}(t)\langle \alpha^{*}(0)\alpha(t)\rangle + \langle B_{+}(t)\alpha(t)\rangle - \langle B_{-}(t)\alpha(t)\rangle, \tag{8.15}$$

from which follows

$$\langle \alpha^{2}(t)\rangle = \langle B_{+}^{2}(t)\rangle - 2\langle B_{+}(t) B_{-}(t)\rangle + \langle B_{-}^{2}(t)\rangle. \tag{8.16}$$

By applying Eq. (3.161), one can then see that

$$\langle B_{\pm}^{2}(t)\rangle = \frac{1}{4} \int_{0}^{t} \int_{0}^{t} e^{-\frac{\lambda_{\pm}}{2}(2t-t'-t'')} \langle f_{\pm}(t') f_{\pm}(t'')\rangle dt' dt'', \tag{8.17}$$

which can also be rewritten as

$$\langle B_{\pm}^{2}(t)\rangle = \frac{\varepsilon}{2} \int_{0}^{t} \int_{0}^{t} e^{-\frac{\lambda_{\pm}}{2}(2t-t'-t'')} \delta(t'-t'') dt' dt''. \tag{8.18}$$

After that, carrying out the integration yields at steady state

$$\langle B_{\pm}^{2}(t)\rangle_{ss} = \frac{\varepsilon}{2\lambda_{\pm}}. \tag{8.19}$$

Following similar procedure, one can verify that

$$\langle B_{-}(t) B_{+}(t)\rangle_{ss} = 0. \tag{8.20}$$

The correlation as a result can be expressed as

$$\langle \alpha^{2}(t)\rangle_{ss} = \frac{\kappa\varepsilon}{\kappa^{2} - 4\varepsilon^{2}}. \tag{8.21}$$

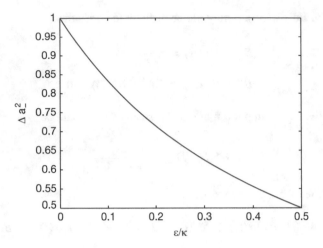

Fig. 8.3 Plot of the minus quadrature variance of the cavity radiation generated by a degenerate parametric oscillator at steady state

Combination of Eqs. (8.7), (8.14) and (8.21) then leads to

$$\Delta a_{\pm}^2 = \frac{\kappa(\kappa \pm 2\varepsilon)}{\kappa^2 - 4\varepsilon^2}. \tag{8.22}$$

One can see from Fig. 8.3 that the cavity radiation would be in squeezed state for a minus (or phase) quadrature. As one can verify from Eq. (8.22), and also confirm from Fig. 8.3, Δa_{-}^2 would not be less than 0.5; which entails that the degree of squeezing of the generated radiation cannot be greater than 50% under normal condition. One can also see that the degree of squeezing increases with the amplitude with which the crystal is pumped. Nonetheless, one cannot increase the strength of the pumping radiation indefinitely under normal operation due to threshold condition; and so the restriction on the degree of squeezing. Comparison of Eqs. (8.14) and (8.22) may also reveal that the photon number would be very large when the degree of squeezing is large. This result may envisage as a degenerate parametric oscillator can be a source of an intense light with reasonably good quantum feature.

It is also possible to appraise the quantum features of the radiation that able to come out of the cavity upon using the input-output relation (see Sect. 3.1.3). In relation to this proposal, the variances of the quadrature operators for the radiation coming out of the cavity can be expressed with the aid of Eqs. (3.62) and (3.74) as

$$\Delta a_{\pm(out)}^2 = 1 + 2\langle \alpha_{out}^*(t)\alpha_{out}(t)\rangle \pm \left[\langle \alpha_{out}^{*2}(t)\rangle + \langle \alpha_{out}^2(t)\rangle\right], \tag{8.23}$$

where

$$\alpha_{out}(t) = \sqrt{\kappa}\alpha(t) - \frac{1}{\sqrt{k}}f_R(t) \tag{8.24}$$

in which $f_R(t)$ is the noise force related to the reservoir modes whose correlations are given by Eqs. (3.60) and (3.61).

It might be good stressing that the stochastic noise force $f(t)$ is the aggregation of the contribution of the cavity and reservoir; and hence, can be expressed as

$$f(t) = f_c(t) + f_R(t), \tag{8.25}$$

where $f_c(t)$ is the noise force associated with the cavity in the absence of damping. Since the noise force for the reservoir does not correlate with $f_c(t)$ and system variables at earlier time, it is not difficult to verify at steady state for a vacuum reservoir that (see Eq. (3.81))

$$\langle \alpha_{out}^*(t)\alpha_{out}(t)\rangle_{ss} = \kappa \langle \alpha(t)\alpha(t)\rangle_{ss}, \qquad \langle \alpha_{out}^2(t)\rangle_{ss} = \kappa \langle \alpha^2(t)\rangle_{ss}, \tag{8.26}$$

from which follows

$$\langle \alpha_{out}^*(t)\alpha_{out}(t)\rangle_{ss} = \frac{2\kappa\varepsilon^2}{\kappa^2 - 4\varepsilon^2}, \tag{8.27}$$

$$\langle \alpha_{out}^2(t)\rangle_{ss} = \frac{\varepsilon\kappa^2}{\kappa^2 - 4\varepsilon^2}. \tag{8.28}$$

It might be possible to see from Eqs. (8.14) and (8.27) that the mean of the photon number that escapes the cavity is less than the amount available in the cavity. This outcome suggests that the intensity or the brightness of the light accessible for application could be reduced from what one can generate.

One can also see that the combination of Eqs. (8.23), (8.27) and (8.28) yields

$$\Delta a_{\pm(out)}^2 = 1 \pm \frac{2\varepsilon\kappa(\kappa \pm 2\varepsilon)}{\kappa^2 - 4\varepsilon^2}. \tag{8.29}$$

One can infer from Eq. (8.29) or comparison of Figs. 8.4 and 8.5 that the output degree of squeezing increases with κ. Even then, since $\kappa < 1$, one can see that the degree of squeezing of the output radiation is less than 50% and when $\kappa = 1$, the two figures are completely equal to each other since there is no barrier in this case. The unbiased noise fluctuations arising while the radiation crosses the boundary or the coupler mirror is believed to be the reason for the reduction of the degree of squeezing for the output radiation. Just as for cavity radiation, the degree of squeezing for the output radiation also increases with the amplitude of the driving coherent radiation since the nonlinear process that leads to the squeezing increases with the strength of pumping.

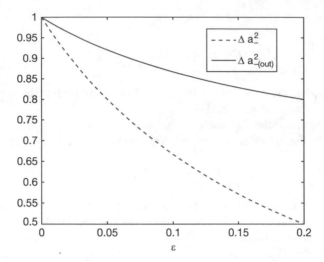

Fig. 8.4 Plots of the minus quadrature variance of the cavity and output radiations for a degenerate parametric oscillator at steady state for $\kappa = 0.4$

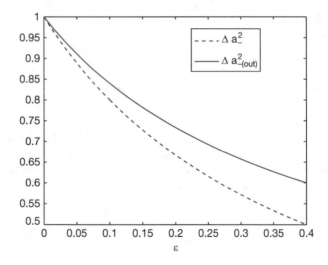

Fig. 8.5 Plots of the minus quadrature variance of the cavity and output radiations for a degenerate parametric oscillator at steady state for $\kappa = 0.8$

8.1.2 Second-Harmonic Generator

In the same way as the process of parametric oscillation, the second-harmonic generation occurs when the response of the atoms of the nonlinear medium depends quadratically on the strength of the applied optical field. This proposal insinuates that the intensity of the light generated at the second-harmonic frequency tends to increase as the square of the intensity of the applied laser light. When two light

Fig. 8.6 Schematic
representation of a
second-harmonic oscillator in
which a nonlinear medium is
placed inside a cavity driven
by a resonant coherent field
of amplitude Ω, and after
interaction a photon of 2ω
frequency is emitted
(compare with Figs. 3.6
and 8.1)

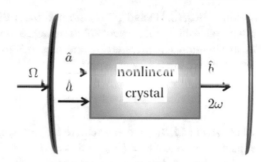

beams with frequency ω are impinged onto a nonlinear crystal, a light with 2ω
frequency would be produced; the reverse process of parametric oscillation [9].
Markedly, at least one of the terms in the nonlinear polarization (8.3) exemplifies
the sum-frequency generation that accounts for the second-harmonic generation of
the radiation at frequency 2ω (see Fig. 8.6).

The second-harmonic generation process can generally be perceived as two
photons at frequency ω successively excite the cascade three-level atom, and as
the photon with frequency 2ω would be subsequently emitted. The interaction
Hamiltonian that describes such a process can be written as

$$\hat{H} = i\Omega(\hat{a}^\dagger - \hat{a}) + i\frac{\lambda}{2}(\hat{a}^{\dagger 2}\hat{b} - \hat{a}^2\hat{b}^\dagger),\tag{8.30}$$

where $\hat{a}$ ($\hat{b}$) is the annihilation operator for the light mode of frequency ω (2ω), λ
is the coupling constant and Ω is the parameter proportional to the amplitude of the
coherent driving field (compare with Eq. (3.234) and note also that $\hat{a}$ is interchanged
with $\hat{b}$).

Following the approach outlined in Sect. 3.1.2, one can verify that the quantum
Langevin equations would be

$$\frac{d\hat{a}}{dt} = \Omega + \lambda\hat{a}^\dagger\hat{b} - \frac{\kappa}{2}\hat{a} + \hat{F}_1(t),\tag{8.31}$$

$$\frac{d\hat{b}}{dt} = -\frac{\lambda}{2}\hat{a}^2 - \frac{\kappa}{2}\hat{b} + \hat{F}_2(t),\tag{8.32}$$

where κ is assumed to be the same for both modes and $\hat{F}_{1,2}(t)$ are the noise operators
for each mode. For two independent vacuum reservoirs:

$$\langle\hat{F}_i(t)\rangle = \langle\hat{F}_j^\dagger(t)\hat{F}_i(t')\rangle_{i\neq j} = \langle\hat{F}_j(t)\hat{F}_i(t')\rangle_{i\neq j} = \langle\hat{F}_i(t)\hat{F}_i(t')\rangle = \langle\hat{F}_i^\dagger(t)\hat{F}_i(t')\rangle = 0,\tag{8.33}$$

$$\langle\hat{F}_i(t)\hat{F}_i^\dagger(t')\rangle = \kappa\delta(t - t'),\tag{8.34}$$

with $i, j = 1, 2$ (relate with Eqs. (3.237)–(3.239)).

Since Eqs. (8.31) and (8.32) are nonlinear differential equations, it may not be possible to obtain their exact solutions. One may thus resorted to use some sort of approximation scheme, that is, one may opt to utilize the linearization procedure introduced in Sect. 3.2.4 in which

$$\hat{a}(t) = \alpha + \hat{A}(t), \qquad \hat{b}(t) = \beta + \hat{B}(t), \tag{8.35}$$

where $\hat{A}(t)$ and $\hat{B}(t)$ are assumed to be very small variations about the steady state expectation values: $\alpha = \langle \hat{a} \rangle_{ss}$ and $\beta = \langle \hat{b} \rangle_{ss}$.

Upon taking the statistical average of Eqs. (8.31) and (8.32) and applying the semiclassical approximation in which at steady state the modes are assumed to be uncorrelated, one can obtain the relations between different parameters. To this end, one can afterwards begin with

$$\lambda \alpha^* \beta - \frac{\kappa}{2} \alpha + \Omega = 0, \tag{8.36}$$

$$\lambda \alpha^2 + \kappa \beta = 0. \tag{8.37}$$

Equations (8.36) and (8.37) stand for the steady state semiclassical expressions generated from Eqs. (8.31) and (8.32) by treating the system operators as c-numbers and also neglecting the noise operators (see Appendix 3).

After that, with the aid of Eqs. (8.31), (8.32), (8.36) and (8.37), one finds following the assumption that $\hat{A}$ and $\hat{B}$ are quite small;

$$\frac{d}{dt} \hat{A}(t) = \varepsilon_1 \hat{B}(t) + \varepsilon_2 \hat{A}^\dagger(t) - \frac{\kappa}{2} \hat{A}(t) + \hat{F}_1(t), \tag{8.38}$$

$$\frac{d}{dt} \hat{B}(t) = -\varepsilon_1 \hat{A}(t) - \frac{\kappa}{2} \hat{B}(t) + \hat{F}_2(t), \tag{8.39}$$

from which follows

$$\frac{d}{dt} \hat{A}_\pm(t) = -a_\pm \hat{A}_\pm(t) + \varepsilon_1 \hat{B}_\pm(t) + \hat{E}_\pm(t), \tag{8.40}$$

$$\frac{d}{dt} \hat{B}_\pm(t) = -\varepsilon_1 \hat{A}_\pm(t) - \frac{\kappa}{2} \hat{B}_\pm(t) + \hat{F}_\pm(t), \tag{8.41}$$

where

$$a_\pm = \frac{\kappa \mp 2\varepsilon_2}{2}, \tag{8.42}$$

$$\hat{A}_\pm = \hat{A}^\dagger \pm \hat{A}, \qquad \hat{B}_\pm = \hat{B}^\dagger \pm \hat{B}, \tag{8.43}$$

$$\hat{E}_\pm(t) = \hat{F}_1^\dagger(t) \pm \hat{F}_1(t), \qquad \hat{F}_\pm(t) = \hat{F}_2^\dagger(t) \pm \hat{F}_2(t) \tag{8.44}$$

in which

$$\varepsilon_1 = \lambda\alpha, \qquad \varepsilon_2 = \lambda\beta \tag{8.45}$$

(compare with Eq. (8.245) and note also the difference of the values of ε_1 and ε_2, and their relationship).

One can see that Eqs. (8.40) and (8.41) are coupled differential equations that can be solved via constructing matrix equation. To solve these differential equations, it proves advantageous introducing a matrix equation of the form

$$\frac{d}{dt}\mathcal{A}(t) = -\mathcal{B}\mathcal{A}(t) + \mathcal{C}(t), \tag{8.46}$$

where

$$\mathcal{A}(t) = \begin{pmatrix} \hat{A}_\pm(t) \\ \hat{B}_\pm(t) \end{pmatrix}, \qquad \mathcal{B} = \begin{pmatrix} a_\pm & -\varepsilon_1 \\ \varepsilon_1 & \frac{\kappa}{2} \end{pmatrix}, \qquad \mathcal{C}(t) = \begin{pmatrix} \hat{E}_\pm(t) \\ \hat{F}_\pm(t) \end{pmatrix}. \tag{8.47}$$

Once the dynamical equations are set, it might be straightforward to see that one can obtain the solution following the approach provided in the appendix of Chap. 3 (see also Appendix 1).

In view of the proposal that the properties of the radiation can be studied by applying the provided solution and derivation presented in the appendix (see also Sect. 3.2.4), the pertinent quadratures are found to be

$$\hat{a}_+ = 2\alpha + \hat{A}_+, \qquad \hat{a}_- = -i\hat{A}_- \tag{8.48}$$

whose variances can be written as

$$\Delta a_\pm^2 = \pm\left[\langle \hat{A}_\pm^2(t)\rangle - \langle \hat{A}_\pm(t)\rangle^2\right]. \tag{8.49}$$

We next proceed to determine the expectation values in Eq. (8.49).

To do so, upon applying Eq. (8.143), one can have

$$\langle \hat{A}_\pm(t)\rangle = \left(\frac{1\pm1}{\lambda}\right)\left[\varepsilon_1 b_\pm(t) + \varepsilon_2 c_\pm(t)\right] - b_\pm(t)\left[\langle \hat{a}^\dagger(0)\rangle \pm \langle \hat{a}(0)\rangle\right]$$
$$- c_\pm(t)\left[\langle \hat{b}^\dagger(0)\rangle \pm \langle \hat{b}(0)\rangle\right] - \langle g_\pm(t)\rangle - \langle f_\pm(t)\rangle. \tag{8.50}$$

Upon taking the cavity modes to be initially in vacuum state and employing the fact that the expectation value of the noise force is zero, one then finds

$$\langle \hat{A}_\pm(t)\rangle = \left(\frac{1\pm1}{\lambda}\right)\left[\varepsilon_1 b_\pm(t) + \varepsilon_2 c_\pm(t)\right], \tag{8.51}$$

from which follows at steady state

$$\langle \hat{A}_\pm(t) \rangle_{ss} = 0. \tag{8.52}$$

It is possible to see in a similar manner that

$$\langle \hat{A}_\pm^2(t) \rangle = \left(\frac{1 \pm 1}{\lambda} \right) \left[\varepsilon_1 \langle \hat{b}_\pm(t) \hat{A}_\pm(t) \rangle + \varepsilon_2 \langle \hat{c}_\pm(t) \hat{A}_\pm(t) \rangle \right]$$
$$- \langle \hat{g}_\pm(t) \hat{A}_\pm(t) \rangle - \langle \hat{f}_\pm(t) \hat{A}_\pm(t) \rangle$$
$$- b_\pm(t) \left[\langle \hat{a}^\dagger(0) \hat{A}_\pm(t) \rangle \pm \langle \hat{a}(0) \hat{A}_\pm(t) \rangle \right]$$
$$- c_\pm(t) \left[\langle \hat{b}^\dagger(0) \hat{A}_\pm(t) \rangle \pm \langle \hat{b}(0) \hat{A}_\pm(t) \rangle \right], \tag{8.53}$$

which can be rewritten with the aid of Eqs. (8.143), (8.145), (8.146), (8.148) and (8.149) along with the fact that the noise force at later time does not correlate with the system variables at earlier time and boson commutation relation as

$$\langle \hat{A}_\pm^2(t) \rangle = \left(\frac{1 \pm 1}{\lambda} \right) \left[\varepsilon_1 b_\pm(t) \langle \hat{A}_\pm(t) \rangle + \varepsilon_2 c_\pm(t) \langle \hat{A}_\pm(t) \rangle \right] \pm \left[b_\pm(t)^2 + c_\pm^2(t) \right]$$
$$+ \frac{1}{4} \int_0^t \int_0^t \left[(1 \mp p)^2 e^{-\pm \eta_\pm(2t-t'-t'')} + (1 \pm p)^2 e^{-\lambda_\pm(2t-t'-t'')} \right.$$
$$+ 2(1 - p^2) e^{-(\eta_\pm(t-t') + \lambda_\pm(t-t''))} \Big] \langle \hat{E}_\pm(t') \hat{E}_\pm(t'') \rangle dt' dt''$$
$$+ \frac{q^2}{4} \int_0^t \int_0^t \left[e^{-\pm \eta_\pm(2t-t'-t'')} + e^{-\lambda_\pm(2t-t'-t'')} \right.$$
$$- 2 e^{-(\eta_\pm(t-t') + \lambda_\pm(t-t''))} \Big] \langle \hat{F}_\pm(t') \hat{F}_\pm(t'') \rangle dt' dt''. \tag{8.54}$$

On the basis of Eq. (8.43), one on the other hand can see that

$$\langle \hat{E}_\pm(t') \hat{E}_\pm(t'') \rangle = \langle \hat{F}_1^\dagger(t') \hat{F}_1^\dagger(t'') \rangle + \langle \hat{F}_1(t') \hat{F}_1(t'') \rangle$$
$$\pm \left[\langle \hat{F}_1^\dagger(t') \hat{F}_1(t'') \rangle + \langle \hat{F}_1(t') \hat{F}_1^\dagger(t'') \rangle \right] \tag{8.55}$$

so that taking the correlations (8.33) and (8.34) of the noise forces into account leads to

$$\langle \hat{E}_\pm(t') \hat{E}_\pm(t'') \rangle = \pm \kappa \delta(t' - t''). \tag{8.56}$$

In the same way, one can check that

$$\langle \hat{F}_\pm(t') \hat{F}_\pm(t'') \rangle = \pm \kappa \delta(t' - t''). \tag{8.57}$$

After that, insertion of Eqs. (8.56) and (8.57) into Eq. (8.54) yields

$$
\langle \hat{A}_\pm^2(t) \rangle = \left(\frac{1 \pm 1}{\lambda} \right) \left[\varepsilon_1 b_\pm(t) \langle \hat{A}_\pm(t) \rangle + \varepsilon_2 c_\pm(t) \langle \hat{A}_\pm(t) \rangle \right] \pm \left[b_+(t)^2 + c_\pm^2(t) \right]
$$

$$
\pm \frac{k}{4} \int_0^t \int_0^t \left[(1 \mp p)^2 e^{-\pm \eta_\pm (2t - t' - t'')} + (1 \pm p)^2 e^{-\lambda_\pm (2t - t' - t'')} \right.
$$

$$
\left. + 2(1 - p^2) e^{-(\eta_\pm(t - t') + \lambda_\pm(t - t''))} \right] \delta(t' - t'') dt' dt''
$$

$$
\pm \frac{\kappa q^2}{4} \int_0^t \int_0^t \left[e^{-\pm \eta_\pm (2t - t' - t'')} + e^{-\lambda_\pm (2t - t' - t'')} \right.
$$

$$
\left. - 2 e^{-(\eta_\pm(t - t') + \lambda_\pm(t - t''))} \right] \delta(t' - t'') dt' dt'', \tag{8.58}
$$

where carrying out the integrations results

$$
\langle \hat{A}_\pm^2(t) \rangle = \pm \frac{\kappa}{4} \left[\frac{1 \mp 2p + p^2}{2\eta_\pm} \left(1 - e^{-2\eta_\pm t}\right) + \frac{2(1 - p^2)}{\eta_\pm + \lambda_\pm} \left(1 - e^{-(\eta_\pm + \lambda_\pm)t}\right) \right.
$$

$$
+ \frac{1 \pm 2p + p^2}{2\lambda_\pm} \left(1 - e^{-2\lambda_\pm t}\right) \right] \pm \frac{q^2 \kappa}{4} \left[\frac{1}{2\eta_\pm} \left(1 - e^{-2\eta_\pm t}\right) \right.
$$

$$
\left. - \frac{2}{\eta_\pm + \lambda_\pm} \left(1 - e^{-(\eta_\pm + \lambda_\pm)t}\right) + \frac{1}{2\lambda_\pm} \left(1 - e^{-2\lambda_\pm t}\right) \right]. \tag{8.59}
$$

It may not be hard to see at steady state

$$
\langle \hat{A}_\pm^2(t) \rangle_{ss} = \pm \frac{\kappa}{4} \left[\frac{1 + p^2 + q^2 \mp 2p}{2\eta_\pm} + \frac{1 + p^2 + q^2 \pm 2p}{2\lambda_\pm} + \frac{2(1 - p^2 - q^2)}{\eta_\pm + \lambda_\pm} \right], \tag{8.60}
$$

which can also be put in a more appealing manner as

$$
\langle \hat{A}_\pm^2(t) \rangle_{ss} = \pm \frac{\kappa}{4} \left[\frac{(\eta_\pm + \lambda_\pm)^2 + 4\eta_\pm \lambda_\pm}{2\eta_\pm \lambda_\pm (\eta_\pm + \lambda_\pm)} + (p^2 + q^2) \frac{(\eta_\pm + \lambda_\pm)^2 - 4\eta_\pm \lambda_\pm}{2\eta_\pm \lambda_\pm (\eta_\pm + \lambda_\pm)} \right.
$$

$$
\left. \mp p \frac{(\lambda_\pm - \eta_\pm)}{\lambda_\pm \eta_\pm} \right], \tag{8.61}
$$

in which

$$
\frac{(\eta_\pm + \lambda_\pm)^2 + 4\eta_\pm \lambda_\pm}{8\eta_\pm \lambda_\pm (\eta_\pm + \lambda_\pm)} = \frac{2\kappa^2(\kappa^2 \pm 2\varepsilon_1^2) + \varepsilon_1^2(\varepsilon_1^2 + 4\kappa^2)}{2\kappa(\kappa^2 \pm \varepsilon_1^2)(\kappa^2 \pm 2\varepsilon_1^2 + 4\varepsilon_1^2)}, \tag{8.62}
$$

$$\frac{(\eta_\pm + \lambda_\pm)^2 - 4\eta_\pm\lambda_\pm}{8\eta_\pm\lambda_\pm(\eta_\pm + \lambda_\pm)} = \frac{\varepsilon_1^2(\varepsilon_1^2 - 4\kappa^2)}{2\kappa(\kappa^2 \pm \varepsilon_1^2)(\kappa^2 \pm 2\varepsilon_1^2 + 4\varepsilon_1^2)}, \qquad (8.63)$$

$$\frac{\lambda_\pm - \eta_\pm}{4\eta_\pm\lambda_\pm} = \frac{\varepsilon_1\sqrt{\varepsilon_1^2 - 4\kappa^2}}{\kappa(\kappa^2 \pm 2\varepsilon_1^2 + 4\varepsilon_1^2)}, \qquad (8.64)$$

$$p^2 + q^2 = \frac{\varepsilon_1^2 + 4\kappa^2}{\varepsilon_1^2 - 4\kappa^2}. \qquad (8.65)$$

Hence upon making use of Eqs. (8.49), (8.52), (8.61)–(8.65), one can reach at

$$\Delta a_\pm^2 = \frac{\kappa^2(\kappa^2 \pm \varepsilon_1^2 + 4\varepsilon_1^2)}{(\kappa^2 \pm \varepsilon_1^2)(\kappa^2 \pm 2\varepsilon_1^2 + 4\varepsilon_1^2)}. \qquad (8.66)$$

One may see from Fig. 8.7 that the pumping radiation, which is often taken as a strong classical radiation, turns out to exhibit squeezing as a result of a nonlinear up conversion process. It seems that the coherence engraved in the driving radiation leads to unequal allocation of noise between the quadratures while combined due to the ensued interaction with atoms of the nonlinear medium to form a new more energetic light. To get a more sensible understanding, it might be advisable relating with the down conversion process where the unconverted classical pumping radiation exhibits squeezing after interacting with a nonlinear medium (see Fig. 3.7).

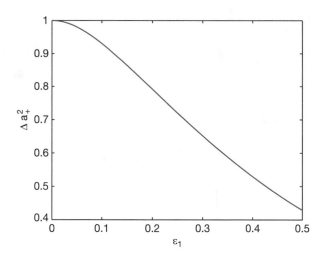

Fig. 8.7 Plot of the plus quadrature variance of the driving radiation of a second-harmonic generator at steady state for $\kappa = 0.5$

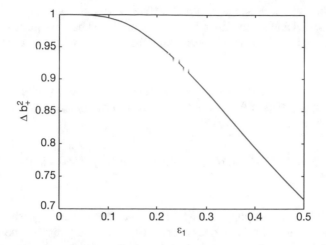

Fig. 8.8 Plot of the plus quadrature variance of the up converted radiation of the second-harmonic generator at steady state for $\kappa = 0.5$

It is also possible to verify following a similar approach that

$$\Delta b_{\pm}^2 = \frac{\kappa^4 \pm 3\kappa^2\varepsilon_1^2 + 4\varepsilon_1^2\kappa^2 + 2\varepsilon_1^4}{(\kappa^2 \pm \varepsilon_1^2)(\kappa^2 \pm 2\varepsilon_1^2 + 4\varepsilon_1^2)}, \qquad (8.67)$$

which is cited to attest to the perception that the second-harmonic radiation can be in squeezed state. One may see from Fig. 8.8 that the up converted radiation can also exhibit squeezing. In this case as well, it is the conversion process that is responsible for the emergence of the nonclassical properties of the radiation. For clarity, it might be insightful comparing what is observed here with the result obtained for the degenerate parametric oscillator when the driving radiation is not treated classically (Sect. 3.2.4). It might also be tempting to compare the parametric up and down conversion processes to get a foresight while developing optical schemes for application.

8.1.3 Multi-Wave Mixer

In the same way as earlier cases, the idea of applying the interaction of electromagnetic field with the nonlinear medium to produce additional field is central to the physical interpretation of the multi-wave mixing process [10]. In the multi-wave mixing procedure, the first input field is presumed to instigate the oscillation of the polarization of the medium to re-radiate with some phase shift. Then the application of the second field drives the polarization so that the interference of the two waves triggers the onset of harmonics in the sum and difference frequencies. In the same way, the application of the third field drives the polarization so that it beats with both input fields as well as the sum and difference frequencies, and then gives rise to the fourth field and so on. Since each of the produced beat frequencies can act as a new source field, in principle, a myriad number of fields with a significant amount of correlation among them can be generated by such a process.

The third-order nonlinear susceptibility responsible for a multi-wave mixing process is actually related to the polarization by $\mathbf{P} = \chi^{(3)}\mathbf{E}^3$, where $\chi^{(3)}$ is fourth-rank susceptibility tensor. For example, a general form of polarization for four-wave mixing can be expressed as

$$P(\omega_4, \mathbf{r}) = \frac{1}{2}\chi^{(3)} E_1(\omega_1) E_2^*(\omega_2) E_3(\omega_3) \exp[i(\mathbf{k}_1 - \mathbf{k}_2 + \mathbf{k}_3).\mathbf{r} - i\omega_4 t],$$

(8.68)

which describes the coupling among four waves each with its own direction of propagation, polarization and frequency. One may recall that efficient coupling between the four waves occurs when energy and momentum are conserved: $\omega_4 = \omega_1 - \omega_2 + \omega_3$ and $\mathbf{k}_4 = \mathbf{k}_1 - \mathbf{k}_2 + \mathbf{k}_3$. These wave and phase matching conditions emanate from the fact that the energy transfer is coherent process; and hence, the four waves must maintain constant phase relative to each other in order to avoid destructive interference.

We now opt to concentrate on the highlight of a multi-wave mixing as source of light with nonclassical properties. Ideal nondegenerate four-wave mixer can be described by interaction Hamiltonian of the form

$$\hat{H} = \chi^{(3)}\big(E^2\hat{a}^\dagger\hat{b}^\dagger + E^{*2}\hat{a}\hat{b}\big),$$

(8.69)

where E stands for the amplitude of the external pumping with frequency ω_p, and $\hat{a}$ ($\hat{b}$) is boson operators for the generated radiation with frequencies $\omega_p + \epsilon$ ($\omega_p - \epsilon$) in which ϵ is a mid-way frequency. The two pump photons are annihilated and at the same time two photons with frequency $\omega_p \pm \epsilon$ are created where the pair of created photons can acquire nonlinear correlations. The emerging nonlinear correlations are often evident in the fluctuations of the quadrature fields defined by $\hat{X}_\phi = \hat{a}e^{-i\phi} + \hat{b}^\dagger e^{i\phi}$, where ϕ is the phase to be fixed based on the corresponding practical requirements.

The variance of the fluctuations in these quadratures as well can be calculated by using

$$V(X_\phi) = \frac{1}{2}\Big[\langle\hat{X}_\phi\hat{X}_\phi^\dagger\rangle + \langle\hat{X}_\phi^\dagger\hat{X}_\phi\rangle\Big] - \langle\hat{X}_\phi^\dagger\rangle\langle\hat{X}_\phi\rangle.$$

(8.70)

Observe that the nonclassical properties of the generated radiation that includes the squeezing can be studied by applying the pertinent Hamiltonian [11].

8.1.4   Correlated Emission Laser

In the process of the interaction of multi-level atoms with radiation, the atomic coherence that can be induced by coupling the atomic energy levels by a strong radiation and/or preparing the atoms initially in the coherent superposition of the energy levels between which direct transition is dipole forbidden is taken to be responsible for various important aspects of the emitted light (see Chap. 6). On the basis of this assertion, we intend to concentrate on the study of the nonclassical features of the radiation generated by cascade three-level scheme with the hope that the approach to be developed can be extended to other schemes. To begin with, the cascade three-level atom can be conveniently construed from the situation in which the intermediate energy level is presumed to have a different parity from the upper and lower energy levels, that is, when a direct transition between the upper and lower energy levels is dipole forbidden (see Fig. 6.16). In case two photons with different frequency are generated, the resulting configuration would be dubbed as nondegenerate, or else degenerate.

Imagine that nondegenerate three-level atoms in a cascade configuration and initially prepared in the coherent superposition of the upper and lower energy levels are injected at a constant rate into the cavity, and removed after the atoms decay spontaneously to energy levels other than the intermediate or the lower (see Fig. 8.9). Pertaining to the presence of the cavity, successively emitted radiation is compelled to oscillate back and forth between the coupler mirrors: the process that leads to the amplification, and so constitutes the lasing mechanism.

On the other hand, owing to the presumed coherent superposition, when the atom decays from the upper to the lower energy level via intermediate, note that the emerging two photons are not emitted independently—the idea that invokes the notion of correlated emission laser. A nondegenerate three-level cascade laser[3] so happens to be a genuine correlated two-photon device having an ability to suppress the noise in one of the quadratures below the vacuum limit. In the nutshell, it is such an unequal partition of the quantum noise between the two quadrature components which is responsible for witnessing the quantum features of the superposed radiation.

The interaction of a three-level cascade atom—whose upper and lower energy levels are initially prepared in arbitrary coherent superposition and also coupled by external coherent radiation—with a two-mode cavity radiation can be described in the interaction picture by the Hamiltonian of the form

$$\hat{H} = ig\big[\hat{a}|a\rangle\langle b| - |b\rangle\langle a|\hat{a}^\dagger + \hat{b}|b\rangle\langle c| - |c\rangle\langle b|\hat{b}^\dagger\big] + i\frac{\Omega}{2}\big[|c\rangle\langle a| - |a\rangle\langle c|\big],$$

$$(8.71)$$

---

[3] Although the atoms can be externally pumped by a coherent light that couples the upper and lower energy levels, and the cavity as a whole can be coupled to various reservoirs, we are not going to deal with such options in details (interested reader can see for instance [12]).

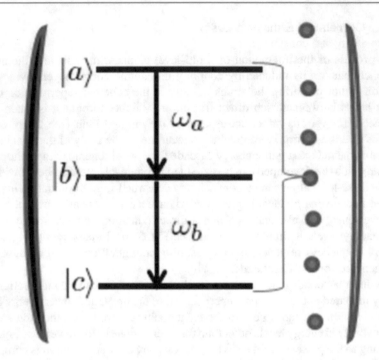

Fig. 8.9 Schematic representation of a nondegenerate three-level cascade laser in which atoms in a three-level cascade configuration are injected into the cavity at a constant rate, and removed after they undergone transition

where Ω is the parameter proportional to the amplitude of the coherent radiation, g is the coupling constant taken to be the same for both transitions, and $\hat{a}, \hat{b}$ are the annihilation operators for the two cavity modes.

Note that the initial state of cascade three-level atom can be taken to be

$$|\Psi_A(0)\rangle = C_a(0)|a\rangle + C_c(0)|c\rangle, \tag{8.72}$$

where

$$C_a(0) = \langle a|\Psi_A(0)\rangle, \qquad C_c(0) = \langle c|\Psi_A(0)\rangle \tag{8.73}$$

are the probability amplitudes for the atom to occupy initially the upper and lower energy levels. The initial density operator therefore can be expressed as

$$\hat{\rho}_A(0) = \rho_{aa}^{(0)}|a\rangle\langle a| + \rho_{ac}^{(0)}|a\rangle\langle c| + \rho_{ca}^{(0)}|c\rangle\langle a| + \rho_{cc}^{(0)}|c\rangle\langle c|, \tag{8.74}$$

where

$$\rho_{aa}^{(0)} = |C_a|^2, \qquad \rho_{ac}^{(0)} = C_a C_c^*, \qquad \rho_{ca}^{(0)} = C_c C_a^*, \qquad \rho_{cc}^{(0)} = |C_c|^2. \tag{8.75}$$

To invoke the lasing, the prepared three-level atoms are assumed to be injected into the cavity at a constant rate r_a, and removed after some time τ which is long enough for the atoms to spontaneously decay as shown in Fig. 8.9. The density operator for the cavity radiation plus a single atom injected into the cavity at time t_j can be denoted by $\rho_{AR}(t, t_j)$ in which $t - \tau \leq t_j \leq t$. The density operator for the atoms in the cavity plus the cavity radiation at time t can then be taken as $\hat{\rho}_{AR}(t) = r_a \sum_j \hat{\rho}_{AR}(t, t_j)\Delta t_j$, where $r_a \Delta t_j$ represents the number of atoms injected into the cavity in the time interval Δt_j. Assuming that the atoms are continuously injected into the cavity, and taking the limit $\Delta t_j \to 0$, the summation over j can be converted into integration with respect to t':

$$\hat{\rho}_{AR}(t) = r_a \int_{t-\tau}^{t} \hat{\rho}_{AR}(t, t')dt', \tag{8.76}$$

where difftiation with respect to t yields

$$\frac{d}{dt}\hat{\rho}_{AR}(t) = r_a\left[\rho_{AR}(t, t) - \hat{\rho}_{AR}(t, t - \tau)\right] + r_a \int_{t-\tau}^{t} \frac{\partial}{\partial t}\hat{\rho}_{AR}(t, t')dt'. \tag{8.77}$$

It might worth noting that $\hat{\rho}_{AR}(t, t)$ stands for the density operator for the atom plus the cavity radiation when the atom is injected into the cavity and $\hat{\rho}_{AR}(t, t - \tau)$ when the atom is removed from the cavity. Since the atomic and radiation variables are not correlated at the instant the atoms are injected into or removed from the cavity, one can propose that $\hat{\rho}_{AR}(t, t) = \hat{\rho}_A(0)\hat{\rho}(t)$ and $\hat{\rho}_{AR}(t, t - \tau) = \hat{\rho}_A(t - \tau)\hat{\rho}(t)$. It is thus possible to write

$$\frac{d}{dt}\hat{\rho}_{AR}(t) = r_a\left[\hat{\rho}_A(0) - \hat{\rho}_A(t - \tau)\right]\hat{\rho}(t) + r_a \int_{t-\tau}^{t} \frac{\partial}{\partial t}\hat{\rho}_{AR}(t, t')dt', \tag{8.78}$$

which can be expressed upon citing Liouvelle-Neumann relation $\frac{\partial \hat{\rho}_{AR}(t)}{\partial t} = -i[\hat{H}, \hat{\rho}_{AR}(t)]$ as

$$\frac{d}{dt}\hat{\rho}_{\alpha\beta}(t) = r_a\langle\alpha|\hat{\rho}_A(0)|\beta\rangle\hat{\rho} - r_a\langle\alpha|\hat{\rho}_A(t - \tau)|\beta\rangle\hat{\rho} - i\langle\alpha|[\hat{H}, \hat{\rho}_{AR}(t)]|\beta\rangle - \gamma\hat{\rho}_{\alpha\beta}, \tag{8.79}$$

where the last term accounts for the decay of the atoms due to spontaneous emission in which $\hat{\rho}_{\alpha\beta} = \langle\alpha|\hat{\rho}_{AR}|\beta\rangle$ with $\alpha, \beta = a, b, c$ and γ is the atomic decay rate taken to be the same for the three atomic energy levels (the case when the atomic damping constant and the dephasing rate are taken to be different can also be very interesting [13]).

Upon making use of Eqs. (8.71), (8.74) and (8.79), one can verify that

$$\frac{d}{dt}\hat{\rho}_{\alpha\beta}(t) = r_a[\rho_{aa}^{(0)}\delta_{\alpha a}\delta_{a\beta} + \rho_{ac}^{(0)}\delta_{\alpha a}\delta_{c\beta} + \rho_{ca}^{(0)}\delta_{\alpha c}\delta_{a\beta} + \rho_{cc}^{(0)}\delta_{\alpha c}\delta_{c\beta}]\hat{\rho}(t)$$

$$- g[\hat{a}^\dagger\hat{\rho}_{\alpha\beta}\delta_{\alpha b} + \hat{b}^\dagger\hat{\rho}_{b\beta}\delta_{\alpha c} - \hat{a}\hat{\rho}_{b\beta}\delta_{\alpha a} - \hat{b}\hat{\rho}_{c\beta}\delta_{\alpha b} + \hat{\rho}_{\alpha a}\hat{a}\delta_{b\beta} + \hat{\rho}_{\alpha b}\hat{b}\delta_{c\beta}$$

$$- \hat{\rho}_{\alpha b}\hat{a}^\dagger\delta_{a\beta} - \hat{\rho}_{\alpha c}\hat{b}^\dagger\delta_{b\beta}]$$

$$- \frac{\Omega}{2}[\hat{\rho}_{c\beta}\delta_{a\alpha} - \hat{\rho}_{\alpha\beta}\delta_{c\alpha} - \hat{\rho}_{\alpha a}\delta_{c\beta} + \hat{\rho}_{\alpha c}\delta_{a\beta}] - \gamma\hat{\rho}_{\alpha\beta}, \qquad (8.80)$$

where $\langle\alpha|\hat{\rho}_A(t-\tau)|\beta\rangle = 0$ based on physical consideration. It is not thus difficult to see that

$$\frac{d}{dt}\hat{\rho}_{aa}(t) = r_a\rho_{aa}^{(0)}\hat{\rho}(t) + g(\hat{a}\hat{\rho}_{ba} + \hat{\rho}_{ab}\hat{a}^\dagger) - \frac{\Omega}{2}(\hat{\rho}_{ac} + \hat{\rho}_{ca}) - \gamma\hat{\rho}_{aa}, \qquad (8.81)$$

$$\frac{d}{dt}\hat{\rho}_{bb}(t) = -g(\hat{a}^\dagger\hat{\rho}_{ab} + \hat{\rho}_{ba}\hat{a} - \hat{b}\hat{\rho}_{cb} - \hat{\rho}_{bc}\hat{b}^\dagger) - \gamma\hat{\rho}_{bb}, \qquad (8.82)$$

$$\frac{d}{dt}\hat{\rho}_{cc}(t) = r_a\rho_{cc}^{(0)}\hat{\rho}(t) - g(\hat{b}^\dagger\hat{\rho}_{bc} + \hat{\rho}_{cb}\hat{b}) + \frac{\Omega}{2}(\hat{\rho}_{ac} + \hat{\rho}_{ca}) - \gamma\hat{\rho}_{cc}, \qquad (8.83)$$

$$\frac{d}{dt}\hat{\rho}_{ab}(t) = g(\hat{a}\hat{\rho}_{bb} - \hat{\rho}_{aa}\hat{a} + \hat{\rho}_{ac}\hat{b}^\dagger) - \frac{\Omega}{2}\hat{\rho}_{cb} - \gamma\hat{\rho}_{ab}, \qquad (8.84)$$

$$\frac{d}{dt}\hat{\rho}_{ac}(t) = r_a\hat{\rho}_{ac}^{(0)}\hat{\rho} + g(\hat{a}\hat{\rho}_{bc} - \hat{\rho}_{ab}\hat{b}) - \frac{\Omega}{2}(\hat{\rho}_{cc} - \hat{\rho}_{aa}) - \gamma\hat{\rho}_{ac}, \qquad (8.85)$$

$$\frac{d}{dt}\hat{\rho}_{cb}(t) = -g(\hat{\rho}_{ca}\hat{a} - \hat{\rho}_{cc}\hat{b}^\dagger + \hat{b}^\dagger\hat{\rho}_{bb}) + \frac{\Omega}{2}\hat{\rho}_{ab} - \gamma\hat{\rho}_{cb}. \qquad (8.86)$$

Confining to a linear analysis that amounts to dropping terms containing g in Eqs. (8.81)–(8.83) and (8.85), and applying the adiabatic approximation,[4] one can affirm that

$$\hat{\rho}_{aa} = \frac{r_a\rho_{aa}^{(0)}}{\gamma}\hat{\rho}(t) - \frac{\Omega}{2\gamma}(\hat{\rho}_{ac} + \hat{\rho}_{ca}), \qquad (8.87)$$

---

[4]In the good cavity limit ($\gamma \gg \kappa$), the cavity mode variables are assumed to change slowly when compared to the atomic variables: the atomic variables are taken to reach steady state in a relatively short time, that is, the time derivative can be set to zero while keeping the remaining atomic and cavity mode variables at time t. It is such consideration that is usually referred to as adiabatic approximation.

$$\hat{\rho}_{cc} = \frac{r_a \rho_{cc}^{(0)}}{\gamma} \hat{\rho}(t) + \frac{\Omega}{2\gamma}(\hat{\rho}_{ac} + \hat{\rho}_{ca}), \tag{8.88}$$

$$\hat{\rho}_{...} = \frac{r_a \rho_{...}^{(0)}}{\gamma} \hat{\rho}(t) - \frac{\Omega}{2\gamma}(\hat{\rho}_{...} + \hat{\rho}_{...}), \tag{8.89}$$

in which $\hat{\rho}_{ac} = \hat{\rho}_{ca}$ in case one assumes that $\rho_{ac}^{(0)} = \rho_{ca}^{(0)}$ and $\hat{\rho}_{bb} = 0$. In view of Eqs. (8.87)–(8.89), one can obtain by applying the adiabatic approximation once again that

$$\hat{\rho}_{ab} = -\frac{g r_a \hat{\rho}}{(4\gamma^2 + \Omega^2)(\gamma^2 + \Omega^2)} \Big[\hat{a}[(4\gamma^2 + \Omega^2)\rho_{aa}^{(0)} - 6\gamma\Omega\rho_{ac}^{(0)} + 3\Omega^2\rho_{cc}^{(0)}]$$
$$+ \hat{b}^\dagger \Big[-\frac{\Omega(2\gamma^2 - \Omega^2)}{\gamma}\rho_{aa}^{(0)} + 2(\Omega^2 - 2\gamma^2)\rho_{ac}^{(0)} + \frac{\Omega(\Omega^2 + 4\gamma^2)}{\gamma}\rho_{cc}^{(0)} \Big] \Big], \tag{8.90}$$

$$\hat{\rho}_{cb} = \frac{g r_a \hat{\rho}}{(4\gamma^2 + \Omega^2)(\gamma^2 + \Omega^2)} \Big[\hat{a}\Big[-\frac{\Omega(4\gamma^2 + \Omega^2)}{\gamma}\rho_{aa}^{(0)} - 2(2\gamma^2 - \Omega^2)\rho_{ac}^{(0)}$$
$$- \frac{\Omega(\Omega^2 - 2\gamma^2)}{\gamma}\rho_{cc}^{(0)} \Big] + \hat{b}^\dagger \Big[3\Omega^2\rho_{aa}^{(0)} + 6\Omega\gamma\rho_{ac}^{(0)} + (\Omega^2 + 4\gamma^2)\rho_{cc}^{(0)} \Big] \Big]. \tag{8.91}$$

On the other hand, note that inserting Eq. (8.71) into

$$\frac{d\hat{\rho}(t)}{dt} = -iTr_A\big[\hat{H},\ \hat{\rho}_{AR}(t)\big] \tag{8.92}$$

results

$$\frac{d\hat{\rho}(t)}{dt} = g\big[\hat{\rho}_{ab}\hat{a}^\dagger - \hat{a}^\dagger\hat{\rho}_{ab} - \hat{b}^\dagger\hat{\rho}_{bc} + \hat{\rho}_{bc}\hat{b}^\dagger + \hat{a}\hat{\rho}_{ba} - \hat{\rho}_{ba}\hat{a} + \hat{b}\hat{\rho}_{cb} - \hat{\rho}_{cb}\hat{b}\big]. \tag{8.93}$$

Upon taking this into account, it might not be difficult to obtain

$$\frac{d\hat{\rho}}{dt} = \frac{AC}{2B}\big[2\hat{a}^\dagger\hat{\rho}\hat{a} - \hat{\rho}\hat{a}\hat{a}^\dagger - \hat{a}\hat{a}^\dagger\hat{\rho}\big] + \frac{AD}{2B}\big[2\hat{b}\hat{\rho}\hat{b}^\dagger - \hat{\rho}\hat{b}^\dagger\hat{b} - \hat{b}^\dagger\hat{b}\hat{\rho}\big]$$
$$+ \frac{AE}{2B}\big[\hat{a}^\dagger\hat{\rho}\hat{b}^\dagger - \hat{\rho}\hat{b}^\dagger\hat{a}^\dagger + \hat{b}\hat{\rho}\hat{a} - \hat{a}\hat{b}\hat{\rho}\big]$$
$$+ \frac{AF}{2B}\big[\hat{a}^\dagger\hat{\rho}\hat{b}^\dagger - \hat{b}^\dagger\hat{a}^\dagger\hat{\rho} + \hat{b}\hat{\rho}\hat{a} - \hat{\rho}\hat{a}\hat{b}\big], \tag{8.94}$$

where

$$A = \frac{2r_a g^2}{\gamma^2}, \tag{8.95}$$

is the linear gain coefficient and

$$B = \left(1 + \frac{\Omega^2}{\gamma^2}\right)\left(1 + \frac{\Omega^2}{4\gamma^2}\right), \tag{8.96}$$

$$C = \rho_{aa}^{(0)}\left(1 + \frac{\Omega^2}{4\gamma^2}\right) - \rho_{ac}^{(0)}\frac{3\Omega}{2\gamma} + \rho_{cc}^{(0)}\frac{3\Omega^2}{4\gamma^2}, \tag{8.97}$$

$$D = \rho_{aa}^{(0)}\frac{3\Omega^2}{4\gamma^2} + \rho_{ac}^{(0)}\frac{3\Omega}{2\gamma} + \rho_{cc}^{(0)}\left(1 + \frac{\Omega^2}{4\gamma^2}\right), \tag{8.98}$$

$$E_\pm = -\rho_{aa}^{(0)}\frac{\Omega}{2\gamma}\left(1 \mp \frac{\Omega^2}{2\gamma^2}\right) - \rho_{ac}^{(0)}\left(1 - \frac{\Omega^2}{2\gamma^2}\right) + \rho_{cc}^{(0)}\frac{\Omega}{\gamma}\left(1 \pm \frac{\Omega^2}{4\gamma^2}\right) \tag{8.99}$$

The time evolution of the density operator for a two-mode cavity radiation coupled to a two-mode squeezed vacuum reservoir via a single-port mirror has been provided elsewhere (see Eq. (3.49)). In line with this, the master equation for the cavity radiation of a nondegenerate three-level cascade laser can be expressed as

$$\frac{d\hat{\rho}(t)}{dt} = \frac{\kappa(N+1)}{2}[2\hat{a}\hat{\rho}\hat{a}^\dagger - \hat{a}^\dagger\hat{a}\hat{\rho} - \hat{\rho}\hat{a}^\dagger\hat{a}]$$

$$+ \frac{1}{2}\left(\frac{AC}{B} + \kappa N\right)[2\hat{a}^\dagger\hat{\rho}\hat{a} - \hat{a}\hat{a}^\dagger\hat{\rho} - \hat{\rho}\hat{a}\hat{a}^\dagger]$$

$$+ \frac{1}{2}\left(\frac{AD}{B} + \kappa(N+1)\right)[2\hat{b}\hat{\rho}\hat{b}^\dagger - \hat{b}^\dagger\hat{b}\hat{\rho} - \hat{\rho}\hat{b}^\dagger\hat{b}]$$

$$+ \frac{\kappa N}{2}[2\hat{b}^\dagger\hat{\rho}\hat{b} - \hat{b}\hat{b}^\dagger\hat{\rho} - \hat{\rho}\hat{b}\hat{b}^\dagger] + \frac{AE_+}{2B}[\hat{a}^\dagger\hat{\rho}\hat{b}^\dagger - \hat{a}\hat{b}\hat{\rho} - \hat{\rho}\hat{a}^\dagger\hat{b}^\dagger + \hat{b}\hat{\rho}\hat{a}]$$

$$+ \frac{AE_-}{2B}[\hat{a}^\dagger\hat{\rho}\hat{b}^\dagger - \hat{a}^\dagger\hat{b}^\dagger\hat{\rho} - \hat{\rho}\hat{a}\hat{b} + \hat{b}\hat{\rho}\hat{a}]$$

$$- M\kappa[\hat{a}^\dagger\hat{\rho}\hat{b}^\dagger + \hat{b}^\dagger\hat{\rho}\hat{a}^\dagger + \hat{a}\hat{\rho}\hat{b} + \hat{b}\hat{\rho}\hat{a} - \hat{a}^\dagger\hat{b}^\dagger\hat{\rho} - \hat{a}\hat{b}\hat{\rho} - \hat{\rho}\hat{a}^\dagger\hat{b}^\dagger - \hat{\rho}\hat{a}\hat{b}]. \tag{8.100}$$

The stochastic differential equations associated with the normal ordering for the cavity mode variables can be generated after that by using Eq. (8.100):

$$\frac{d\alpha(t)}{dt} = -\frac{\kappa B - AC}{2B}\alpha(t) + \frac{AE_+}{2B}\beta^*(t) + f_a(t), \tag{8.101}$$

$$\frac{d\beta^*(t)}{dt} = -\frac{\kappa B + AD}{2B}\beta^*(t) - \frac{AE_-}{2B}\alpha(t) + f_b^*(t) \tag{8.102}$$

in which $f_a(t)$ and $f_b^*(t)$ are the noise forces with correlation properties:

$$\langle f_a(t)\rangle = \langle f_b^*(t)\rangle = \langle f_a(t')f_a(t)\rangle = \langle f_a^*(t')f_a(t)\rangle = \langle f_b(t')f_b(t)\rangle = 0, \tag{8.103}$$

$$\langle f_a(t')f_a^*(t)\rangle = \left(\frac{AC}{B} + \kappa N\right)\delta(t - t'), \qquad \langle f_b(t')f_b^*(t)\rangle = \kappa N\delta(t - t'), \tag{8.104}$$

$$\langle f_b(t')f_a(t)\rangle = -\left(\frac{AE_-}{2B} - \kappa M\right)\delta(t - t'), \tag{8.105}$$

which can be obtained following the procedure outlined in Sect. 3.2.2 (see Appendix 2 and also [12]).

It nowadays becomes evident that introducing a new variable

$$\eta = 1 - 2\rho_{aa}^{(0)} \tag{8.106}$$

with $-1 \leq \eta \leq 1$ is helpful in writing the resulting expressions in a more compact form. It is worth noting that η is the negative of the population inversion (5.36), and so can conveniently represent initially prepared population inversion. As a result, for three-level atoms initially prepared in a coherent superposition of the top and bottom levels, one can write that

$$\rho_{cc}^{(0)} = \frac{1 + \eta}{2}, \qquad \rho_{ac}^{(0)} = \frac{\sqrt{1 - \eta^2}}{2}, \tag{8.107}$$

in case $\rho_{ac}^{(0)} = \rho_{ac}^{*(0)}$.

8.2 Illustrations of Nonclassical Properties of Light

To illustrate some of the nonclassical features of light highlighted in Chap. 7, we choose to make use of a nondegenerate three-level cascade laser without external pumping ($\Omega = 0$) as a toy model hoping to enjoy the relative simplicity of the rigor and its richness [14]. Owing to the coherence induced by preparing the atoms initially in the coherent superposition of the upper and lower energy levels that creates the population transfer pathways and leads to the correlated two-photon emission, this quantum system is believed to be the source of a nonclassical light characterized by a strong correlation of the emitted two photons at two frequencies placed symmetrically at either side of the central frequency (see Sect. 6.5.2).

8.2.1　Two-Mode Squeezing

We first and foremost seek to explore the squeezing property of the generated two-mode cavity radiation when the system is coupled to ordinary two-mode vacuum reservoir ($N = M = 0$). One may recall that the squeezing property of a two-mode cavity radiation can be described by $\hat{c} = (\hat{a} + \hat{b})/\sqrt{2}$ (see Sect. 2.2.3), where the quadrature operators are defined by

$$\hat{c}_+ = \hat{c}^\dagger + \hat{c}, \qquad \hat{c}_- = i(\hat{c}^\dagger - \hat{c}). \tag{8.108}$$

With this information, the variances of these quadrature operators are found to be

$$\Delta c_\pm^2 = 1 \pm \left[\langle \hat{c}^{\dagger^2} \rangle + \langle \hat{c}^2 \rangle \pm 2\langle \hat{c}^\dagger \hat{c} \rangle + \langle \hat{c}^\dagger \rangle^2 + \langle \hat{c} \rangle^2 \pm 2\langle \hat{c} \rangle \langle \hat{c}^\dagger \rangle \right], \tag{8.109}$$

which can be expressed in terms of the c-number variables as

$$\begin{aligned}
\Delta c_\pm^2 = {} & 1 + \langle \alpha^* \alpha \rangle + \langle \beta^* \beta \rangle + \langle \alpha^* \beta \rangle + \langle \alpha \beta^* \rangle - \langle \alpha^* \rangle \langle \beta^* \rangle - \langle \alpha \rangle \langle \beta \rangle \\
& \pm \left[\langle \alpha^* \beta^* \rangle + \langle \alpha \beta \rangle - \langle \alpha^* \rangle \langle \alpha \rangle - \langle \beta^* \rangle \langle \beta \rangle - \langle \alpha^* \rangle \langle \beta \rangle - \langle \alpha \rangle \langle \beta^* \rangle \right] \\
& - \frac{1}{2} \left[\langle \alpha^* \rangle^2 + \langle \alpha \rangle^2 + \langle \beta^* \rangle^2 + \langle \beta \rangle^2 \mp \left(\langle \alpha^{*^2} \rangle + \langle \alpha^2 \rangle + \langle \beta^{*^2} \rangle + \langle \beta^2 \rangle \right) \right].
\end{aligned} \tag{8.110}$$

Upon assuming the two cavity modes to be initially in two-mode vacuum state, one can verify with the help of Eqs. (8.161) and (8.162) that

$$\langle \alpha(t) \rangle = \langle \beta(t) \rangle = 0, \tag{8.111}$$

and thus follows

$$\begin{aligned}
\Delta c_\pm^2 = {} & 1 + \langle \alpha^*(t) \alpha(t) \rangle + \langle \beta^*(t) \beta(t) \rangle + \langle \alpha^*(t) \beta(t) \rangle + \langle \alpha(t) \beta^*(t) \rangle \\
& \pm \left[\langle \alpha^*(t) \beta^*(t) \rangle + \langle \alpha(t) \beta(t) \rangle + \frac{1}{2} \left(\langle \alpha^{*^2}(t) \rangle + \langle \alpha^2(t) \rangle + \langle \beta^{*^2}(t) \rangle + \langle \beta^2(t) \rangle \right) \right].
\end{aligned} \tag{8.112}$$

We next strive to obtain the expectation values in Eq. (8.112) term by term. To this end, upon taking Eq. (8.161) into account, one can write

$$\begin{aligned}
\langle \alpha^*(t) \alpha(t) \rangle = {} & A_+(t) \langle \alpha^*(t) \alpha(0) \rangle + B_+(t) \langle \alpha^*(t) \beta^*(0) \rangle \\
& + \langle \alpha^*(t) F_+(t) \rangle + \langle \alpha^*(t) G_+(t) \rangle.
\end{aligned} \tag{8.113}$$

In the same way, by applying the complex conjugate of Eq. (8.161), it is possible to see that

$$\langle \alpha^*(t)\alpha(0)\rangle \quad A_+(t)\langle \alpha^*(0)\alpha(0)\rangle + B_+(t)\langle \alpha(0)\beta(0)\rangle$$
$$+ \langle \alpha(0)F_+^*(t)\rangle + \langle \alpha(0)G_+^*(t)\rangle. \tag{8.114}$$

With the aid of the assumption that the noise force at time t does not meaningfully related to the cavity mode variables at earlier times, and taking the cavity modes to be initially in a vacuum state, one gets

$$\langle \alpha^*(t)\alpha(0)\rangle = 0. \tag{8.115}$$

It should also be possible to verify in the same way:

$$\langle \alpha^*(t)\beta(0)\rangle = 0, \tag{8.116}$$
$$\langle \alpha^*(t)F_+(t)\rangle = \langle F_+^*(t)F_+(t) + \langle G_+^*(t)F_+(t), \tag{8.117}$$
$$\langle \alpha^*(t)G_+(t)\rangle = \langle F_+^*(t)G_+(t) + \langle G_+^*(t)G_+(t). \tag{8.118}$$

Equation (8.113) can be consequently rewritten as

$$\langle \alpha^*(t)\alpha(t)\rangle = \langle F_+^*(t)F_+(t)\rangle + \langle G_+^*(t)G_+(t)\rangle + \langle F_+^*(t)G_+(t)\rangle + \langle G_+^*(t)F_+(t)\rangle. \tag{8.119}$$

It then turns out to be appropriate and necessary evaluating the correlations in Eq. (8.119). So on the basis of Eq. (8.165), one can have

$$\langle F_+^*(t)F_+(t)\rangle = \frac{1}{4\eta} \int_0^t \int_0^t [(\eta-1)e^{-(\kappa+A\eta)(t-t')/2} + (\eta+1)e^{-\kappa(t-t')/2}]$$
$$\times [(\eta-1)e^{-(\kappa+A\eta)(t-t'')/2} + (\eta+1)e^{-\kappa(t-t'')/2}]$$
$$\times \langle f_a(t'')f_a^*(t')\rangle dt' dt'', \tag{8.120}$$

from which follows using Eqs. (8.104) and (8.106)

$$\langle F_+^*(t)F_+(t)\rangle = \frac{A(1-\eta)}{2\eta^2}\left[\frac{(\eta-1)^2(1-e^{-(\kappa+A\eta)t})}{4(\kappa+A\eta)} + \frac{(\eta+1)^2(1-e^{-\kappa t})}{4\kappa} \right.$$
$$\left. + \frac{(\eta^2-1)(1-e^{-(2\kappa+A\eta)t/2})}{(2\kappa+A\eta)} \right]. \tag{8.121}$$

It may not be difficult to verify in a similar manner that

$$\langle G_+^*(t)G_+(t)\rangle = 0, \tag{8.122}$$

$$\langle F_+^*(t)G_+(t)\rangle = \langle G_+^*(t)F_+(t)\rangle = \frac{A(1-\eta^2)}{4\eta^2}\left[\frac{(\eta-1)(1-e^{-(\kappa+A\eta)t})}{(4\kappa+A\eta)}\right.$$

$$\left.-\frac{(\eta+1)(1-e^{-\kappa t})}{4\kappa}+\frac{(1-e^{-(2\kappa+A\eta)t/2})}{(2\kappa+A\eta)}\right]. \tag{8.123}$$

After that, applying Eqs. (8.119), (8.121)–(8.123), one can see that

$$\langle \alpha^*(t)\alpha(t)\rangle = -\frac{A(1-\eta)^2}{4\eta(\kappa+A\eta)}(1-e^{-(\kappa+A\eta)t})+\frac{A(1-\eta^2)}{2\eta(2\kappa+A\eta)}(1-e^{-(2\kappa+A\eta)t}). \tag{8.124}$$

Following a similar procedure, one can see that

$$\langle \beta^*(t)\beta(t)\rangle = -\frac{A(1-\eta^2)}{4\eta(\kappa+A\eta)}(1-e^{-(\kappa+A\eta)t})+\frac{A(1-\eta^2)}{2\eta(2\kappa+A\eta)}(1-e^{-(2\kappa+A\eta)t}), \tag{8.125}$$

$$\langle \alpha(t)\beta(t)\rangle = -\frac{A(1-\eta)\sqrt{1-\eta^2}}{4\eta(\kappa+A\eta)}(1-e^{-(\kappa+A\eta)t})+\frac{A\sqrt{1-\eta^2}}{2\eta(2\kappa+A\eta)}(1-e^{-(2\kappa+A\eta)t}), \tag{8.126}$$

$$\langle \alpha^2(t)\rangle = \langle \beta^2(t)\rangle = \langle \alpha^*(t)\beta(t)\rangle = 0. \tag{8.127}$$

It is not difficult to see the quadrature variances of the two-mode cavity radiation (8.110) to be

$$\Delta c_\pm^2 = 1 - \frac{A(1-\eta)(1\pm\sqrt{1-\eta^2})}{2\eta(\kappa+A\eta)}(1-e^{-(\kappa+A\eta)t})$$

$$+\frac{A(1-\eta^2\pm\sqrt{1-\eta^2})}{\eta(2\kappa+A\eta)}(1-e^{-(2\kappa+A\eta)t}), \tag{8.128}$$

which takes at steady state the form

$$\Delta c_\pm^2 = 1 + A(1-\eta)\left[\frac{4\kappa(\kappa+A\eta)+(A^2+A(2\kappa+A\eta))(1\pm\sqrt{1-\eta^2})}{4\kappa(2\kappa+A\eta)(\kappa+A\eta)}\right]$$

$$\pm A\sqrt{1-\eta^2}\left[\frac{4\kappa(\kappa+A\eta)+A^2(\eta^2-1\mp\sqrt{1-\eta^2})}{4\kappa(2\kappa+A\eta)(\kappa+A\eta)}\right]. \tag{8.129}$$

Pertaining to threshold condition, mark that the interpretation of Eq. (8.129) should be restricted at steady state to $\eta > -\frac{\kappa}{A}$.

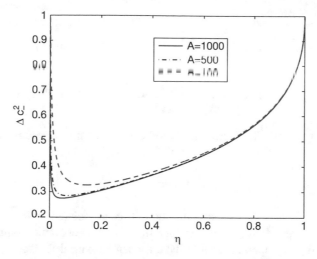

Fig. 8.10 Plots of the minus quadrature variance of the two-mode cavity radiation generated by a nondegenerate three-level cascade laser at steady state for $\kappa = 0.5$ and different values of A

One can see from Fig. 8.10 that the two-mode squeezing manifests for values of η between 0 and 1 that corresponds to the case when the atoms are initially prepared in such a way that there are more atoms that occupy the lower energy level: the regime designated as lasing without population inversion. The strongest degree of squeezing would be observed when initially nearly half of the atoms are in the upper energy level. One can also see that the degree of squeezing increases with the linear gain coefficient that can be interpreted as the rate at which the atoms are injected into the cavity in case the atomic damping rate is taken to be constant. This result entails that the more the number of atoms injected into the cavity, of course without directly interacting with each other, the more would be the chance for producing correlated radiation. Even then, the squeezing feature ceases to exist when the atoms are initially prepared with a maximum ($\eta = 0$; $\rho_{aa}^{(0)} = \rho_{cc}^{(0)} = \rho_{ac}^{(0)} = 1/2$) or a minimum ($\eta = 1$; $\rho_{aa}^{(0)} = \rho_{ac}^{(0)} = 0$ and $\rho_{cc}^{(0)} = 1$) atomic coherence.

8.2.2 Bipartite Entanglement

It is worth noting that the variances of the quadrature operators incorporated in the Duan-Geidke-Cirac-Zoller type entanglement quantification criteria (see Eq. (7.111)) are already obtained. As a result, based on the fact that for a nondegenerate three-level cascade laser $\langle \alpha^2(t) \rangle = \langle \beta^2(t) \rangle = \langle \alpha^*(t)\beta(t) \rangle = 0$, the variances of the quadrature operators (see Eqs. (7.113), (7.114) and (8.127)) can be expressed as

$$\Delta u^2 = \Delta v^2 = 1 + \langle \alpha^*(t)\alpha(t) \rangle + \langle \beta^*(t)\beta(t) \rangle - 2\langle \alpha(t)\beta(t) \rangle, \qquad (8.130)$$

from which follows

$$\Delta u^2 + \Delta v^2 = 2 - \frac{A(1-\eta)(1-\sqrt{1-\eta^2})}{\eta(\kappa + A\eta)}(1 - e^{-(\kappa + A\eta)t})$$

$$+ \frac{2A(1-\eta^2 - \sqrt{1-\eta^2})}{\eta(2\kappa + A\eta)}(1 - e^{-(2\kappa + A\eta)t}). \qquad (8.131)$$

Comparison of Eqs. (8.128) and (8.131) indicates that for the system under consideration[5]

$$\Delta u^2 + \Delta v^2 = 2\Delta c_-^2. \qquad (8.132)$$

Recall also that the inseparability condition associated with a logarithmic negativity can be inferred from the fact that for the separable composite state, the product density operator should have a positive partial transpose, that is, the entanglement would be manifested when E_n (logarithmic negativity) is positive (see Eqs. (7.124) and (7.125)). To attest to the proposal that a nondegenerate three-level cascade laser can be the source of entanglement as quantified by this measure, one can therefore express the c-number variables associated with the normal ordering as given by Eqs. (7.121)–(7.123) using the fact that $\langle \alpha(t) \rangle = \langle \beta(t) \rangle = \langle \alpha^2(t) \rangle = \langle \beta^2(t) \rangle = \langle \alpha(t)\beta^*(t) \rangle = 0$ as

$$m = 1 + 2\langle \alpha^*(t)\alpha(t) \rangle, \quad n = 1 + 2\langle \beta^*(t)\beta(t) \rangle, \quad c = \langle \alpha(t)\beta(t) \rangle + \langle \alpha^*(t)\beta^*(t) \rangle. \qquad (8.133)$$

With this information, it might not be hard to see from Eq. (7.120) that

$$\det A = m^2, \qquad \det B = n^2, \qquad \det C = -c^2, \qquad (8.134)$$

and as a result follows

$$\det A = 1 + 4\langle \alpha^*(t)\alpha(t) \rangle [\langle \alpha^*(t)\alpha(t) \rangle + 1], \qquad (8.135)$$

$$\det B = 1 + 4\langle \beta^*(t)\beta(t) \rangle [\langle \beta^*(t)\beta(t) \rangle + 1], \qquad (8.136)$$

$$\det C_{AB} = -4\langle \alpha(t)\beta(t) \rangle^2, \qquad (8.137)$$

---

[5]In relation to the assumption that a direct interaction between the atoms is neglected, some correlations for a nondegenerate three-level cascade laser are found to be zero (see Eq. (8.127)), which might be responsible for obtaining such similarity between the squeezing and entanglement. Since there is no concrete rationale to anticipate that these correlations would vanish for other systems as well, the direct relation between the degree of quadrature squeezing and entanglement should not be perceived as something generally true.

$$\det\Xi = 16\left[\frac{1}{4} + \frac{\langle\alpha^*(t)\alpha(t)\rangle + \langle\beta^*(t)\beta(t)\rangle}{2}\right.$$

$$\left. \mid \langle\alpha^*(t)\alpha(t)\rangle\langle\beta^*(t)\beta(t)\rangle - \langle\alpha(t)\beta(t)\rangle^2\mid^2\right]^2. \tag{8.138}$$

Taking this into account, the logarithmic negativity E_n (7.124) at steady state is plotted.

One can see from Fig. 8.11 that the generated radiation exhibits entanglement except when all atoms initially occupy the lower energy level (when $\eta = 1$). Assuming that a larger E_n corresponds to a stronger degree of entanglement, one can assert that the degree of entanglement increases with the rate at which the atoms are injected into the cavity. This suggestion agrees well with the result one expects for instance from the available degree of squeezing (see Fig. 8.10). Motivated by this observation, we also opt to compare the degree of entanglement that can be quantified by using EPR type quadratures with logarithmic negativity. One can then see from Fig. 8.12 that a substantial degree of entanglement can be witnessed when $\eta = 0$ contrary to what one learns from the EPR type quadratures. Even then, it might be worthwhile noting that the values of η for which a maximum degree of entanglement is exhibited is very close to each other. One may by and large conclude that the logarithmic negativity and EPR type quadratures quantify entanglement in the same way except for the difference near maximum coherence.

One may also recall that the two-mode Gaussian composite state is said to be entangled when one of the eigenvalues of the symplectic matrix is less than one: $V_s < 1$, that is, the degree of entanglement can be quantified by directly applying the correlations in Eq. (7.116). With this in mind, one can see from Fig. 8.13

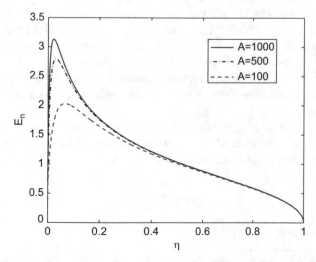

Fig. 8.11 Plots of the logarithmic negativity of the cavity radiation at steady state for $\kappa = 0.5$ and different values of A

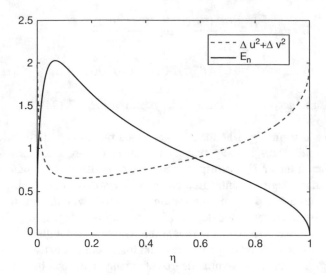

Fig. 8.12 Plots of the logarithmic negativity and quadrature variances of the cavity radiation at steady state for $\kappa = 0.5$ and $A = 100$

that a significant degree of entanglement can be captured by the criterion of the logarithmic negativity except when $\eta = 1$, and the degree of entanglement increases with the rate at which the atoms are injected into the cavity. Besides, comparison of Figs. 8.10 and 8.13 reveals that there is a substantial similarity between the nature of the entanglement predicted by the logarithmic negativity and EPR type quadratures. Despite the perceived striking similarities, one may infer from Fig. 8.14 that the smallest eigenvalue of the symplectic matrix can capture entanglement when the EPR type quadratures failed. It is also possible to see that the criterion of logarithmic negativity can predict a higher degree of entanglement than EPR type quadratures based criterion. With the result for $\eta = 0$ and owing to the fact that the initially prepared coherent superposition would be perceived as a source of the nonclassical features of radiation, it arguably appears more suitable employing the logarithmic negativity to quantify the entanglement of the radiation generated by correlated emission laser.[6]

In addition, the photon number correlation of the two modes of the cavity radiation can also be utilized to study the quantum features of radiation (see Sect. 4.2). With this observation, one can write

$$g(n_a, n_b) = \frac{\langle \hat{n}_a \hat{n}_b \rangle}{\langle \hat{n}_a \rangle \langle \hat{n}_b \rangle} \tag{8.139}$$

---

[6]The difference in the degree of quantifiable entanglement can be related to the assumption and the context from which each measure is derived. Since these criteria are only sufficient conditions, it is not advisable in practical context to solely rely on any one of them.

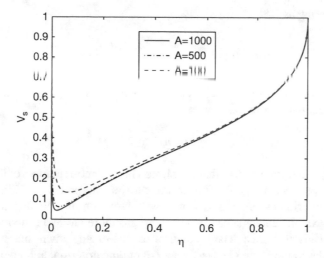

Fig. 8.13 Plots of the smallest eigenvalue of symplectic matrix of the cavity radiation at steady state for $\kappa = 0.5$ and different values of A

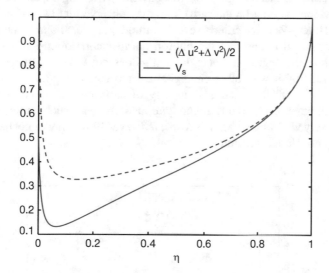

Fig. 8.14 Plots of the smallest eigenvalue of symplectic matrix and the quadrature variances at steady state for $\kappa = 0.5$ and $A = 100$

in which $\langle \hat{n}_a \hat{n}_b \rangle = \langle \hat{a}^\dagger(t)\hat{a}(t)\hat{b}^\dagger(t)\hat{b}(t) \rangle$ is the correlation of the photon numbers, and $\langle \hat{n}_a \rangle = \langle \hat{a}^\dagger(t)\hat{a}(t) \rangle$ and $\langle \hat{n}_b \rangle = \langle \hat{b}^\dagger(t)\hat{b}(t) \rangle$ are the photon numbers (compare with Eq. (4.199)). It may not be hard to see that the operators in Eq. (8.139) are in the normal order.

So in view of the fact that $\alpha(t)$ and $\beta(t)$ are Gaussian variables of zero mean, Eq. (8.139) can be expressed in terms of c-number variables associated with the normal ordering as [15]

$$g(n_a, n_b) = 1 + \frac{\langle \alpha(t)\beta(t)\rangle^2}{\langle \alpha^*(t)\alpha(t)\rangle\langle \beta^*(t)\beta(t)\rangle}. \tag{8.140}$$

In light of this, the Hillery-Zubairy entanglement measure (7.131) can be equated for the system under consideration to

$$g(n_a, n_b) > 2. \tag{8.141}$$

One may see from Fig. 8.15 that the photon number correlation would be greater than 2 except when $\eta = 0$. This result can be interpreted as Hillery-Zubairy entanglement criterion predicts that the cavity radiation of a nondegenerate three-level laser exhibits entanglement when there are more atoms in the lower energy level as already remarked. The prediction of this entanglement measure is more or less similar to the Duan-Geidke-Cirac-Zoller type criterion, although the photon number correlation is found to decrease with the linear gain coefficient. Nonetheless, in case one assumes that a larger $g(n_a, n_b)$ envisages a stronger quantumness of the radiation (which may not be the case), one may observe that Duan-Geidke-Cirac-Zoller and Hillery-Zubairy criteria reveal contradicting results in some respect. It might be insightful noting that the photon number correlation—which does not strictly capture the degree of entanglement—does not have the upper limit, and so directly relating it with other entanglement measures may not be appropriate. Even then, owing to the relative simplicity of carrying out photon coincidence measurement when compared to homodyne measurement, such an approach may found to be very vital when the aim is restricted to verify whether or not the radiation does exhibit nonclassical features.

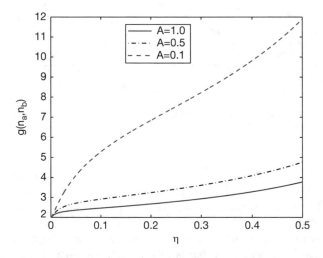

Fig. 8.15 Plots of the photon number correlation of the cavity radiation at steady state for $\kappa = 0.5$ and different values of A

8.2.3 Quantum Discord

We now choose to limit the discussion to continuous variable bipartite system that can be described by a two-mode Gaussian state, and studied within the domain of the general Gaussian formalism (see Sect. 7.4). It should be clear that the practical implementation can be achieved with the help of passive and active linear optics as in homodyne detection setup [16]. In this context, to examine the nature of the quantum discord—epitomized by $D_A > 0$ (see Eq. (7.176))—of the radiation emitted by a correlated emission laser, one can begin with the observation that a two-mode Gaussian state is fully specified up to a local displacement by its covariance matrix (2.182) whose symplectic eigenvalues are

$$v_\pm = \pm\sqrt{\frac{\Delta \pm \sqrt{\Delta^2 - 4\det\Xi}}{2}}, \tag{8.142}$$

where $\Delta = \det A + \det B + 2\det C_{AB}$ (relate with Eqs. (2.183) and (7.189)). One may note that the required parameters are already calculated and given by Eqs. (8.135)–(8.138), and the analysis can be done by using Eq. (7.198).

The result depicted in Fig. 8.16 indicates that the cavity radiation of the correlated emission laser exhibits a strong quantum feature that can be quantified by quantum discord as expected. Markedly, D_A is found to increase with the rate at which the atoms are injected into the cavity. Assuming that D_A large corresponds to a stronger nonclassical behavior, it may not be difficult to realize that the degree of nonclassical features increases with the number of atoms initially occupy the upper energy level. However, the quantum discord turns out to be zero for $\eta = 1$ irrespective of the

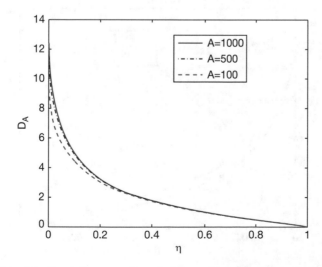

Fig. 8.16 Plots of the forward quantum discord of the cavity radiation at steady state for $\kappa = 0.5$ and different values of A

values of the linear gain coefficient. This outcome asserts that there is no way for the atoms to emit the radiation capable of forging nonclassical correlations if they are initially prepared in the lower energy level. It is also shown that quantum discord is almost independent of the rate of injection of the atoms except when they are initially prepared in a relatively stronger coherent superposition.

One may also note that a qualitative comparison between the quantum discord and quadrature entanglement evinces that the quantum discord can capture the quantumness of the radiation specially when the quadrature entanglement criterion failed. Quantum discord particularly predicts a maximum degree of nonclassical feature for $\eta = 0$ while the criterion of Duan-Geidke-Cirac-Zoller type predicts its absence. On account of this, it might be worthy contemplating that although it is not possible to directly compare the detectable degree of quantumness due to the openness of criterion (7.176) from above, there could be a situation in which the degree of quantumness that can be captured by the quantum discord can be less than entanglement.

Moreover, it may not be beyond comprehension to apprehend that inferring about a system A based on the measurement on system B can be different from the reverse process, that is, the backward quantum discord D_B should also be looked into to get the full picture of the involved nonclassical correlations. Since the mean photon numbers for a nondegenerate three-level cascade laser in the two modes are different (compare Eqs. (8.124) with (8.125), it is reasonable to expect difference between the two versions of the quantum discord (asymmetric discord). To attest to this claim, the forward and backward quantum discords are compared based on the discussion provided in Sect. 7.4. There is indeed unmistakable difference between the two versions of the quantum discord as shown in Fig. 8.17. Detailed scrutiny may

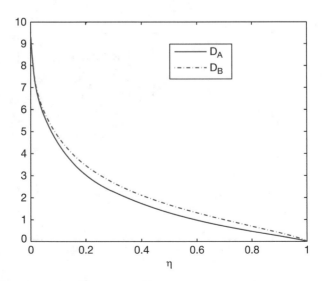

Fig. 8.17 Plots of the forward and backward quantum discord of the cavity radiation at steady state for $\kappa = 0.5$ and $A = 100$

also reveal that the difference between the two would be significant when one has a small number of atoms in the cavity, that is, when one expects relatively smaller degree of quantumness. With the fact that the photon number has also the same nature, it appears reasonable to associate the field of the quantum discord with the intensity of the radiation, even though the quantum feature is directly related to the correlation.

8.2.4   Quantum Steering

As observed in Sect. 7.5, there is a suggestion that quantum steering can be captured in continuous variable regime by Eq. (7.206), which is admissible whenever $S < 1$.

It is clearly shown in Fig. 8.18 that the radiation generated by nondegenerate cascade three-level laser exhibits quantum steering in the regime of lasing without population inversion where the degree of steering increases with the rate at which the atoms are injected into the cavity, and would be maximum when the atoms are prepared in such a way that nearly half of them are in the upper energy level. Since for a cascade three-level laser the mean photon numbers of the two modes are different, the degree of forward and backward quantum steering is found to be different. As one may also see from Fig. 8.19, the chance of steering the state in the receiving side would be greater if one utilizes the light generated during transition from the upper to the intermediate energy level as steerer.

It might also be important comparing quantum steering with other quantum features of the radiation under similar condition. For instance, Fig. 8.20 indicates that the degree of nonclassicality quantified by quantum steering and logarithmic

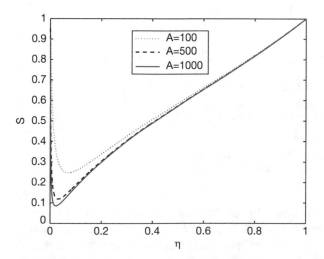

Fig. 8.18 Plots of the forward quantum steering of the cavity radiation at steady state for $\kappa = 0.5$ and different values of A

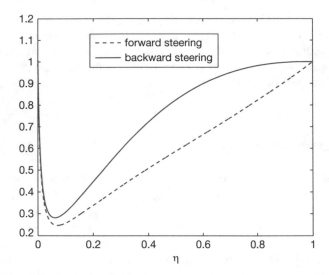

Fig. 8.19 Plots of the forward and backward quantum steering of the cavity radiation at steady state for $\kappa = 0.5$ and $A = 100$

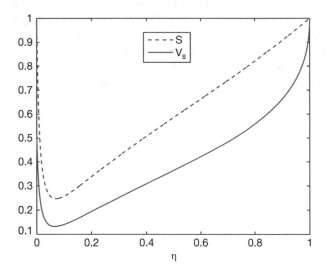

Fig. 8.20 Plots of the forward quantum steering and smallest eigenvalue of symplectic matrix of the cavity radiation at steady state for $\kappa = 0.5$ and $A = 100$

negativity as captured by the smallest eigenvalue of the symplectic matrix shows nearly the same behavior. Critical scrutiny may reveal that quantum steering predicts lesser degree of quantumness when compared to logarithmic negativity but greater when compared to quadrature entanglement as capture by Duan et al. type criterion for some range of atomic inversion (see Fig. 8.21). Since quantum steering is

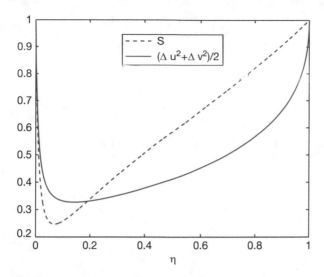

Fig. 8.21 Plots of the forward quantum steering and quadrature variances of the cavity radiation at steady state for $\kappa = 0.5$ and $A = 100$

directly related to photon number correlation, it is expected to be less complicated to measure. This might be advantageous when compared to quadrature entanglement that requires homodyne measurement.

Appendix 1: Dynamical Equation of Linear Second-Harmonic Oscillator

As suggested in the text, the solution of the coupled differential Eqs. (8.40) and (8.41) is found following the approach provided in the appendix of Chap. 3;

$$\hat{A}_{\pm}(t) = \left(\frac{1 \pm 1}{\lambda}\right)\left[\varepsilon_1 b_{\pm}(t) + \varepsilon_2 c_{\pm}(t)\right] - b_{\pm}(t)\left[\hat{a}^{\dagger}(0) \pm \hat{a}(0)\right]$$
$$- c_{\pm}(t)\left[\hat{b}^{\dagger}(0) \pm \hat{b}(0)\right] - g_{\pm}(t) - f_{\pm}(t), \tag{8.143}$$

$$\hat{B}_{\pm}(t) = -\left(\frac{1 \pm 1}{\lambda}\right)\left[\varepsilon_1 c_{\pm}(t) - \varepsilon_2 d_{\pm}(t)\right] + c_{\pm}(t)\left[\hat{a}^{\dagger}(0) \pm \hat{a}(0)\right]$$
$$- d_{\pm}(t)\left[\hat{b}^{\dagger}(0) \pm \hat{b}(0)\right] + f_{\pm}(t) - h_{\pm}(t), \tag{8.144}$$

where

$$b_{\pm}(t) = \frac{1}{2}\left[(1 \mp p)e^{-\eta_{\pm}t} + (1 \pm p)e^{-\lambda_{\pm}t}\right], \tag{8.145}$$

$$c_{\pm}(t) = \frac{q}{2}\left[e^{-\eta_{\pm}t} - e^{-\lambda_{\pm}t}\right], \tag{8.146}$$

$$d_{\pm}(t) = \frac{1}{2}\left[(1 \pm p)e^{-\eta_{\pm}t} + (1 \mp p)e^{-\lambda_{\pm}t}\right], \tag{8.147}$$

$$f_{\pm}(t) = \frac{q}{2}\int_{0}^{t}\left[e^{-\lambda_{\pm}(t-t')} - e^{-\eta_{\pm}(t-t')}\right]\hat{F}_{\pm}(t')dt', \tag{8.148}$$

$$g_{\pm}(t) = \frac{1}{2}\int_{0}^{t}[(1 \mp p)\,e^{-\eta_{\pm}(t-t')} + (1 \pm p)\,e^{-\lambda_{\pm}(t-t')}]\hat{E}_{\pm}(t')dt', \tag{8.149}$$

$$h_{\pm}(t) = \frac{1}{2}\int_{0}^{t}[(1 \pm p)\,e^{-\eta_{\pm}(t-t')} + (1 \mp p)\,e^{-\lambda_{\pm}(t-t')}]\hat{F}_{\pm}(t')dt', \tag{8.150}$$

in which

$$\lambda_{\pm} = \left(\frac{\kappa^2 \pm \varepsilon_1^2}{2\kappa}\right) + \frac{\varepsilon_1}{2\kappa}\sqrt{\varepsilon_1^2 - 4\kappa^2}, \tag{8.151}$$

$$\eta_{\pm} = \left(\frac{\kappa^2 \pm \varepsilon_1^2}{2\kappa}\right) - \frac{\varepsilon_1}{2\kappa}\sqrt{\varepsilon_1^2 - 4\kappa^2}, \tag{8.152}$$

$$p = \frac{\varepsilon_1}{\sqrt{\varepsilon_1^2 - 4\kappa^2}}, \tag{8.153}$$

$$q = \frac{2\kappa}{\sqrt{\varepsilon_1^2 - 4\kappa^2}}. \tag{8.154}$$

These solutions correspond to physically consistent discussion for $\lambda_{\pm}$ and $\eta_{\pm}$ positives; the restriction referred to as threshold condition.

Appendix 2: Dynamical Equation of Nondegenerate Three-Level Cascade Laser

Taking Eq. (8.106) into consideration, one can see from Eqs (8.101) and (8.102) that

$$\frac{d}{dt}\alpha(t) = -a_+\alpha(t) - b_+\beta^*(t) + f_a(t), \tag{8.155}$$

$$\frac{d}{dt}\beta^*(t) = -a_-\beta^*(t) - b_-\alpha(t) + f_b^*(t), \tag{8.156}$$

where

$$a_\pm = \frac{\kappa}{2} + \frac{A}{4B}\left[\frac{3\Omega}{2\gamma}\sqrt{1-\eta^2} + \eta\left(1 - \frac{\Omega^2}{2\gamma^2}\right) \mp \left(1 + \frac{\Omega^2}{\gamma^2}\right)\right], \tag{8.157}$$

$$b_\pm = -\frac{A}{4B}\left[\frac{\Omega}{2\gamma}\left(1 + \frac{\Omega^2}{\gamma^2}\right) \pm \left[\frac{3\eta\Omega}{2\gamma} - \sqrt{1-\eta^2}\left(1 - \frac{\Omega^2}{2\gamma^2}\right)\right]\right]. \tag{8.158}$$

To solve these coupled differential equations, one may introduce a matrix equation of the form

$$\frac{d}{dt}\mathcal{A}(t) = -\mathcal{B}\mathcal{A}(t) + \mathcal{C}(t), \tag{8.159}$$

where

$$\mathcal{A}(t) = \begin{pmatrix} \alpha(t) \\ \beta^*(t) \end{pmatrix}, \qquad \mathcal{B} = \begin{pmatrix} a_+ & b_+ \\ b_- & a_- \end{pmatrix}, \qquad \mathcal{C}(t) = \begin{pmatrix} f_a(t) \\ f_b^*(t) \end{pmatrix}. \tag{8.160}$$

Then following the approach provided in Appendix 4 of Chap. 3, one can obtain

$$\alpha(t) = A_+(t)\alpha(0) + B_+(t)\beta^*(0) + F_+(t) + G_+(t), \tag{8.161}$$

$$\beta(t) = A_-(t)\beta(0) + B_-(t)\alpha^*(0) + F_-(t) + G_-(t), \tag{8.162}$$

where

$$A_\pm(t) = \frac{1}{2}\left[(1 \pm p)e^{-\lambda_2 t} + (1 \mp p)e^{-\lambda_1 t}\right], \tag{8.163}$$

$$B_\pm(t) = \frac{q_\pm}{2}[e^{-\lambda_1 t} - e^{-\lambda_2 t}], \tag{8.164}$$

$$F_+(t) = \frac{1}{2} \int_0^t [(1+p)e^{-\lambda_2(t-t')} + (1-p)e^{-\lambda_1(t-t')}] f_a(t')dt', \qquad (8.165)$$

$$F_-(t) = \frac{1}{2} \int_0^t [(1-p)e^{-\lambda_2(t-t')} + (1+p)e^{-\lambda_1(t-t')}] f_b(t')dt', \qquad (8.166)$$

$$G_+(t) = \frac{q_+}{2} \int_0^t [e^{-\lambda_1(t-t')} - e^{-\lambda_2(t-t')}] f_b^*(t')dt', \qquad (8.167)$$

$$G_-(t) = \frac{q_-}{2} \int_0^t [e^{-\lambda_1(t-t')} - e^{-\lambda_2(t-t')}] f_a^*(t')dt', \qquad (8.168)$$

with

$$\lambda_1 = \left(\frac{a_+ + a_-}{2}\right) + \sqrt{b_+ b_- + \left(\frac{a_+ - a_-}{2}\right)^2}, \qquad (8.169)$$

$$\lambda_2 = \left(\frac{a_+ + a_-}{2}\right) - \sqrt{b_+ b_- + \left(\frac{a_+ - a_-}{2}\right)^2}. \qquad (8.170)$$

$$p = \frac{1 + \frac{\Omega^2}{\gamma^2}}{\left[\left(1 + \frac{\Omega^2}{\gamma^2}\right)^2 \left(1 + \frac{\Omega^2}{4\gamma^2}\right) - \left(\frac{3\eta\Omega}{2\gamma} - \sqrt{1 - \eta^2}\left(1 - \frac{\Omega^2}{2\gamma^2}\right)\right)^2\right]^{\frac{1}{2}}}, \qquad (8.171)$$

$$q_\pm = \frac{-\frac{\Omega}{2\gamma}\left(1 + \frac{\Omega^2}{\gamma^2}\right) \mp \left[\frac{3\eta\Omega}{2\gamma} - \sqrt{1 - \eta^2}\left(1 - \frac{\Omega^2}{2\gamma^2}\right)\right]}{\left[\left(1 + \frac{\Omega^2}{\gamma^2}\right)^2 \left(1 + \frac{\Omega^2}{4\gamma^2}\right) - \left(\frac{3\eta\Omega}{2\gamma} - \sqrt{1 - \eta^2}\left(1 - \frac{\Omega^2}{2\gamma^2}\right)\right)^2\right]^{\frac{1}{2}}}. \qquad (8.172)$$

The obtained solution would be physically meaningful provided that λ_1 and λ_2 are positive: the restriction taken as threshold condition.

Appendix 3: Proof That $\alpha = \alpha^*$ and $\beta = \beta^*$

Multiplication of Eqs. (8.36) and (8.37) by λ yields

$$\varepsilon_1^* \varepsilon_2 - \frac{\kappa}{2} \varepsilon_1 + \lambda \Omega = 0, \tag{8.173}$$

$$\varepsilon_1^2 = -\kappa \varepsilon_2, \tag{8.174}$$

where $\varepsilon_1 = \lambda \alpha$ and $\varepsilon_2 = \lambda \beta$. Then upon multiplying Eq. (8.173) by ε_1^* throughout and using Eq. (8.174), one gets

$$-\kappa \varepsilon_2^* \varepsilon_2 + \frac{\kappa}{2} \varepsilon_1^* \varepsilon_1 - \lambda \Omega \varepsilon_1^* = 0. \tag{8.175}$$

After that, upon subtracting Eq. (8.175) from its complex conjugate, one can arrive at

$$\lambda \Omega (\varepsilon_1 - \varepsilon_1^*) = 0, \tag{8.176}$$

that is, $\varepsilon_1 = \varepsilon_1^*$, as result $\varepsilon_2 = \varepsilon_2^*$, which leads to $\alpha = \alpha^*$ and $\beta = \beta^*$.

References

1. S.L. Braunstein, P. van Loock, Rev. Mod. Phys. **77**, 513 (2005)
2. N.M. Kroll, Phys. Rev. **127**, 1207 (1962); J.K. Oshman, S.E. Harris, IEEE J. Quant. Elect. **4**, 491 (1968); M.T. Raiford, Phys. Rev. A **2**, 1541 (1970); D. Stoler, Phys. Rev. Lett. **33**, 1397 (1974); U.L. Andersen et al., Phys. Scr. **91**, 053001 (2016)
3. K.F. Hulme, Rep. Prog. Phys. **36**, 497 (1973); R.W. Boyd, *Nonlinear Optics* (Elsevier, Oxford, 2008)
4. M.O. Scully, M.S. Zubairy, *Quantum Optics* (Cambridge University Press, Cambridge, 1997)
5. W. Brunner, H. Paul, Prog. Opt. **15**, 1 (1977); I. Breunig, D. Haertle, K. Buse, App. Phys. B **105**, 99 (2011)
6. L. Gilles et al., Phys. Rev. A **55**, 2245 (1997); S. Chaturvedi, K. Dechoum, P.D. Drummond, ibid. **65**, 033805 (2002); S. Tesfa, Eur. Phys. J. D **46**, 351 (2008) and J. Phys. B **41**, 065506 (2008)
7. J. Hald et al., Phys. Rev. Lett. **83**, 1319 (1999); Y. Zhang et al., Phys. Rev. A **62**, 023813 (2000); C. Simon, D. Bouwmeester, Phys. Rev. Lett **91**, 05360 (2003)
8. S. Tesfa, Nonlinear. Opt. Quant. Opt. **38**, 39 (2008)
9. P.A. Franken et al., Phys. Rev. Lett. **7**, 118 (1961)
10. Z. Zuo et al., Phys. Rev. Lett. **97**, 193904 (2006)
11. R.E. Slusher et al., Phys. Rev. Lett. **55**, 2409 (1985)
12. M.O. Scully, Phys. Rev. Lett. **55**, 2802 (1985); M.O. Scully, M.S. Zubairy, Phys. Rev. A **35**, 752 (1987); N.A. Ansari, ibid. **48**, 4686 (1993); H. Xiong, M.O. Scully, M.S. Zubairy, Phys. Rev. Lett. **94**, 023601 (2005); S. Tesfa, A nondegenerate three-level cascade laser coupled to a two-mode squeezed vacuum reservoir. Ph.D. thesis, Addis Ababa University, 2008
13. S. Tesfa, Phys. Rev A **79**, 033810 and ibid. 063815 (2009)

14. S. Tesfa, Phys. Rev. A **74**, 043816 (2006); J. Phys. B **41**, 055503 (2008); ibid. **42**, 215506 (2009) and Opt. Commun. **285**, 830 (2012)
15. S.M. Barnett, P.M. Badmore, *Methods in Theoretical Quantum Optics* (Oxford University Press, New York, 1997); S. Tesfa, J. Phys. B **41**, 145501 (2008) and ibid. **42**, 215506 (2009)
16. R. Auccaise et al., Phys. Rev. Lett. **107**, 070501 (2011); G. Passante et al., Phys. Rev. A **84**, 044302 (2011); A. Chiuri et al., ibid., 020304 (2011)

Part IV

Exploring Quantum Optical Processes

Optical Qubits and Quantum Logic Operations 9

One of the interesting features of double-slit experiment is the likelihood of casting the distribution of the intensity on the screen when the photon is capable of interfering with itself due to the inaccessibility of the information of which-path (see Chap. 4). One may recall that the moment the which-path question is answered, the interference pattern disappears. The phenomenon of quantum superposition can also be associated with the situation in which the quantum property of the state of the particle happens to be manifested in not only one state, but also in all possible states at once (for historical account and introduction; see for instance [1]). Superposition as a result can be taken as the most important attribute of quantum theory that harbors the perceived pertinent mystery. For example, it is the indistinguishability of the amplitudes that leads to the interference event deployed to exemplify the complementarity inherent in the foundation of quantum theory (see Sect. 4.1 and also [2]).

The phenomenon of superposition can be epitomized by the condition that if the system can be in two or more configurations, it can also be in the superimposed state, and the content of each configuration in the superposition can be specified by a complex number. If $|\Psi_a\rangle$ and $|\Psi_b\rangle$ denote the state of the photons going through slits a and b, the state that represents the coherent superposition can be denoted for instance by

$$|\Psi\rangle = \frac{1}{\sqrt{2}}\left[|\Psi_a\rangle + |\Psi_b\rangle\right]. \tag{9.1}$$

In line with this, the information about the quantum property can be encoded onto the qubit of a two-state quantum system, where the states can be denoted by $|0\rangle$ and $|1\rangle$.

© The Author(s), under exclusive license to Springer Nature Switzerland AG 2020
S. Tesfa, *Quantum Optical Processes*, Lecture Notes in Physics 976,
https://doi.org/10.1007/978-3-030-62348-7_9

It should then be admissible to construe that any quantum system with at least two quantum states can serve as a qubit (see Chaps. 11 and 12). One may also contend that the most vital property of a quantum state is the coherent superposition

$$|Q\rangle = \alpha|0\rangle + \beta|1\rangle \tag{9.2}$$

in which α and β are complex probability amplitudes with the condition that $|\alpha|^2 + |\beta|^2 = 1$. There is no way for someone to know which state the particle occupies since $|\alpha|^2$ and $|\beta|^2$ designate the probabilities for the combined state to be in respective state—the main trait of coherent superposition.

It might also be necessary stressing that superposition embodies the correlation between the subsystems that gives rise to the distinction between quantum correlation and its classical counterpart (see Sect. 4.2). With this delineation, the superimposed states that display quantum interconnectedness are referred to as entangled states. We thus seek to explore the entangled states that can be constructed from atomic energy levels with the help of external coherent light (see Sect. 9.2 and also [3]), and from the polarization and quadratures of light (see Sect. 9.3).

Quantum logic gate on the other hand is one of the quantum architectures that operates on a small number of qubits (entangled states) and considered as the building block of quantum circuitry [4]. A logic gate is thus a physical system with a capacity to implement a variety of operations based on the properties of the control and input variables while subjected to a random error (see Sect. 9.4). In classical scenario, to compute a function, one utilizes a two-bit algebra to designate the inputs and outputs while the parameters that denote the function, let us say, $a_i's$ are binary variables: 0 or 1. So arbitrary large function such as $f(a_1, a_2, \ldots, a_n)$ is often described by n-bit Boolean function applied in succession to simulate the prescribed function on a small part of the inputs chosen from the set of $\{a_i\}$.

In quantum regime however the action of such a gate on the inputs can be expressed as

$$|\Psi\rangle_{out} = \hat{U}|\Psi\rangle_{in}, \tag{9.3}$$

where $\hat{U}$ is the evolution operator. It might worth noting that a quantum gate can also be construed as a networked scheme that can be used to construct quantum circuitry; and referred to as universal if its replicas can be joined together to make a circuit that can compute a function and at the same time establish a desired unitary transformation on a set of quantum variables that involve a reversible process.[1] Based on the unitary nature of quantum operation, the peculiar traits of

---

[1]One of the principal concerns in classical computation is related to logical and physical reversibility that refers to the ability to reconstruct the input from the output via manipulating the gate function [6]. But in the context of thermodynamics, the change in entropy due to the loss of one bit of information corresponds to the energy increase of $kT\ln2$, where k is Boltzman's constant and T is the temperature. The heat that dissipates during the process is taken as a signature for physical irreversibility: the situation in which the microscopic physical state of the system cannot be restored exactly (compare with the notion of decoherence).

quantum state such as superposition, interference, entanglement, nonlocality and the accompanying measurement process are expected to enhance the power of manipulation. One may note that a quantum logic gate can also be portrayed in the context of quantum optics as a device whose inputs and outputs are taken to be optical polarizations, which can be extended to conditions available regime when required [5].

9.1 General Background

Imagine the source capable of emitting two particles with opposite momentum. Due to the perceived difference in the momentum, let us say, when particle 1 emerges in the left side, particle 2 would emerge towards right, or if particle 1 is found in the upper beam, particle 2 would be in the lower. In this kind of delineation, the two particles carry different bit values: particle 1 is either in state $|0\rangle$ or $|1\rangle$ while particle 2 takes the remaining. A two-party superposed state can be designated in quantum theory by

$$|\Psi\rangle = \frac{1}{\sqrt{2}}\left[|0\rangle_1 \otimes |1\rangle_2 + |1\rangle_1 \otimes |0\rangle_2\right]. \tag{9.4}$$

Such a state is christened as entangled since it cannot be separated.

The most interesting and puzzling property of this entangled state is neither of the two qubits carries definite value. Nonetheless, as soon as one of them is subjected to measurement where the result is random, the other is immediately found to carry the opposite value irrespective of the locations of the individual particles. It is consequently possible to state that one of the widely studied examples of the bipartite entangled states is the four sets of Bell states commonly known as EPR states:

$$|\Phi_\pm\rangle = \frac{1}{\sqrt{2}}\left[|00\rangle \pm |11\rangle\right], \tag{9.5}$$

$$|\Psi_\pm\rangle = \frac{1}{\sqrt{2}}\left[|01\rangle \pm |10\rangle\right]. \tag{9.6}$$

These states are maximally entangled in the sense that if one discards the information pertaining to one of the qubits, the measurement performed on the other pair would yield completely random result.

For the source that emits three particles at $120°$, the superposition of the quantum states can be expressed in the same way as

$$|\Psi\rangle = \frac{1}{\sqrt{2}}\big[|0\rangle_1 \otimes |0\rangle_2 \otimes |0\rangle_3 + |1\rangle_1 \otimes |1\rangle_2 \otimes |1\rangle_3\big], \qquad (9.7)$$

which is a tripartite entangled state, and is usually dubbed as Greenberger-Horne-Zeilinger (GHZ) state. Note that none of the three qubits carries information by its own, and so does not have a definite bit value. Even then, as soon as one of the three qubits is measured, the other two will assume a well defined value as long as the measurement is performed in $\{0, 1\}$ bases. This understanding holds irrespective of the spatial separation between the three measurements; the notion that leads to the phenomenon of quantum teleportation.

Even though there is no standard measure that quantifies multipartite entanglement, many measures qualify the GHZ state as a maximally three-party entangled state. The GHZ state also leads to the violation of Bell's inequality, and so exhibits a strong nonclassical correlation. The GHZ state can also be expressed in a compact mathematical form as

$$|GHZ\rangle_{\pm} = \frac{1}{\sqrt{2}}\big[|000\rangle \pm |111\rangle\big]. \qquad (9.8)$$

It is possible to infer from Eq. (9.8) as $|GHZ\rangle_{\pm}$ states account for the case when there is one particle in each compartment and none in the corresponding state.

There is also another three-party entangled state: the W state that can be expressed as

$$|W\rangle = \frac{1}{\sqrt{3}}\big[|100\rangle + |010\rangle + |001\rangle\big]. \qquad (9.9)$$

Markedly, $|W\rangle$ state is not necessarily equivalent to the $|GHZ\rangle$ state. A simple manipulation for example may reveal that the entanglement in the $|W\rangle$ state is robust against the loss of one qubit while the $|GHZ\rangle$ state reduces to the product of two qubits for the same case. In light of this, tracing over one of the three qubits in the $|GHZ\rangle$ state leads to unentangled mixture state: $|00\rangle\langle00| + |11\rangle\langle11|$ but tracing out one qubit in the $|W\rangle$ state leads to $\sqrt{\frac{2}{3}}|\psi^+\rangle\langle\psi^+| + \sqrt{\frac{1}{3}}|00\rangle\langle00|$—a maximally entangled state.

One may note that even though the $|GHZ\rangle$ state is introduced for three qubits, it can also be interpreted as the superposition of many particle states in their respective basis state $|0\rangle$ and $|1\rangle$. This thinking can be extended to more qubits, and so may lead to the possibility of expressing the $|GHZ\rangle$ state in a more general form as

$$|GHZ\rangle = \frac{1}{\sqrt{2}}\big[|000...0\rangle + |111...1\rangle\big]. \qquad (9.10)$$

It should also be possible to generalize for $|W\rangle$ state as

$$|W\rangle = \frac{1}{\sqrt{N}}\big[|000\ldots1\rangle + |000..10\rangle + \ldots + |010\ldots0\rangle + |100\ldots0\rangle\big]. \qquad (9.11)$$

One can see that $|W\rangle$ state of N qubits consists of equally weighed superposition of the N states each of which has exactly one qubit in state $|1\rangle$ and all the others in $|0\rangle$ while $|GHZ\rangle$ state[2] stands for the quantum superposition of all subsystems being in state 0 with the rest being in state 1.

9.2 Light Induced Atomic Qubits

Based on earlier mathematical depiction of a qubit, it may not be hard to conceive that two remote atomic states that constitute the qubit can also be entangled with exchange of quantized light. Such a proposal can be ascertained for instance upon injecting a two-mode squeezed light into two spatially separated cavities that contain a single two-level atom each and when the atom in each cavity is presumed to resonantly interact with the pertinent radiation in the cavity. Owing to the nonclassical properties of squeezed light, the pair of atoms are expected to coherently interact despite the accompanying spatial separation. The interaction of arbitrary two-level atom with radiation can also lead to entanglement of the atom with the radiation as long as the coherent superposition of the energy levels of the atom can be induced (see Chap. 6 and also [7]). One can thus see that in case the decaying process of the atom to the energy states participating in the interaction is prolonged or dephasing is minimized, the presumed induced entanglement can survive, and so can be transferred to the second atom via exchange of quantized radiation.

It is a well established fact that in the good cavity limit the photon emitted by one of the atoms can be almost immediately absorbed by the other in which the exchange of the energy or radiation between the atoms is believed to be responsible for inducing the sought for entanglement. Since the strong coupling can reduce the loss of quantum properties via decoherence, the inhibition of damping can be realized; and as a result, in the strong coupling regime, the established entanglement could be robust. The major challenge in practical utilization of the entangled atomic states however can be related to the difficulty of isolating the atoms from the surrounding environment since the interaction of atomic system with the surrounding environment gives rise to decoherence that leads to the loss of quantum information stored in the system (see Sect. 10.3).

---

[2]Although we are content with this way of describing entangled state, it is rewarding to note that other set of entangled state can also be introduced following a similar technique; and present discussion can be extended to systems with a continuous variable as we shall discuss in the following.

It might be sufficient to be restricted to two two-level atoms that can be designated by the set of bare atomic states $\{|a_1\rangle \otimes |a_2\rangle,\ |a_1\rangle \otimes |b_2\rangle,\ |a_2\rangle \otimes |b_1\rangle,\ |b_1\rangle \otimes |b_2\rangle\}$ to illustrate this idea, where the numbers represent the atoms and the letters the atomic levels. In case the dipole-dipole interaction between the atoms is inferred, the combined product states may form two superposed states usually turn out to be entangled. To put this in context, imagine that the two atoms are identical and atomic eigenstates that describe the two two-atom system as

$$|\pm\rangle = \frac{1}{\sqrt{2}}\left[|a_1\rangle \otimes |b_2\rangle \pm e^{i\phi}|b_1\rangle \otimes |a_2\rangle\right], \tag{9.12}$$

where ϕ is the phase factor [8]. Note that these states can capture the effects of the underlying dipole-dipole interaction; and at the same time, can be taken as atomic Bell states since one can generate a maximally entangled state by fixing the accompanying phase, which constitute the sought for atomic entangled states.

It is found out to be rewarding foreseeing that there are some interesting entangled states that can be constructed from two two-level atoms. For instance, a two-photon entangling mechanism in which the superposed sates of the atoms, when both or neither of them are excited would be dressed by the two-photon coherent interaction, can be thought of (see Sect. 6.3). That means, superposed states induced by a two-photon excitation process encompassing nonclassical correlations can transfer population from the two-atom lower state to the upper state without necessarily populating the intermediate energy level. This consideration can make possible the construction of a GHZ type two-party entangled state from the atoms coupled to a two-mode squeezed state.

It should also be straightforward to see that a three-level atom can be taken as another physical resource that can be exploited to construct atomic entangled states, that is, for a simple system composed of two atomic dipole transitions, it is possible to introduce superposed states such as

$$\hat{E}_s = p\hat{S}_1^+ + q\hat{S}_2^+, \tag{9.13}$$

$$\hat{E}_a = q\hat{S}_1^+ - p\hat{S}_2^+, \tag{9.14}$$

where p and q are complex transformation coefficients, and $\hat{S}_1$ and $\hat{S}_2$ are the atomic operators. For the corresponding transformation to be unitary, these coefficients need to form a complete and an orthonormal set: $|p|^2 + |q|^2 = 1$.

In case the superposition is induced by spontaneous emission as in the cascade scheme, these coefficients can be chosen as

$$p = \sqrt{\frac{\gamma_1}{\gamma_1 + \gamma_2}}, \qquad q = \sqrt{\frac{\gamma_2}{\gamma_1 + \gamma_2}}, \tag{9.15}$$

where γ_1 and γ_2 are the spontaneous decay rates of the two dipole transitions, and if the superposition is induced by external pumping;

$$p = \sqrt{\frac{\Omega_1}{\Omega_1 + \Omega_2}}, \qquad q = \sqrt{\frac{\Omega_2}{\Omega_1 + \Omega_2}}, \tag{9.16}$$

where Ω_1 and Ω_2 are the Rabi frequencies of the laser fields.

In addition, the superposed states denoted by the subscripts a and s (that stand for asymmetric and symmetric) that correspond to different configurations can be evaluated in terms of the actual atomic energy levels participating in the transition. For example, for $\vee$-type configuration: $|+\rangle = p|u\rangle + q|b\rangle$ and $|-\rangle = q|u\rangle - p|b\rangle$; for $\wedge$-type configuration $|+\rangle = p|b\rangle + q|l\rangle$ and $|-\rangle = q|b\rangle - p|l\rangle$; and for cascade type $|+\rangle = p|u\rangle + q|l\rangle$ and $|-\rangle = q|u\rangle - p|l\rangle$, where $|u\rangle$, $|b\rangle$ and $|l\rangle$ denote the upper, intermediate and lower atomic energy levels (see Sect. 6.5).

In case, $\gamma_1 = \gamma_2$ and $\Omega_1 = \Omega_2$, one can readily see that

$$|\pm\rangle = \frac{1}{\sqrt{2}}\big[|u\rangle \pm |b\rangle\big], \qquad \text{for } \vee \text{ configuration;} \tag{9.17}$$

$$|\pm\rangle = \frac{1}{\sqrt{2}}\big[|b\rangle \pm |l\rangle\big], \qquad \text{for } \wedge \text{ configuration;} \tag{9.18}$$

$$|\pm\rangle = \frac{1}{\sqrt{2}}\big[|u\rangle \pm |l\rangle\big], \qquad \text{for } \Xi \text{ configuration.} \tag{9.19}$$

Upon taking these representations as basic qubits, it should be possible in principle to construct entangled state for two three-level atoms. By extending the number of atomic energy levels involved in the interaction, it should be allowable to construct in the same way a multi-party entangled state.

Moreover, in view of blockade scenario, one can argue that the atom-radiation entangled states should also be thought of following the same reasoning (see Sect. 6.6 and also [7]). Since the interaction can be taken as exchange of energy between the electronic excitation and radiation field, the atom-radiation coupled system can be expressed by the atomic dressed states as

$$|\pm\rangle = \frac{1}{\sqrt{2}}\big[c_g|0, n + 1\rangle \pm c_e|1, n\rangle\big], \tag{9.20}$$

where $c_{g,e}$ are probability amplitudes (compare with Eqs. (6.13) and (6.243)). One may see that Eq. (9.20) represents the process equivalent to stimulated absorption and emission where the spontaneous emission signifies the emitted photon while the atom cascades down the ladder, and stands for a simple form of atom-radiation entangled state.

To extend this discussion to a multi-level atom, one may begin with the fact that the superposition of $N + 1$ number states can be constructed from the vacuum state by injecting N appropriately prepared two-level atoms into the cavity, and detecting all of them in the ground state, or by passing one suitably prepared multi-level atom through the cavity [9]. The latter scenario in particular is consistent with the framework of the generalized Jaynes–Cummings model that involves a multi-level atom interacting with a single-mode cavity field. Let us hence consider $(N+1)$-level atom with a ladder configuration in which the single-photon transitions connect adjacent levels where the atom is initially taken to be in the nth level and the cavity field in the vacuum state.

The initial state of the whole system in this consideration would be $|n, 0\rangle$, and the pertinent subspace would be spanned by $\{|n, 0\rangle, \ |n - 1, 1\rangle, \ldots, \ |j, n - j\rangle, \ \ldots, |1, n - 1\rangle\}$, where $n - j$ is the photon number of the cavity field and j is a positive number lesser or equal to n. Then imagine that the atom initially in its lowest energy state $|1\rangle$ is sent to N classical fields. Assume also that nth classical field is tuned to the transition between the nth and $(n + 1)$th energy levels. One can further assume that the transition frequency between each pair of adjacent energy levels is different from those between other pairs of adjacent energy levels.

During the passage of the atom through the nth classical field, only the transition from $|n\rangle$ to $|n + 1\rangle$ energy states is important, and so other transitions can be neglected. After tuning of the first classical field, one can reasonably expect

$$|\Phi\rangle = a_1|1\rangle + \sqrt{1 - |a_1|^2}|2\rangle, \tag{9.21}$$

where coefficient a_1 can be controlled by adjusting the amplitude and phase of the classical field. It is also required that N transitions occur during the interaction with the cavity mode so that the transition frequency differences should not be very large. This process can be understood as the way of generating arbitrary superposition of the first $N + 1$-Fock states of the cavity field from the vacuum state by sending suitably prepared $(N + 1)$-level atom with a ladder configuration through the cavity and detecting it in the lowest state. It is therefore arguably possible to construct practically feasible atom-radiation entangled state.

9.3   Photonic Qubits

On the basis of the proposal that a qubit can be constructed from a two-party entangled system, it becomes imperative extending the previous discussion on a discrete system to a continuous variable quantum system, specifically, to light. Since photons have only two eigenvalues of their propagation or wave vectors that correspond to the right and left-handed circular polarizations, it might be insightful noting that linearly polarized photon can be considered as a genuine two-party quantum system. The characterization of a photonic qubit in the most part is based on the perception that quantum information can be encoded on the quantum states of certain degrees of freedom such as the polarization of individual photon. The possibility of performing quantum information processing tasks with the help of photons depends at fundamental level on the fact that nonclassical properties of the photon can be manipulated either by using simple optical elements such as interferometer or by letting them to interact with a matter at the atom-photon interface (see Chap. 6 and also [10]). Owing to the interaction between photonic and atomic qubits, it may not be difficult to foresee the transfer of quantum information from the flying photons to the localized atoms, and vice versa. Such a simple information transferring mechanism might be very important to invoke a discussion on quantum information processing (see Chap. 12).

9.3.1 Discrete Photonic Qubits

In relation to earlier discussion, arbitrary photonic qubit can be generally expressed as $\alpha|H\rangle + \beta|V\rangle$, where $|H\rangle$ and $|V\rangle$ stand for the horizontal and vertical polarization states, and α and β are constants with $|\alpha|^2 + |\beta|^2 = 1$ (compare with Eq. (9.2)). Existing experience entails that these states can be created and manipulated with a relative ease and high precision by making use of simple linear optical elements such as polarizing beam splitters, polarizers and wave plates (see Chap. 11 and also [11]). There are also other degrees of freedom of the photon that can be exploited to encode information on qubit. Spatial qubits of a single photon for example can be constructed from two different spatial modes of the light or the path the photon might follow ($\alpha|a\rangle + \beta|b\rangle$). This may occur when a single photon exits a beam splitter with two output modes designated by $|a\rangle$ and $|b\rangle$ that can be prepared by making use of suitable phase shifters and beam splitters (see Sect. 4.3.2).

It might not be hard to conceive that a two-state qubit with the same spectral shape but time shifted by much more than the coherence time can be realized by sending a single photon via unbalanced Mach–Zehnder interferometer (see Sect. 4.5 and also Fig. 4.1). The wave packet in this case would be split by the first beam splitter with transmission and reflection coefficients $|\alpha|^2$ and $|\beta|^2$ into two coherent parts. Imagine the transmitted part to propagate along the short arm while the reflected along the long arm. In case the wave packet extension is shorter than the arm length difference, the output from the ports of the second 50:50 beam splitter would be two wave packets separated in time. If no photon is registered in one of the output ports, in the other output one would have a single photon in a coherent superposition of $\alpha|E\rangle + \beta|L\rangle$, where E and L stand for early and late (see Fig. 9.1).

Its such a procedure of separating the light into distinct two states that would be taken as the underlying recipe for studying the photonic entangled states. To establish the scheme of a photonic qubit, it would be advisable starting with a well defined spatial mode that propagates, let us say, along z-axis. So it would be reasonable taking its vertical and horizontal polarization states to point along x- and y-axes that uniquely define the required two basis states of the qubit according to

Fig. 9.1 Schematic representation of the setup that can separate the light impinged onto beam splitter into two different states such as early and late, and later recombine them on another beam splitter so that an entangled photonic state is formed

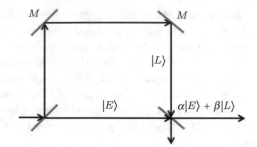

the transformation:[3] $|V\rangle \leftrightarrow |0\rangle$ and $|H\rangle \leftrightarrow |1\rangle$, where the pertinent orthogonal polarization modes can be described by the creation and annihilation operators $\hat{a}_V^\dagger$, $\hat{a}_V$, $\hat{a}_H^\dagger$ and $\hat{a}_H$ that obey the commutation relations: $[\hat{a}_k, \hat{a}_j] = [\hat{a}_k^\dagger, \hat{a}_j^\dagger] = 0$ and $[\hat{a}_k, \hat{a}_j^\dagger] = \delta_{kj}\hat{I}$ in which $k, j = V, H$ [11].

Consistent with earlier discussion, being a quantum object, photonic qubit is expected to offer a far better option for manipulation than classical bits. In this respect, it should be possible to begin with the entangled states construed from the vertical and horizontal polarization states:

$$|D^\pm\rangle = \frac{1}{\sqrt{2}}\big[|H\rangle \pm |V\rangle\big], \tag{9.22}$$

$$|R^\pm\rangle = \frac{1}{\sqrt{2}}\big[|H\rangle \pm i|V\rangle\big], \tag{9.23}$$

$$|P\rangle = p|H\rangle + e^{i\phi}q|V\rangle, \tag{9.24}$$

where $p^2 + q^2 = 1$ and ϕ is arbitrary phase. This kind of choice provides additional degrees of freedom such as displacement, rotation and phase shift that can be used in the field of continuous variable quantum information processing.

One of the interesting aspects of such representation might be the possibility of rotating the polarization angle, for instance by 45°, and generate the quantum state of the form

$$|E\rangle = |H\rangle \otimes |H\rangle + |H\rangle \otimes |V\rangle + |V\rangle \otimes |H\rangle + |V\rangle \otimes |V\rangle, \tag{9.25}$$

---

[3] A polarized light beam can be taken in quantum realm as being in state $|H\rangle$ if in some time interval there is a unit probability for one photon to be detected at the horizontal output of the polarizing beam splitter placed in the beam's path, and $|V\rangle$ if found at the vertical output. Note that a polarized beam splitter is the device that splits the incident beam into two beams having different linear polarization based on the appropriate choice of the thickness of the crystal, the wavelength of light and the variation of the index of refraction.

which envisages equal superposition of the various compositions. Repetition of similar approach or rotating across the accessible polarization angle can lead to a higher-order bit values; as a result, one may foresee the possibility for composition. Doubtless, one should note that quantum theory allows to form superpositions of the vertical and horizontal polarization states with maximally entangled Bell states

$$|\Phi^{\pm}\rangle = \frac{1}{\sqrt{2}}\left[|H\rangle_1 \otimes |H\rangle_2 \pm |V\rangle_1 \otimes |V\rangle_2\right], \tag{9.26}$$

$$|\Psi^{\pm}\rangle = \frac{1}{\sqrt{2}}\left[|H\rangle_1 \otimes |V\rangle_2 \pm |V\rangle_1 \otimes |H\rangle_2\right] \tag{9.27}$$

(note the striking similarity with Eqs. (9.5) and (9.6)).

As illustration, one may begin with the fact that the number state has no classical analogue, and so expected to exhibit certain peculiar statistical properties where the relevant example includes the superposition of Fock states,[4] which can lead to nonclassical features such as photon anti-bunching and quadrature squeezing. The interest of generating a superposed quantum states of light fields can be linked to cavity quantum electrodynamics in which the beams of atoms are made to strongly interact with the cavity mode [12].

The interaction of a two-level atom with an external coherent light can be described in this case, and in the dressed picture, by states

$$|+\rangle = \cos\theta_n |u, n-1\rangle - \sin\theta_n |l, n\rangle, \tag{9.28}$$

$$|-\rangle = \sin\theta_n |u, n-1\rangle + \cos\theta_n |l, n\rangle, \tag{9.29}$$

where u and l represent the upper and lower energy levels, n the number of the photons in the cavity and θ_n the atom-radiation interaction (relate with Eqs. (6.24) and (6.25)). Equations (9.28) and (9.29) can also be rewritten when the atom is initially prepared in a certain coherent superposition by adjusting the involved parameters as

$$|n, \pm\rangle = \frac{1}{\sqrt{2}}\left[|u, n\rangle \pm i|l, n+1\rangle\right], \tag{9.30}$$

which epitomizes a maximally entangled state; and hence, can be taken as a standard qubit (compare with Eq. (9.12)).

One may note that Eq. (9.30) describes the entanglement between atomic energy levels and the photon number in the cavity via successive emission-absorption events. To generate the Fock states in the cavity regime, a variety of ways such

---

[4]When the information of the which-path of a light is ambiguous, the correlation between the horizontal and vertical polarization would be maximum; and the emerging wave function could be replaced by incoherent mixture of the states resulting from the suppression of the interference.

as a single two-level atom coupled to a cavity mode radiation can be thought of. Since such an interaction with low noise fluctuations can facilitate the generation of noise fluctuations below vacuum level, it might be required during practical implementations.

To gain a better insight to the manipulation based on Fock states (see Sect. 9.2), one may begin with the quantum state of the form

$$|\Psi_n\rangle = A_0|0\rangle + A_n|n\rangle, \tag{9.31}$$

with $|A_0|^2 + |A_n|^2 = 1$. Note that Eq. (9.31) represents one of the photonic qubits. For a one photon system, it might be straightforward to write

$$|\Psi_1\rangle = A_0|0\rangle + A_1|1\rangle. \tag{9.32}$$

One can also consider the state such as

$$|\Psi_2\rangle = A_0|0\rangle + A_2|2\rangle. \tag{9.33}$$

9.3.2   Continuous Variable Photonic Qubits

One can also extend previous discussion to arbitrary eigenstates of the photon number as

$$|\Psi\rangle_N = \frac{1}{\sqrt{N+1}} \sum_{n=0}^{N} |n\rangle_a \otimes |N-1\rangle_b. \tag{9.34}$$

This depiction suggests that if one successively employs a boson creation operator on the coherent state, that is, if one adds photons to the coherent light, one can expect a nonclassical state of light (see Sect. 4.4). In light of this, if m photons are assumed to be added to the coherent light, one can get entangled state of the form

$$|\alpha, m\rangle = \frac{e^{-\frac{\alpha^*\alpha}{2}}}{\sqrt{\sum_{n=0}^{m} \frac{(\alpha^*\alpha)^n (m!)^2}{(n!)^2 (m-n)!}}} \sum_{n=0}^{\infty} \frac{\alpha^n \sqrt{(m+n)!}}{n!} |n\rangle \otimes |m\rangle. \tag{9.35}$$

Due to the emerging strong entanglement, it should be possible in principle to construct a desired photonic qubits following this approach.

One can also cite as alternative scheme the process by which a series of photon subtractions or additions accompanied by a suitable displacement can lead to the generation of arbitrary superposition of Fock states [13]. One may note that a finite number quantum state can be realized as superposition of Fock states with any desired accuracy. The way in which a single photon detection generates arbitrary superposition of Fock states for example can be understood by considering a two-mode squeezed vacuum state that can be generated from a nondegenerate parametric process.

9.3.2.1 For Light with Finite Number

In the basis of the photon number states, one can generally write

$$|\Phi\rangle_{s,i} \propto \sum_n q^n |n\rangle_s \otimes |n\rangle_i, \tag{9.36}$$

where the quantity q ($0 \leq q < 1$) relies on the pump power and nonlinear coefficient of the crystal. After that, if one can split, let us say, the idler mode into three modes using a nonlinear photon interaction, and if the generated idler modes are assumed to be displaced by coherent amplitudes of β_1, β_2, and β_3, in the limit of a small pump power and displacement, the projection process can be expressed as

$$|\Phi\rangle_s \propto \langle 0|_i \left(\frac{\hat{a}}{\sqrt{3}} + \beta_1\right) \left(\frac{\hat{a}}{\sqrt{3}} + \beta_2\right) \left(\frac{\hat{a}}{\sqrt{3}} + \beta_3\right) |\Phi\rangle_{i,s}, \tag{9.37}$$

where $\hat{a}$ is the annihilation operator acting on the idler mode and $1/\sqrt{3}$ accounts for the splitting of the idler mode into three.

In practical setting, with the aid of linear optical components, it is possible to split the idler mode into several modes, and displace each of these modes by a coherent amplitudes β_i, where the idler modes can be measured by a single photon detectors. Then when the coincidence occurs, the signal mode can be projected onto the desired superposed state. In the limit of a small pump power and displacements, one can observe that

$$|\Phi\rangle = \beta_1\beta_2\beta_3|0\rangle + \frac{q}{\sqrt{3}}(\beta_1\beta_2 + \beta_2\beta_3 + \beta_3\beta_1)|1\rangle$$

$$+ \frac{\sqrt{2}}{3}q^2(\beta_1 + \beta_2 + \beta_3)|2\rangle + \frac{\sqrt{2}}{3}q^3|3\rangle. \tag{9.38}$$

This discussion may evince the possibility of generating entangled Fock state via splitting the radiation with the help of nonlinear interaction.

It also appears possible to extend this discussion to a large number of suitable idler modes, which may allow us to generate entangled state of a multi-mode Fock state. One notable example is the state usually referred to as path entangled number state or NOON state: an entangled two-mode state in which N photons are in one of the two paths of the interferometer, and zero in the other (see the mechanism shown in Sect. 4.3.2 and also [14]).

Suppose the light that passes through the phase shifter is denoted by $\hat{c}$ and the other by $\hat{d}$ (for a general; scheme see Fig. 9.2, and for the rationale and emerging

Fig. 9.2 Schematic representation of the setup used to mix two light modes at different arms of the beam splitter and resolve the photon number falling on separate detectors

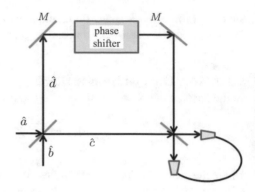

expressions Sect. 4.1.5). Taking the effect of the beam splitter and phase shifter into account, the radiation falling on the screen can be designated by

$$\hat{c} = \frac{1}{\sqrt{2}}(\hat{a} - i\hat{b})e^{i\phi}, \qquad \hat{d} = \frac{1}{\sqrt{2}}(\hat{b} - i\hat{a}), \tag{9.39}$$

where $\hat{a}$ and $\hat{b}$ designate the input radiations falling on the beam splitter. The transformations between the incoming modes ($\hat{a}$ and $\hat{b}$) and the outgoing modes ($\hat{c}$ and $\hat{d}$) on the polarized beam splitter can be described by $a_H \to \hat{c}_H$, $b_H \to \hat{d}_H$, $a_V \to i\hat{d}_V$ and $b_V \to i\hat{c}_V$. With this denotation, the radiation that superimposed on the screen can be expressed as

$$\hat{c} + \hat{d} = \frac{1}{\sqrt{2}}\left[(e^{i\phi} - i)\hat{a} + (1 - ie^{i\phi})\hat{b}\right]. \tag{9.40}$$

Imagine also that the radiation emitted by a nondegenerate parametric oscillator (with one photon in each part) is impinged onto the beam splitter as shown in Fig. 9.3 (relate also with Fig. 9.1). The simplest choice to represent the nonclassical photon number product in such a case is the state $|\Psi\rangle = |1\rangle_a \otimes |1\rangle_b$, which is the natural output of a single-photon parametric down conversion. When two indistinguishable photons are impinged onto a 50:50 beam splitter in both input arms, the phase relation can be made to work in such a way that the output modes would always be in the state $|2002\rangle$, which is the operational description of the Hong-Ou-Mandel interferometer as sketched in Fig. 9.3 (see also [15]).

In regards to present discussion, the two photons from different input modes of the beam splitter cannot trigger a twofold detection coincidence at the output modes. So one of the corresponding interesting ideas would be: the interference effects that emerge as a result of the passage of the radiation through symmetric and lossless beam splitter leads to the product number state

$$|2002\rangle = \frac{1}{\sqrt{2}}\left[|0\rangle_a|2\rangle_b + |2\rangle_a|0\rangle_b\right]. \tag{9.41}$$

Fig. 9.3 Schematic representation of the mechanism that used to entangle the photons generated by a single source, and appear at one of the detectors (Hong-Ou-Mandel interferometer). Note that in one of the pathes, the radiation is made to pass through nonlinear medium

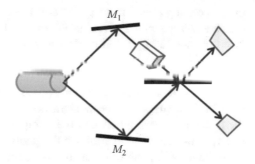

The entanglement appears to be between the photon number and the path. Technical issues aside, since there are two indistinguishable paths, the amplitude associated with the paths is expected to interfere; the rational for expecting the combined state to be entangled.

This presumption can hence be extended to the case when $N/2$ photons are sent through the beam splitter from both ports of the scheme represented by Fig. 9.1. The output radiation in this case constitutes the state often designated as a path entangled number state or NOON state

$$|\text{NOON}\rangle = \frac{1}{\sqrt{2}}\left[|N\rangle_a \otimes |0\rangle_b + |0\rangle_a \otimes |N\rangle_b\right], \tag{9.42}$$

which can also be expressed as

$$|\text{NOON}\rangle = \frac{1}{\sqrt{2N!}}\left[\hat{a}_1^{\dagger N} + \hat{a}_2^{\dagger N}\right]|0,0\rangle = \frac{1}{\sqrt{2N!}}\left[\hat{a}_1^N + \hat{a}_2^N\right]|N,N\rangle. \tag{9.43}$$

Equation (9.43) may indicate that the NOON state can be generated by a successive application of either photon addition or subtraction mechanism. Following the same reasoning, this discussion can be extended to the case when one has a beam splitter with a partial transmission in which the output state can be described in terms of the entangled state as

$$|n,m\rangle = \frac{1}{\sqrt{2}}\left[|n\rangle_a \otimes |m\rangle_b + |m\rangle_a \otimes |n\rangle_b\right], \tag{9.44}$$

where n and m are the number of photons in respective modes, and characterize a maximally entangled number state.

9.3.2.2 For Coherent Light

On the basis of the fact that unlike the number state, the coherent state is typified by infinite photon number but with a definite phase, while constructing a photonic qubit from a coherent state, the emphasis should be on the phase manipulations rather than the photon number as in the earlier cases. One may note that the process of constructing a qubit from a coherent state can be achieved by sending a laser light across a nonlinear medium at least in one of the arms of the pertinent interferometer (see Figs. 4.1 and 9.3), and then manipulate the emerging superimposed light with the aid of linear optical elements such as beam splitters, mirrors and phase shifters. Besides, it might worth noting that the pertinent superposition of the coherent states generated with the help of linear optical setup should conserve the total photon number. The suggested mechanism however might have a potential to constrain the phase relationship between the components of the multi-mode coherent state superposition required to establish the photonic qubits.

To begin with, one can consider the process of mixing the coherent state with the vacuum state at a beam splitter that may yield entangled coherent state. But since the pertinent coherence is extremely sensitivity to environment, the realization of the superposition of coherent state is not an easy task. Despite such inherent drawbacks, soon after the introduction of a single-mode superposition of coherent state or Schrödinger kitten, entangled coherent state or superposition of a multi-mode coherent state becomes significantly important [16]. One can particularly cite the characterization of the entangled coherent state as superposition of a two-mode coherent state with equal amplitude but opposite optical phase, and allowed to have an arbitrary phase relationship between the two components of the bipartite system; the perception that may lead to even and odd states that provide the binary states required to conceive the qubit. Entangled coherent state can therefore be outright treated as a bimodal state; but later, can be generalized as superposition of a multi-mode coherent state.

The generalization to a multi-mode scenario insinuates the intricacies of the multipartite entanglement to be manifested in the entangled coherent states such as GHZ and W types of states [17]. To illustrate such a proposal, one may begin with the fact that the superposition of N coherent states can be expressed as

$$|\Psi\rangle_\alpha = \frac{1}{\sqrt{N + 2\sum_{k>j}^N e^{\frac{1}{2}|\alpha_j - \alpha_k|^2} \cos\left[\theta_j - \theta_k - \mathrm{Im}(\alpha_j^* \alpha_k)\right]}} \sum_{j=1}^N e^{i\theta_j} |\alpha_j\rangle,$$

(9.45)

which reduces for even and odd number coherent states to

$$|\Psi\rangle_{2\alpha} = \frac{1}{\sqrt{2(1 + e^{-2\alpha^* \alpha} \cos\theta)}} \left[|\alpha\rangle + e^{i\theta} |-\alpha\rangle\right],$$

(9.46)

where θ stands for the phase difference between the two coherent states.

In the context of continuous variable, the phase responsible for witnessing the nonclassical features is often related to the local oscillator that needs to be mode matched with the classical beam. As one can see, the phase shift can be incorporated as in Mach–Zehnder interferometer by increasing or decreasing the optical path length of the beam when compared to the local oscillator by adding transparent material of tailored depth in the accompanying interpretation (see Sect. 4.3.2 and also [18]).

To explore certain quantum features in terms of a continuous variable, it appears reasonable to introduce the asymmetric entangled coherent state of the type

$$|\Psi\rangle_\pm = \frac{1}{\sqrt{2}\sqrt{1 \pm e^{-2\alpha^*\alpha - 2\beta^*\beta}}} \big[|\alpha\rangle_a \otimes |\beta\rangle_b \pm |-\alpha\rangle_a \otimes |-\beta\rangle_b\big], \qquad (9.47)$$

where $|\Psi\rangle_+$ ($|\Psi\rangle_-$) can be taken to possess even (odd) sums of photons. When similar coherent light is impinged onto the beam splitter, the resulting entangled coherent state would look like

$$|\Psi\rangle_\pm = \frac{1}{\sqrt{2}\sqrt{1 \pm e^{-4\alpha^*\alpha}}} \big[|\alpha\rangle_a \otimes |\alpha\rangle_b \pm |-\alpha\rangle_a \otimes |-\alpha\rangle_b\big]. \qquad (9.48)$$

These states are usually dubbed as even and odd coherent state. The even-odd terminology refers to the situation in which the photon number distribution is nonzero for even (odd) photon number in case of the even (odd) coherent state. The detection of the coherent entangled states can be realized by optical homodyne detection mechanism, where the marginal distribution for the conjugate quadrature exhibits interference fringes that yield information on the strength of the coherence of the resulting superimposed state.

In the same token, imagine that entangled coherent state can be produced by feeding a coherent light and the vacuum state into Mach–Zehnder type interferometer that contains a nonlinear medium in one of or both arms [19]. Assume also that two coherent modes denoted by coherent states $|\alpha\rangle$ and $|\beta\rangle$ are injected into the interferometer composed of two beam splitters with a nonlinear Kerr medium in one of the arms (see Fig. 9.2). The pertinent unitary operator in such a case can be denoted by

$$\hat{U} = \hat{B}\hat{K}\hat{B}, \qquad (9.49)$$

where the beam splitter Hamiltonian is given by $\hat{B} = \exp\big[\frac{i\pi}{4}\big(\hat{a}^\dagger\hat{b} + \hat{a}\hat{b}^\dagger\big)\big]$ and the Kerr medium transformation by $\hat{K} = \exp\big[-i\chi\big(\hat{a}^\dagger\hat{a}\big)^2\big]$ in which χ represents the nonlinearity.

If one then assumes that the initial state is the product coherent state $|\alpha\rangle \otimes |\beta\rangle$, the output state would evolve as

$$|\Psi\rangle_{out} = \hat{U}|\alpha\rangle \otimes |\beta\rangle. \qquad (9.50)$$

One may recall that the beam splitter transforms the product state as

$$\hat{B}|\alpha\rangle_1 \otimes |\beta\rangle_2 = \frac{1}{\sqrt{2}}\big[|\alpha + i\beta\rangle_1 \otimes |\beta + i\alpha\rangle_2\big], \qquad (9.51)$$

where the numbers denote the arms of the interferometer (see Chap. 2). Note that the output state is also a direct product state of the emerging output states, and so does not lead to entanglement.

Contrary to this, the insertion of nonlinearity in one of the arms of the interferometer can transform the coherent state into a superposition of two macroscopically distinct coherent states [20]. In support of this, it might be possible to see that the Kerr medium can transform the coherent state as

$$\hat{K}|\alpha\rangle = e^{-\frac{\alpha^*\alpha}{2}} \sum_{n=0}^{\infty} \frac{\alpha^n}{\sqrt{n!}} e^{-i\chi n^2}|n\rangle, \tag{9.52}$$

and for $\chi = \pi/2$

$$\hat{K}|\alpha\rangle = \frac{1}{\sqrt{2}} \left[e^{-i\pi/4}|\alpha\rangle + e^{i\pi/4}|-\alpha\rangle \right]. \tag{9.53}$$

Equation (9.53) represents a special case of the superposition of the coherent states with equal complex field amplitudes and phase separations which is a maximally entangled state.

Then combination of the effect of the Kerr medium and beam splitter indicates that

$$\hat{U}|\alpha\rangle \otimes |\beta\rangle = \frac{1}{\sqrt{2}} \left[e^{-i\pi/4}|i\beta\rangle \otimes |i\alpha\rangle + e^{i\pi/4}|-\alpha\rangle \otimes |\beta\rangle \right]. \tag{9.54}$$

The resulting state would be the superposition of the coherent state $|\alpha\rangle$ in one mode and $|\beta\rangle$ in the other in case the phase is disregarded. In view of this understanding, if the coherent light is mixed at the beam splitter; and if what comes out in one of the arms interacts with the nonlinear medium on its way to the screen, the output radiation in the weak interaction limit and in the number state representation can be written as

$$|\psi\rangle = e^{\frac{-\alpha^*\alpha}{2}} \sum_{n=0}^{\infty} \frac{\alpha^2}{\sqrt{n!}} |n\rangle \otimes |-i\Omega\chi n\rangle, \tag{9.55}$$

where Ω stands for the strength of the pumping and χ for the strength of nonlinearity. This discussion may entail that a Gaussian state can be decomposed into thermal states, which makes thermal state or number state one of the fundamental blocks for constructing a Gaussian entangled photonic states.

9.3.2.3 For Squeezed Light
Let us now cite one of practically realizable setups in which upon applying the squeezing operator on the vacuum state, one can generate a squeezed vacuum state

that can be described in terms of a number state as

$$|r, r\rangle = \frac{1}{\sqrt{\cosh r}} \sum_{n=0}^{\infty} \frac{\sqrt{(2n)!}}{n! 2^n} \tanh^n r |2n\rangle \qquad (9.56)$$

and in the same way a two-mode squeezed vacuum state often dubbed as optical EPR state,

$$|EPR\rangle = |r, r\rangle = \sqrt{1 - \tanh^2(r)} \sum_{n}^{\infty} \tanh^n(r) |n\rangle_a \otimes |n\rangle_b, \qquad (9.57)$$

can be generated. Note that a maximally entangled $|EPR\rangle$ state can be obtained for a large squeezing ($r \to \infty$) [21].

This consideration can also be extended further based on the fact that the efficiency of a quantum information processing task significantly depends on the degree of entanglement (nonlocality) of the quantum state shared by the participating parties. In some cases, such dependence can be demonstrated by using the states constructed by mixing a maximally entangled state with a separable mixed state in the way that

$$\hat{\rho} = p|\psi\rangle\langle\psi| + \frac{(1 - p)}{d^2} \hat{I}_1 \otimes \hat{I}_2; \qquad 0 \le p \le 1, \qquad (9.58)$$

where

$$|\psi\rangle = \frac{1}{\sqrt{d}} \sum_{j=1}^{d} |j\rangle_1 \otimes |j\rangle_2 \qquad (9.59)$$

is taken to be the maximally entangled state and d is the dimension of the Hilbert space. The state defined by Eq. (9.58) is characterized by the probability of the maximally entangled state; and taken as entangled if $p > 1/(1 + d)$.

Even though the quest for generating a pure maximally entangled states is tempting, it is somewhat counterbalanced by the increasing emergence of the mixed states and non-maximally entangled states. Because of the unavoidable effects of losses and decohering interactions, these states are reasonably taken as the basic constituents of modern quantum information processing tasks as they determine the performance of the available protocols. Then upon comparing the maximum fidelity achievable for the classical communication and local operation with the average fidelity of this state, it is suggested that

$$\hat{\rho} = \frac{1}{1 + d} \left[\frac{1}{\sqrt{d}} \sum_{j=1}^{\infty} |j\rangle_1 \langle j|_2 + \frac{1}{d} \hat{I}_1 \otimes \hat{I}_2 \right], \qquad (9.60)$$

which stands for entangled state.

One can also imagine that such a representation can be extended to continuous variable system. With this in mind, let us pick the state of light generated by a nondegenerate optical parametric amplifier:

$$\hat{\rho}_{\text{NOPA}} = \left[1 - \tanh^2 r\right] \sum_{n,m}^{\infty} \tanh^{n+m} r |m, m\rangle \otimes |n, n\rangle. \tag{9.61}$$

Although the two-mode radiation is in an entangled $|\text{NOPA}\rangle$ state, each mode separately is in a thermal state. In line with this proposal, one may note that

$$\hat{\rho}_{\text{thermal}} = \left[1 - \tanh^2 s\right]^2 \sum_{n,m}^{\infty} \tanh^{2(n+m)} s |m\rangle \langle m|n\rangle \langle n|, \tag{9.62}$$

where $s = \sinh^{-1} \sqrt{\langle \hat{n} \rangle}$.

Combination of these descriptions of the light can be expressed as

$$\hat{\rho} = p \hat{\rho}_{\text{NOPA}} + (1 - p) \hat{\rho}_{\text{thermal}}. \tag{9.63}$$

Markedly, such a state which is often referred to as Werner state is worthy example of entangled optical state [22]. Depending on the weight of p, note that the two-qubit Werner states would be entangled when $p > 1/3$.

9.4 Quantum Optical Logic Operations

It is common knowledge that a quantum gate can be efficaciously epitomized by a matrix, that is, the quantum gate that acts on k qubits can be denoted by a $2^k \otimes 2^k$ unitary matrix, where a single-qubit quantum gate for example can be designated by a $2 \otimes 2$ matrix. The idea of representing a quantum gate by a square matrix entails that the number of qubits in the input and output parts of the process should be equal to accomplish a meaningful operation, that is, the action of the gate on the quantum state (or computation; see Sect. 12.1.3) can be expressed in terms of the multiplication of the vector (that represents the state) by the matrix (that stands for the gate) as defined by Eq. (9.3). In the context of quantum optics, logical operations can be implemented by the action of boson operator on various optical states [5, 11, 23].

9.4.1 Single-Qubit Optical Gates

One can generally express a quantum logic gate constructed from discrete states $|0\rangle$ and $|1\rangle$[^5] in terms of the unitary matrix

$$U = \begin{pmatrix} \alpha & \gamma \\ \beta & \delta \end{pmatrix}, \tag{9.64}$$

where α, β, γ, δ are space holders that take either 0 or 1. It is therefore possible to introduce as demonstration the unity or identity gate with $\alpha = \delta = 1$ and $\beta = \gamma = 0$ as

$$I = \begin{pmatrix} 1 & 0 \\ 0 & 1 \end{pmatrix}. \tag{9.65}$$

Upon choosing different logical values of the elements of the matrix (9.64), one can obviously think of various implementable gates.

9.4.1.1 Pauli Gates

Pauli X: This gate can be considered as the quantum equivalent of the classical NOT gate, and equated to the rotation of the Bloch sphere by π radians around x-axis.[^6] It generally maps $|0\rangle \to |1\rangle$ and $|1\rangle \to |0\rangle$; and can be represented by

$$X = \begin{pmatrix} 0 & 1 \\ 1 & 0 \end{pmatrix}. \tag{9.66}$$

Pauli Y: This is the gate that can be captured by the rotation around y-axis of the Bloch sphere by π radians, and so taken to map $|0\rangle \to i|1\rangle$ and $|0\rangle \to -i||0\rangle$;

$$Y = \begin{pmatrix} 0 & -i \\ i & 0 \end{pmatrix}. \tag{9.67}$$

Pauli Z: This is a special class of phase shift gate (with π phase shift), and equated to the rotation of the Bloch sphere by π radians around z-axis. It leaves $|0\rangle$ unchanged and maps $|1\rangle \to -|1\rangle$;

$$Z = \begin{pmatrix} 1 & 0 \\ 0 & -1 \end{pmatrix}. \tag{9.68}$$

[^5]: In most cases, one can use the transformation: $|0\rangle \leftrightarrow \begin{pmatrix} 1 \\ 0 \end{pmatrix}$ and $|1\rangle \leftrightarrow \begin{pmatrix} 0 \\ 1 \end{pmatrix}$ (mark the typeface).

[^6]: Such interpretation stems from the observation that the eigenstates of the Pauli operators can be identified with the points on the Bloch sphere and the line through these points defines the axis of rotation.

These mathematical expressions are also dubbed as Pauli matrices denoted by σ_x, σ_y and σ_z (see Sect. 5.1.2). This observation envisages that a Pauli operation can be implemented by making use of atomic or spin system. So on account of the properties of atomic operators, it is not difficult to verify that the Pauli matrices anti-commute. This feature of the Pauli gates can be portrayed by eigenvalue equations: for Pauli X gate, $X|\pm\rangle = \pm|\pm\rangle$ with $|\pm\rangle = \frac{1}{\sqrt{2}}[|0\rangle \pm |1\rangle]$; for Pauli Y gate, $Y|L\rangle = |L\rangle$ and $Y|R\rangle = -|R\rangle$ with $|L\rangle = \frac{1}{\sqrt{2}}[|0\rangle + i|1\rangle]$ and $|R\rangle = \frac{1}{\sqrt{2}}[|0\rangle - i|1\rangle]$; and for Pauli Z gate, $Z|0\rangle = |0\rangle$ and $Z|1\rangle = -|1\rangle$.

Moreover, for a two-mode light, one can denote the computational bases as

$$|0\rangle = \hat{a}_1^\dagger |0, 0\rangle, \qquad |1\rangle = \hat{a}_2^\dagger |0, 0\rangle, \tag{9.69}$$

where the indices are inserted to able to distinguish the two modes. Besides, the combined beam splitting and phase shifting operations can be characterized in terms of the Hamiltonian

$$\hat{H} = \theta e^{i\phi}\hat{a}_1^\dagger \hat{a}_2 + \theta e^{-i\phi}\hat{a}_2^\dagger \hat{a}_1, \tag{9.70}$$

where θ denotes the reflectivity of the beam splitter and ϕ the phase shift (see Sect. 4.3.2).

It may not be hard then to verify following the procedure outlined in Chap. 2 that

$$\hat{a}_1 \longrightarrow \cos\theta \, \hat{a}_1 - i e^{i\phi} \sin\theta \, \hat{a}_2, \tag{9.71}$$

$$\hat{a}_2 \longrightarrow \cos\theta \, \hat{a}_2 - i e^{-i\phi} \sin\theta \, \hat{a}_1. \tag{9.72}$$

In case such an operation is presumed to be applied on the vacuum state, the computational bases (9.69) transform as

$$|0\rangle \longrightarrow \cos\theta|0\rangle + i e^{-i\phi} \sin\theta|1\rangle, \tag{9.73}$$

$$|1\rangle \longrightarrow \cos\theta|1\rangle + i e^{i\phi} \sin\theta|0\rangle. \tag{9.74}$$

Owing to such transformation and upon suitably choosing the values of θ and ϕ, it should be possible to construct the eigenstates of the Pauli matrices: $|\pm\rangle$, $|L\rangle$ and $|R\rangle$; which can be taken as justification for the possibility of implementing Pauli gates in the regime of continuous variable with the help of passive beam splitters and phase shifters [24].

9.4.1.2 Hadamard Gate

Hadamard transformation on the other hand can be taken as one of the most basic gate operations that offers elegant and versatile utilities in the area of quantum information processing due to its simple symmetric form:

$$H = \frac{1}{\sqrt{2}} \begin{pmatrix} 1 & 1 \\ 1 & 1 \end{pmatrix}.$$

(9.75)

It is possible to note from Eq. (9.75) that if one multiplies a two-point vector by the Hadamard matrix, one will get the sum and difference of the two points, which institutes the procedure for summation and subtraction. It is also possible to see that the product of two Hadamard gates will be unity: $HH = I$.

Note that application of this gate operation on a pure quantum state invokes transformation of the form

$$H|0\rangle \longrightarrow \frac{1}{\sqrt{2}}\big[|0\rangle + |1\rangle\big], \qquad H|1\rangle \longrightarrow \frac{1}{\sqrt{2}}\big[|0\rangle - |1\rangle\big].$$

(9.76)

To reinforce its applicability, one can relate the Hadamard gate with Pauli matrices as

$$HXH = Z, \qquad HZH = X, \qquad HYH = -Y$$

(9.77)

and express as

$$H = \frac{1}{\sqrt{2}}\big[X + Z\big].$$

(9.78)

Equation (9.78) may indicate that the Hadamard gate exemplifies a coherent combination of two Pauli gates; and so, one may envision the quantum nature of Hadamard gate.

To extend this consideration to the regime of continuous variable, one can begin with interchanging the position and momentum bases of a continuous variable quantum state by $\pi/2$ rotation in phase space that can be enforced by replacing the Hadamard rotation in the Bloch sphere with Fourier transformation. Notably, the continuous variable version of the Hadmard gate can be designated by a Fourier transformation [25]

$$F(\ell)|x\rangle = \frac{1}{\ell\sqrt{\pi}} \int dy \, e^{\frac{i2xy}{\ell^2}} |y\rangle,$$

(9.79)

where ℓ is the scaled length. Such transformation can be used while going from the position to the momentum basis upon setting $\ell = \sqrt{2}$. With this denotation, one can attest to

$$F(\ell)F^\dagger(\ell)|x\rangle = F^\dagger(\ell)F(\ell)|x\rangle = |x\rangle,$$

(9.80)

where $F(\ell)F^\dagger(\ell) = F^\dagger(\ell)F(\ell) = I$.

In the process of connecting such description with light, one may perceive that the consideration that the light incident from above (below) on a 50:50 beam splitter has 50% probability to emerge either in the above or below output beams without loss of particle ultimately imposes a constraint on the phase of the beam splitter (see Chap. 4). One of the approaches to overcome such a limitation is to fix the phase in such a way that the beam splitter can be represented by the Hadamard transformation, which suggests that a Hadamard gate can be designated by a beam splitting procedure.

As illustration, think of the incident state as the general qubit

$$|Q\rangle_{in} = \alpha|0\rangle_{in} + \beta|1\rangle_{in}, \tag{9.81}$$

and subsequently construing the beam splitter as the Hadamard gate operation is tantamount to setting

$$H|Q\rangle_{in} = \frac{1}{\sqrt{2}}\big[(\alpha + \beta)|0\rangle_{out} + (\alpha - \beta)|1\rangle_{out}\big], \tag{9.82}$$

where $\alpha + \beta$ is the measure of the probability for finding the beam in the outgoing upper arm (basis state $|0\rangle_{out}$) and $\alpha - \beta$ in the outgoing lower beam (basis state $|1\rangle_{out}$).

When $\alpha = 0$ or $\beta = 0$, one can see that there is an equal probability for the beam to be in the upper and lower outgoing states. But for $\alpha = \beta$, the outgoing beam would be only in the upper outgoing state. Besides, in case in one of the input arms, let us say, the lower arm there is no beam ($\alpha = 1$ and $\beta = 0$), the outgoing beam would be in the upper output arm, which envisages that the interference of the two amplitudes incident on the final beam splitter leads to having the beam with certainty in one of the outgoing arms (see also Sect. 11.4).

It might be instructive to consider a sequence of similar action of the beam splitter to constitute repeated Hadamard transformations. For example, it is possible to replicate the Mach–Zehnder interferometer with two identical beam splitters (see Sect. 4.3.2), that is,

$$HH|Q\rangle_{in} = |Q\rangle_{in}. \tag{9.83}$$

Equation (9.83) exemplifies that a double Hadamard transformation can be equated to the identity operation. One can hence assert that the output qubit of the empty Mach–Zehnder interferometer takes a definite value provided that the input qubit also has a definite value since between the two Hadamard transformations the value of the qubit would be maximally undefined.

One can also consider the case when the computational basis states—coherent states $|\pm\alpha\rangle$—are expressed in terms of the diagonal or entangled basis states. One can then imagine the input state of the type

$$|\Psi\rangle_{in} = \frac{1}{\sqrt{2(1 \pm e^{-2\alpha^*\alpha})}}\left[|u\rangle \perp | - u\rangle\right] \tag{9.81}$$

taken as the even and odd coherent entangled state or qubit in which the corresponding Hadamard gate operation relies on the supply of a superposed coherent state assumed to have the same amplitude as the coherent state of the computational bases (see Sect. 9.3).

The pertinent gate works by displacing arbitrary coherent entangled input state followed by nondistinguishable subtraction of a single photon from either the displaced input or the resource state. This process can be realized by reflecting a small portion of either state using asymmetric beam splitters; interfering the resulting beams on a beam splitter with transmissivity T and reflectivity R; and then detecting the photon at the output with a single-photon detector. Such a mechanism can be captured by the operator $R\hat{a} + T\hat{b}$, where $\hat{a}$ and $\hat{b}$ are annihilation operators that stand for the subtraction of a photon from the displaced input and the coherent state superposition resource.

Once the beam splitting is established, the resulting two-mode state can be projected onto a single-mode quadrature eigenstate $|x\rangle$ with the help of homodyne detection, where the emerging output state can be expressed as

$$|\Psi\rangle_{out} = \frac{1}{\sqrt{2(1 \pm e^{-2\alpha^*\alpha})}}$$
$$\times \left\{ u\left[|\alpha\rangle + | - \alpha\rangle\right] + \frac{T}{2R}\left(u + v\frac{\langle x|0\rangle}{\langle x|2\alpha\rangle}\right)\left[|\alpha\rangle - | - \alpha\rangle\right]\right\}. \tag{9.85}$$

Afterwards, upon utilizing the beam splitter with $T \ll R$ and setting the quadrature such that

$$\langle x|0\rangle \gg \langle x|2\alpha\rangle, \qquad T\sqrt{2(1 + e^{-2\alpha^*\alpha})}\langle x|0\rangle = 2R\sqrt{2(1 - e^{-2\alpha^*\alpha})}\langle x|2\alpha\rangle, \tag{9.86}$$

the Hadmard transformation can be implemented [26].

The success of this scheme however depends on the joint measurement of the optical process in terms of the quadrature with value x. Since the even coherent state superposition with a small amplitude can be taken as the reminiscent of a squeezed vacuum state, it might be advantageous replacing the ideal resource with the squeezed vacuum state. The transformed state in this regard has the form

$$|\Psi\rangle_{out} = u\hat{S}(r)|0\rangle - \frac{T\sinh(r)}{2R\alpha}\left(u + v\frac{\langle x|0\rangle}{\langle x|2\alpha\rangle}\right)\hat{a}^\dagger|0\rangle. \tag{9.87}$$

9.4.1.3 Phase Gate

One can also introduce a phase in the setup of beam splitting to the amplitude of one of the two beams by using a phase shifter or a nonlinear medium in one of the arms of the Mach–Zehnder type interferometer. Pertaining to the ensuing difference between the emerging two beams, one can construe a quantum phase gate that can be described by the transformation of the type

$$\Phi|j\rangle_1 \otimes |l\rangle_2 \longrightarrow e^{i\phi_{jl}}|j\rangle_1 \otimes |l\rangle_2, \qquad (9.88)$$

where $j, l = 0, 1$ are the basis states of the two qubits and 1 (2) stands for control (target) qubit such as vertical, horizontal, circularly left or right polarizations. The action of the intended phase shifter can be expressed in terms of the computational bases and unitary transformations of the type

$$\Phi|0\rangle = |0\rangle, \qquad \Phi|1\rangle = e^{i\phi}|1\rangle, \qquad (9.89)$$

which invoke the notion of phase gate;

$$\Phi = \begin{pmatrix} 1 & 0 \\ 0 & e^{i\phi} \end{pmatrix}. \qquad (9.90)$$

It might also be required to apply the phase gate along with the Pauli gates:

$$\Phi X \Phi^\dagger = Y, \qquad \Phi Y \Phi^\dagger = -X, \qquad \Phi Z \Phi^\dagger = Z. \qquad (9.91)$$

Based on the proposal that such realization insinuates as if the output qubit can be generated by a successive application of gate operations, one can exploit the peculiar aspects of each process in the combined phase and Hadamard gate operations [27].

As a testament, one can see with the help of Eq. (9.81) that

$$H\Phi H|Q\rangle_{in} = \frac{\alpha}{2}\left[\left(e^{i\phi} + 1\right)|0\rangle + \left(e^{i\phi} - 1\right)|1\rangle\right]$$
$$+ \frac{\beta}{2}\left[\left(e^{i\phi} - 1\right)|0\rangle + \left(e^{i\phi} + 1\right)|1\rangle\right]. \qquad (9.92)$$

One can deduce from Eq. (9.92) that successive application of the Hadamard and phase gates can substantially change the nature of the input qubit. For example, for $\alpha = 1$ and $\beta = 0$, one can see that

$$H\Phi H|Q\rangle_{in} = \frac{1}{2}\left[\left(e^{i\phi} + 1\right)|0\rangle + \left(e^{i\phi} - 1\right)|1\rangle\right]. \qquad (9.93)$$

The phase shifter can switch the state of the output beam between the computational bases $|0\rangle$ and $|1\rangle$ (check for $\phi = 0$ and $\phi = \pi$).

The phase shifter can also be expressed in terms of the momentum and position variables following the same trend as

$$\Phi_q(\theta) - e^{i\theta\hat{q}^2}, \qquad \Phi_p(\theta) - e^{i\theta\hat{p}^2} \qquad (9.94)$$

where θ accounts for the angle of rotation. Such characterization entails that the phase gate can be constructed from the combination of squeezing and rotation of the quadratures. So it should be possible to initiate a transformation from the position to momentum quadrature;

$$\Phi_p = e^{i\frac{\pi}{2}(\hat{q}^2+\hat{p}^2)} \Phi_q e^{-i\frac{\pi}{2}(\hat{q}^2+\hat{p}^2)} \qquad (9.95)$$

in which θ is set to $\pi/2$ that can be realized by making use of a pumped second-order nonlinear material such as parametric oscillator and phase shifter (see Sect. 8.1).

To construct the phase gate from a coherent light, one can also begin with the observation that the phase shifting mechanism can be incorporated via a coherent displacement that can be realized by applying optical displacement operator. In connection to this, imagine arbitrary qubit in the coherent state basis,

$$|\Psi\rangle_{in} = A|\alpha\rangle + B|-\alpha\rangle, \qquad (9.96)$$

to be displaced by a constant amplitude β

$$|\Psi'\rangle_{in} \longrightarrow \hat{D}(\beta)|\Psi\rangle_{in} \qquad (9.97)$$

that can be implemented by mixing the signal beam with auxiliary strong coherent field on highly unbalanced beam splitter or homodying (see Sect. 2.2.1).

After a single photon is subtracted from the state, and if the inverse displacement of β has been performed, one can come up with

$$|\Psi\rangle_{out} = \hat{D}^\dagger(\beta)\,\hat{a}\,\hat{D}(\beta)|\Psi\rangle_{in} = A(\beta+\alpha)|\alpha\rangle + B(\beta-\alpha)|-\alpha\rangle \qquad (9.98)$$

(see Sect. 4.4). Then upon setting $\beta - \alpha$ as $(\beta + \alpha)e^{i\phi}$, one can see that

$$|\Psi\rangle_{out} = (\beta+\alpha)\big[A|\alpha\rangle + Be^{i\phi}|-\alpha\rangle\big], \qquad (9.99)$$

where Eq. (9.99) epitomizes the phase gate up to certain global phase factor.

By making use of the usual creation and annihilation operators, the optical phase shifting process can also be expressed in terms of

$$\hat{a}_{out}^\dagger = e^{i\phi(t)\hat{a}_{in}^\dagger\hat{a}_{in}t}\,\hat{a}_{in}^\dagger\,e^{-i\phi(t)\hat{a}_{in}^\dagger\hat{a}_{in}t} \qquad (9.100)$$

in which the time dependence is absorbed in the phase. Note that the phase shifter can be modeled as a slab of transparent material with the index of refraction different from free space, and can be designated upon parameterizing the probability amplitudes as $\cos\theta$ and $\sin\theta$ by

$$\hat{a}_{out}^\dagger = \cos\theta\hat{a}_{in}^\dagger + ie^{-i\phi}\sin\theta\hat{b}_{in}^\dagger, \qquad \hat{b}_{out}^\dagger = ie^{i\phi}\sin\theta\hat{a}_{in}^\dagger + \cos\theta\hat{b}_{in}^\dagger. \qquad (9.101)$$

It is worth noting that the relative phase shift $ie^{\pm i\phi}$ is inserted to ensure the transformation to be unitary and the beam splitter to be calibrated by adjusting θ and ϕ.

Markedly, the Hamiltonian of the beam splitter can be expressed as

$$\hat{H} = \theta e^{i\phi}\hat{a}_{in}^\dagger\hat{b}_{in} + \theta e^{-i\phi}\hat{a}_{in}\hat{b}_{in}^\dagger \qquad (9.102)$$

when required (relate with Eq. (9.70)) since the same mathematical description can be used to effect the ensued evolution due to the polarization rotation achievable by utilizing quarter and half wave plates. Instead of $\hat{a}_{in}$ and $\hat{b}_{in}$, the two incoming modes are expected to have different polarizations.

With such delineation, one can invoke a variable transformation such as $\hat{a}_{in} \rightarrow \hat{a}_x$ and $\hat{b}_{in} \rightarrow \hat{a}_y$ for some orthogonal set of spatial coordinates x and y. The parameters θ and ϕ in this scenario stand for the angle of rotations with

$$\hat{a}_{x'}^\dagger = \cos\theta\hat{a}_x^\dagger + ie^{-i\phi}\sin\theta\hat{a}_y^\dagger, \qquad \hat{a}_{y'}^\dagger = ie^{i\phi}\sin\theta\hat{a}_x^\dagger + \cos\theta\hat{a}_y^\dagger \qquad (9.103)$$

that are believed to be sufficient to implement a photonic single-qubit operation. If the polarized beam splitter is cut to horizontal and vertical polarization directions, the transformation of the incoming modes $\hat{a}_{in}$ and $\hat{b}_{in}$ may yield

$$\hat{a}_{in,H} \rightarrow \hat{a}_{out,H}, \qquad \hat{a}_{in,V} \rightarrow \hat{b}_{out,V}, \qquad \hat{b}_{in,H} \rightarrow \hat{b}_{out,H}, \qquad \hat{b}_{in,V} \rightarrow \hat{a}_{out,V}. \qquad (9.104)$$

9.4.2 Two-Qubit Optical Gates

Even though the underlying physical process is relatively simpler to manipulate, single-qubit operations alone cannot unlock the entire perceived power of a collection of qubits that can be prepared in the computational bases $|0\rangle$ and $|1\rangle$. It might be worthy therefore to include a multi-qubit operations that can be described in terms of the computational bases: $|00\rangle$, $|01\rangle$, $|10\rangle$, $|11\rangle$. Note that a two-qubit optical gate in which the evolution of one of the qubits is conditioned upon the state of the other is taken to be significantly important in continuous processing quantum information since they can provide the vital entanglement [28].

To begin with, one can introduce a two-qubit operation that has a potential to swap the logical value of the target with the control such as the swapping of the atomic qubit with photonic qubit, and denoted by 4×4 matrix;

$$W_{SW} = \begin{pmatrix} 1 & 0 & 0 & 0 \\ 0 & 0 & 1 & 0 \\ 0 & 1 & 0 & 0 \\ 0 & 0 & 0 & 1 \end{pmatrix}. \tag{9.105}$$

It is also found admissible to swap unknown qubit state $|\psi\rangle$ with a state $|0\rangle$ by carrying out a successive CNOT gate operations (to be introduced later) as

$$CN_{21} CN_{12}|\psi\rangle_1 \otimes |0\rangle_2 = |0\rangle_1 \otimes |\psi\rangle_2. \tag{9.106}$$

It is thus possible to construct a general two-qubit SWAP gate from CNOT gates as

$$SW_{12}|j\rangle_1 \otimes |l\rangle_2 = CN_{12} CN_{21} CN_{12}|j\rangle_1 \otimes |l\rangle_2 = |l\rangle_1 \otimes |j\rangle_2, \tag{9.107}$$

where $j, l = 0, 1$ are the basis states of the two qubits and 1 (2) stands for control (target) qubit. Following the same reasoning, the SWAP gate can be constructed from the Pauli gates as

$$SW_{12} = \frac{1}{2}\left[1 + X_1 X_2 + Y_1 Y_2 + Z_1 Z_2\right]. \tag{9.108}$$

Expressing the SWAP gate in the regime of continuous variable,

$$SW_{12}|x\rangle_1 \otimes |p\rangle_2 = |p\rangle_1 \otimes |x\rangle_2, \tag{9.109}$$

perhaps can be somehow tricky. Even then, without going into details, we opt to list some possible schemes:

$$SW_{12} = NOT_1 NOT_2 CN_{12}^- CN_{21}^- CN_{12}^-, \tag{9.110}$$

$$SW_{12} = NOT_2 CN_{12}^- CN_{21}^- CN_{12}^+ = e^{i\hat{x}_1 \hat{p}_2} NOT_1 e^{i\hat{x}_2 \hat{p}_1} e^{-i\hat{x}_1 \hat{p}_2}, \tag{9.111}$$

$$SW_{12} = NOT_2 e^{\frac{i}{2}(\hat{x}_1 \hat{p}_2 - \hat{x}_2 \hat{p}_1)}, \tag{9.112}$$

where $NOT_j = e^{i\hat{a}_j^\dagger \hat{a}_j}$ and $j = 1, 2$. Upon taking Eq. (9.112) into account, it is also possible to write the SWAP gate in terms of boson operators as

$$SW_{12} = e^{i\hat{a}_2^\dagger \hat{a}_2} e^{\frac{1}{2}(\hat{a}_1^\dagger \hat{a}_2 - \hat{a}^\dagger \hat{a}_1)}. \tag{9.113}$$

It is not difficult to conceive that this discussion can be extended to arbitrary two-qubit gate by introducing a 4×4 unitary matrix commonly dubbed as controlled-U

$$W_{CU} = \begin{pmatrix} 1 & 0 & 0 & 0 \\ 0 & 1 & 0 & 0 \\ 0 & 0 & \alpha & \gamma \\ 0 & 0 & \beta & \delta \end{pmatrix}, \tag{9.114}$$

where the 2×2 matrix with elements $\{\alpha, \beta, \gamma, \delta\}$ is a place holder and designated as U-matrix. One can after that take the operation that more often required in application and known as the controlled NOT (CNOT) gate, which is applicable to disentangle EPR states.

A fundamental working principle of the CNOT gate amounts to negating the value of the target qubit when the control qubit has a logical value 1, but in all other options the logical value of the control qubit would not be changed, that is,

$$|0\rangle_c |0\rangle_t \longrightarrow |0\rangle_c |0\rangle_t, \qquad |1\rangle_c |0\rangle_t \longrightarrow |1\rangle_c |1\rangle_t;$$

$$|0\rangle_c |1\rangle_t \longrightarrow |0\rangle_c |1\rangle_t, \qquad |1\rangle_c |1\rangle_t \longrightarrow |1\rangle_c |0\rangle_t, \tag{9.115}$$

where c (t) stands for control (target) qubit. The CNOT gate particularly flips the second (target) qubit when the first (control) qubit is 1. Beside, note that a CNOT gate is a two-qubit gate in which one qubit is flipped conditioned on the state of another qubit and can be expressed as

$$CN_{12}|j\rangle_1 \otimes |l\rangle_2 = |j\rangle_1 \otimes |j \oplus l\rangle_2, \tag{9.116}$$

where $\oplus$ denotes addition modulo 2.

A CNOT gate so can be represented in case $\alpha = \delta = 0$ and $\beta = \gamma = 1$;

$$W_{CN} = \begin{pmatrix} 1 & 0 & 0 & 0 \\ 0 & 1 & 0 & 0 \\ 0 & 0 & 0 & 1 \\ 0 & 0 & 1 & 0 \end{pmatrix}. \tag{9.117}$$

Owing to the symmetry of matrix (9.117), the manipulation of the CNOT gate turns out to be considerably easier. For example, the operation of the CNOT gate on the state

$$|\Psi\rangle = C_{00}|00\rangle + C_{01}|01\rangle + C_{10}|10\rangle + C_{11}|11\rangle \tag{9.118}$$

leads to

$$W_{CN}|\Psi\rangle = \left(C_{00}, C_{01}, C_{11}, C_{10} \right)^T \tag{9.119}$$

in which the coefficients of the basis states $|10\rangle$ and $|11\rangle$ are interchanged. One may also see that a W_{CN} gate can be employed in generating EPR type states, that is,

$$W_{CN}|00\rangle \rightarrow \frac{1}{\sqrt{2}}[|00\rangle + |11\rangle], \tag{9.120}$$

where the same kind of outcomes are implied for the other sets as well.

A two-qubit CNOT gate operation (9.116) can also be expressed as

$$CN_{12}^{\pm}|x\rangle_1 \otimes |p\rangle_2 = |x\rangle_1 \otimes |x \pm p\rangle_2, \tag{9.121}$$

where

$$CN_{12}^{+} = e^{-i\hat{x}_1\hat{p}_2}, \qquad CN_{12}^{-} = \text{NOT}_2 e^{i\hat{x}_1\hat{p}_2} = e^{-i\hat{x}_1\hat{p}_2}\text{NOT}_2 \tag{9.122}$$

in which $\text{NOT} = (-1)e^{\hat{a}^\dagger\hat{a}}$ with $\text{NOT}|x\rangle = |-x\rangle$, $\text{NOT}|p\rangle = |-p\rangle$, and $|x\rangle$ ($|p\rangle$) is the eigenstate of the position (momentum) operator $\hat{x}$ ($\hat{p}$) [29]. One can also write

$$CN_{12}^{\pm}|p\rangle_1 \otimes |q\rangle_2 = |p\rangle_1 \otimes |p \pm q\rangle_2, \tag{9.123}$$

where

$$CN_{12}^{+} = e^{i\hat{q}_2\hat{p}_1}, \qquad CN_{12}^{-} = \text{NOT}_2 e^{-i\hat{q}_2\hat{p}_1} = e^{i\hat{q}_2\hat{p}_1}\text{NOT}_2. \tag{9.124}$$

To illustrate such a logical qubit, one can designate the value of each qubit by the photon available in one of the two paths of the interferometer or two optical fibers in which one of the paths denote the logical value **0** while the other **1**. The logic operations encompass the resulting interference between the optical paths that can be very sensitive to the pertinent phase shift. The polarization inspired qubits in which the values of the qubits are accounted by the polarization states of the photon are often used to setup the required logical operation. Although the process of choosing appropriate polarization of the light may diminish the degree of interference between the separate optical paths, it can have an advantage during applications that can be implemented by impinging two input qubits onto a polarizing beam splitter [30].

A controlled phase gate that can be designated in a concise form by the transformation

$$|q_1, q_2\rangle \longrightarrow (-1)^{q_1 q_2}|q_1, q_2\rangle, \tag{9.125}$$

where q_k are presumed to take logical values 0 and 1 can also be constructed by replacing the place holder U-matrix by the phase matrix Φ (9.90). To begin with, consider the general phase and time space transitions of the form:

$$\hat{X}(s) = e^{is\hat{p}}, \qquad \hat{Z}(t) = e^{it\hat{q}}, \tag{9.126}$$

where s and t are real continuous value parameters, and $\hat{p}$ and $\hat{q}$ are momentum and position quadrature operators.

Recall that when $\hat{X}(s)$ acts on a continuously indexed basis state $|q\rangle_q$, it transforms the state as

$$\hat{X}(s)|q\rangle_q = |q + s\rangle_q, \tag{9.127}$$

and the eigenstate of the momentum also transforms as

$$\hat{Z}(t)|p\rangle_p = |p + t\rangle_p \tag{9.128}$$

which can be carried out by making use of Fourier transformation.

Owing to similar generic transformation, it is possible to realize a continuous variable two-qubit gate operations, that means, a controlled-U gate operations can be made to include additional options by properly choosing the elements of the place holder matrix. For example, by replacing the space holder by the Pauli X and Z matrices, one can come up with controlled-X and controlled-Z gate operations. Such quantum gate operations can be implemented in the regime of continuous variable by displacing the target qubit on the phase space by the amount determined by the position eigenvalue of the control qubit [31]:

$$C_X = e^{-i\hat{q}_c \otimes \hat{p}_t}, \qquad C_Z = e^{i\hat{q}_c \otimes \hat{q}_t}. \tag{9.129}$$

It would be imperative stressing that a logic operation is inherently nonlinear. Note that while using a simple linear optical element to perform logical operation, the required nonlinearity can be induced via mixing the input photons with various ancilla photons with the help of linear optical elements. The nonlinearity can enter via the accompanying measurement process since a single-photon detector either records a photon or not, and projects out the desired logical output state provided that some result is obtained from the planned measurement (relate with the discussion in Sect. 4.4).

References

1. M.A. Nielsen, I.L. Chuang, *Quantum Computation and Quantum Infromation*, 10th Anniversary edn. (Cambridge University, Cambridge, 2010)
2. T.B. Pittman et al., Phys. Rev. Lett. **77**, 1917 (1996); D.V. Strekalov, T.B. Pittman, Y.H. Shih, Phys. Rev. A **57**, 567 (1998); X.F. Qian et al., Phys. Rev. Lett. **117**, 153901 (2016).

3. J.M. Raimond, M. Brune, S. Haroche, Rev. Mod. Phys. **73**, 565 (2001)
4. A. Barenco et al., Phys. Rev. A **52**, 3457 (1995); S. Lloyd, Phys. Rev. Lett. **75**, 346 (1995); H.F. Hofmann, Phys. Rev. Lett. **94**, 160504 (2005); P. Lambropoulos, D. Petrosyan, *Fundamentals of Quantum Optics and Quantum Information* (Springer, Berline, 2007)
5. H. Yonezawa, A. Furusawa, P. van Loock, Phys. Rev. A **76**, 032305 (2007); S. Yokoyama et al., ibid. **90**, 012311 (2014)
6. A. Peres, Phys. Rev. A **32**, 3266 (1985)
7. M. Saffman, T.G. Walker, K. Molmer, Rev. Mod. Phys. **82**, 2313 (2010)
8. Z. Ficek, R. Tanas, Phys. Rep. **372**, 369 (2002); T. Wilk et al., Phys. Rev. Lett. **104**, 010502 (2010)
9. S.B. Zheng, Opt. Commun. **154**, 290 (1998)
10. L.H. Pedersen, K. Molmer, Phys. Rev. A **79**, 012320 (2009); J.W. Pan et al., Rev. Mod. Phys. **84**, 777 (2012)
11. P. Kok et al., Rev. Mod. Phys. **79**, 135 (2007); H.A. Bachor, T.C. Ralph, *A Guide to Experiments in Quantum Optics*, 2nd revised and elarged edn. (Wiley-VCH Verlag GmbH and Co. KGaA, Weinheim, 2004)
12. T. Liu et al., Sci. Rep. **6**, 32004 (2016)
13. M. Dakna et al., Phys. Rev. A **59**, 1658 (1999); J. Fiurasek, R.G. Patron, N.J. Cerf, ibid. **72**, 033822 (2005); M. Yukawa et al., Opt. Express **21**, 5529 (2013)
14. A.N. Boto et al., Phys. Rev. Lett. **85**, 2733 (2000); S.D. Huver, C.F. Wildfeuer, J.P. Dowling, Phys. Rev. A **78**, 063828 (2008)
15. C.K. Hong, Z.Y. Ou, L. Mandel, Phys. Rev. Lett. **59**, 2044 (1987); P. Kok, H. Lee, J.P. Dowling, Phys. Rev. A **65**, 052104 (2002); Y. Israel et al., Phys. Rev. A **85**, 022115 (2012)
16. E.V. Mikheev et al., Sci. Rep. **9**, 14301 (2019)
17. X. Wang, B.C. Sanders, Phys. Rev. A **65**, 012303 (2001); H. Jeong, N.B. An, ibid. **74**, 022104 (2006); B.C. Sanders, J. Phys. A **45**, 244002 (2012)
18. J.S.N. Nielsen et al., Phys. Rev. Lett. **97**, 083604 (2006); K. Huang et al., ibid. **115**, 023602 (2015)
19. N. Imoto, H.A. Haus, Y. Yamamoto, Phys. Rev. A **32**, 2287 (1985); B.C. Sanders, G.J. Milburn, ibid. **39**, 694 (1989); B.C. Sanders, ibid. **45**, 6811 (1992); B.C. Sanders, D.A. Rice, ibid. **61**, 013805 (1999)
20. B. Yurke, D. Stoler, Phys. Rev. Lett. **57**, 13 (1986)
21. H.P. Yuen, Phys. Rev. A **13**, 2226 (1976)
22. R.F. Werner, Phys. Rev. A **40**, 4277 (1989); L. Mista, R. Filip, J. Fiurasek, ibid. **65**, 062315 (2002); M. Barbieri et al., Phys. Rev. Lett. **92**, 177901 (2004)
23. Z. Zhoa et al., Phys. Rev. Lett. **94**, 030501 (2005); Y. Wang et al., Phys. Rev. A **81**, 022311 (2010); S. Slussarenkoa, G.J. Prydeb, Appl. Phys. Rev. **6**, 041303 (2019); O. Pfister, J. Phys. B **53**, 012001 (2020)
24. P. Kok, B.W. Covett, *Introduction to Optical Quantum Information Processing* (Cambridge University, New York, 2010)
25. S. Parker, S. Bose, M.B. Plenio, Phys. Rev. A **61**, 032305 (2000)
26. A. Tipsmark et al., Phys. Rev. A **84**, 050301 (2011); S.A. Podoshvedov, Opt. Commun. **285**, 3896 (2012); S.E. Nigg, Phys. Rev. A **89**, 022340 (2014)
27. H.F. Hofmann, S. Takeuchi, Phys. Rev. A **66**, 024308 (2002); C. Ottaviani et al., ibid. **73**, 010301 (2006); P. Marek, J. Fiurasek, ibid. **82**, 014304 (2010); R. Blandino et al., New J. Phys. **14**, 013017 (2012)
28. X. Wang, J. Phys. A **34**, 9577 (2001)
29. T.C. Ralph et al., Phys. Rev. A **65**, 062324 (2002); R. Filip, ibid. **69**, 052313 (2004); R. Okamoto et al., Phys. Rev. Lett. **95**, 210506 (2005)
30. T.B. Pittman, B.C. Jacobs, J.D. Franson, Phys. Rev. Lett. **88**, 257902 (2002); T.B. Pittman et al., Phys. Rev. A **68**, 032316 (2003); J.L. O'Brien et al., Nature **426**, 264 (2003); S. Gasparoni et al., Phys. Rev. Lett. **93**, 020504 (2004); N.K. Langford et al., ibid. **95**, 210504 (2005)
31. R. Wu, R. Chakrabarti, H. Rabitz, Phys. Rev. A **77**, 052303 (2008); R. Ukai et al., Phys. Rev. Lett. **107**, 250501 (2011)

Implications of the Quantumness

10

While dealing with schemes that encompass applications in quantum versus classical scenario, two contradicting options need to be considered: some operations that can be implemented in the regime of classical physics but may not be repeated in the quantum realm such as signaling, cloning, broadcasting, deleting, and that cannot be done based on principle of classical physics but may be implemented in the quantum realm such as entanglement swapping, quantum steering, quantum dense coding and quantum teleportation. It is such perceived advantage (disadvantage) along with unavoidable circumstance of losing the quantumness to the environment and associated measurement problem that facilitate (hamper) technological advance and understanding, which is taken as the notion of the implication of exploiting quantum operations or processes as a resource for application. One can cite for instance the idea of no-signaling or no-communication theorem which states that the instantaneous collapse of the wave function of an entangled state by the measurement on the conjugate state at a remote location cannot be used to send a useful information [1].

It might be worthy of noting that prior to taking up the discussion on quantum optical processes and relevant applications as in next chapters (see Chaps. 11 and 12), there are still issues that we need to recap such as measurement problem (see Sects. 10.1 and also [2]), fragility of the quantumness (see Sect. 10.3 and also [3]) and inability to perfectly copy arbitrary quantum state (see Sect. 10.2 and also [4]). According to orthodox quantum theory for example all measurements cannot be carried out at the same space and time for conjugate variables since they do not commute with each other which leads to inherent uncertainty.

Pertaining to the notion of joint measurability, this principle is linked most recently to entanglement steering that may be utilized in implementing remote manipulation of quantum process [5]. In other words, quantum measurement outcomes are usually jointly unpredictable or they are incompatible, which can be taken as a quantum resource in generating random numbers in quantum key distribution (see Chap. 4 and also [6]). Besides, there is more or less an agreement

© The Author(s), under exclusive license to Springer Nature Switzerland AG 2020
S. Tesfa, *Quantum Optical Processes*, Lecture Notes in Physics 976,
https://doi.org/10.1007/978-3-030-62348-7_10

nowadays on the idea that quantum theory appears random in the sense that experimental results can be predicted and understood in the framework that quantum measurements are fundamentally random. Even then, one may note that there are a significant number of quantum processes or procedures that draw there working principle from quantum measurement theories specially from von Neumann weak measurement and positive operator value measure [7].

The other facet of quantumness is the manifestation of coherent superposition among quantum states, which is taken to be the epitome of the corresponding peculiar features; and yet, susceptible to loss or quantum decoherence [8]. Although it is prone to loss and difficult to observe directly, quantum superposition is found to be one of the major resources for quantum information processing (see Chap. 9). On top of this, measurement (or interaction) converts the state of the interacting systems to an eigenstate of the measured observable, which by itself is an entangling process; and so kills the meager chance that exists for directly accessing the dynamics of the interaction (or the result of quantum information processing). Such consideration may envisage that after a series of interactions as one may face in realistic quantum information processing, it would not be straightforward to assure the reliability of the generated information. The main aim of this chapter is thus geared towards drawing the attention of the reader to how significantly the quantum aspects of a certain operation or mechanism or principle can alter the classical perception; and consequently, the interpretation of the outcome of certain measurement or process or procedure in the realm of quantum theory, and also bringing to light the overall idea or strategy on how one may plan to overcome the inherent challenges and setbacks in this regard.

10.1   Recap of Quantum Measurement Theory

It is a well established fact that quantum formulation describes the dynamics of the system in terms of deterministic unitary evolution of the wave function which has never been observed. Nevertheless, the wave function is frequently used to compute or determine the probability by which the pertinent macroscopic events would be observed (see Eq. (3.1)). It is such dichotomy between the wave function model and observed macroscopic events that leads to interpretational conundrum (see Sect. 4.1). It should also be evident by now that even though the mathematical model can convey about what one seeks to observe in classical physics, in quantum theory, the mathematical model does not entail observation by itself—one is hence obliged to resort to interpretation to decipher the content of the measurement. It might be imperative to underscore at this moment as this is not just a philosophical issue, but also as it is the glaring truth or reality under which the forthcoming quantum optical processes and corresponding applications should be understood.

As already highlighted, one may often need to model quantum state of certain system as the superposition of two or more possible outcomes that describe the accompanying states although quantum superposition leads to interference like entwined effects. One of the main theoretical questions is then how superposition of different possibilities emerges as some particular observation; and in practical context, how one can extract the sought for information from the resulting entanglement. One may note that this quantum aspect, which is perceived as quantum measurement problem, affects not only how one can analyze some experimental results such as Bell inequality but also how one can readout the information or the valuable message after certain manipulation or implementation is carried out. It appears that various setbacks can transpire from an apparent conflict in the working principles or various interpretations of quantum measurement theory. The linear dynamics of quantum theory particularly seems to be in conflict with nonlinear collapse of the wave function during measurement, that is, the postulate of the collapse of the wave function appears to be right about what happens when one makes measurement but the dynamics whenever one is not making.

Such understanding can also be framed as the wave function evolves into a linear superposition of different states according to Schrödinger equation while practical measurement actually proclaims the measured system to be in a definite state—which unfortunately has a significant bearing on the realization of subsequent dynamical evolution as required in quantum information processing. In other words, the wave function evolves deterministically, that is, knowing the wave function at one moment implies that the wave function at any later time can be determined by using Schrödinger equation. So while aspiring to utilize quantum aspects of the system for instance light, it might be compelling to know whether it is possible to predict precisely the results of the required manipulations or how much the result of certain manipulation is reliable. One might also need to figure out how it is possible to establish a seamless relationship between quantum outcomes and the corresponding classical reality or clicks of the detector.

On the other hand, the working principle of several quantum operations, processes and implementations are drawn from the postulates of quantum measurement theories such as projective measurement, von Neumann weak measure, positive operator value measure and breaking of joint measurability [9]. In addition, in the context of quantum optics, it is difficult to think of certain quantum information processing without the mechanism of wave mixing and photon detection such as beam splitting including phase beam splitting, interferometric setups most prominently Match-Zehnder interferometry, direct and homodyne detections, tomography and coincidence detection (see Chap. 4). It is therefore uncanny not to seriously consider the implication a measurement procedure might have on the result and interpretation of practical manipulations while processing information with aid of quantum features of light as intended to be explored in the remaining chapters.

10.2 Quantum No-Cloning

It is common knowledge that a classical state of a given system can be casted onto many copies upon measuring the emerging copies as accurately as possible, which helps to unravel the content of the information imprinted onto the state. This everyday experience however could be drastically altered in the quantum world since it is impossible to know with certainty the information content of a quantum state. This assertion stems from the observation that while one measures the quantum system, its state collapses—having different measurement values of the quantum state makes the information to be written on the copy unreliable (see Sects. 4.1 and 10.1). To overcome such a setback, it might be required to activate procedure such as quantum tomography that encompasses reconstructing the quantum state or density matrix by carrying out a measurement on the system coming from the source overtime to get the information of the probabilities associated with each quantum state. It should not be beyond comprehension to foresee that the situation could be worse when one seeks to copy unknown quantum state for a simple reason that it is not possible to make a perfect copy of unknown quantum state. This idea, which is a direct consequence of the principle of quantum theory and forbids the creation of identical copies of arbitrary unknown quantum state, is then the crux of quantum no-cloning.

Contrary to such assertion, it should be stressed that a perfect copying can be achieved in the quantum regime as well for the case when the states are orthogonal with the copier exclusively built for that set of states. Since it is not reasonable to aspire to uniquely know the right copier for a general case without prior knowledge of the state to be copied, one cannot practically copy unknown quantum state perfectly. Such an argument can be taken as the rationale thinking underlying the notion of quantum no-cloning theorem: no quantum operation exists that can perfectly and deterministically duplicates a pure state.[1]

One can thus naively argue that if it were admissible to clone a quantum state, it could have been possible to violate the uncertainty principle. To justify this proposal, one can begin with the proposal that if one can clone a particle A's state to particle B's, one can then measure A's position and B's momentum with desired precision. Since the uncertainty principle is known to impose limitation on what is to be known about the particle's position and momentum, it is envisaged that quantum cloning is not realizable at least in some respect.

To attest to such proposal, one can begin with the assumption that the state of the quantum system expected to be copied to be $|\Psi\rangle_A$ and the desirability of having a blank state with initial state $|c\rangle_B$ to make the copy. Pertaining to the presumption that the experimenter does not have a prior knowledge of the quantum state to be copied, the initial (blank) state can be taken to be independent of $|\Psi\rangle_A$. As a result of the interaction that prompts the copying process, the composite system in this case can be represented by the tensor product of the two states $|\Psi\rangle_A \otimes |c\rangle_B$. Note that there could be two ways to manipulate the composite system. For one, it is possible to carry out a measurement that irreversibly collapses the system into some eigenstates of the observable; and in the process, corrupt the information contained in the quantum state. This is not obviously what one seeks to achieve. One on the other hand can control the Hamiltonian of the system until the pertinent unitary

---

[1]Since deleting can be taken as the reverse process of copying, one can propose that deleting unknown quantum state is also not possible.

evolution is realized, that is, until the usual quantum dynamics emerges. If things go well as expected, the corresponding evolution operator acts as the copier operation for arbitrary $|\Psi\rangle$ and $|\phi\rangle$ provided that

$$\hat{U}|\Psi\rangle_A \otimes |c\rangle_B = |\Psi\rangle_A|\Psi\rangle_B, \qquad \hat{U}|\phi\rangle_A \otimes |c\rangle_B = |\phi\rangle_A|\phi\rangle_B, \qquad (10.1)$$

It is a well known fact that the unitary operation preserves the inner product, that is,

$$\langle c|_B \otimes \langle \phi|_A \hat{U}^\dagger \hat{U}|\Psi\rangle_A \otimes |c\rangle_B = \langle \phi|_B \langle \phi|_A |\Psi\rangle_A |\Psi\rangle_B, \qquad (10.2)$$

which reduces to

$$\langle \phi|\Psi\rangle = \langle \phi|\Psi\rangle^2. \qquad (10.3)$$

Equation (10.3) clearly shows that either $|\phi\rangle = |\Psi\rangle$ or $\langle \phi|\Psi\rangle = 0$ (the two states are orthogonal); which cannot be necessarily true for a general case. To verify this, it suffices to consider maximally entangled states $|\phi\rangle = \frac{1}{\sqrt{2}}\big[|0\rangle + |1\rangle\big]$ and $|\Psi\rangle = \frac{1}{\sqrt{2}}\big[|0\rangle - i|1\rangle\big]$ (see Sect. 9.1). Taking the tensor product of these states may reveal that the prescribed evolution operation cannot clone the general quantum state. This exercise therefore could be perceived as a tactical proof of the no-cloning theorem. One can also assert that the no-cloning theorem does not luck generality since arbitrary quantum operation can be implemented via introducing ancillary state, and performing a suitable unitary evolution: no-clonability should be understood as a unique aspect of arbitrary sets of quantum states.

It is also possible to describe quantum no-clonability from another perspective. To do so, one can begin with the idea that if one knows that the system's state is one of the states in some set but if one does not know which, can one prepare another system in the same state in case the elements of the set are pairwise orthogonal? For such a set, note that there exists a measurement procedure that can capture the exact state of the system without disturbing it. However, in case the set contains two elements that are not pairwise orthogonal (the set of all quantum states possesses such pairs), a similar situation leads to opposite result. Such an argument entails that even though one can narrow down the state of the quantum system to just two possibilities, one still cannot clone unless the states happen to be orthogonal.[2]

---

[2]Even in the classical context, given only the result of one flip of a coin, one cannot simulate the second independent toss of the same coin. The assertion of this statement follows from the linearity of classical probability, and has exactly the same structure as the argument leading to quantum no-cloning theorem, which entails that if one wishes to insist that no-cloning is a uniquely quantum aspect some care should be taken in formulating the idea. One way of doing this task encompasses restricting the states to pure states since classical pure states are pairwise orthogonal but quantum pure states are not.

For now, imagine that a cloning machine can duplicate arbitrary pure state according to

$$\hat{U}|A\rangle|B\rangle \otimes |a\rangle = |A\rangle|A\rangle \otimes |a(A)\rangle, \tag{10.4}$$

where $|A\rangle$ stands for the original arbitrary pure state, $|B\rangle$ for initial blank state, $|a\rangle$ for initial auxiliary state and $|a(A)\rangle$ for the auxiliary state after cloning operation. It seems that this scheme has a potential to copy indefinite or unknown number of original state. Let us then check the validity of the prescribed operation of this machine for arbitrary state in a sense that the amplitudes α and β are not known: $|\Psi\rangle = \alpha|0\rangle + \beta|1\rangle$.

According to the cloning recipe given by Eq. (10.4), one can see that

$$\big[\alpha|0\rangle + \beta|1\rangle\big]|B\rangle \otimes |a\rangle = \alpha|00\rangle \otimes |a(0)\rangle + \beta|11\rangle \otimes |a(1)\rangle. \tag{10.5}$$

Owing to the assumption that $|\Psi\rangle$ is a pure state, the same transformation can also lead to

$$\big[\alpha|0\rangle + \beta|1\rangle\big]|B\rangle \otimes |a\rangle = \big[\alpha^2|00\rangle + \beta^2|11\rangle + \alpha\beta|01\rangle + \beta\alpha|10\rangle\big] \otimes |a(\Psi)\rangle. \tag{10.6}$$

Since the right side of Eqs. (10.5) and (10.6) are not the same, the assumption that there is a universal copier machine should be wrong. This outcome in other words ensures that a perfect quantum cloning machine does not exist for arbitrary pure state, which is the consequence of the linearity of quantum theory and casts a significant bearing on the corresponding application.

In the context of quantum optics, although one can build a hypothetical cloning machine that produces several clones of the horizontal and vertical polarization states of the incoming photon, circularly polarized states cannot yield circularly polarized clones. Suppose one has a quantum cloning machine (see also Sect. 11.4)

$$|H\rangle \otimes |QCM\rangle \longrightarrow |H, H\rangle|C(H)\rangle, \qquad |V\rangle \otimes |QCM\rangle \longrightarrow |V, V\rangle|C(V)\rangle. \tag{10.7}$$

Let us now transform the left and right polarized states (see Sect. 9.3),

$$|L\rangle = \frac{1}{\sqrt{2}}\big[|H\rangle + i|V\rangle\big], \qquad |R\rangle = \frac{1}{\sqrt{2}}\big[|H\rangle - i|V\rangle\big], \tag{10.8}$$

using the same machine:

$$|L\rangle \otimes |QCM\rangle \longrightarrow \frac{1}{\sqrt{2}}\big[|H, H\rangle|C(H)\rangle + i|V, V\rangle|C(V)\rangle\big] \neq |L, L\rangle|C(L)\rangle, \tag{10.9}$$

$$|R\rangle \otimes |QCM\rangle \longrightarrow \frac{1}{\sqrt{2}}\big[|H, H\rangle|C(H)\rangle - i|V, V\rangle|C(V)\rangle\big] \neq |R, R\rangle|C(R)\rangle. \tag{10.10}$$

One can see from Eqs. (10.9) and (10.10) that cloning of a circularly polarized state fails even when $|C(H)\rangle = |C(V)\rangle$.

The phenomenon of quantum no-cloning can also be manifested when a light with a definite polarization encountered an excited atom on its way. In this case as well, some non vanishing probability is expected although the atom can emit the second photon via stimulated emission designed to have the same polarization as the original photon. It may also be vital to inquire whether it is possible to amplify the quantum state (to produce several copies of polarized photon each having the same state as the original) by this scheme. It should be clear by now that such amplification cannot be realized: this is the usual quantum no-cloning theorem put in different context [10].

One may be tempted to object to such a view point based on the fact that in classical telecommunication the information that travels in an optical fiber is encoded in the state of light, and it is amplified several times from the source to the receiver which otherwise should have been degraded. This mechanism appears to allow the phenomenon of quantum cloning. Nonetheless, to understand the misrepresentation in this regard, one needs to emphasize that the telecom signal is made up of a large number of photons prepared in the same quantum state. The seemingly prevalent amplification amounts to producing some new copies of the state out of the available similar states. Even then, it is worth noting that quantum no-cloning theorem does still apply since spontaneous emission is always available in amplifiers whatever small its proportion may be.

It might be necessary therefore to deal with implications of quantum no-cloning theorem on practical application at some point. To begin with, the no-clonability of the state of the quantum system happens to invalidate classical error correction techniques (see Sect. 12.1.1). This understanding heralds that it is not possible to create a backup copy of a given state in the middle of quantum computation, and apply it to correct subsequent errors as usually done in classical counterpart. The limitation imposed by quantum no-clonability in contrast can be a vital ingredient in forbidding eavesdropper from creating copies of the transmitted quantum crypto-graphic key, and subsequent manipulation that can lead to deciphering the encoded message, that means, no-clonability has a potential to enhance the security of quantum communication (see Sect. 12.1.2). The no-clonability of the quantum state can also preserve one of the pillars of quantum theory—the uncertainty principle. Otherwise, as mere intuition may reveal, if one can clone unknown quantum state, surely one can make as many copies as required; and consequently, measure each dynamical variable with arbitrary precision that may allow to bypass the restriction imposed by uncertainty principle and then annul the corresponding application.

In addition, it may not be difficult to foresee that the no-cloning theorem prohibits broadcasting of pure quantum states [11] since it is required to put the two systems in the product state such as cloning is possible to broadcast it. The notion of quantum no-broadcasting so can be taken as the direct consequence of quantum no-cloning. One can simply state: the claim that a pure quantum state cannot be copied leads to the presumption that it cannot be broadcasted. Even though the no-cloning theorem forbids to create two copies of the state given a single copy, it might worth noting that a mixed state no-cloning theorem is not sufficient to ascertain the validity of no-broadcasting since there can be many conceivable ways to broadcast a mixed state without the pertinent joint state necessarily being in the product state.

The idea of quantum no-broadcasting, therefore, can be understood as having a single copy of the state does not necessarily lead to creating the state such that its parts are the same as the original state: given an initial state with a density operator $\hat{\rho}$, it is impossible to create a composite state with density operator $\hat{\rho}_{AB}$ such that $Tr_{A,B}(\hat{\rho}_{AB}) = \hat{\rho}$. Let us now consider a broadcasting machine that consists of a source system whose unknown state is to be broadcasted ($\hat{\sigma}_s$), the system onto which the source state should be copied ($\hat{\sigma}_t$) and auxiliary system or the machine ($\hat{\sigma}_m$) that interacts with the source and target systems via unitary evolution.

With such labeling, the broadcasting process can be designate by the transformation

$$\hat{\rho}^{in} = \hat{\sigma}_s \otimes \hat{\sigma}_t \otimes \hat{\sigma}_m \rightarrow \hat{\rho}^{out}, \tag{10.11}$$

where $Tr_{t,m}(\hat{\rho}^{out}) = Tr_{s,m}(\hat{\rho}^{in}) = \hat{\sigma}$ is the density of the source system. One can hence deduce that the main essence of quantum no-broadcasting theorem is to show that there is no unitary transformation that allows such a process to exist for arbitrary source. Even though such observation seems to have a sever limitation on quantum communication, it can have a disguised importance such as initiating a one-to-one direct communication; and hence, can lead to a more reliable and securer information communication.

10.3   Fragility of the Quantumness

It might be clear by now that interference phenomenon in which the double-slit experiment provides a standard example is the main ingredient of some interesting features of quantum theory and resulting process (see Chaps. 7, 9 and 11). In practical setting, whatever small may be, one cannot disregard the interference term that depends on the superposition of the amplitude of the wave functions of the particles that made to overlap. Note that such assertion upholds consistent description of the particle in terms of the wave function (quantum theory; relate with the discussion in Chap. 4). Contrary to this long standing presumption, there are situations in which the interference term would not be manifested, that is, there are cases when the classical probability formulae is sufficient to explain the measurement result. This might make the study of the spontaneous interaction between the system and its environment very interesting. One prominent example is when one seeks to implement a detection measurement at the slits or when there are sufficiently many stray particles scattered off the particles supposed to be suitably interact with the wave in the space between the slits and screen. The interference term in such a case disappears due to the entanglement of the state of the strayed particles with that of the experimental particles. It is this kind of spontaneous suppression of interference via interaction with the environment that is often dubbed as quantum decoherence [12].

One may note that there is a tradeoff between the interference (phase information) and the which-path information: the better one can distinguish the two possible paths, the less visible the interference pattern would become. Pecoherence can also be perceived as the process that arises from continuous monitoring of the system by the environment. This kind of proposal in another words may connote as the environment performs nondemolition measurement on the system, and delineate decoherence as quantum effect that should be distinguished from classical dissipation and stochastic fluctuations or noise. Since the state of the object and the environment can be in the superposition of a lot of very well localized terms, the notion of quantum decoherence might be more complex than perceiving it as the entanglement of measuring apparatus with the measured system as insinuated earlier. One can also characterize decoherence as the quantum nature of the system is leaked into the environment, which leads to superposition of the system and environment states. In light of this, one may take the emerging mixture as a proper quantum ensemble in the context of measurement, and leads to the realization of a precisely one state in the ensemble.

When a quantum system interacts with its environment, it would become entangled with environmental degrees of freedom (see Chap. 4). It is this entanglement that profoundly influences what one can locally observe while measuring the system, that is, quantum interference effects can become effectively suppressed and so making them exorbitantly difficult to observe in most cases of practical interest. One may thus take entanglement as a nonlocal property of a composite system that would be significantly affected by the environment;[3] and as a result, can be perceived as the main obstacle in the realization of quantum information processing (see Chaps. 11 and 12). In relation to this, several quantum aspects of interest can be cited: the suppression of the interference can be extremely fast process, and so can easily defy instantaneous observation; the environment may tend to be coupled to, and suppress interference between preferred set of states that can be robust; and while the system are expected to be entangled with the environment, the states between which the interference is suppressed are the ones that gets least entangled with the environment.

The preferred states in case of decoherence are Schrödinger-type waves localized in both position and momentum that are the coherent states of the system that approximate the corresponding classical world. Decoherence can then be viewed in some aspect as a dynamical filter on the space of the quantum states that has a potential to single out those states that can be steadily prepared and maintained while effectively excluding other states. This proposition may herald that the superposed states necessary for quantum information processing tasks are those most susceptible to decoherence. This can be taken on the other hand as the

---

[3]There has been a considerable effort to model the environment responsible for decoherence as thermal reservoir, and explore its effect on the quantum features of the cavity radiation [13].

proposal that qubit systems should be engineered in the way that the interaction with environment can be optimized, which is detrimental to the preparation and longevity of the desired superposed states; and at the same time, they should remain sufficiently open to allow for their control.

Recall that the quantum system in the cavity is unavoidably coupled to the fluctuations in the external environment due to its contact with the walls of the container (see Sect. 3.1). As a result of such interaction, the quantum state invariably losses its purity. It is the interaction between the environment modes and the quantum system of interest that leads to the loss of coherence via unitary evolution. It has become evident over the years that entangled states lose their coherence rather so quickly since the intended interaction almost automatically collapses the quantum state, that is, the available entanglement as a nonlocal property of a composite quantum system is extremely fragile against the influence of the environment;[4] and should be taken into consideration while designing a pertinent application.

To look into the origin of the loss of quantum features, one can begin with the assumption that the coupling of the quantum state with environment modes induces a joint unitary time evolution of the form

$$\hat{U}(t)|\psi\rangle \otimes |E\rangle \longrightarrow |\psi\rangle \otimes |E_\psi(t)\rangle, \tag{10.12}$$

where $|E\rangle$ is some fixed initial states of the environment that can be addressed by the evolution operator $\hat{U}(t)$ responsible for the dynamics. Equation (10.12) entails that the environment acts as some kind of measurement procedure that captures the content of the information of the quantum state. For a general mixed initial state, note that the unitary time evolution can be expressed as

$$\hat{U}(t)\big[\alpha|0\rangle + \beta|1\rangle\big] \otimes |E\rangle \longrightarrow \alpha|0\rangle \otimes |E_0(t)\rangle + \beta|1\rangle \otimes |E_1(t)\rangle. \tag{10.13}$$

The density operator for the emerging mixed state can then be traced over the state of the environment, and one obtains

$$\rho(t) = \begin{pmatrix} \alpha^*\alpha & \alpha\beta^*\langle E_1|E_0\rangle \\ \alpha^*\beta\langle E_0|E_1\rangle & \beta^*\beta \end{pmatrix}, \tag{10.14}$$

where $\rho(t) = Tr_E(\hat{\rho}_{SE})$.

---

[4]The unfortunate situation in which the entanglement disappears while the quantum system unitarily evolves is often dubbed as entanglement sudden death. To be able to grasp the details of the dynamics of the quantum, the study of the effect of decoherence on entanglement evolution is a must despite the availability of numerous strategies such as error correction and decoherence free subspaces that can be utilized to bypass it [14].

It might not be difficult to see that, in case $|E_0(t)\rangle$ and $|E_1(t)\rangle$ are not orthogonal, the information that leaks to the environment can be designated as

$$\langle F_0(t)|F_1(t)\rangle = e^{-\Gamma(t)} \tag{10.15}$$

in which $\Gamma(t)$ (the rate of damping of coherence) is taken to be a function of time and its specific form depends on the details of the coupling between the quantum state and environment (see Sect. 3.1 and also [15]). Such interaction curtails the influence coming from the off-diagonal matrix elements that account for the coherence overtime; and hence, the reason for relating it with decoherence—doing so without significantly affecting the dynamics of the density operator is referred to as dephasing.

Imagine a double-slit experiment in which the passage of the particle through slits 1 and 2 is denoted by $|s_1\rangle$ and $|s_2\rangle$, and the environment by $|E\rangle$. According to the recipe given by Eq. (10.12), for initial superimposed state $\alpha|s_1\rangle + \beta|s_2\rangle$, the final state would be

$$|\psi\rangle^{out} = \alpha|s_1\rangle \otimes |E_1\rangle + \beta|s_2\rangle \otimes |E_2\rangle, \tag{10.16}$$

where $|E_{1,2}\rangle$ register the changes as a result of the system-environment interaction.

The possible local measurement on the system can thus be expressed as

$$\hat{\rho} = |\alpha|^2|s_1\rangle\langle s_1| + |\beta|^2|s_2\rangle\langle s_2| + \alpha\beta^*|s_1\rangle\langle s_2|\langle E_2|E_1\rangle + \alpha^*\beta|s_2\rangle\langle s_1|\langle E_1|E_2\rangle. \tag{10.17}$$

For instance, the probability density of measuring the position of the particle would be

$$P(x) = |\alpha|^2\langle x|s_1\rangle^2 + |\beta|^2\langle x|s_2\rangle^2 + 2\mathrm{Re}\left[\alpha\beta^*\langle x|s_1\rangle\langle x|s_2\rangle\langle E_1|E_2\rangle\right]. \tag{10.18}$$

The last term accounts for the interference term where the visibility of the interference pattern is quantifiable by the overlap $\langle E_1|E_2\rangle$. The idea is: the coherence would be retained for the case when the interaction between S and E is in such a way that E is completely unable to resolve the paths of the particle, that is, when $|E_1\rangle$ and E_2 are indistinguishable.

One may recall in addition that the structure and amount of the information that the environment encodes about the system can be quantified by the measure of classical or quantum mutual information (see Sect. 7.4 and also [16]). A classical mutual information in this sense is quantifiable based on the choice of particular observable of the system S and the environment E, and characterizes how well one can predict the outcome of the measurement of a given observable of S by measuring some observable on the fraction of E. Quantum mutual information can also be expressed as $H(S) = S(\rho_S) + S(\rho_E) - S(\rho_{SE})$, where S (E) stands for system (environment) and $S(X) = Tr(X\log X)$ is the von Neuman entropy associated with X. Quantum mutual information markedly quantifies the degree of quantum correlations between S and E. The classical and quantum mutual information can thus give rise to similar result since the difference between the two measures (quantum discord) disappears when decoherence is sufficiently effective to select a well defined classical pointer basis of the measuring device (relate with the discussion in Sect. 4.1.4).

Note that quantum measurement theory stipulates that the dynamics of the system treated in an isolation is irreversible. In other words, since the physical system is invariably linked to the surrounding environment, it is not possible to get rid of decoherence; and hence, looking for the way of protecting a given state of the system (or part of) it from decoherence, namely, entanglement purification and distillation could be very interesting [17]. So entanglement purification or distillation schemes would have a paramount importance if one wishes to communicate or compute within quantum regime.

To overcome the effects of decoherence on the realization of certain quantum implementation, one might be tempted to look for corresponding specified space: the Hilbert space of the system in which every state in the subspace is immune to decoherence [18]. One important condition for this to happen is the selection of the preferred state

$$\hat{S}_\alpha |s_i\rangle = \lambda_i^\alpha |s_i \qquad \text{for all } \alpha \text{ and } i, \qquad (10.19)$$

which form orthonormal bases of the subspace where the eigenvalues λ_i^α are independent of i and $\hat{S}_\alpha$ is the operator of the environment perceived to monitor the system.

This condition entails that the action of $\hat{S}_\alpha$ must be the same for the basis states $|s_i\rangle$ of the decoherence free subspace, and the existence of the decoherence free subspace thus corresponds to a dynamical symmetry in the structure of the system-environment interaction. The necessary condition for such a symmetry is the absence of the terms in the interaction Hamiltonian that act jointly on the system and environment in a nontrivial manner. For arbitrary state, a decoherence free subspace can be represented by a recipe that, if

$$\hat{U} |\psi\rangle \otimes |E(0)\rangle \longrightarrow |\psi\rangle \otimes |E(t)\rangle, \qquad (10.20)$$

the state $|\psi\rangle$ is not entangled with the environment, and so immune to decoherence (relate with Eq. (10.12)).

In the context of quantum optics, the system-environment system can be described by the interaction Hamiltonian of the form

$$\hat{H} = i \sum_j g_j \big(\hat{a}\hat{b}_j^\dagger - \hat{a}^\dagger \hat{b}_j\big), \qquad (10.21)$$

where $\hat{b}_j$ are the bath operators and $\hat{a}$ is the system operator (compare with Eq. (3.28)). By applying the standard techniques, the master equation that describes the interaction of the system with environment in Born-Markov approximation can be expressed as

$$\frac{d\hat{\rho}}{dt} = -i[\hat{H}, \hat{\rho}] + \frac{1}{2}\sum_{k,j}\gamma_{kj}\{[\hat{S}_k, \hat{\rho}\hat{S}_j^\dagger] + [\hat{S}_k\hat{\rho}, \hat{S}_j^\dagger]\}, \tag{10.22}$$

where γ_{kj} carries the information about the decoherence (dissipation as well; compare with Eq. (6.163)). Once the master equation is known, it should be clear by now that the effects of the decoherence can be included into the analysis [13, 19].

References

1. G.C. Ghirardi, A. Rimini, T. Weber, Lett. Nuovo Cim. **27**, 293 (1980); D. Dieks, Phys. Lett. A **92**, 271 (1982); T.F. Jordan, Phys. Lett. A **94**, 264 (1983)
2. M. Born, P. Jordan, Z. Phys. **34**, 858 (1925); W. Heisenberg, Z. Phys. **43**, 172 (1927) (English version: W.A. Fedaka, J.J. Prentis, Am. J. Phys. **77**, 128 (2009))
3. T. Baumgratz, M. Cramer, M.B. Plenio, Phys. Rev. Lett. **113**, 140401 (2014); J. Xu, Phys. Rev. A **93**, 032111 (2016); A. Streltsov, G. Adesso, M.B. Plenio, Rev. Mod. Phys. **89**, 041003 (2017)
4. M. Schlosshauer, Rev. Mod. Phys. **76**, 1267 (2005)
5. R. Uola, T. Moroder, O. Guhne, Phys. Rev. Lett. **113**, 160403 (2014)
6. T. Heinosaari et al., J. Phys. A **48**, 435301 (2015)
7. C.H. Bennett, Phys. Rev. Lett. **68**, 3121 (1992); J.M. Renes, Phys. Rev. A **70**, 052314 (2004); N. Gisin, Phys. Lett. A **210**, 151 (1996); E.S. Gomez et al., Phys. Rev. Lett. **117**, 260401 (2016)
8. P.W. Shor, Phys. Rev. A **52**, 2493 (1995)
9. S. Izumi et al., Sci. Rep. **8**, 2999 (2018); U. Khalid, J. Rehman, H. Shin, Sci. Rep. **10**, 2443 (2020)
10. W.K. Wootters, W.H. Zurek, Nature **299**, 802 (1982)
11. H. Barnum et al., Phys. Rev. Lett. **76**, 2818 (1996); S. Bandyopadhyay, G. Kar, Phys. Rev. A **60**, 3296 (1999); H. Barnum et al., Phys. Rev. Lett. **99**, 240501 (2007); A. Kalev, I. Hen, Phys. Rev. Lett. **100**, 210502 (2008); S. Luo, N. Li, X. Cao, Phys. Rev. A **79**, 054305 (2009)
12. W.H. Zurek, Phys. Rev. D **24**, 1516 (1981); W.H. Zurek, Rev. Mod. Phys. **75**, 715 (2003)
13. S. Tesfa, J. Phys. B **40**, 2373 (2007)
14. T. Hiroshima, Phys. Rev. A **63**, 022305 (2001); T. Yu, J.H. Eberly, Phys. Rev. B **66**, 193306 (2002); Q. Chen, M. Feng, Phys. Rev. A **82**, 052329 (2010)
15. W.H. Zurek, Phys. Rev. D **26**, 1862 (1982); K. Hornberger, J.E. Sipe, Phys. Rev. A **68**, 012105 (2003)
16. H. Ollivier, D. Poulin, W.H. Zurek, Phys. Rev. A **72**, 042113 (2005); R.B. Kohout, W.H. Zurek, Phys. Rev. A **73**, 062310 (2006)
17. C.H. Bennett et al., Phys. Rev. Lett. **76**, 722 (1996); L.M. Duan et al., Phys. Rev. Lett. **84**, 4002 (2000); J.W. Pan et al., Nature **410**, 1067 (2001)
18. P. Zanardi, M. Rasetti, Phys. Rev. Lett. **79**, 3306 (1997); D.A. Lidar, I.L. Chuang, K.B. Whaley, Phys. Rev. Lett. **81**, 2594 (1998); D. Bacon et al., Phys. Rev. Lett. **85**, 1758 (2000)
19. S. Tesfa, J. Mod. Opt. **55**, 1587 (2008)

Quantum Optical Processes

11

The aspiration to harness the nonclassical properties of light to improve the way we communicate and analyze data emanates from the expectation that the strong correlation shared among entangled states, or interconnectedness established by the entanglement, can enhance the efficiency of the exchange of information between nodes or sites. Despite such claim, as already noted, while one attempts to implement certain quantum operations, there are things that one may not able to do such as perfectly copying quantum state (see Sect. 10). It might be as a result essential to devise a means of mitigating the challenges imposed by quantum no-cloning by the way of instituting alternative quantum copying procedure. That means, contrary to the ramification of quantum no-cloning, implementation of the procedure of copying a quantum state should be expedient whenever there is an inclination to distribute quantum information among parties with minimum measurement disturbance and a need to enhance transmission fidelity over a lossy quantum channel. In relation to the observation that quantum principles do not preclude partial or imperfect copying of the quantum state, we explore the situation near perfect or optimal copying of the quantum state: the idea connoted as quantum cloning (see Sect. 11.4).

One can also argue that the change made by the experimenter on one of the entangled parties as in EPR steering can alter the properties of the other without necessarily obliging them to communicate due to the nonlocal character (see Chap. 7). This deceptively simple consideration can be taken as the starting point to explain much of the quantum optical processes we intend to discuss. For instance, when each particle in the composition is made to entangle with its partner, one may recall that appropriate measurement such as Bell state analyzer collapses the state of the remaining particle onto the entangled state (see Chap. 4). It is such a striking outcome of the projection measurement that instigates for example the conception of entanglement swapping (see Sect. 11.1).

It might be alluring nowadays to consider the content of quantum information to be transferred from the sending to receiving site with the aid of entanglement

S. Tesfa, *Quantum Optical Processes*, Lecture Notes in Physics 976,
https://doi.org/10.1007/978-3-030-62348-7_11

between the particles as resource. Since the process of information communication and the accompanying local operation cannot destroy the established entanglement, one can propose that in case the initial quantum state is part of the entangled state, the outcome as well is part of the entangled state even when these sites are presumed to be far apart. The rationale in this kind of thinking grows out of the understanding that two systems that do not directly interact can be entangled with the help of entanglement swapping. Such a process can constitute the notion of quantum teleportation: the phenomenon that can be perceived as if the unknown quantum state is destroyed at the sending site while its replica appears at receiving site; and also of quantum telecloning if used along with quantum cloning (see Sects. 11.2 and 11.5).

In case the sender (Alice) and the receiver (Bob) initially obtain one particle of a pair of maximally entangled Bell states, based on the content of the message she wishes to send, Alice should be able to perform one of the four local operations on the shared entangled state, and then send her qubit to Bob. Bob on the other hand can decode the desired message by performing the projective measurement in the Bell bases on the two-party system (see Fig. 11.3). One may note that the four manipulations lead to four orthogonal Bell states that can be taken as four distinguishable messages. This process entails that two bits of information can be sent via Bob's two-state particle to Alice who reads the encoded information via manipulating the Bell state of the two-party system. Upon identifying each Bell states with different messages, one can incidentally encode two bits of message via manipulating the state of one of the two particles. This procedure envisages enhancement of the information capacity of the transmission channel to two bits: this kind of doubling is usually alluded to as quantum dense (super) coding (see Sect. 11.3 and also [1]).

In the context of quantum optics, much of these quantum operations can be implemented by exploiting polarization based photonic qubits where the bosonic nature of the electromagnetic field is exploited within the scope of two-photon quantum interference (see Sect. 9.3). Nonetheless, in the quest of developing quantum information processing schemes in the realm of Gaussian continuous variable, it is found compelling using variables with a continuous spectrum of the amplitude of electromagnetic field as ingredient (see Sect. 2.4). Such consideration can be promising since one can manipulate for example the squeezing by using linear optical elements such as beam splitters and phase shifters [2]. The main theme of this chapter is thus set to introduce quantum optical processes that can be realized by using a coherent light as a source and expected to lead to viable schemes with a relative advantage in implementing quantum information processing tasks.

11.1   Entanglement Swapping

The observation that when a part of the states of the particles in a given composition are made to entangle with each other, the remaining states of the particles can also be made to entangle with the aid of appropriate measurement leads to the notion of entanglement swapping. Imagine that two experimenters share an entangled pair of quantum states; and one of them, let us say, Bob sends his half to the third party (Charlie) via quantum channel. The manifestation of the phenomenon of quantum swapping in this framework amounts to the realization of the combined state of Alice and Charlie taken to be initially uncorrelated to be entangled. Entanglement swapping process therefore encompasses preparing two independent pairs of entangled particles and to subject one particle from each pair to a joint measurement that can project the other formerly independent two particles onto entangled state. One can also assume that they share quantum nonlocal correlations when the outcome of the measurement permits.

One can begin as illustration with the proposal that for a binary variable u_k and its complement v_k (usually taken as $1 - u_k$) the Bell state to be expressed as

$$|\Phi^{\pm}\rangle_{k,j} = |u_k, u_j\rangle \pm |v_k, v_j\rangle, \tag{11.1}$$

where $|u_k\rangle$ and $|v_k\rangle$ are presumed to form an orthogonal set (compare with definition in Sect. 9), and the state that describes the participating four particles as

$$|\Phi^+\rangle_{1,2} \otimes |\Phi^+\rangle_{3,4} = |u_1, u_2, u_3, u_4\rangle + |v_1, v_2, u_3, u_4\rangle + |u_1, u_2, v_3, v_4\rangle$$
$$+ |v_1, v_2, v_3, v_4\rangle. \tag{11.2}$$

One can see from Eq. (11.2) that particles 1 and 2, and 3 and 4 are mutually entangled. If one then performs a Bell measurement on particles 2 and 3, that is, if one projects states 2 and 3 onto the Bell state as described in Sect. 9, the joint state of the four particles becomes

$$|\Phi^{\pm}\rangle = \big[|u_2, u_3\rangle \pm |v_2, v_3\rangle\big] \otimes \big[|u_1, u_4\rangle \pm |v_1, v_4\rangle\big], \tag{11.3}$$

$$|\Psi^{\pm}\rangle = \big[|u_2, v_3\rangle \pm |v_2, u_3\rangle\big] \otimes \big[|u_1, v_4\rangle \pm |v_1, u_4\rangle\big]. \tag{11.4}$$

One can infer from Eqs. (11.2)–(11.4) that prior to measurement, the Bell pairs are 1 and 2, and 3 and 4, but after measurement 2 and 3, and 1 and 4 turn out to be entangled.

To adopt the same procedure for light, one may begin with the assumption that there are two entangled photon pairs (1, 2) and (3, 4) where such designation can be related to the signal and idler modes of two nondegenerate parametric oscillators [3]. Intuitive consideration may suggest that photons 1 and 4 can be entangled when the joint measurement of photons 2 and 3 in the Bell basis is carried out and provided that the result of the measurement is communicated to 1 and 4. This embodies the main recipe that leads to the realization of entanglement swapping; which can also be framed as the entanglement induced between the two photons via simultaneous emission by the same source can be transferred or swapped to two photons originated from different sources formerly completely independent.

Take for example two EPR type light sources that simultaneously emit a pair of entangled photons, where the combined system can be described by a state

$$|\Psi\rangle = \frac{1}{2}\big[|H\rangle_1|V\rangle_2 - |V\rangle_1|H\rangle_2\big] \otimes \big[|H\rangle_3|V\rangle_4 - |V\rangle_3|H\rangle_4\big]. \tag{11.5}$$

Suppose one photon from each pair, let us say, photons 2 and 3 are subjected to Bell state measurement in the way shown in Fig. 11.2. According to the scheme of entanglement swapping (see Fig. 11.1), it is expected that the measurement projects the other two outgoing photons (1 and 4) onto entangled state. Nonetheless, it is possible to see from Eq. (11.5) that photons 1 and 2, and 3 and 4 are entangled in anti-symmetric polarized state, that is, the state of the pair 1 and 2 is factorizable from the state of the pair 3 and 4. This observation entails that there is no entanglement of photons 1 (or 2) with photons 3 (or 4). This outcome may indicate that having a set of two entangled pairs alone does not necessarily guarantee entanglement swapping.

Consider the case in which the joint Bell state measurement is performed on photons 2 and 3, that is, photons 2 and 3 are projected onto one of the four Bell states that form a complete bases for the combined state of photons 2 and 3 as shown in Fig. 11.2:

$$|\Psi^\pm\rangle_{23} = \frac{1}{\sqrt{2}}\big[|H\rangle_2|V\rangle_3 \pm |V\rangle_2|H\rangle_3\big], \tag{11.6}$$

$$|\Phi^\pm\rangle_{23} = \frac{1}{\sqrt{2}}\big[|H\rangle_2|H\rangle_3 \pm |V\rangle_2|V\rangle_3\big] \tag{11.7}$$

(see Eqs. (9.26) and (9.27), and also Fig. 9.1).

Fig. 11.1 Schematic representation of entanglement swapping designed from two EPR type light sources pumped by a laser light (LL) from one side and reflected light (RL) from the other, two phase beam splitters (PBS) and a mirror (M) (see also [4])

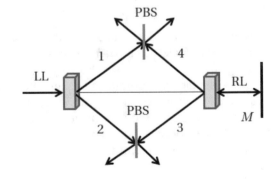

Fig. 11.2 Schematic representation of Bell-state analyzer designed by using a 50:50 beam splitter and two phase beam splitters (PBS)

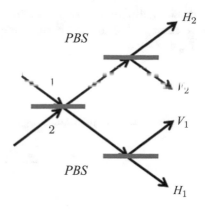

With the aid of the projection measurement corresponding to photons 1 and 4, it then appears straightforward to propose that

$$|\Psi\rangle_{1234} = \frac{1}{2}\left[|\Psi^+\rangle_{14}|\Psi^+\rangle_{23} + |\Psi^-\rangle_{14}|\Psi^-\rangle_{23} + |\Phi^+\rangle_{14}|\Phi^+\rangle_{23} + |\Phi^-\rangle_{14}|\Phi^-\rangle_{23}\right].$$
(11.8)

One can surmise from Eq. (11.8) that photons 1 and 4 emerge entangled despite the fact that they have never interacted with one another in the past, and the same argument holds for photons 2 and 3. This understanding can also be extended to N-party scenario.[1]

The idea of entanglement swapping between photon pairs can be taken as a fundamental building block in a mechanism that utilizes quantum relays and quantum repeaters such as quantum networks (see Sect. 12.1.2). For a photonic quantum communication in particular, since the distance between the sending and receiving sites is significantly limited by the decoherence that arises from coupling to the environment, one needs to divide the distance into smaller sections over which entanglement can be distributed where the created separate sections are meant to be bridged by entanglement swapping. The multistage entanglement swapping is, therefore, expected to be helpful in speeding up the distribution of entanglement and in improving the protection against certain errors [6].

One may then note that by taking the assertion that a continuous variable entanglement swapping can be dealt with in the framework of Gaussian matrix (7.120) into consideration, a continuous variable entanglement swapping can be implemented by exploiting the nonclassical correlations in the amplitudes of the electromagnetic field such as squeezing in the same way as for a discrete case (see also Sect. 2.4 and also [7]). To do so, using the setup such as Fig. 11.1 with the numbers

---

[1]Care should be taken when one wishes to contrast entanglement swapping with quantum monogamy: the phenomenon that forbids entangling one system strongly with other two systems at the same time [5].

denoting different bosonic operators, and when the optical modes are presumed to be combined on a 50:50 beam splitter (instead of the phase beam splitter), one can see that

$$\hat{a}_u = \frac{1}{\sqrt{2}}(\hat{a}_2 - \hat{a}_3), \qquad \hat{a}_v = \frac{1}{\sqrt{2}}(\hat{a}_2 + \hat{a}_3) \tag{11.9}$$

with the pertinent quadratures $\hat{q}_{u,v}$ and $\hat{p}_{u,v}$ defined by

$$\hat{q}_2 \to \frac{\hat{q}_u + \hat{q}_v}{\sqrt{2}}, \qquad \hat{q}_3 \to \frac{\hat{q}_v - \hat{q}_u}{\sqrt{2}}, \tag{11.10}$$

$$\hat{p}_2 \to \frac{\hat{p}_u + \hat{p}_v}{\sqrt{2}}, \qquad \hat{p}_3 \to \frac{\hat{p}_v - \hat{p}_u}{\sqrt{2}}, \tag{11.11}$$

which can be measured via a homodyne detection (see Sect. 4.5.2).

Depending on the covariance matrix of the input state (2.182), the state after the Bell measurement can be shown to be the entangled state on modes 1 and 4. In order to obtain the optimal outputs, the displacements need to be weighed by gain factors consistent with the output of the form:

$$\hat{q}_{1out} = \hat{q}_1 - \sqrt{2}g_1\hat{q}'_u, \qquad \hat{q}_{4out} = \hat{q}_4 + \sqrt{2}g_4\hat{q}'_u, \tag{11.12}$$

$$\hat{p}_{1out} = \hat{p}_1 + \sqrt{2}g_1\hat{p}'_v, \qquad \hat{p}_{4out} = \hat{p}_1 + \sqrt{2}g_4\hat{p}'_v. \tag{11.13}$$

The idea is: the Bell measurement yields $q_u = q'_u$ and $p_v = p'_v$, and the covariance matrix would not be changed, although the values of the first moments can be different.[2] To attest to whether the output states exhibit a nonclassical correlation, one may need to displace the conditional states so that a Gaussian state with vanishing first moment is obtained; identify the input states; optimize various gains; and test if modes 1 and 4 are coherently correlated (entangled).

For many practical cases, the displacement can be adjusted so that $g_1 = g_4 = c^2/(n+m)$, that is,

$$\sigma_{opt} = \begin{pmatrix} m - \frac{c^2}{n+m} & 0 & \frac{c^2}{n+m} & 0 \\ 0 & m - \frac{c^2}{n+m} & 0 & -\frac{c^2}{n+m} \\ \frac{c^2}{n+m} & 0 & n - \frac{c^2}{n+m} & 0 \\ 0 & -\frac{c^2}{n+m} & 0 & n - \frac{c^2}{n+m} \end{pmatrix}, \tag{11.14}$$

---

[2] See for example Eq. (2.166) and recall also that for $\theta = \pi/4$;

$$\hat{q}_j = \frac{1}{\sqrt{2}}[\hat{q}_j(0) + \hat{q}_k(0)], \qquad \hat{q}_k = \frac{1}{\sqrt{2}}[\hat{q}_j(0) - \hat{q}_k(0)].$$

where n and m are the mean photon numbers of the respective modes, and c stands for the correlation between the modes (compare with Eq. (2.181) and see also [8]). The logarithmic negativity can thus be taken as a good figure of merit to justify entanglement swapping. It is encouraging to see in practical aspect that the ᵣₑ₎ ₕᵤₗₚᵢₙ ₒₚᵣᵢ ₕₕᵣₒᵤgₕ.₁ₙ.₁ₗ .ₛᵥₙₗₗₚᵢₙg can be exploited to entangle radiation with an ensemble of atoms placed at different sites that can be very much handy in storing, and then transferring quantum data between to sites [9, 10].

11.2   Quantum Teleportation

The assertion that a lack of a precise simultaneous measurement of non-commuting observable precludes a complete transfer of a quantum state to other location entails that it is unpractical to aspire to get the information required to completely describe the quantum state at receiving site (see Sects. 4.1 and 10.1). It is expected despite such claim that quantum teleportation—the phenomenon that can be perceived as if an unknown quantum state is destroyed at the sending site while its perfect replica appears at receiving site—can overcome such restriction when the sender and receiver opt to consult a helper that shares nonlocal quantum correlation with them. As a result of presumed appearing and disappearing of the quantum state at the same time and different sites, quantum teleportation is believed to be one of the major techniques applicable to transfer quantum information to arbitrary distant location reliably.

One may note that by invoking appropriate joint measurements on the state to be transferred or teleported and the helper state, the sender can transmit classical measurement outcomes that enable the receiver to reconstruct the quantum state upon carrying out a unitary transformation on the helper state (see Fig. 11.3). There are two crucial ingredients in this consideration: the source of entangled states for encoding information at the sender site and Bell state analyzer for reconstructing the quantum state at receiving site (see Fig. 11.2). Supported by the processes of encoding and state reconstruction, it is hence expected that one can transfer a complete state of the particle to another particle placed at some other place where the original system losses its quantum state in the process, and there is no transportation of the particle or physical part of the system as the name might suggest.

It might be worthy to note that quantum information cannot be transmitted reliably via a classical channel alone as this would allow to replicate the signal in violation of quantum no-cloning theorem (see Sect. 10.2). Teleportation of unknown quantum state can thus be conceived as transportation of its defining characteristics through classical channel followed by its reconstitution at receiving site with the help of entanglement analyzer, where the collaborators do not require to know each other's location as prescribed by the phenomenon of quantum nonlocality. To teleport a part of the information encoded in the unknown input state via quantum correlation between the two separated subsystems, a classical information is required to be sent via conventional classical channel for confirmation. Owing to the fact that the classical information broadcasted over the classical channel is minuscule when compared to the infinite amount of the content of information required to specify the unknown quantum state, it is fair to perceive quantum teleportation as disembodied transport. Even then, the notion of teleportation constitutes the situation in which the unknown state goes in and the same state comes out: perfect teleportation—the feat that cannot be achieved in a real life [11].

To quantify teleportation, one may need to formulate appropriate criteria. To this end, there are certain issues that require quantitative treatment: the states supplied by the third party should be unknown to the sender and receiver; entanglement should be a verifiable resource in the sense that the output is close to the input but not close enough as the information would be sent through classical channel; each and every trial should achieve the output sufficiently close to the input (unconditional teleportation); and the quality of the teleportation should be good enough to transfer the quantum entanglement instead of a small subset of the unknown quantum state.

Intuitive consideration on the other hand may reveal that the quantum state does not exist independently of what one knows; rather, it captures the best information available about how the quantum system reacts to the existing circumstances (see Sect. 4.1.4). This idea may provoke a careful rethinking of what is actually transported. Teleporting a quantum state in the way described may point to the circumstance in which someone is lurking in the background for instance the third party who is the keeper of the knowledge about how the system is prepared. The goal of quantum teleportation is thus to transfer what the third party can say about the system placed in the sender's possession onto the system in the receiver's possession. This kind of interpretation may lead to the obvious conclusion that it is the information in its purest sense that is teleported. In present model, when the sender and receiver declare that the teleportation process is realized, the third party (the verifier) should be able to know that the original system is finally in the receiver's possession.

11.2.1 Basic Framework

On the basis of the resourcefulness of entanglement in the implementation of quantum teleportation, there might be a need to know the critical value of the degree of entanglement that can be captured in terms of the transferred content of information. During the realization of quantum teleportation, imagine that an original state $|\Psi_{in}\rangle$ at sender's site is put into the process while a different state with density $\hat{\rho}^{out}$ comes out at receiver's site. It can be presumed that these states embody what the third party or the verifier can say about the given operation. The question that should be addressed outright is when or in what circumstance do $|\Psi_{in}\rangle$ and $\hat{\rho}^{out}$ become close enough to each other as required by teleportation.

It turns out to be appropriate and appealing to gauge the closeness between $|\Psi_{in}\rangle$ and $\hat{\rho}^{out}$ with the aid of fidelity

$$F = \langle \Psi_{in}|\hat{\rho}^{out}|\Psi_{in}\rangle. \tag{11.15}$$

This measure of merit has a nice property that it equals to 1 if $\hat{\rho}^{out} = |\Psi_{in}\rangle\langle\Psi_{in}|$ and 0 if the input and output states are distinguishable with certainty. Quantum teleportation is then said to be successful when the fidelity exceeds the classical or no-cloning limit, that is, when $\Gamma_c = 1/2$—the best achievable value without entanglement.

It would be unwise to consent to the presumption that Alice and Bob do not try to get rid of entanglement and use classical channel instead. To avoid such mishaps, Victor (the verifier) requires certain measure that helps him to assess whether the similarity between the teleported and the input states surpasses the boundary exceedable by utilizing the resource associated with entanglement. Assume also that Alice and Bob who are in contact (or Victor who has sent one qubit of the entangled state to each) produced two particles with maximally entangled Bell state

$$|EPR\rangle = \frac{1}{\sqrt{2}}\big[|0\rangle_1|0\rangle_2 + |1\rangle_1|1\rangle_2\big]. \tag{11.16}$$

If Alice acquires one part of this particle (to be sent to Bob) and makes Bell's measurement beforehand; the idea is that the original state of the particle would be lost at the sending site but readily teleported to Bob's particle due to the established entanglement.

For now, imagine that if Alice has particle 1 in a certain quantum state but unknown to her, the emerging qubit will be $|\Psi\rangle_1 = \alpha|0\rangle_1 + \beta|1\rangle_1$. Suppose Alice wants to communicate the unknown quantum bit to Bob. Assume also that they have a classical telephone and a pair of entangled qubits at their disposal. One way to carry out teleportation at the sending site encompasses: measuring the qubit; guessing the state based on the outcome of the measurement; and describing it to Bob via the telephone. Although this approach seems a natural thing to do, the state would be transferred (if at all) with very poor fidelity. Compelled by the ensuing obstacles associated with quantum no-clonability, Alice might be forced to send the qubit to Bob at the price of simultaneously erasing it at her site. Such action, which is odd to do in classical regime, can be taken as the foundation of the idea of teleportation: the quantum state can be transferred from one place to another without necessarily being copied.

To implement quantum teleportation, Alice needs to measure her qubit and part of maximally entangled state, which is an entangling processing as it projects the state onto Bell states

$$|\Psi^\pm\rangle = \frac{1}{\sqrt{2}}[|0\rangle|1\rangle \pm |1\rangle|0\rangle], \qquad |\Phi^\pm\rangle = \frac{1}{\sqrt{2}}[|0\rangle|0\rangle \pm |1\rangle|1\rangle] \tag{11.17}$$

(relate with Eqs. (9.5) and (9.6)) [12]. She afterwards needs to communicate a qubit in state $|q\rangle = a|0\rangle + b|1\rangle$ on system A with the aid of a singlet state residing on her system A' (part of the maximally entangled state) and Bob's system B.

In line with this, the total initial state can be written as

$$|\Psi\rangle_{AA'B} = |q\rangle_A \otimes \frac{1}{\sqrt{2}}[|0\rangle|0\rangle + |1\rangle|1\rangle]_{A'B}, \tag{11.18}$$

can be rewritten as

$$|\Psi\rangle_{AA'B} = \frac{1}{\sqrt{2}}\big[a|0\rangle \otimes (|0\rangle|0\rangle + |1\rangle|1\rangle)_{A'B} + b|1\rangle \otimes (|0\rangle|0\rangle + |1\rangle|1\rangle)_{A'B}\big]. \tag{11.19}$$

It is possible to see from Eq. (11.19) that the first two qubits are at Alice's site and the last at Bob's.

Let Alice then applies the CNOT gate transformation to her two qubits with the control qubit being the qubit to be teleported to Bob (see Sect. 9.4);

$$|\Psi'\rangle_{AA'B} = \frac{1}{\sqrt{2}}\big[a|0\rangle \otimes (|0\rangle|0\rangle + |1\rangle|1\rangle)_{A'B} + b|1\rangle \otimes (|1\rangle_{A'}|0\rangle_B + |0\rangle_{A'}|1\rangle_B)\big]. \tag{11.20}$$

Suppose she also applies the Hadamard gate transformation to the first qubit

$$|\Psi''\rangle_{AA'B} = \frac{1}{2}\big[a(|0\rangle + |1\rangle) \otimes (|0\rangle|0\rangle + |1\rangle|1\rangle)_{A'B} + b(|0\rangle - |1\rangle) \otimes (|1\rangle_{A'}|0\rangle_B$$
$$+ |0\rangle_{A'}|1\rangle_B)\big], \tag{11.21}$$

from which follows

$$|\Psi''\rangle_{AA'B} = \frac{1}{2}\big[|00_{A'}\rangle \otimes (a|0\rangle + b|1\rangle)_B + |01_{A'}\rangle \otimes (a|1\rangle + b|0\rangle)_B$$
$$+ |10_{A'}\rangle \otimes (a|0\rangle - b|1\rangle)_B + |11_{A'}\rangle \otimes (a|1\rangle - b|0\rangle)_B\big], \tag{11.22}$$

or

$$|\Psi\rangle_{AA'B} = \frac{1}{\sqrt{2}}\big[|\Phi^+\rangle_{AA'} \otimes (a|0\rangle + b|1\rangle)_B + |\Phi^-\rangle_{AA'} \otimes (a|0\rangle - b|1\rangle)_B$$
$$+ |\Psi^+\rangle_{AA'} \otimes (a|0\rangle + b|1\rangle)_B + |\Psi^+\rangle_{AA'} \otimes (a|0\rangle - b|1\rangle)_B\big] \tag{11.23}$$

depending on how one opts to look into the matter.

One can gather from such a result that when Alice measures her system, she induces four states in Bob's system with equal proportion (relate with Fig. 11.2). Observe that the resulting states in Bob's system are similar to the state of the qubit $|q\rangle$ that Alice sought to send to him. Their mixture is unfortunately equal to the initial state of the system at Bob's site and so Bob cannot get a valuable information instantly. The output shown in Eqs. (11.22) and (11.23) can be utilized when Alice wants to convey her version of the outcome of the measurement to Bob via classical channel. In the context of quantum teleportation, after Bob received the classical code—depending on the two bits of information encoded in the Bell states—he is ⲁⲡⲗⲉ ⲧⲟ ⲧⲁⲗ ⲧⲉ ⲣⲟⲇⲟⲇ ⲗⲓⲧⲉ ⲧⲉⲓⲧ ⲗⲩ ⲧⲉⲧ ⲟⲓ ⲧⲉ ⲣⲟⲧ Pⲓⲉⲗⲓ ⲧⲣⲁⲇ ⲟⲣⲙⲁⲧⲓⲟⲛⲧ (ⲛⲛⲛ Cⲏⲟⲡ. 1ⲧ and also [13]). Markedly, Bob should obtain state $|q\rangle$ at the end of each rotation at his site while Alice possesses just one of the Bell states, and so no information about the state $|q\rangle$ would be left with her.

Imagine also that Alice measures the two qubits in her possession, and communicate the result to Bob with two classical bits of information encoded on four possible states: $|00\rangle$, $|01\rangle$, $|10\rangle$ and $|11\rangle$. One may envisage that the measurement outcome 00 entails that the state of Bob's qubit is equivalent to the teleported state. If Bob receives this classical code from Alice, he then knows for sure that the teleported state is the same as whatever associated with $|00\rangle$ state in Eq. (11.22). But if for instance he receives 01, he should perform the NOT-gate transformation of his qubit to find out the teleported state. Likewise, if he obtains a message such as 10 or 11, he can know the teleported state by carrying out σ_z or σ_z-NOT Pauli transformation of his state. Such exercise may indicate that, in the quantum teleportation, a single qubit in arbitrary quantum state can be teleported using just two bits of information sent from one spatial location to another via classical communication channel (relate with the discussion in Sect. 11.3).

11.2.2 Coherent State Teleportation

It might be acceptable to envision quantum teleportation of continuous variable to rely on the entangled state of the electromagnetic field as in the original EPR conjecture (see Sects. 2.4 and 7.2). With this in mind, the entangled optical state can be sent to Alice and Bob in the same way as the discrete case in which Alice is expected to couple the mode she seeks to teleport with her part of the EPR mode on the beam splitter. In this context, it is the Bell measurement of the position quadrature at one of the beam splitter outputs and of the momentum at the other that gives rise to the classical values to be sent to Bob via classical channel. One can as a result propose that once the light is represented in terms of optical polarization states, quantum teleportation can be implemented following the already outlined conceptual framework (see Fig. 11.2).

One can also think of distributing the available entanglement by sending photon pairs via optical fibers. In this case, once the quantum teleportation channel is established, one can remove the classical channel, that is, the sender needs to perform Bell state measurement between the photon from the entangled pair and

---

[3]Note that $|00\rangle = \frac{1}{\sqrt{2}}[\Phi^+\rangle + |\Phi^-\rangle]$; $|01\rangle = \frac{1}{\sqrt{2}}[\Psi^+\rangle + |\Psi^-\rangle]$;
$|11\rangle = \frac{1}{\sqrt{2}}[\Phi^+\rangle - |\Phi^-\rangle]$; $|10\rangle = \frac{1}{\sqrt{2}}[\Psi^+\rangle - |\Psi^-\rangle]$.

photonic qubit that carries the quantum state to be teleported. The general procedure required to revive the communicated information thus encompasses that Alice informs Bob of the result of her Bell state measurement, and Bob on the other performs outcome dependent unitary rotation on his system. One may realize that the size of the classical information sent by Alice is smaller than what is required to provide the classical description of the teleported quantum state.

In a certain teleportation scheme, the entangled (EPR) state can be created from two independent squeezed fields α_1 and α_2, and half of this entangled state, let us say, α_1 is sent to Alice who in turn combines it on a 50:50 beam splitter with unknown input state α_{in} (which is unknown to both) intended for teleportation [14, 15]. Alice in the process measures the position and momentum quadratures of the two output fields α_a and α_b from the beam splitter, which can yield the continuous variable analogue of the Bell state measurement. One may note that Alice gains no information about the input state in the limit of perfect EPR correlations. Even then, the output photocurrents from Alice's two quadrature measurements G_x and G_p are presumed to be transmitted to Bob via classical channel. Bob then applies this information to perform continuous phase space displacement on the second EPR beam α_2, and thereby generates the teleported output state α_{out}. The teleported state is expected to acquire certain fidelity with the original unknown input state as verified by Victor: who generates the original input, and also measures the teleported output (see Fig. 11.3).

One can after that begin with the assumption that two independently generated squeezed vacuum modes are superimposed on a 50:50 beam splitter. As highlighted in Chap. 2 and Sect. 4.3.2, the squeezing operation can transform the quadratures as

$$\hat{q}(t) \to \hat{S}(r)\hat{q}(0)\hat{S}^\dagger(r) = \hat{q}(0)e^r, \qquad \hat{p}(t) \to \hat{S}(r)\hat{p}(0)\hat{S}^\dagger(r) = \hat{p}(0)e^{-r} \tag{11.24}$$

and the 50:50 beam splitting as

$$\hat{q}_j = \frac{1}{\sqrt{2}}(\hat{q}_j(0) + \hat{q}_k(0)), \qquad \hat{q}_k = \frac{1}{\sqrt{2}}(\hat{q}_j(0) - \hat{q}_k(0)). \tag{11.25}$$

Since the involved canonical momentum would not be altered by such transformations, one can see that the two output quadrature modes transform as

$$\hat{x}_\pm = \frac{1}{\sqrt{2}}[e^r \hat{x}_1^{(0)} \pm e^{-r}\hat{x}_2^{(0)}], \qquad \hat{p}_\pm = \frac{1}{\sqrt{2}}[e^{-r}\hat{p}_1^{(0)} \pm e^r\hat{p}_2^{(0)}], \tag{11.26}$$

where the superscript denotes the initial vacuum modes and r is the squeeze parameter. In the limit of infinite squeezing ($r \to \infty$), the output modes become infinitely noisy: $\hat{x}_+ - \hat{x}_- \to 0$ and $\hat{p}_+ + \hat{p}_- \to 0$.

Suppose mode 1 (denoted by $\hat{x}_+$ and $\hat{p}_+$) is sent to Alice and mode 2 (denoted by $\hat{x}_-$ and $\hat{p}_-$) to Bob, and let Alice's mode is superimposed on a 50:50 beam splitter

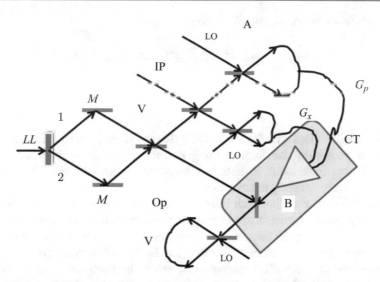

Fig. 11.3 Schematic representation of quantum teleportation designed from two-mode EPR type light source. The two-mode EPR type light source is pumped by laser light (LL) and the output is directed by two mirrors to be superimposed on a 50:50 beam splitter, the entangle light is superimpose with the input (IP) on a 50:50 beam splitter at Vector's site, where the resulting quadratures are measured by two homodyne detectors at Alice's site, the classical data of this measurement is transmitted through classical channel (CT) to Bob's where it is combine with the other part of the entangled light received via quantum channel at a 50:50 beam splitter and the data is finally superimposed on another 50:50 beam splitter with local oscillator and the output (OP) quadratures are measured at Victor's site (see also [16])

with the input mode as shown in Fig. 11.3;

$$\hat{x}_{u(v)} = \frac{1}{\sqrt{2}}\left(\hat{x}^{in} - (+)\hat{x}_a\right), \qquad \hat{p}_{u(v)} = \frac{1}{\sqrt{2}}\left(\hat{p}^{in} - (+)\hat{p}_a\right), \qquad (11.27)$$

from which Bob's mode follows

$$\hat{x}_b = \hat{x}^{in} + \hat{x}_b - \hat{x}_a - \sqrt{2}\hat{x}_u, \qquad \hat{p}_b = \hat{p}^{in} + \hat{p}_a + \hat{p}_b - \sqrt{2}\hat{p}_v. \qquad (11.28)$$

After receiving Alice's classical results x_u and p_u via classical transmission, let us say, Bob displaces his mode as

$$\hat{x}_b \rightarrow \hat{x}^{out} = \hat{x}_b + \sqrt{2}G_x\hat{x}_u, \qquad \hat{p}_b \rightarrow \hat{p}^{out} = \hat{p}_b + \sqrt{2}G_p\hat{p}_v, \qquad (11.29)$$

where $\hat{x}_u$ and $\hat{p}_v$ are displaced elements and $G_{x,p}$ is the corresponding gain [17].
For $G_x = G_p = 1$ for example the teleported fields become

$$\hat{x}^{out} = \hat{x}^{in} - \sqrt{2}e^{-r}\hat{x}_b^{(0)}, \qquad \hat{p}^{out} = \hat{p}^{in} + \sqrt{2}e^{-r}\hat{p}_a^{(0)}, \qquad (11.30)$$

and for arbitrary but equal gain $(G_x = G_p = G)$

$$\hat{x}^{out} = G\hat{x}^{in} - \frac{G-1}{\sqrt{2}}e^r\hat{x}_a^{(0)} - \frac{G+1}{\sqrt{2}}e^{-r}\hat{x}_b^{(0)}, \tag{11.31}$$

$$\hat{p}^{out} = G\hat{p}^{in} + \frac{G-1}{\sqrt{2}}e^r\hat{p}_b^{(0)} + \frac{G+1}{\sqrt{2}}e^{-r}\hat{p}_a^{(0)}. \tag{11.32}$$

One can see from Eqs. (11.31) and (11.32) that for $G = 1$ and infinite squeezing, there is a possibility for implementing perfect teleportation. Nonetheless, the squeezing in realistic setup is finite and the detectors are inefficient, which make such a proposal practically unattainable.

In the hope of resolving such inherent inconsistency, one may seek to derive quantum teleportation criteria upon employing the least noisy model in which Victor is assumed to prepare the initial input drawn from a fixed set of states, and passes half of it to Alice and the other half to Bob. Let Bob, after carrying out the required transformation to accomplish teleportation, sends the teleported part of his state back to Victor so that he judges whether teleportation has taken place.

For the input quadratures at Alice's site and the output quadratures at Bob's site, the least noisy linear model—the case in which Alice and Bob are classically communicating—can be represented by the transformation of the form

$$\hat{x}^{out,j} = G_x\hat{x}^{in} + \frac{G_x}{s_a}\hat{x}_a^{(0)} + \frac{1}{s_{b,j}}\hat{x}_{b,j}^{(0)}, \tag{11.33}$$

$$\hat{p}^{out,j} = G_p\hat{p}^{in} - G_p s_a\hat{p}_a^{(0)} + s_{b,j}\hat{p}_{b,j}^{(0)}, \tag{11.34}$$

where j stands for the label of the copy, $G_{x,p}$ for the gain of x, p measurement and s_a (s_b) for the parameter given by Alice's (Bob's) measurement strategy and accounts for the noise penalty due to her (his) homodyne detection. One may note that the gains $G_{x,p}$ as well as the parameter s_b can be manipulated by Bob as long as the output quadrature operators satisfy the usual commutation relation.

In case the input denotes pure Gaussian states with coherent amplitudes

$$\hat{x}^{in} = \langle\hat{x}^{in}\rangle + \frac{1}{s_v}\hat{x}^{(0)}, \qquad \hat{p}^{in} = \langle\hat{p}^{in}\rangle + s_v\hat{p}^{(0)}, \tag{11.35}$$

where Victor can choose in each trial the coherent amplitude and to what extent the input is squeezed via parameter s_v. If Victor finds overlapping amplitude in all trials, he can look at the excess noise in each trial at least within some error margin. One can hence define the normalized variances as

$$V_x^{out,in} = \frac{\langle\Delta(\hat{x}^{out} - \hat{x}^{in})^2\rangle}{\Delta x_{vac}^2}, \qquad V_p^{out,in} = \frac{\langle\Delta(\hat{p}^{out} - \hat{p}^{in})^2\rangle}{\Delta p_{vac}^2}. \tag{11.36}$$

It should be consequently possible to verify for a unit gain that

$$V_r^{out,in} V_n^{out,in} = \frac{(s_a^2 + s_b^2)^2}{n_a^2 n_b^2}. \tag{11.37}$$

Equation (11.37) would be minimum for $s_a = s_b$ with the optimum value of 4 that corresponds to a classical teleportation. One can thus write the fundamental requirement that reveals whether a given state with arbitrary coherent amplitude is teleported, that is,

$$V_x^{out,in} V_p^{out,in} \geq 4. \tag{11.38}$$

During a genuine quantum teleportation, Victor is expected to observe the violation of this inequality. Verification of teleportation is believed to boost Victor's confidence in Alice's and Bob's honesty that they indeed utilized entanglement as a resource while communicating.

One of the challenges related to the condition epitomized by Eq. (11.38) is that, the variances are not directly measurable since the input state would be destroyed by the teleportation process. Victor can however combine for a Gaussian input state his knowledge of the input variances V^{in} with the detected variances V^{out} to infer $V^{out,in}$. As illustration, one can take arbitrary s_v (set of input states that contains the coherent and squeezed states) for which

$$V_x^{out} V_p^{out} = \frac{(s_v^2 + s_a^2 + s_b^2)^2}{s_v^2 s_a^2 s_b^2}. \tag{11.39}$$

Equation (11.39) would also be minimized for $s_v = s_a = s_b = 1$, that is,

$$V_x^{out} V_p^{out} \geq 9, \tag{11.40}$$

the violation of which can be taken as testament to the implementation of quantum teleportation. Note that there are still various conditions that can be derived based on different considerations, where the degree of the overlap can be determined by calculating the fidelity [18].

11.3   Quantum Dense Coding

If Alice wants to send the content of certain quantum information to Bob using quantum bits, she needs to encode the information onto the qubit, and then send the classical code. Since nonorthogonal quantum states cannot be reliably distinguished, one may guess that Alice cannot achieve better than one bit per qubit: there may not be gained advantage in opting to employ quantum approach. However, available works clearly show that in case Alice and Bob share a maximally entangled state, two classical bits per qubit can be communicated—the idea commonly known as quantum dense coding [19]. Although there is a considerable effort to distinguish between dense and super dense codings depending on whether the capacity of the channel can be doubled or not, we lump them together for ease of presentation.[4] Quantum dense coding so can be perceived as the way of communicating two bits of information via manipulation of one of the two entangled states that can individually carry a bit of information in which the essential aspect of this phenomenon is the presumption of shared entangled state that can be mapped onto another state via local manipulation.

One can thus begin with the assumption as one of the Bell states, let us say,

$$|\Psi^+\rangle = \frac{1}{\sqrt{2}}\big[|0\rangle_A \otimes |1\rangle_B + |1\rangle_A \otimes |0\rangle_B\big] \tag{11.41}$$

is shared between Alice and Bob with the condition that Alice can transform the composite system into one of the other Bell states by manipulating her part (see Sect. 11.2). If Alice does nothing, the system remains in the same state. But if she manages to send her particle via unitary gates that can be described by one of the Pauli matrices, the two-party entangled system is expected to change. For example, when initially Alice and Bob each obtains half of the entangled pair in the state $|\Psi^+\rangle$, Bob can perform one out of four possible unitary transformations on his state, that is, Bob can accomplish four transformations (see Chap. 9): identity operation that yields the original state $|\Psi^+\rangle$; state exchange ($|0\rangle_B \leftrightarrow |1\rangle_B$ and $|1\rangle_B \leftrightarrow |0\rangle_B$) that leads to $|\Phi^+\rangle$; state dependent phase shift (differing $|0\rangle_B$ and $|1\rangle_A$ by π) that leads to transforming the original state to $|\Psi^-\rangle$; and state exchange and phase shift that lead to the state $|\Phi^-\rangle$ (see Eqs. (9.5) and (9.6)). It is this kind of simple manipulations allowable by quantum theory that result in the doubling of the channel capacity.

It might also be imperative introducing the idea of distributed dense coding and be able to classify mixed states as quantum or classical according to their measure of dense coding. One may need to first reaffirm that the phenomenon of dense coding arises from the perception that the shared entanglement can make feasible the exchange of more information content than classical setup. The situation can also be seen from different angles: for one, the sender and receiver can be perceived to be at a considerable distant place from each other and so are not allowed to communicate with each other; for the other, they can use local operations and classical communication if required and so are allowed to perform global operations. In such arrangement, for a single receiver, there is an exact quantum dense coding capacity that cannot be increased via communication between the senders, that means, their joint operations can be discussed in relation to quantum monogamy. This assertion may imply that in case there are more than one receivers, there is an

---

[4]There is a suggestion recently that a classical superposition of the electric field can also lead to the situation similar to quantum dense coding [20].

upper bound for the quantum dense coding capacities. For instance, for a bipartite case, the amount of the classical information that can be sent via a d-dimensional quantum system would be at most $\log_2 d$ bits [21]. One can recall that it is not the qubit in one of the four orthogonal states that should be sent but a pair of entangled qubits.

To look into how this mechanism works, one may need to stick to the assumption that Alice and Bob share one of the entangled singlet states. If Alice then wants to inform Bob about one of the four events denoted by j, she needs to rotate her qubit according to Pauli's transformation (see Sect. 9.4). The singlet state rotated by a Pauli matrix $\hat{\sigma}_j$ on Alice's qubit corresponds to $|\Psi_j\rangle = [\hat{\sigma}_j]_A \otimes \hat{I}_B |\Psi_0\rangle$: Bell state is orthogonal to $|\Psi_{j'}\rangle = [\hat{\sigma}_{j'}]_A \otimes \hat{I}_B |\Psi_0\rangle$ for $j \neq j'$ since Bell states are mutually orthogonal. The idea is: if Bob gets Alice's half of the entangled state after carrying out the required rotation, he can discriminate between four Bell states and infer j.

In this way, the notion that Alice sends one qubit would give Bob $\log_2^4 = 2$ bits of information—which is the essence of quantum dense coding. The perceived entanglement between the sender and receiver is thus expected to extend the capacity of the channel beyond classically admissible bound. If the sender and receiver share for example entangled bipartite state in $\hat{d}_A \otimes \hat{d}_B$ dimensional phase space, the sender may able to send more than $\log_2 d_A$ bits to the receiver, that is, more than the maximal information content of her subsystem when no entanglement is shared. Even then, Alice certainly cannot send more than $\log_2 d_A + \log_2 d_B$ bits to Bob.

Given previously shared state $\hat{\rho}^{AB}$ in dimension $\hat{d}_A \otimes \hat{d}_B$, a general dense coding scheme consists of two steps. At first, Alice performs a local unitary transformation with probability p_i on her part of $\hat{\rho}^{AB}$, that is, she transforms the state $\hat{\rho}^{AB}$ to the ensemble $\{p_j, \hat{\rho}^{AB}\}$ with

$$\hat{\rho}_j^{AB} = \hat{U}_j \otimes \hat{I}_{d_B} \hat{\rho}^{AB} U_j^\dagger \otimes \hat{I}_{d_B}, \tag{11.42}$$

where $\hat{I}_{d_B}$ is the identity operator on the Bob's Hilbert space; after that, she sends her part of the ensemble state to Bob. Bob on the other extracts the maximal information about the index j from the ensemble $\{p_j, \hat{\rho}^{AB}\}$ by performing projective measurements.

One may see that the maximum amount of information Bob can gather from his measurement is bounded from above by

$$S(\hat{\rho}') = \sum_j p_j S(\hat{\rho}_j^{AB}) + \sum_j p_j S(\hat{\rho}_j^{AB} | \hat{\rho}), \tag{11.43}$$

where $S(\hat{\rho})$ is the von Neumann entropy, $S(\hat{\rho}|\hat{\rho}') = Tr(\hat{\rho}\log_2\hat{\rho} - \hat{\rho}\log_2\hat{\rho}')$ is the relative entropy and $\hat{\rho}' = \sum_j p_j \hat{\rho}_j^{AB}$.

It is proposed that this bound can be attained asymptotically so that the capacity of dense coding would be

$$C_d = \max \sum_j p_j S(\hat{\rho}_i^{AB}|\hat{\rho}'), \tag{11.44}$$

where the maximization procedure is carried out overall sets of unitaries performed by Alice. The capacity of dense coding for a given shared state thus turns out to be

$$C_d = \log_2 d_A + S(\hat{\rho}^B) - S(\hat{\rho}^{AB}) \tag{11.45}$$

(see Sect. 7.4 and also [22]). One can infer from Eq. (11.45) that the quantity C_d can be increased when Alice and Bob are allowed to locally operate on the shared state. An increase in C_d via filtering however requires a classical communication between them; they may not be allowed to perform classical communication to effect the change in the shared state.

Since a classical procedure that does not require a shared quantum state can be used by Alice to send at most $\log_2 d_A$ bits of classical information, the shared quantum state turns out to be useful to achieve dense coding with the capacity more than $\log_2 d_A$. One may note that such states are precisely those for which

$$S(\hat{\rho}^B) > S(\hat{\rho}^{AB}). \tag{11.46}$$

Equation (11.46) exemplifies states that are mixed more locally than globally, and never be satisfied for separable states.

This observation suggests that even states that are entangled, but not distillable, cannot be used for dense coding operation since it is impossible to obtain a maximally entangled state from them by local operations and classical communication. It is also possible to put forward that entangled states can be used to transfer classical bits encoded in a quantum state beyond the classical limit and prepare known quantum state at remote location. Even though such a procedure is initially introduced for the case of a single sender and receiver, to seek viable application of such communication schemes, it happens to be compelling to consider information transmission network that engages several senders and receivers where the knowledge of correlations among separated physical systems is required [23].

11.3.1 Dense Coding with Polarized Light

In light of the operational interpretation of quantum dense coding, the input to the interferometric setup can be one or half of the pair of entangled photonic qubits in the Bell basis state while the remaining qubit is sent unchanged to Bob (see Fig. 11.3). After processing the former photonic qubit in one of the four ways using phase beam splitters, Alice is expected to send it to Bob who measures the two qubits and obtains two classical bits. So the outcome would be: Bob receives two classical bits that match those Alice sent by manipulating her part.

In another but equivalent approach, Bob is perceived to be the owner of the source of information, that is he sends one part of the entangled photonic qubit to Alice and keeps one for himself. After that, Alice is expected to manipulate her part of the qubit and send to Bob those over which she has control, and discard those over which she does not have. The underpinning principle is: no one other than Bob should obtain the information from the communicated photon which would be possible in case the message is encoded in the Bell states of the twin photons but not in the local state of the communicated photon. The nonlocal character of the quantum state hence guarantees that Alice can encode two bits of information by preparing one of the four Bell states through local operation on her photonic qubit.

The joint measurement of the twin photons that can institute the entanglement with Bell state analyzer would be one of the ways to decode the two bits of information (see Fig. 11.2). In case Alice has an optical system that generates

$$|\Psi^+\rangle = \frac{1}{\sqrt{2}}\left[|H\rangle_A|V\rangle_B + |V\rangle_A|H\rangle_B\right] \tag{11.47}$$

as twin beam, she can utilize two of the Bell states $|\Psi^+\rangle$ and $|\Psi^-\rangle$ to perform the outlined procedure (see Eqs. (9.26) and (9.27), and also Fig. 11.2). It might worth noting that she can change the Bell state from $|\Psi^+\rangle$ to $|\Psi^-\rangle$ and thus encode a bit of information in the state by operating the phase shift on the vertical polarization component of her photon. Then Alice sends her photon to Bob where the communicated photon and Bob's photon are superimposed on a 50:50 beam splitter. After that, a circular polarization beam splitter can be inserted in each output direction of the beam splitter where the output photons are counted by single-photon detectors. At last, Bob can decode the information encoded in the entangled states via coincidence counting. One may then infer from Fig. 11.4 that two photons can be directed to the same output port with the same circular polarization state for $|\Psi^+\rangle$, while at the same time, they are directed to different output ports with different circular polarization states for $|\Psi^-\rangle$.

Fig. 11.4 Schematic representation of quantum dense coding designed from EPR type light source pumped by a laser light (LL), phase beam splitters (PBS), phase shifter (Φ) and mirrors (compare with Fig. 11.3)

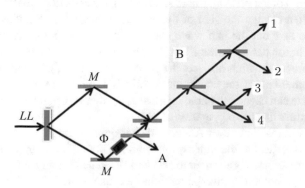

A bit of information can be encoded in this way in the Bell states through local operation on the photon, and can be decoded later by the joint measurement since the four manipulations (identity, polarization flip, phase shift and phase shift plus rotation) result in the four orthogonal Bell states that correspond to four distinguishable messages. This mechanism unequivocally indicates that two bits of information can be sent via Bob's two-state particle to Alice who finally reads the encoded information from the obtained Bell state of the two-particle system [24]. One can see for instance that linear optics implementation of dense coding can reach up to three messages of its theoretical two-bit (four messages) while one considers the Bell states. This is mainly because, the recognition of two Bell states $|\Psi^\pm\rangle$ can be achieved while the other two $|\Phi^\pm\rangle$ cannot be told apart: both are taken to send one and the same message.

To overcome such predicament, let us begin with four bases:

$$|\chi^{1(2)}\rangle = \frac{1}{\sqrt{2}}\big[|H\rangle_1 \otimes |V\rangle_2 + (-)|V\rangle_1 \otimes |H\rangle_2\big], \qquad (11.48)$$

$$|\chi^3\rangle = |H\rangle_1 \otimes |H\rangle_2, \qquad (11.49)$$

$$|\chi^4\rangle = |V\rangle_1 \otimes |V\rangle_2. \qquad (11.50)$$

The general dense coding process in this representation can be achieved by sending the photons to polarizing beam splitters, and then detect the output using photon number resolving detectors (see Sect. 4.3.2). When one sends two parallel polarization photons to a beam splitter from opposite sides, the photons that emerge from the same side would bunch together. This approach can allow one to discriminate between $|\chi^3\rangle$ and $|\chi^4\rangle$, and from $|\chi^{1(2)}\rangle$ with the aid of photon number resolving detectors. Although this technique can discriminate the vertical and horizontal polarizations, it cannot do so with the mixture $|H\rangle \otimes |V\rangle$. Bell state measurement, on the other hand, can distinguish between $|\chi^1\rangle$ and $|\chi^2\rangle$. It is hence instructive to exploit these approaches simultaneously to discriminate among the four computational bases, and subsequently the corresponding encoded messages.

In practical setting, Alice can start with $|\Psi^+\rangle$ (generated by a spontaneous parametric down conversion process). To send $|\chi^1\rangle$, she needs to put nothing in the path of her photon. But to send $|\chi^2\rangle$, she needs to put half wave plate in the path, and then change the sign of the vertical polarization. Besides, to send $|\chi^3\rangle$, she needs to take out the half wave plate and put in 45^0 polarizer oriented horizontally ($|H\rangle$ photon passes through but $|V\rangle$ photon is reflected), and turns $|\Psi^+\rangle$ into $|\Phi^-\rangle$ in the process, which projects the photons to state $|H\rangle$ in half of the occurrences. In the other half of the occurrences, Alice's photon is reflected from the polarizer, and so she can have both photons in state $|V\rangle$, that is, a pair in state $|\chi^4\rangle$. It might be good to stress that the preparation of $|\chi^3\rangle$ and $|\chi^4\rangle$ should pertain to the physics of entangled systems: whenever Alice sends her qubit through a polarizer oriented horizontally or vertically, the other qubit from the entangled pair (originally in the state $|\Phi^+\rangle$) immediately goes over to $|H\rangle$ and $|V\rangle$ states for any subsequent measurement along horizontal or vertical directions (see also Sect. 7.2.1).

11.3.2 Continuous Variable Dense Coding

In the realm of continuous variable, the entanglement resource that should be shared by Alice and Bob is a pair of EPR beams having quantum correlations between canonically conjugate quadrature variables that can be efficiently generated by a nonlinear optical process such as parametric oscillator. The corresponding quantum communication scheme can be thought of as one of the entangled pairs is fed into Alice's station while the message is encoded as quantum state by a simple phase space offset by way of the displacement operator. The state pertinent to the displacement of the EPR type beam constitutes the quantum signal and communicated along the quantum channel to Bob's station where it is decoded with the aid of the second component of the EPR pair of the beam and homodyne detector (see Fig. 11.5). The resulting suitably normalized photocurrents typify the message received by Bob in which the finite correlations inherent in the EPR beams are expected to enable quantum dense coding with enhanced channel capacity [25].

Note that there could be two main motivations for in depth study of quantum dense coding in the realm of continuous variables: the information capacity is expected to beat a single-mode coherent (and squeezed) state communication, and the emerging quantum dense coding is expected to be highly immune to unauthorized eavesdropping since the signal-to-noise ratio in the channel is very small. It might hence be reasonable to anticipate that the emerging quantum dense coding can have a significant potential for application (see Chap. 12).

Imagine that Alice and Bob share a two-mode state having correlations between $\hat{x}_1$ and $\hat{x}_2$, and $\hat{p}_1$ and $\hat{p}_2$ with condition that $\langle \Delta \hat{x}_- \Delta \hat{p}_+ \rangle = 0$, where $\hat{x}_- = \hat{x}_1 - \hat{x}_2$ and $\hat{p}_+ = \hat{p}_1 + \hat{p}_2$: there is no correlation among respective variances. Since for a two-mode squeezed light the variances are the same, Alice can encode the same amount of information on both quadratures. From technical point of view, after Alice sends her mode to Bob, let us say, he measures the corresponding quadratures. One may deduce from earlier discussion that the conditional probability distribution

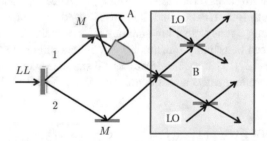

Fig. 11.5 Schematic representation of continuous variable quantum dense coding designed from EPR type light source pumped by laser light, beam splitters, mirrors, amplitude and/or phase modulator at Alice's site (compare with Figs. 11.3 and 11.4)

for Bob to read value β_j when Alice encodes α_j is necessarily Gaussian;

$$P(\beta|\alpha) = \mathcal{N} \exp\left[\frac{(\beta_x - \frac{\alpha_x}{\sqrt{2}})}{V_{x-}} - \frac{(\beta_p - \frac{\alpha_p}{\sqrt{2}})}{V_{p+}}\right], \tag{11.51}$$

where V_{x-} (V_{p+}) are the variances of the maximum (minimum) quadrature projections and $\mathcal{N}$ is the normalization constant.

Let us now turn the attention to the capacity of arbitrary Gaussian quantum channel with the emphasis that a continuous variable quantum dense coding can be quantified in terms of the available quadrature spectral variances. One may note that such variances can be directly measurable with the help of linear optical elements and homodyne detection (see Fig. 11.5). It is a well established fact that the Shannon capacity of a given communication channel with Gaussian noise of power (variance) N and Gaussian distributed signal power S operating at the bandwidth limit can be expressed as

$$C = \frac{1}{2} \log_2\left[1 + \frac{S}{N}\right]. \tag{11.52}$$

Markedly, Eq. (11.52) can be applicable to calculate the channel capacity of quantum states with Gaussian probability distributions such as coherent and squeezed states, and S/N can be interpreted as a signal-to-noise ratio. Besides, the average photon number of a beam of light can be expressed as

$$\bar{n} = \frac{1}{4}\left(V_{x-} + V_{p+}\right) - \frac{1}{2}. \tag{11.53}$$

Various degrees of and options for quantum dense coding can hence be studied [26].

One may also consider the case when a classical information is presumed to be encoded by the choice of one particular quantum state from among predefined ensemble of quantum states (continuous variable) by Alice for communication over quantum channel to Bob. If Alice and Bob are allowed to communicate via a one-way exchange along a noisy quantum channel, the optimal amount of the classical information that can be reliably transmitted can be determined in the same way as earlier approaches [27]. That means, if a classical signal α taken from the ensemble of P_a is to be communicated as a quantum state $\hat{\rho}_\alpha$, Holevo's bound would be bounded by

$$H(A:B) \leq S(\hat{\rho}) - \int d^2\alpha \, P_\alpha S(\hat{\rho}_\alpha) \leq S(\hat{\rho}), \tag{11.54}$$

where $\hat{\rho} = \int d^2\alpha \, P_\alpha \hat{\rho}_\alpha$ and the equality would be achievable when Alice sends a pure orthogonal states.

It is worth noting that the roles of the classical and quantum channels in quantum dense coding and quantum teleportation are interchanged (see Sect. 11.2), that is, instead of reliably transferring a quantum information through a classical channel using entanglement as in teleportation, the amount of classical information communication sent from Alice to Bob would be increased when Alice sends her half of the entangled state through a quantum channel. But like quantum teleportation, quantum dense coding also relies significantly on the shared entanglement. If Alice and Bob share quantum resource, quantum theory should hence allow for the realization of a practical procedure that circumvents the classical bound on the channel capacity.

Suppose Alice wants to send the message as letter a from her set of alphabets $\{a_j\}$. To do so, let us say, she produces a quantum state of density $\hat{\rho}_a$ with probability p_a; the state of the channel can be expressed as

$$\hat{\rho} = \sum_a p_a \hat{\rho}_a. \qquad (11.55)$$

Under the same condition, let Bob employs generalized positive operator value measurement $\hat{E}_b$ to try to extract information about the letter Alice attempts to send. When he acquires the state $\hat{\rho}_a$ representing letter a from Alice's alphabet, suppose Bob obtains instead letter b from his own alphabet with a conditional probability from which one may compute the mutual information (see Sect. 7.4).

Intuitive consideration may reveal that the inherent upper bound on the obtained mutual information is independent of Bob's measurement strategy; the achievable upper bound can so help to determine the channel capacity for communicating classical information with the aid of quantum states. When the states employed to send information are from infinite dimensional Hilbert space such as a single-mode bosonic field, it is required to attach some constraint to the channel usage to get a finite value for the pertinent capacity.

In this respect, one may assume that there is a constraint on the mean number of photon that can pass down the channel per usage: $\langle \hat{n} \rangle = \bar{n}$. For this constraint to happen, the maximum entropy may be interpreted as the channel capacity achieved when Alice uses alphabet of number states distributed according to thermal distribution. The channel capacity in this case can be expressed as

$$C = (1 + \bar{n}) \ln(1 + \bar{n}) - \bar{n} \ln \bar{n}. \qquad (11.56)$$

One can see that Eq. (11.56) reduces for large $\bar{n}$ to $1 + \ln \bar{n}$. The capacity when Alice applies the strategy that involves coherent state and squeezed state alphabet for example is found to be [28]

$$C^{coh} = \ln \bar{n}, \qquad C^{sq} = \ln 2 + \ln \bar{n}. \qquad (11.57)$$

In quantum dense coding, Alice and Bob need to communicate via two channels but Alice needs to modulate only one of them while the second channel is used to communicate the other half of the entangled state to Bob (see Fig. 11.5). Note that Alice's local action would be to span the system with the square of the dimension of Hilbert space of the piece she holds. Since information is the logarithm of the number of distinguishable states, $\log(n^2) = 2\log n$, one can expect the doubling of the capacity of the channel.

It might not be difficult to see that this discussion can be extended to arbitrary two-mode Gaussian state [29]. We can as a result forward the condition that aids to detect two-mode Gaussian states that can be expressed in terms of the variances of two correlated quadratures;

$$V_{x-}V_{p+} < C, \tag{11.58}$$

where C is the constant that depends on the selected scheme in which $C = \left(\frac{1}{2e}\right)^2$ for squeezed state and $C = \frac{1}{16}$ for number state. One ought to note that to exhibit entanglement, $C < \frac{1}{16}$.

11.4   Quantum Cloning

The inability to perfectly clone unknown quantum state can lead to the assessment that the lack of a precise simultaneous measurement of a non-commuting observable precludes a complete transfer of quantum state to other location but can be exploited in securing the secrecy of the information (see Chap. 10). In the midst of such seemingly conflicting propositions, imagine that each of the sender's photon can be prepared at random in one of the four possible combinations of horizontal and vertical polarization states (see Fig. 11.2). Based on the fact that quantum principles do not preclude imperfect or partial copying of the quantum state, one may opt to proclaim as quantum cloning operation is the existing best way to make copies of quantum information.

It might be compelling to inquire how close the copy can be to the original state, and what happens to the original state after the copying process had taken place. It is due to such basic concerns and daunting questions along with the perceived applications that one needs to look beyond the no-cloning theorem, and to consider the mechanism of quantum (approximate) cloning. Although there is a practical interest to copy the aspect of the quantum state including entanglement, the scope of the discussion in this section is restricted to the way of copying a portion of the qubit approximately due to the underlying challenge. The main reason to do so stems from the presumption that such an action does not negate the fundamental principle of the no-cloning theorem in the sense that there is still some part of the quantum state that may not be copied. Such a copy—whatever its amount (degree; fidelity) be—can guide one to delve into some information of the original state [30].

Quantum cloning can be interpreted in the layman terms as the process of taking arbitrary unknown quantum state, and make the copy without altering the original state in a conceivable way;[5]

$$\hat{U} |\Psi\rangle_A |c\rangle_B = |\Psi\rangle_A |\Psi\rangle_B, \qquad (11.59)$$

where $\hat{U}$ is the cloner operator, $|\Psi\rangle_A$ is the state to be cloned and $|c\rangle_B$ is the place holder of the copy or initial blank state (see also Eq. (10.1)). Note that the discussion on quantum cloning schemes that geared towards achieving optimal cloning transformation compatible with quantum no-cloning theorem can be highly required; continuous variable quantum cloning specially is expected to be attractive due to the relative ease in preparing and manipulating the respective quantum states with the help of elements of linear optics (see Sect. 9.3).

Imagine that there is a quantum state $|s\rangle$ that one would like to copy and can be expressed in terms of the basis states $|0\rangle_a$ and $|1\rangle_a$, where the subscript a stands for the system to be copied. Assume also that the state $|s\rangle_a$ that can be constructed from a linear combination of the same basis states is fed into the quantum copying machine. This machine is thus expected to yield $|s\rangle_a$ as output in which the copy is presumed to be identical to the one fed into it (quantum state $|s\rangle$). This can be taken as if one state is put into the machine, and then two quantum states are generated—ideal copying. Consider as illustration the transformation of the form

$$|s\rangle_a \otimes |QCM\rangle_x \longrightarrow |s\rangle_a |s\rangle_b \otimes |QCM'\rangle_x, \qquad (11.60)$$

where $|s\rangle_a$ is the input state, $|s\rangle_b$ is the sought for copy, $|QCM\rangle_x$ is the initial state of the machine and $|QCM'\rangle_x$ is its final state (compare with Eq. (10.4)). Note that the input state of the copy is not specified but can be taken without much effort as the vacuum state just like a blank paper in the copier machine.

One may in practice require a cloning machine that can achieve equal fidelity for every output state—universal quantum cloning machine. Since there is no specific information regarding the input state beforehand, such presumption can be perceived to be equivalent to the process of distributing information to different receivers, and so the notion of universal quantum cloning machine envisages the performance to be the same for every input state. In the simplest case scenario, one qubit is taken to be cloned to yield two copies that can be identical to each other, that is, they can be symmetric; and yet, different from the original input state. The two copies can also be different; and yet, both of them can be similar to the original

---

[5]It might be compelling to inquire whether quantum entanglement is cloneable. Existing studies suggest that it is possible to clone in some cases a part of the original entanglement much in the same way as a quantum state can be cloned imperfectly. For a maximally entangled input state, the cloning machine produces two qubit pairs with the same and highest amount of entanglement. Even though quantum operation can perfectly duplicate the entanglement of a maximally entangled states, it does not necessarily preserve separability, that is, some separable quantum states can become entangled after cloning [31].

input state but with different similarities, that is, they can be asymmetric. With this characterization, a quantum cloning machine can be delineated as universal, if it copies equally well the states (if the fidelity is independent of the initial state); symmetric, if at the output all the clones have the same fidelity; and optimal, if for a given fidelity of the originals, the fidelities of the clones are the maximum allowable by quantum theory.

The main task of quantum copying procedure is therefore to produce at least two identical states $|s\rangle_a$ and $|s\rangle_b$ as output although quantum theory does not allow the anticipated transformation for arbitrary input state (see Sect. 10.2). For clarity, consider the copying process designated by the transformations:

$$|0\rangle_a|QCM\rangle_x \rightarrow |0\rangle_a|0\rangle_b|QCM_0\rangle_x, \tag{11.61}$$

$$|1\rangle_a|QCM\rangle_x \rightarrow |1\rangle_a|1\rangle_b|QCM_1\rangle_x. \tag{11.62}$$

One can see from the unitary nature of the transformation process and the orthonormality of the basis states that the states of the copying machine are normalized

$$\langle QCM|QCM\rangle = \langle QCM_0|QCM_0\rangle = \langle QCM_1|QCM_1\rangle = 1. \tag{11.63}$$

The quantum copying machine can be technically recognized in such a way that the basis vectors $|0\rangle_a$ and $|1\rangle_a$ are ideally copied provided that a similar kind of transformation can be invoked. Since the superposition of these basis states can be written as $|s\rangle_a = \alpha|0\rangle_a + \beta|1\rangle_a$, one can see for $\langle QCM_0|QCM_1\rangle = 0$ that

$$|s\rangle_a|QCM\rangle_x \rightarrow \alpha|0\rangle_a|0\rangle_b|QCM_0\rangle_x + \beta|1\rangle_a|1\rangle_b|QCM_1\rangle_x = |s\rangle_{abx}^{out}. \tag{11.64}$$

The reduced density operator that represents the state of the system after the copying process has taken place can be expressed as

$$\hat{\rho}_{ab}^{(out)} = Tr_x[\hat{\rho}_{abx}^{(out)}] = \alpha^2|00\rangle\langle00| + \beta^2|11\rangle\langle11|, \tag{11.65}$$

where

$$|00\rangle \equiv |0\rangle_a \otimes |0\rangle_b, \qquad |11\rangle \equiv |1\rangle_a \otimes |1\rangle_b. \tag{11.66}$$

Once the combined density operator is known, the reduced density operators for each state can be obtained;

$$\hat{\rho}_a^{(out)} = Tr_b[\hat{\rho}_{ab}^{(out)}] = \alpha^2|0\rangle_{aa}\langle0| + \beta^2|1\rangle_{aa}\langle1|, \tag{11.67}$$

$$\hat{\rho}_b^{(out)} = Tr_a[\hat{\rho}_{ab}^{(out)}] = \alpha^2|0\rangle_{bb}\langle0| + \beta^2|1\rangle_{bb}\langle1|. \tag{11.68}$$

Equations (11.67) and (11.68) may indicate that both the original and the copy at the output are identical. Even though this can be taken as a good news, the original state at the output turns out to be mixed state which is a bad news since the off-diagonal elements are not copied but destroyed.

11.4.1 Quantification of Quantum Cloning

Since copying a quantum state is highly demanded, it might be imperative aspiring to quantify the degree or amount of the clonability of a given state. To judge how good a certain copying procedure could be, one needs to look for a way of comparing the obtained output with what the corresponding ideal output should be, that is, one requires a way of comparing the pertinent density matrices. With this in mind, it is possible to utilize the square of the Hilbert-Schmidt norm of the difference between the two density matrices that can be defined for operator $\hat{A}$ as

$$||\hat{A}|| = \sqrt{Tr(\hat{A}^\dagger \hat{A})} \tag{11.69}$$

with the property that $|Tr(\hat{A}^\dagger \hat{B})| \leq ||\hat{A}|| ||\hat{B}||$.

One can also express the distance between the density matrices as

$$D = ||\hat{\rho}_1 - \hat{\rho}_2||^2. \tag{11.70}$$

Then the norm of the distance between the input and output density operators of the original mode can be written as

$$D_a = Tr[\hat{\rho}_a^{(id)} - \hat{\rho}_a^{(out)}]^2, \tag{11.71}$$

where the input density operator of the original mode is denoted by $\hat{\rho}_a^{(id)}$. When viewed in relation to this measure of merit

$$\hat{\rho}_a^{(id)} = \alpha^2 |0\rangle_a \langle 0| + \alpha\beta |0\rangle_a \langle 1| + \alpha\beta |1\rangle_a \langle 0| + \beta^2 |1\rangle_a \langle 1|, \tag{11.72}$$

from which the Hilbert-Schmidt norm of the difference between the density operators follows

$$D_a = 2\alpha^2 \beta^2 = 2\alpha^2 (1 - \alpha^2). \tag{11.73}$$

Equation (11.73) reflects that the states $|0\rangle_a$ (for $\alpha = 1$) and $|1\rangle_a$ (for $\alpha = 0$) are copied perfectly while the superimposed states $|s^\pm\rangle_a = \frac{1}{\sqrt{2}}[|0\rangle_a \pm |1\rangle_a]$ are copied the worst [30].

One may also seek to look for a more qualitative approaches that allow to weigh by how much the initial state can be copied to the blank state. In line with this suggestion, the symmetric cloning—the situation in which $\hat{\rho}_1 = \hat{\rho}_2$—can be studied by making use of the local fidelity

$$F = \langle \Phi | \hat{\rho}_{1,2} | \Phi \rangle \tag{11.74}$$

that quantifies the degree of similarity of the clones to the input, where the fidelity can be taken as a figure of merit that qualifies each clone independently [32]. For a system having more than two states, one of the valid figure of merits that can quantify the degree of quantum cloning is a single copy fidelity defined for each of the outputs $j = 1, ..., M$ of the cloning machine as the overlap between j and the initial state:

$$F_j = \langle \Phi | \hat{\rho}_j | \Phi \rangle, \tag{11.75}$$

where j is the state of the clone.

For the operation epitomized by Eq. (11.60), the cloning process can be expressed as

$$\hat{U} | \Psi \rangle_a | \Psi \rangle_b \otimes | QCM \rangle_x \rightarrow | \Psi \rangle_{abx}^{out} \tag{11.76}$$

such that the partial traces on the original qubit and the clone satisfy

$$\hat{\rho}_a = \hat{\rho}_b = F | \Psi \rangle \langle \Psi | + (1 - F) | \Psi^\perp \rangle \langle \Psi^\perp |, \tag{11.77}$$

where F is the fidelity independent of the initial state. The main idea is: for arbitrary normalized input state, since quantum theory is linear, one can obtain the copies by invoking unitary operation. After tracing out the ancillary state, the density operator[6] turns out to be

$$\hat{\rho}^{out} = \frac{2}{3} | \Psi \rangle \langle \Psi | + \frac{1}{6} \left[| \Psi \rangle | \Psi^\perp \rangle + | \Psi^\perp \rangle | \Psi \rangle \right] \otimes \left[\langle \Psi | \langle \Psi^\perp | + \langle \Psi^\perp | \langle \Psi | \right], \tag{11.78}$$

which also reduces to

$$\hat{\rho}_1 = \hat{\rho}_2 = \frac{2}{3} | \Psi \rangle | \Psi \rangle + \frac{1}{6} \hat{I}. \tag{11.79}$$

---

[6]For d-dimensional case, note that

$$\hat{\rho}_1 = \hat{\rho}_2 = \eta | \Psi \rangle | \Psi \rangle + \frac{1 - \eta}{d} \hat{I},$$

where η is the shrink factor [33]. In line with this, there is a proposal that the information about the direction in space can be better encoded onto orthogonal qubits than in two identical qubits [34]. One may note that the worst possible fidelity for the cloning of a d-dimensional quantum system is d^{-1} that can be obtained for the maximally mixed state.

One can thus see that the pertinent fidelity would be

$$F = \langle \Psi | \hat{\rho}_1 | \Psi \rangle = \frac{5}{6}, \tag{11.80}$$

and the density operator that describes the optimal cloning can be expressed as

$$\hat{\rho} = \frac{5}{6} | \Psi \rangle \langle \Psi | + \frac{1}{6} | \Psi^\perp \rangle \langle \Psi^\perp |, \tag{11.81}$$

where $| \Psi^\perp \rangle$ is the state orthogonal to $| \Psi \rangle$. The fidelity determined in this way provides a measure of similarity between the states that can be described by the density operator of the original input state.

If this figure of merit happens to be one, the two states are completely the same; but if zero, they are orthogonal. Since a completely mixed state contains nothing about the input state, the corresponding fidelity of 1/2 should be the farthest distance between the two quantum states. This type of fidelity does not depend on the input state, that means, the copies exhibit state independent characteristics, and so the cloning machine can be taken as universal.

One can also observe that the result exemplified by Eq. (11.80) is the highest fidelity that can be achieved for a cloning machine, and thus stands for the optimal case. For initial state $| \Psi \rangle = \alpha | 0 \rangle + \beta | 1 \rangle$ and if cloning machine is in initial state $| C \rangle$, one can see that

$$|0\rangle \otimes |C\rangle \rightarrow |\Psi_0\rangle, \qquad |1\rangle \otimes |C\rangle \rightarrow |\Psi_1\rangle \tag{11.82}$$

with the final states belonging to the product of the Hilbert spaces of the clones. So based on the linearity of quantum theory, one can propose that

$$|\Psi\rangle \otimes |C\rangle \rightarrow \alpha |\Psi_0\rangle + \beta |\Psi_1\rangle \equiv |\Psi^{out}\rangle \tag{11.83}$$

in which the fidelity of the clones can be expressed as

$$F_A = \langle \Psi | Tr_{BC}(\hat{\rho}^{out}) | \Psi \rangle, \qquad F_B = \langle \Psi | Tr_{AC}(\hat{\rho}^{out}) | \Psi \rangle. \tag{11.84}$$

Under the condition that the fidelity of the clones is equal: $F_A = F_B$, the thesis would be quantum theory permits the existence of a cloning transformation with fidelity $F^{\text{Uni}} = \frac{5}{6}$ [35].

Besides, the transformation of a universal cloning machine can be epitomized by

$$|\Psi_0\rangle = \sqrt{\frac{2}{3}} |00\rangle_{AB} \otimes |0\rangle_C + \sqrt{\frac{1}{3}} |\Psi^+\rangle_{AB} \otimes |1\rangle_C, \tag{11.85}$$

$$|\Psi_1\rangle = \sqrt{\frac{2}{3}} |11\rangle_{AB} \otimes |1\rangle_C + \sqrt{\frac{1}{3}} |\Psi^+\rangle_{AB} \otimes |0\rangle_C, \tag{11.86}$$

where $|\Psi^+\rangle = \left[|01\rangle + |10\rangle|\right]/\sqrt{2}$, and $|0\rangle_C$ and $|1\rangle_C$ denote the two orthogonal states of the cloning machine (see Sect. 9). Upon tracing over the cloning machine, the two clones of the input $|0\rangle$ are afterwards left in the joint state

$$\hat{\rho}_{AB} = Tr_C(\hat{\rho}^{out}) = \frac{2}{3}|00\rangle\langle00| + \frac{1}{3}|\Psi^+\rangle\langle\Psi^+|, \tag{11.87}$$

and the input state becomes a projector onto $|\Psi\rangle^{\otimes2}$ while the second term is some maximally entangled state

$$\hat{\rho}_A = \hat{\rho}_B = \frac{2}{3}|\Psi\rangle\langle\Psi| + \frac{1}{6}\hat{I} \tag{11.88}$$

(compare with Eq. (11.79)).

Such consideration can also be generalized for the case when N particles are assumed to carry the pure initial state to be copied to $M - N$ particles that represent the copies (blank states) and interpreted as $N \to M$ cloning:

$$|\Psi\rangle^{\otimes N} \otimes |R\rangle^{\otimes(M-N)} \otimes |QCM\rangle \longrightarrow |\Psi\rangle^{out}, \tag{11.89}$$

where $|R\rangle$ is arbitrarily chosen reference state. So upon using an axiomatic approach in the context of earlier discussion, the cloning map that yields the optimal $N \to M$ cloning for d-dimensional states is found to be [36]

$$F_{N \to M} = \frac{N(d-1) + M(N+1)}{M(N+d)}, \tag{11.90}$$

which reduces for $d = 2$ to

$$F_{N \to M} = \frac{N + M(N+1)}{M(N+2)}. \tag{11.91}$$

One may at some point intend to clone a qubit designated by the polarization state of the photon. To do so, imagine population inverted atomic medium whose initial state and interaction Hamiltonian with the electromagnetic field are taken to be invariant under a given polarization transformation so that it can emit photons of various polarizations with the same probability. If a given light is presumed to impinged onto such a medium, it can stimulate the emission of a light of the same polarization, wherein the photons in the final state can be considered as clones of the original incoming photons. But whenever there is a stimulated emission, there is also spontaneous emission of photons of the wrong polarization that limits the achievable fidelity of the cloning. It might be interesting thus to look into how small the damaging effect of the spontaneous emission can be on the optimal cloning in the physical process that involves stimulated emission.

For a single two-level atom with dipole moment μ, it is a well established fact that the amplification of the incoming photon of polarization vector **u** depends on their scalar product [37]. In case the polarization vector of the incoming photon is parallel to the dipole moment, the state $|1\rangle_{\parallel}$ after some interaction time will evolve into the state containing the desired two photon state $|2\rangle_{\parallel}$ due to stimulated emission. But if **u** is orthogonal to μ, the two-photon component of the resulting state corresponds to $|1\rangle_{\parallel}|1\rangle_{\perp}$, where $\parallel$ ($\perp$) stands for polarization vector parallel (orthogonal) to **u**. This happens due to the spontaneous emission that spoils the amplification since one of the two photons has the wrong ($\perp$) polarization. If one considers an amplifier that comprises of two such atoms with orthogonal dipole moments μ_1 and μ_2, it is possible to amplify the photon independently of its polarization although the process suffers from unavoidable noise originating from spontaneous emission. Assuming that the two atoms interact in a similar manner with the incoming photon, one can intuitively notice that if one atom amplifies the photon well (when **u** is close to μ_1), the second atom will amplify it poorly since **u** is by then approximately orthogonal to μ_2. The balance between these two effects results in amplification that does not depend on μ.

In this regard, by filtering out the resulting two-photon component, one may get

$$|1\rangle_{\parallel} \otimes |0\rangle_{\perp} \longrightarrow \frac{2}{3}|2\rangle_{\parallel}\langle 2| \otimes |0\rangle_{\perp}\langle 0| + \frac{1}{3}|1\rangle_{\parallel}\langle 1| \otimes |1\rangle_{\perp}\langle 1| \tag{11.92}$$

irrespective of the direction of the dipole moment. Mark that cloning of the polarization via stimulated emission can be attained with probability 2/3 while the spontaneous emission blurs the measure of polarization by probability of 1/3. Let us now take the cloning of a general qubit that can be represented by the polarization state of the photon. Owing to the pertaining physical condition, there is always spontaneous emission of photons of wrong polarization, that means, although one seeks to have stimulated emission for photon polarization to reach universality, there is inevitably spontaneous emission that limits the obtainable fidelity. The fidelity of obtaining a stimulated emission (cloning) can then be expressed as

$$F_c = \sum_{k=0}^{N+1} \sum_{j=0}^{N} \frac{k}{k+j} \, p(k, j), \tag{11.93}$$

where $p(k, j)$ is the probability for finding k right and j wrong photons.

In the same token, imagine a three-level atomic system with lower energy level $|l\rangle$ and two upper energy levels $|u_1\rangle$ and $|u_2\rangle$ connected by two electromagnetic field modes $\hat{a}_1$ and $\hat{a}_2$ (see Sect. 6.5). One may aspire in this case to clone the superposed states $(\alpha\hat{a}_1 + \beta\hat{a}_2)|0\rangle$, where α and β are constants. Recall that the interaction Hamiltonian that describes the atomic system has the form

$$\hat{H} = ig \sum_{k=1}^{N} \left[|u_1\rangle\langle l|_j \hat{a}_1 + |u_2\rangle\langle l|_j \hat{a}_2 \right] + H.c, \tag{11.94}$$

where j refers to the jth atom and g to the strength of the coupling between the atom and radiation. It might also be important for each atom to be initially prepared in the mixed state

$$\rho_0 = \frac{1}{2}\big[|u_1\rangle\langle u_1| + |u_2\rangle\langle u_2|\big], \qquad (11.95)$$

which is invariant under unitary transformation. Note that the invariance of the Hamiltonian and initial state ensures the universality of the cloning process.

One may realize that in such a scheme the atoms do not only act as the photon source but also play the role of ancilla. Markedly, in case the atomic states are rewritten in the bases related to the original by the same unitary transformation, the interaction Hamiltonian and initial state of the atoms look exactly the same as in the original bases, which suggests that it is sufficient to analyze the performance of the cloner for one basis.[7] The time development of the density of the combined atom-photon system can be obtained by successive application of the evolution operator in certain time step starting with

$$\hat{\rho} = \frac{1}{2^N}\big[|u_1\rangle\langle u_1| + |u_2\rangle\langle u_2|\big]^{\otimes N} \otimes \hat{a}|0\rangle\langle 0|\hat{a}^\dagger. \qquad (11.96)$$

The optimal fidelity of finding N photons as perfect output clones thus turns out[8] to be

$$F_{1 \to N} = \frac{2N + 1}{3N}. \qquad (11.97)$$

11.4.2 Continuous Variable Quantum Cloning

It may be imperative looking for continuous variable quantum cloning transformation where the two copies obey a general transformation relation in which the product of the position error variance on the first copy times the momentum error variance on the second is zero [40]. In light of this, we now strive to extend the

---

[7]Discussion in Chap. 6 indicates that the atom describable by such Hamiltonian can be de-excited via emission of a photon in the pertinent mode (1 or 2), and excited via absorption of a photon of the appropriate type, that is, the number of photons plus excited atoms would be a conserved quantity.

[8]One can describe the required linear amplifier as the ensemble of atoms initially in the excited state and can emit photons polarized along a given direction as in Kerr medium. The atoms are assumed to be pumped with the photon of a suitable energy, let us say, polarized along the vertical direction. At the exit from the amplifier, it is possible to select the case in which one and only one additional photon has been emitted, and then analyze the output in the (H, V) bases. One can after that look for the probability that the additional photon is polarized along H (spontaneous emission) and V (stimulated emission). Once these probabilities are determined, one can find the emerging cloning fidelity [38, 39].

discussion on quantum cloning to the notion of universal cloner of position and momentum variables.

To do so, one can begin with the position basis whose state is normalized: $\langle x|x'\rangle - \delta(x - x')$ in which the momentum eigenstates can take the form

$$|p\rangle = \frac{1}{\sqrt{2\pi}} \int e^{ipx}|x\rangle \, dx \tag{11.98}$$

(see Sect. 2.4). Note that the pertinent maximally entangled states can be expressed as

$$|\Psi(x, p)\rangle = \frac{1}{\sqrt{2\pi}} \int_{-\infty}^{\infty} e^{ipx'}|x'\rangle_1 \otimes |x' + x\rangle_2 \, dx', \tag{11.99}$$

where the subscripts 1 and 2 stand for the variables that represent the two particles, and x and p are real parameters. Incidentally, applying some unitary operator on one of the entangled variables may allow to invoke a unitary transformation.

It is also possible to express the corresponding displacement operator as

$$\hat{D}(x, p) = \int_{-\infty}^{\infty} e^{ipx}|x + x'\rangle\langle x'| \, dx', \tag{11.100}$$

which represents the momentum shift of p followed by the position shift of x; the process that can be exploited in the implementation of quantum cloning (see Sect. 2.4). Then assume that the input variable of the cloner is initially entangled with the reference variable so that their joint state is $|\Psi(0, 0)\rangle$, that is, if cloning yields a position shift error of x on the copy, the joint state of the reference and copy variables would be $|\Psi(x, 0)\rangle$. One can also argue that a momentum shift error of p would result in $|\Psi(0, p)\rangle$ in which x and p errors occur according to the probability distribution $P(x, p)$. The joint state in this case can be expressed as a mixture: $\int_{-\infty}^{\infty} P(x, p)|\Psi(x, p)\rangle\langle\Psi(x, p)| \, dxdp$.

Imagine that after a cloning machine exemplified by a unitary transformation acting on the input variable (let us say variable 2) supplemented by two auxiliary variables, the blank copy (variable 3) and the ancilla (variable 4). Besides, suppose the reference variable (variable 1)—which is maximally entangled with the cloner input—is introduced to reduce the involved rigor. Assume also that the reference and the input variables are in the joint state $|\Psi(0, 0)\rangle_{1,2}$ while the auxiliary variables 3 and 4 are initially prepared in the state

$$|\phi\rangle_{3,4} = \int \int_{-\infty}^{\infty} f(x, p)|\Psi(x, -p)\rangle_{3,4} \, dxdp, \tag{11.101}$$

where $f(x, p)$ is arbitrary complex function and the minus sign designates the phase conjugation process (see Sect. 11.5.2). After applying unitary transformation, variables 2 and 3 are taken as the two outputs of the cloner while the output system is presumed to traced over variable 4. The ultimate goal is then to obtain the unitary transformation $\hat{U}$ such that the joint state of the reference, the two copies and the ancilla after cloning would be

$$|\Phi\rangle = \int\int_{-\infty}^{\infty} f(x, p)|\Psi(x, p)\rangle_{1,2}|\Psi(x, -p)\rangle_{3,4}\, dxdp. \tag{11.102}$$

The cloning operator that generates this kind of transformation turns out to be

$$\hat{U} = e^{-i(\hat{x}_4 - \hat{x}_3)\hat{p}_2} e^{-i\hat{x}_2(\hat{p}_3 + \hat{p}_4)}. \tag{11.103}$$

Once one able to define the unitary operation, the quantum cloning process can be epitomized by $|\Phi'\rangle = \hat{U}|\Phi\rangle$. It might be thus appealing to state that a similar transformation can be described in the Heisenberg picture by

$$\hat{x}'_k = \hat{U}^\dagger \hat{x}_k \hat{U}, \qquad \hat{p}'_k = \hat{U}^\dagger \hat{p}_k \hat{U} \tag{11.104}$$

leaving the state $|\Psi\rangle$ invariant when the primed parameters stand for the clone.

For the cloning transformation to be optimal, certain requirements are expected to be imposed. For instance, the M output quadrature modes should have the mean values

$$\langle x_k \rangle = \langle \alpha | \hat{x}_0 | \alpha \rangle, \qquad \langle p_k \rangle = \langle \alpha | \hat{p}_0 | \alpha \rangle \tag{11.105}$$

for $k = 0, ..., M - 1$ as input modes, and the quadrature variances to be

$$V_k = \langle \hat{q}_k^2 \rangle - \langle \hat{q}_k \rangle^2 = \frac{1}{2}, \tag{11.106}$$

where $\hat{q}_k = c\hat{x}_k + d\hat{p}_k$ in which $|c|^2 + |d|^2 = 1$. The quadrature variances should be left invariant by complex rotations in phase space, and the commutation relations

$$[\hat{x}'_k, \hat{x}'_j] = [\hat{p}'_k, \hat{p}'_j] = 0, \qquad [\hat{x}'_k, \hat{p}'_j] = i\delta_{kj} \tag{11.107}$$

should also be preserved throughout the evolution for $i, j = 0, ..., M - 1$ and the ancilla.

Pertaining to these requirements, it is possible to write for $1 \rightarrow 2$ cloning:

$$\hat{x}'_0 = \hat{x}_0 + \frac{\hat{x}_1}{\sqrt{2}} + \frac{\hat{x}_z}{\sqrt{2}}, \quad \hat{x}'_1 = \hat{x}_0 - \frac{\hat{x}_1}{\sqrt{2}} + \frac{\hat{x}_z}{\sqrt{2}}, \quad \hat{x}'_z = \hat{x}_0 + \sqrt{2}\hat{x}_z, \tag{11.108}$$

$$\hat{p}'_0 = \hat{p}_0 + \frac{\hat{p}_1}{\sqrt{2}} - \frac{\hat{p}_z}{\sqrt{2}}, \quad \hat{p}'_1 = \hat{p}_0 - \frac{\hat{p}_1}{\sqrt{2}} - \frac{\hat{p}_z}{\sqrt{2}}, \quad \hat{p}'_z = -\hat{p}_0 + \sqrt{2}\hat{p}_z. \tag{11.109}$$

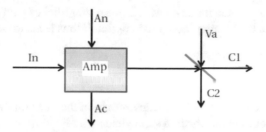

Fig. 11.6 Schematic representation of continuous variable quantum cloner constructed from the input (In), ancillia (An), linear amplifier (Amp), vacuum input (Va) with anti-clone (AC), clone 1 (C1) and clone 2 (C2) as output (see also [41])

Such transformations may indicate that the state in which the ancilla is left after cloning is centered at $(x_0, -p_0)$, that is, in the phase conjugate state. This may entail that a continuous variable cloner can generate an anti-clone together with the two clones (see Fig. 11.6).

It can also be interpreted in the context of quantum optics as a sequence of two canonical transformations:

$$\hat{a}_0' = \sqrt{2}\hat{a}_0 + \hat{a}_z^\dagger, \qquad \hat{a}_z' = \hat{a}_0^\dagger + \sqrt{2}\hat{a}_z \tag{11.110}$$

with phase insensitive amplifier of gain 2, and

$$\hat{a}_0'' = \frac{1}{\sqrt{2}}[\hat{a}_0' + \hat{a}_1], \qquad \hat{a}_1'' = \frac{1}{\sqrt{2}}[\hat{a}_0' - \hat{a}_1] \tag{11.111}$$

in a phase free 50:50 beam splitter, where $\hat{a}_k = (\hat{x}_k + i\hat{p}_k)/\sqrt{2}$ (see Sect. 4.3.2).

In $N \to M$ cloning transformation, energy has to be brought to the $M - N$ blank modes in order to drive them from the vacuum state into a state with desired mean that can be achieved with the aid of linear amplifier. One should also observe that cloning induces noise that emanates from the amplification process and grows with the amplifier gain. The cloning procedure can therefore be technically implemented by symmetrically amplifying the N input modes, and then distributing the output among the M modes, where the concentration and distribution processes can be carried out by a discrete Fourier transform

$$\hat{a}_k' = \frac{1}{\sqrt{N}} \sum_{l=0}^{N-1} \exp\left[\frac{i2\pi kl}{N}\right] \hat{a}_l, \tag{11.112}$$

with $k = 0, \ldots, N-1$. This operation concentrates the energy of the N input modes into one single mode $\hat{a}_0$ and leaves the remaining $N-1$ modes in the vacuum state.

Amplifying the created mode $\hat{a}_0$ with a linear amplifier of gain M/N (this value is chosen to make the cloning mechanism optimal) thus leads to

$$\hat{a}_0' = \sqrt{\frac{M}{N}}\hat{a}_0 + \sqrt{\frac{M-N}{N}}\hat{a}_z^\dagger, \qquad \hat{a}_z' = \sqrt{\frac{M-N}{N}}\hat{a}_0^\dagger + \sqrt{\frac{M}{N}}\hat{a}_z. \qquad (11.113)$$

Then performing a discrete Fourier transformation that acts on M modes between $\hat{a}_0'$ and $M-1$ modes in the vacuum state yields

$$\hat{a}_k'' = \frac{1}{\sqrt{M}} \sum_{l=0}^{M-1} \exp\left[\frac{i2\pi kl}{M}\right]\hat{a}_l', \qquad (11.114)$$

where $k = 0, ..., M-1$ and $\hat{a}_i' = \hat{a}_i$ for $i = N, ..., M-1$. Note that the unitary matrix used to instigate such transformation can be realized with a sequence of beam splitters and phase shifters.[9] This means, $N \to M$ cloning transformation can be implemented with the aid of passive optical elements except for the linear amplifier.

11.4.3 Coherent State Quantum Cloning

The process of impinging two photons simultaneously onto the beam splitter from two different input arms can enhance the probability of inducing the coherence as long as the photons are indistinguishable. So the pertinent quantum cloning machine is anticipated to yield M identical optimal clones from N replicas of the coherent state and its phase conjugate [43], and can be practically implemented with the help of simple optical elements and homodyne detection (see Sect. 9.3).

One may recall that the quality of the cloning machine can be quantified by the fidelity

$$F = \frac{2}{\sqrt{(1+\Delta X_c^2)(1+\Delta P_c^2)}} \exp\left[-\frac{(x_c - x_{in})^2}{2(1+\Delta X_c^2)} - \frac{(p_c - p_{in})^2}{2(1+\Delta P_c^2)}\right] \qquad (11.115)$$

(see the appendix). In case of a unit gain, that is, when $x_c = x_{in}$ and $p_c = p_{in}$;

$$F = \frac{2}{\sqrt{(1+\Delta X_c^2)(1+\Delta P_c^2)}}. \qquad (11.116)$$

---

[9]Linear optical processes such as phase shifting and beam splitting can be described by a linear transformation of the form $\hat{a}_i' = \sum_j U_{jl}\hat{a}_l$, where U_{jl} is the pertinent $N \otimes N$ unitary matrix. Such mathematical construction can envisage that mixing of optical modes that can be described by a unitary matrix can be implemented with the aid of linear optical elements [42].

In addition, $N \to M$ symmetric Gaussian cloner is presumed to be a linear, trace preserving and complete positive map that yields M clones from $N \leq M$ identical replicas of a coherent state $|\alpha\rangle$. One can as a result express the reduced density operator of one of the outputs as

$$\hat{\rho} = \frac{1}{\pi \sigma_{N \to M}^2} \int d^2\beta \, e^{-\sigma_{N \to M}^2 \beta^* \beta} \hat{D}(\beta) |\alpha\rangle \langle \alpha| \hat{D}^\dagger(\beta), \qquad (11.117)$$

where the integration is carried out over complex plane.

For a symmetric Gaussian cloner, the involved variances should be equal: $\sigma_X^2 = \sigma_P^2 = \sigma_{N \to M}^2$ and the fidelity would therefore be

$$F_{N \to M} = \langle \alpha | \hat{\rho} | \alpha \rangle = \frac{1}{1 + \sigma_{N \to M}^2}. \qquad (11.118)$$

Such cloning transformation can be specialized for a Gaussian wave packet that can be described by a coherent state denoted by $\hat{a} = (\hat{x} + i\hat{p})/\sqrt{2}$, where the related displacement is $\hat{D}(\alpha) = e^{\alpha \hat{a}^\dagger - \alpha^* \hat{a}}$ with $\alpha = (x + ip)/\sqrt{2}$. In a coherent state representation (see Sects. 2.2 and 2.4),

$$\hat{D}(\alpha) |\alpha_0\rangle = e^{i \text{Im}(\alpha \alpha_0^*)} |\alpha_0 + \alpha\rangle \qquad (11.119)$$

and in case the input is taken to be a coherent state $|\alpha_0\rangle$, its output would be the mixture

$$\hat{\rho} = \frac{2}{\pi} \int d^2\alpha \, e^{-2\alpha^* \alpha} |\alpha_0 + \alpha\rangle \langle \alpha_0 + \alpha|, \qquad (11.120)$$

and so the fidelity would be

$$F_{1 \to 2} = \langle \alpha_0 | \hat{\rho} | \alpha_0 \rangle = \frac{2}{3}. \qquad (11.121)$$

The quadratures of the two output modes can also be related to the two input quadratures by

$$\hat{X}_{a,b} = g_{X_{a,b}} \hat{X}_{in} + \hat{F}_{X_{a,b}}, \qquad \hat{P}_{a,b} = g_{P_{a,b}} \hat{P}_{in} + \hat{F}_{P_{a,b}}, \qquad (11.122)$$

where g_i's and $\hat{F}_i$'s stand for the linearized gains and noise forces of each channel [44]. Then based on the proposal that the added noises are not correlated with the input signals, one can obtain from Eq. (11.122)

$$[\hat{F}_{X_a}, \hat{F}_{P_b}] = -g_{X_a} g_{P_b} [\hat{X}_{in}, \hat{P}_{in}] \qquad (11.123)$$

and infer that

$$\Delta \hat{F}_{X_a} \Delta \hat{F}_{P_b} \geq |g_{X_a} g_{P_b}| V_0, \tag{11.124}$$

where V_0 is the vacuum noise variance since the two field modes are taken to be different: $[\hat{X}_a, \hat{P}_b] = 0$.

One may note that the pertinent variances can be written as

$$V_{X_j} = \left(\frac{\Delta \hat{X}_j}{|g_{X_j}|}\right)^2 - \left(\Delta \hat{X}_{in}\right)^2 = \left(\frac{\Delta \hat{F}_{X_j}}{|g_{X_j}|}\right)^2, \tag{11.125}$$

where j stands for a and b. It is also possible to obtain symmetrical inequalities such as $V_{X_{a,b}} V_{P_{b,a}} \geq V_0^2$. Existing works show that the duplicator ($1 \rightarrow 2$ cloner) that reaches these limiting cases can be realized with the aid of an amplifier and a 50:50 beam splitter [45].

One can begin with the assumption that there is an optical mode $\hat{c}$ that initially contains unknown quantum state and taken to be cloned. Let the cloning procedure yields duplicates of the input state with modes $\hat{a}$ and $\hat{c}$. Imagine also that the implementation of the cloning requires ancilla mode $\hat{b}$. To demonstrate the cloning procedure, suppose modes $\hat{a}$ and $\hat{b}$ are in vacuum state, and the resulting three-mode state is written as $|0\rangle_a |0\rangle_b |\Psi\rangle_c$ at the input.

In view of the outlined scheme, the output state can be expressed as

$$|\Psi\rangle_{out} = e^{\hat{c}(\hat{a}^\dagger + \hat{b}) - \hat{c}^\dagger(\hat{a} + \hat{b}^\dagger) + \kappa(\hat{a}\hat{b} - \hat{a}^\dagger \hat{b}^\dagger)} |0\rangle_a |0\rangle_b |\Psi\rangle_c^{in}, \tag{11.126}$$

where κ is the parameter used to control the asymmetry of the clones. As one can see from earlier discussion, the cloning transformation can be carried out by the action of $\exp[\hat{c}(\hat{a}^\dagger + \hat{b}) - \hat{c}^\dagger(\hat{a} + \hat{b}^\dagger)]$, where the modes a and b enter the cloning machine in entangled two-mode squeezed state $\exp[\kappa(\hat{a}\hat{b} - \hat{a}^\dagger \hat{b}^\dagger)]|0\rangle_a |0\rangle_b$. To reduce the rigor, assume that $\hat{a}$ and $\hat{b}$ are initially in the vacuum state and define the cloning transformation by evolution operator

$$\hat{U} = \exp[\hat{c}(\hat{a}^\dagger + \hat{b}) - \hat{c}^\dagger(\hat{a} + \hat{b}^\dagger) + \kappa(\hat{a}\hat{b} - \hat{a}^\dagger \hat{b}^\dagger)], \tag{11.127}$$

where the cloning transformation can be epitomized by

$$\hat{O} \rightarrow \hat{U} \hat{O} \hat{U}. \tag{11.128}$$

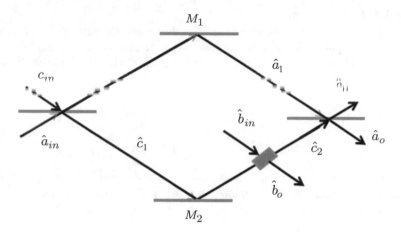

Fig. 11.7 Schematic representation of a coherent state quantum cloner constructed from beam splitters, mirrors and a nonlinear crystal (see also [46])

During a symmetric cloning transformation (see Fig. 11.7), one can find

$$\hat{a}_{out} = \hat{c}_{in} + \frac{1}{\sqrt{2}}[\hat{a}_{in} - \hat{b}_{in}^{\dagger}], \tag{11.129}$$

$$\hat{b}_{out} = \sqrt{2}\hat{b}_{in} - \hat{c}_{in}^{\dagger}, \tag{11.130}$$

$$\hat{c}_{out} = \hat{c}_{in} - \frac{1}{\sqrt{2}}[\hat{a}_{in} + \hat{b}_{in}^{\dagger}]. \tag{11.131}$$

In the same way, assuming the noise induced by the cloning process due to beam splitting to be thermal, the Q-function of the cloned coherent state can be written as

$$Q_{a,c}(\alpha) = \frac{1}{(\bar{n}+1)\pi} \exp\left[-\frac{|\alpha - \beta|^2}{\bar{n}+1}\right] \tag{11.132}$$

(compare with Eq. (2.96)), where $|\beta\rangle$ is the amplitude of the coherent state to be cloned and $\bar{n}$ is the mean photon number of the chaotic light [47].

One may see that the corresponding fidelity would be

$$F_{a,c} = \langle\beta|\hat{\rho}_{out}|\beta\rangle = \pi Q_{a,c}(\beta) = \frac{1}{\bar{n}+1}. \tag{11.133}$$

Note that the intended $1 \rightarrow 2$ cloning can be achieved by using two beam splitters and a squeezer.

11.4.3.1 $1 \rightarrow M$ Cloner

It should be possible to generalize the earlier discussion to illustrate the process of copying one input to M identical outputs [44]. One may assume to ease the rigor that each output channel has a unit gain and the copies to be identical—in the sense that the variances are the same for the outputs—and the pairwise correlation not to depend on the outputs.

The quadratures of the M outputs of the $1 \rightarrow M$ cloner then obey:

$$\hat{X}_l = \hat{X}_{in} + \hat{F}_{X_l}, \qquad \hat{P}_l = \hat{P}_{in} + \hat{F}_{P_l} \tag{11.134}$$

for every $2 \le l \le M$. One can also see that $V_X = \Delta F_{\hat{X}_l}^2$ and $V_P = \Delta F_{P_l}^2$. It would not be hard to generalize the commutation relations for arbitrary number of copies: $[\hat{F}_{X_l}, \hat{F}_{P_j}] = -[\hat{X}_{in}, \hat{P}_{in}]$ for every $l \ne j$ and $[\hat{F}_{X_l}, \hat{F}_{P_l}] = 0$. It might be appealing to define the noise force as

$$\hat{F} = \hat{F}_{X_1} + \gamma \sum_{l=2}^{M} \hat{F}_{X_l}, \tag{11.135}$$

where γ is a real constant.

On account of the properties of the noise of each quadrature, one can verify that

$$\Delta \hat{F} \Delta \hat{F}_{P_1} \ge \gamma (M - 1) V_0. \tag{11.136}$$

It might also be straightforward to see that

$$\Delta F^2 = \Delta F_{X_1}^2 + \gamma^2 \sum_{l=2}^{M} \Delta F_{X_l}^2 + 2\gamma \sum_{l=2}^{M} \langle \hat{F}_{X_1} \hat{F}_{X_l} \rangle + \gamma^2 \sum_{l,j>1}^{l \ne j} \langle \hat{F}_{X_l} \hat{F}_{X_j} \rangle, \tag{11.137}$$

which can be written as

$$\Delta F^2 = [1 + \gamma^2 (M - 1)] V_X + [2\gamma (M - 1) + \gamma^2 (M - 1)(M - 2)] C_X, \tag{11.138}$$

where $C_X = \langle \hat{F}_{X_l} \hat{F}_{X_j} \rangle$ is the correlation of the noise for every $l \ne j$.

One can see that Eq. (11.138) reduces for $\gamma = -2/(M - 2)$ to

$$\Delta F^2 = \frac{M^2}{(M - 2)^2} V_X, \tag{11.139}$$

which leads to what usually designated as $1 \rightarrow M$ cloning limit

$$V_X V_P \ge \left[\frac{2(M - 1)}{M} \right]^2 V_0^2. \tag{11.140}$$

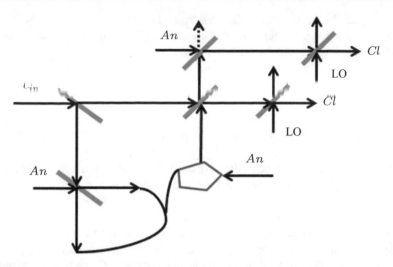

Fig. 11.8 Schematic representation of a coherent state multistage quantum cloner with input c_{in}, ancilla An, local oscillator LO and clones Cl

We now opt to revisit a coherent state cloning in which the cloning transformation is thought to be a sequence of parametric amplifiers and $M - 1$ beam splitters with adjusted transmittance and reflectance (relate with the idea depicted in Figs. 11.7 and 11.8). In this case, one can invoke the following transformations:

$$\hat{c}_1 = \sqrt{M}\hat{c}_{in} - \sqrt{M-1}\hat{b}_{in}^\dagger, \qquad \hat{b}_{out} = \sqrt{M}\hat{b}_{in} - \sqrt{M-1}\hat{c}_{in}^\dagger. \qquad (11.141)$$

If after the squeezer mode $\hat{c}_1$ is mixed with $\hat{a}_{1,in}$ at the first beam splitter and so on, it is possible to see that the output of the successive transformations can take the form

$$\hat{a}_{j,out} = \sqrt{\frac{1}{M-j+1}}\hat{c}_j + \sqrt{\frac{M-j}{M-j+1}}\hat{a}_{j,in}, \qquad (11.142)$$

$$\hat{c}_{j+1} = \sqrt{\frac{M-j}{M-j+1}}\hat{c}_j - \sqrt{\frac{1}{M-j+1}}\hat{a}_{j,in}. \qquad (11.143)$$

It might not be difficult to obtain

$$\hat{a}_{j,out} = \hat{c}_{in} - \sqrt{\frac{M-1}{M}}\hat{b}_{in}^\dagger + \sum_{k=1}^{j}\beta_{jk}\hat{a}_{k,in}, \qquad (11.144)$$

$$\hat{c}_M = \hat{c}_{in} - \sqrt{\frac{M-1}{M}}\hat{b}_{in} + \sum_{k=1}^{M-1}\beta_{Mk}\hat{a}_{k,in}, \qquad (11.145)$$

where β_{jk} can be inferred from the accompanying transformation. $\hat{c}_{in}$ in this case is prepared in a coherent state $|\beta\rangle$; and hence, the Q-function of the clones and the corresponding fidelity turn out be

$$Q(\alpha) = \frac{M}{(2M-1)\pi} \exp\left[-\frac{M}{2M-1}|\alpha - \beta|^2\right], \tag{11.146}$$

$$F_{1\to M} = \frac{M}{2M-1}, \tag{11.147}$$

which is the upper bound (the optimal value) for $1 \to M$ cloner. Since this fidelity reduces for large M to $F_{1\to\infty} = 1/2$, this process does not support quantum cloning.

11.4.3.2 $N \to M$ Cloner

One may note that the previous discussion can be extended to the case in which N coherent state inputs yield M identical clones: $N \to M$ optimal cloning. Let us now turn the attention to look into how one can produce M clones from N original replicas of a coherent state with equal fidelity for $M \geq N$ [33]. In this respect, one can define the intended state as

$$|\Psi\rangle = |\alpha\rangle^{\otimes N} \otimes |0\rangle^{\otimes(M-N)} \otimes |0\rangle_a, \tag{11.148}$$

which denotes the initial joint state of the N input modes prepared in coherent state $|\alpha\rangle$ to be cloned with the additional $M - N$ blank modes and the ancillary mode designated by subscript a, where the blank modes and the ancilla are assumed to be initially prepared in the vacuum state.

To facilitate the task, one can designate a pair of quadrature operators related to each mode by a set of $\{\hat{x}_k, \hat{p}_k\}$, where $k = 0, ..., N - 1$ refer to N original input modes and $k = N, ..., M - 1$ to the additional blank modes. Recall that cloning transformation can be thought of as $|\Psi'\rangle = \hat{U}|\Psi\rangle$ such that the M modes are left in the same state $|\alpha\rangle$ or mixed with a state very close to it. In this mechanism, one part of the interferometer is taken to contain a chain of $N - 1$ beam splitters denoted by $\hat{a}_{j+1}$, and the modes are expected to propagate between the jth and $(j + 1)$th beam splitter.

In a similar way, the pertinent transformation can be expressed as

$$\hat{a}_{j+1} = \sqrt{\frac{j}{j+1}}\hat{a}_j + \sqrt{\frac{1}{j+1}}\hat{c}_{j+1,in}, \quad \hat{c}_{j+1,out} = \sqrt{\frac{1}{j+1}}\hat{a}_j - \sqrt{\frac{j}{j+1}}\hat{c}_{j+1,in}, \tag{11.149}$$

where $j = 1, 2, \ldots, N$, $\hat{d}_1 = \hat{c}_{1,in}$ and $\hat{d}_N = \hat{c}_{1,out}$. The radiation generated by the source is assumed to be amplified by the factor of $\sqrt{M/N}$; and as a result, one can have

$$\hat{u} \quad \sqrt{\frac{M}{N}}\hat{c}_{1,out} \quad \sqrt{\frac{M-N}{N}}\hat{b}_{in}^{\dagger}, \quad \hat{v}_{out} = \sqrt{\frac{M}{N}}\hat{b}_{in} - \sqrt{\frac{M-N}{N}}\hat{c}_{1,out}^{\dagger}. \tag{11.150}$$

If the amplified signal is fed into M-port interferometer, which is the sequence of $M - 1$ beam splitters, and $\hat{e}_{j+1}$ stands for the mode propagating between the jth and $(j + 1)$th beam splitter, the resulting transformation can be expressed in terms of

$$\hat{a}_{j,out} = \sqrt{\frac{1}{M-j+1}}\hat{e}_j + \sqrt{\frac{M-j}{M-j+1}}\hat{a}_{j,in}, \tag{11.151}$$

$$\hat{e}_{j+1} = \sqrt{\frac{M-j}{M-j+1}}\hat{e}_j - \sqrt{\frac{1}{M-j+1}}\hat{a}_{j,in}. \tag{11.152}$$

Since the amplification spoils the signal, each copy is presumed to contain $\bar{n} = (M - N)/(MN)$ chaotic photons. Pertaining to the obtained transformation and earlier discussion, for a coherent input, one can find [39, 47]

$$F_{N \to M} = \frac{MN}{MN + M - N}. \tag{11.153}$$

One can see that Eq. (11.153) reduces for $N = 1$ to $F_{1 \to M} = M/(2M - 1)$.

11.4.3.3 $1 + 1 \to M$ Cloner

In view of earlier delineation, $1 + 1 \to M$ quantum cloning machine is expected to yield M identical optimal clones from two inputs such as a coherent state and its phase conjugate. The mechanism of mixing a coherent light on a beam splitter with the help of linear optical elements and application of the homodyne detection is deemed to be sufficient to demonstrate such a proposal. One may then need to recap the claim that a pair of conjugate Gaussian states can carry more information than using the same states twice (see Sect. 11.4.1), which may envisage that the cloning machine that admits a phase conjugate input modes can acquire enhanced fidelity.

One may consider the input coherent state that can be expressed as

$$|\alpha_{in}\rangle = \frac{1}{2}|x_{in} + ip_{in}\rangle, \quad |\alpha_{in}^*\rangle = \frac{1}{2}|x_{in} - ip_{in}\rangle, \tag{11.154}$$

where x_{in} and p_{in} are the expectation values of the quadratures (see Sect. 2.4 and also [48, 49]). Imagine also the case in which the input coherent state $\hat{c}_{in}$ and its phase conjugate state $\hat{c}_{in}^*$ are prepared by amplitude and phase modulators. One may also conceive the machine that generates many clones of the input state designated by the density operator $\hat{\rho}_c$ and expectation values x_c and p_c. This can be attained by making use of the amplitude and phase modulators in which the amplitude modulated signals are in phase while phase modulated are out of phase.

Assume also that the input mode is split by beam splitter and the vacuum mode is made to enter, and then combined with the phase conjugate state on another 50:50 beam splitter. When the coherent input mode is divided by the beam splitter, the reflected output radiation can be expressed as

$$\hat{c}_r = \sqrt{R}\hat{c}_{in} + \sqrt{T}\hat{F}, \tag{11.155}$$

where $\hat{F}$ stands for vacuum fluctuation modes that appear due to splitting (see Sect. 4.3.2). Let this coherent state is combined with its phase conjugate on a second 50:50 beam splitter, and then a homodyne measurement is performed on the output modes.

Earlier discussion may reveal that the measured quadrature operators can be described by

$$\hat{X} = \frac{1}{\sqrt{2}}[\sqrt{R}\hat{X}_{in} + \sqrt{T}\hat{X}_F + \hat{X}_{pc}], \tag{11.156}$$

$$\hat{P} = \frac{1}{\sqrt{2}}[\sqrt{R}\hat{P}_{in} + \sqrt{T}\hat{P}_F - \hat{P}_{pc}], \tag{11.157}$$

where the subscript pc designates phase conjugate. Let the measured values are used after that to modulate the amplitude and phase of the ancillary coherent state to optimize the gain.

To adhere to the impossibility of getting a perfect quantum cloning, the obtained beam is made to pass through 99:1 beam splitter with the transmitted part of $\hat{c}_{in}$, that is, 99% of the radiation is transmitted. The displaced field in this consideration would take[10] the form

$$\hat{c}_d = \left(\sqrt{1-R} + \frac{g}{\sqrt{2}}\sqrt{R}\right)\hat{c}_{in} - \left(\sqrt{R} - \frac{g}{\sqrt{2}}\sqrt{1-R}\right)\hat{F} + \frac{g}{\sqrt{2}}\hat{c}^{*\dagger}. \tag{11.158}$$

One may stick to the optimization condition that equate to $g = \sqrt{2R/(1-R)}$ to cancel the contribution of the vacuum fluctuations. This choice of the gain leads to

$$\hat{c}_d = \frac{1}{\sqrt{1-R}}\hat{c}_{in} + \sqrt{\frac{R}{1-R}}\hat{c}^{*\dagger}. \tag{11.159}$$

---

[10]The following result is obtained by using $\hat{O} \to \hat{D}^\dagger\hat{O}\hat{D} = \hat{O} + \frac{1}{2}[\hat{X} + i\hat{P}]$ [15].

The proposal that the extension to M replicas can be achieved by utilizing a sequence of beam splitters can be implemented by distributing the displaced field into M clones describable by $\{\hat{c}_l; l = 1, 2, 3, \ldots, M\}$ with the help of a sequence of $M - 1$ beam splitters. Then by carefully adjusting the gain via manipulating the reflectivity and transitivity of the beam splitters, it is found that [39]

$$\Delta^2 X = \Delta^2 P = 1 + \frac{(M - 1)^2}{2M^2}. \tag{11.160}$$

The fidelity of $1 + 1 \to M$ coherent state cloner is thus turned out to be

$$F_{1+1 \to M} = \frac{4M^2}{4M^2 + (M - 1)^2}, \tag{11.161}$$

which reduces for large M ($M \to \infty$) to

$$F_{1+1 \to \infty} = \frac{4}{5}. \tag{11.162}$$

Note that the fidelity shows a significant enhancement when compared to $F_{1 \to \infty} = 1/2$ (see also Eq. (11.147)).

11.4.3.4 $N + N \to M$ Cloner

$N + N \to M$ cloner on the other hand characterizes quantum cloning machine that yields M identical optimal clones from N replicas of a coherent state and N replicas of its phase conjugate. This mechanism is the extension of $1+1 \to M$ cloner, and so can be implemented by combining N replicas of coherent input state and N replicas of the pertinent complex conjugate (phase conjugate).

With this in mind, let us begin with N identically prepared coherent states $|\Phi\rangle$ described by $\{\hat{a}_l; l = 1, 2, 3, \ldots, N\}$ cloned into a single mode $\hat{c}_1$ with amplitude $\sqrt{N}\Phi$, and can be generated by superimposing N input modes in $N - 1$ beam splitters;

$$\hat{c}_1 = \frac{1}{\sqrt{N}} \sum_{l=1}^{N} \hat{a}_l \tag{11.163}$$

and $N - 1$ vacuum modes. The same method can be used to generate the pertinent phase conjugate input mode $\hat{c}_2$ with amplitude $\sqrt{N}\Phi^*$ from N replicas of $|\Phi\rangle$ stored in $\{\hat{b}_l; l = 1, 2, 3, \ldots, N\}$ that can be expressed as

$$\hat{c}_2 = \frac{1}{\sqrt{N}} \sum_{l=1}^{N} \hat{b}_l. \tag{11.164}$$

It might worth noting that $\hat{c}_1$ and $\hat{c}_2$ in the same way can be made to participate in the copying process in which the displaced field can be expressed as [51]

$$\hat{c}_d = \frac{1}{\sqrt{1-R}}\hat{c}_l + \frac{\sqrt{R}}{\sqrt{1-R}}\hat{c}_l^\dagger. \tag{11.165}$$

Adjusting the reflectivity and transitivity of the beam splitters may suggest that

$$\Delta^2 X = \Delta^2 P = 1 + \frac{(M-N)^2}{2MN}. \tag{11.166}$$

The corresponding fidelity then can be put forward as

$$F_{N+N \to M} = \frac{4M^2N}{4M^2N + (M-N)^2}, \tag{11.167}$$

which reduces for large M ($M \to \infty$) to

$$F_{N+N \to \infty} = \frac{4N}{4N+1}. \tag{11.168}$$

One may note that the fidelity is considerably enhanced (nearly perfect fidelity).

11.5   Quantum Telecloning

While teleportation is geared towards communicating quantum information faithfully between the sending and receiving sites, optimal quantum cloning targets to spread the information among the parties in the most efficient way (see Sects. 11.2 and 11.4). The notion of quantum nonlocal cloning (telecloning) consequently can be perceived as a procedure that projects a given quantum input onto a certain number of outputs via combining the process of quantum teleportation and optimal quantum cloning [52]. Quantum telecloning therefore refers to simultaneous distribution of unknown quantum state to distant parties relying on multipartite entangled state as a resource to establish the required correlation among the parties.

Since quantum information is presumed to be copied and transmitted in a single step, the most straightforward scheme available to the experimenter to implement quantum telecloning is to generate optimal clones locally using appropriate quantum network, and then teleport each copy to the recipient by a means of previously shared maximally entangled states. One may observe that the best option is to send the available optimal quantum clones of the original state (see Sect. 10.2). The accompanying task thus encompasses generating M number of the optimal quantum clones of the input, and distribute the replicas (not so perfect) among distant parties. Since this procedure requires M units of initial entanglement and sending of M independent two-bit classical messages, it is tempting to consider quantum telecloning as a generalization of quantum teleportation with multiple receivers.

One may also conceive that without multipartite entanglement, only two-step teleportation is possible in which first the sender makes clones locally, and then sends them to each receiver with bipartite quantum teleportation, that is, teleportation followed by local cloning. The main challenge in this consideration is: the two-step protocol requires a maximal bipartite entanglement for required teleportation with optimal fidelity that corresponds to states with perfect squeezing (infinite energy) [17]. But if three parties Alice, Bob and Clair share appropriate tripartite entangled state, Alice can teleport unknown quantum state to Bob and Clair simultaneously without necessarily following the two-step process—the scheme designated as $1 \rightarrow 2$ quantum telecloning. Quantum telecloning to arbitrary M number of receivers can also be performed in the same way by utilizing a multipartite entanglement.

At the heart of the process of quantum telecloning notably resides the multipartite entanglement shared among the sender and receivers. Contrary to earlier suggestion related to infinite squeezing, a continuous variable telecloning of a coherent state is nevertheless found to require only finite squeezing to achieve the same optimal fidelity (see Sect. 11.5.1). This task seems to insinuate an extensive networking and a complicated setup which is undoubtedly enduring. Even then, when the parties share multipartite entanglement, it should be possible to simultaneously convey M copies of the message by a means of single measurement at the sender site. The sender in this case needs to announce the two bits that determine the measurement result; after that, each recipient is supposed to perform appropriate local rotation conditioned on this information; which is the reminiscent of the teleportation procedure (see Sect. 11.2 and also Fig. 11.3).

To illustrate the main idea, one can begin with the assumption that Alice possesses an unknown state $|\phi\rangle_U$ that she seeks to send to M collaborators who share a multipartite entangled (EPR) state with her. The state should be chosen so that after Alice performs a local measurement, and informs the other parties of its result, the latter can each obtain an optimal copy given by

$$\hat{\rho}^{out} = F|\phi\rangle\langle\phi|_U + (1 - F)|\phi^\perp\rangle\langle\phi^\perp|_U, \qquad (11.169)$$

where $|\phi^\perp\rangle_U$ is the state orthogonal to $|\phi\rangle_U$ and F the fidelity that denotes the imperfection of the copies (see Eqs. (9.58) and (11.77)).

One can also imagine the multipartite entangled state among the sender and receivers be designated by

$$|\text{EPR}\rangle = \sum_{j=0}^{M-1} \sqrt{\frac{(M-1)}{M(M+1)}} \big[|0\rangle_P \otimes |A_j\rangle_A \otimes |\{0, M-j\}\{1, j\}\rangle_C$$

$$+ |1\rangle_P \otimes |A_{M-1-j}\rangle_A \otimes |\{0, j\}, \{1, M-j\}\rangle_C \big], \qquad (11.170)$$

where C denotes the M qubits holding the copies (each of which is held by one of the receivers state although they may be far away from each other), $|A_j\rangle$'s stand for M orthogonal normalized states of the ancillia, $|\{0, M - j\}, \{1, j\}\rangle$ for symmetric and normalized states of M qubits when $M - j$ of them are in state $|0\rangle$ and j of them are in the orthogonal state $|1\rangle$, P stands for a single qubit held by the sender (port qubit) while A for $M - 1$ qubit ancilla taken to be on the sender's side.

One may see that the tensor product of $|EPR\rangle$ with $|\phi\rangle_U$ held by the sender is a $2M + 1$ qubit state:

$$|\Psi\rangle_{UPAC} = \frac{1}{2}\left[|\Phi^+\rangle_{UP} \otimes \left(a|\phi_0\rangle_{AC} + b|\phi_1\rangle_{AC}\right) + |\Phi^-\rangle_{UP} \otimes \left(a|\phi_0\rangle_{AC} - b|\phi_1\rangle_{AC}\right)\right.$$
$$\left. + |\Psi^+\rangle_{UP} \otimes \left(b|\phi_0\rangle_{AC} + a|\phi_1\rangle_{AC}\right) + |\Psi^-\rangle_{UP} \otimes \left(b|\phi_0\rangle_{AC} - a|\phi_1\rangle_{AC}\right)\right]$$
$$(11.171)$$

in which

$$|\phi_0\rangle_{AC} = \sum_{j=0}^{M-1} \sqrt{\frac{2(M-1)}{M(M+1)}} |A_j\rangle_A \otimes |\{0, M - j\}, \{1, j\}\rangle_C, \qquad (11.172)$$

$$|\phi_1\rangle_{AC} = \sum_{j=0}^{M-1} \sqrt{\frac{2(M-1)}{M(M+1)}} |A_{M-1-j}\rangle_A \otimes |\{0, j\}, \{1, M - j\}\rangle_C \ (11.173)$$

(compare with Eq. (11.23)).

Telecloning is thus said to be implemented when Alice performs Bell measurement of qubits U and P, and obtains $|\Phi^+\rangle$; and if the subsystem AC is projected precisely into the optimally cloned state $a|\phi_0\rangle_{AC} + b|\phi_1\rangle_{AC}$. In case one of the other Bell states is obtained, one can still recover the correct state of AC by invoking Pauli transformations.

11.5.1 Continuous Variable Quantum Telecloning

A genuine many party quantum telecloning depends on the correlation induced by the accompanying tripartite entanglement that can be generated for a system of continuous variable by a nonlinear optical process using squeezed vacuum states and two beam splitters where even an infinitesimal squeezing leads to a fully inseparable tripartite states (see Sect. 11.4.3). One may recall that various alternatives of tripartite entanglement can be generated by choosing proper transmittance or reflectivity of the beam splitter and degree of squeezing. A continuous variable analogue of the $|GHZ\rangle$ state for example can be created by combining three squeezed vacuum states with high degree of squeezing on two beam splitters. Nonetheless, the entanglement required for quantum telecloning is expected to comprise both bipartite and tripartite entanglement varieties that should not be necessarily maximally entangled (see

Eq. (9.11)). A continuous variable telecloning mechanism that can broadcast the information of an unknown quantum state without loss from a sender to several spatially separated receivers by exploiting multipartite entanglement as quantum channels can hence be conceived.

The quantum aspect of the unknown state in such a scheme is presumed to be distributed to M optimal clones and $M - 1$ anti-clones using $2M$-partite entanglement [53, 54]. Since the anti-clones might be lost during information distribution, quantum channels do not require maximum entanglement: the emerging telecloning can be regarded as imperfect; and yet, valid. This idea might be connoted as optimal telecloning can be achieved by using a pairwise bipartite entanglement between the sender and receivers, which is not necessarily maximum. Despite earlier observation related to a two-step procedure in a discrete case, a continuous variable phase conjugate input telecloning (with nonlocal clones and local anti-clones) requires just bipartite entanglement as a resource. The bipartite entangled state in this case is a two-mode Gaussian entangled state that can be obtained directly form a nonlinear parametric interaction or indirectly by mixing two independent squeezed beams on a beam splitter (see Chap. 7).

With this information, it is possible to begin with the presumption that one can have two optical parametric oscillators with two squeezed vacuum modes that can be represented by a set of quadrature operators $(\hat{X}_1, \hat{P}_1)$ and $(\hat{X}_2, \hat{P}_2)$. Assume that these beams are combined on a 50:50 beam splitter with $\pi/2$ phase shift. Let then one of the output beams is divided by another beam splitter into (B, C) modes. After that, assume that the three output modes $(\hat{X}_j, \hat{P}_j; \quad j = A, B, C)$ are entangled with arbitrary pairwise squeezing, that is, modes A, B and A, C are pairwise bipartite entangled state. With the aid of the established tripartite entanglement between modes A, B and C, one can argue that the sender can implement quantum telecloning of continuous variable input [55].

In light of this, it is possible to conceive a continuous variable scheme that relies on a finite squeezing resource and enables realization of a symmetric $1 \rightarrow M$ coherent state telecloning with maximum fidelities allowable by quantum theory with the help of $(M + 1)$-partite entangled Gaussian state, beam splitters and homodyne detection (see also Sect. 11.2). In this context, to ensure quantum telecloning, Alice should first perform a joint (Bell) measurement on her entangled mode $(\hat{X}_A, \hat{P}_A)$ and the unknown input mode $(\hat{x}_{in}, \hat{p}_{in})$.

The output of a 50:50 beam splitter in this case can be expressed in terms of quadrature operators as

$$\hat{x} = \frac{1}{\sqrt{2}}[\hat{x}_{in} - \hat{X}_A], \qquad \hat{p} = \frac{1}{\sqrt{2}}[\hat{p}_{in} + \hat{P}_A] \qquad (11.174)$$

(see Sects. 4.3.2 and 9.3). The initial modes of the receivers (Bob and Clair) can also be expressed as

$$\hat{X}_{B,C} = \hat{x}_{in} - (\hat{X}_A - \hat{X}_{B,C}) - \sqrt{2}\hat{x}, \qquad (11.175)$$

$$\hat{P}_{B,C} = \hat{p}_{in} - (\hat{P}_A + \hat{P}_{B,C}) - \sqrt{2}\hat{p} \qquad (11.176)$$

in which Bob's and Clair's modes remain unchanged.

It may not be hard to conceive based on the quantum measurement theory that after Alice's measurement, $\hat{X}$ and $\hat{P}$ collapse to eigenvalues x and p. Then, upon receiving the measurement result from Alice and Bob, Clair can displace their modes as shown in Fig. 11.9;

$$\hat{X}_{B,C} \rightarrow \hat{x}_{1,2} = \hat{X}_{B,C} + \sqrt{2}x, \qquad (11.177)$$

$$\hat{P}_{B,C} \rightarrow \hat{p}_{1,2} = \hat{P}_{B,C} + \sqrt{2}p. \qquad (11.178)$$

Since x and p stand for classical information, they can be duplicated. The modulated beams in practice would be combined with Bob's and Clair's initial modes on 99:1 beam splitter (see Sect. 11.2.2).

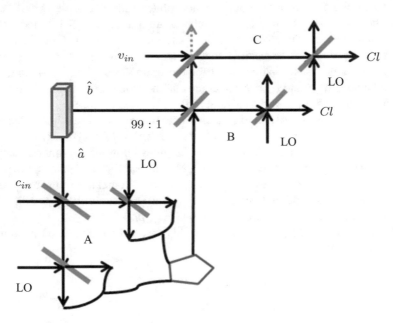

Fig. 11.9 Schematic representation of continuous variable telecloning constructed from 50:50 and 99:1 beam splitters, where LO, v_{in} and Cl stand for local oscillator, vacuum input and clone with coherent input c_{in}

The output modes produced by the telecloning process thus turn out to be

$$\hat{x}_{1(2)} = \hat{x}_{in} + \frac{1}{2}\left(1 - \sqrt{2}\right)\hat{X}_1 - \frac{1}{2}\left(1 + \sqrt{2}\right)\hat{X}_2 + (-)\frac{1}{\sqrt{2}}\hat{X}_v^{(0)}, \quad (11.179)$$

$$\hat{p}_{1(2)} = \hat{p}_{in} + \frac{1}{2}\left(1 + \sqrt{2}\right)\hat{P}_1 - \frac{1}{2}\left(1 - \sqrt{2}\right)\hat{P}_2 + (-)\frac{1}{\sqrt{2}}\hat{P}_v^{(U)}, \quad (11.180)$$

where the subscript v denotes the vacuum input at the second beam splitter and $+$ ($-$) stands for clone 1 (clone 2).

The additional noise $x_v^{(0)}$ and $p_v^{(0)}$ in the nutshell can be minimized by tuning the degree of squeezing where the minimization (optimization) procedure leads to

$$\hat{x}_{1(2)} = \hat{x}_{in} - \frac{1}{2}\left(\hat{X}_1^{(0)} + \hat{X}_2^{(0)}\right) + (-)\frac{1}{\sqrt{2}}\hat{X}_v^{(0)}, \quad (11.181)$$

$$\hat{p}_{1(2)} = \hat{p}_{in} + \frac{1}{2}\left(\hat{P}_1^{(0)} + \hat{P}_2^{(0)}\right) + (-)\frac{1}{\sqrt{2}}\hat{P}_v^{(0)} \quad (11.182)$$

(see Sect. 11.4.2). Note that Eqs. (11.181) and (11.182) represent the sought for optimal clones of continuous variable input mode.

To achieve approximate quantum distribution, one may need to construct a quantum structure that distributes input qubit to M optimal clones and $M - 1$ anti-clones by using $2M$-partite entanglement: $1 \rightarrow M + (M - 1)$ telecloner that can be regarded as perfect nonlocal distributor of quantum information. It should be noted that an optimal telecloning can be achieved by employing a bipartite entanglement between the sender and receivers. With the description that in the loss free quantum distribution both clones and anti-clones are required; for $M = 2$, one can get $1 \rightarrow 2 + 1$ telecloning with the aid of bipartite entangled light. Take also the source of the light to be a two-mode Gaussian entangled state, and the EPR entangled beams to be divided on a 50:50 beam splitter [54].

Then the output radiation can be expressed in coherent state representation as

$$\hat{a}'_{\pm} = \frac{1}{\sqrt{2}}\left(\hat{a}_{E1} \pm \hat{v}_1\right), \qquad \hat{a}_{\pm} = \frac{1}{\sqrt{2}}\left(\hat{a}_{E2} \pm \hat{v}_2\right), \quad (11.183)$$

where $\hat{v}_1$ and $\hat{v}_2$ refer to the vacuum noise induced due to splitting, and the telecloned state are taken to be partitioned in $\{\hat{a}'_+, \hat{a}'_-\}$ and $\{\hat{a}_+, \hat{a}_-\}$: the elements in the same set come from one of the entangled pairs. The idea is: by using quadripartite entangled modes, the sender can distribute a coherent state input to three receivers who can produce two clones and one anti-clone at their sites. To perform $1 \rightarrow 2 + 1$ telecloning, the sender thus first performs Bell measurements, let us say, on the entangled mode $\hat{a}'_+$ and the unknown input mode $\hat{a}_{in}$ in which the measurement consists of a 50:50 beam splitter and two homodyne detectors with outputs of the type

$$\hat{x} = \frac{1}{\sqrt{2}}(\hat{X}'_+ - \hat{X}_{in}), \qquad \hat{p} = \frac{1}{\sqrt{2}}(\hat{P}'_+ - \hat{P}_{in}). \tag{11.184}$$

After receiving this data from the sender, let Bob, Clair and Dan modulate the amplitude and the phase of their data by using auxiliary beam at their sites. Then after combining the modulated beam with their modes ($\hat{a}_1$, $\hat{a}_2$, $\hat{a}_3$, respectively) on 99:1 beam splitter, the output mode that would be produced by the telecloning process—upon setting the gains so that an optimal condition to be achieved—is expected to be of the form

$$\hat{a}_B^{out} = \hat{a}_{in} + \frac{1}{\sqrt{2}}(\hat{a}_{E2} - \hat{a}_{E1}^\dagger) + \frac{1}{\sqrt{2}}(\hat{v}_2 - \hat{v}_1^\dagger), \tag{11.185}$$

$$\hat{a}_C^{out} = \hat{a}_{in} + \frac{1}{\sqrt{2}}(\hat{a}_{E2} - \hat{a}_{E1}^\dagger) - \frac{1}{\sqrt{2}}(\hat{v}_2 + \hat{v}_1^\dagger), \tag{11.186}$$

$$\hat{a}_D^{out} = \hat{a}_{in}^\dagger - \sqrt{2}\hat{v}_1. \tag{11.187}$$

On account of Eqs. (11.185)–(11.187), one can argue that Bob and Clair possess the cloned states while Dan the anti-cloned state. Note that the additional noise acquired by Dan's state is independent of the EPR entanglement, and so represents optimal anti-clone of the input state.

The unknown state in case of a perfect EPR entanglement input is completely unknown not only to Alice but also to anyone engaged in the process of telecloning: the quantum information of the unknown state would be partitioned and distributed completely to Bob, Clair and Dan. The pertinent fidelity can markedly be calculated in the same way as in Sect. 11.4.2 and found to be

$$F_{1\to2+1}^{clone} = \frac{2}{3 + e^{-2r}}, \tag{11.188}$$

$$F_{1\to2+1}^{anticlone} = \frac{1}{2}. \tag{11.189}$$

One may note that this approach can also be extended to the case when there are N inputs, M clones and $M - 1$ anti-clones.

11.5.2 Phase Conjugate Quantum Telecloning

The perception that a pair of conjugate Gaussian states can carry more information than a pair of identical coherent states enables one to achieve a better fidelity with a cloning machine admitting phase conjugate coherent states inputs when compared to a case with identical input copies [48]. Suppose one of the EPR entangled beams,

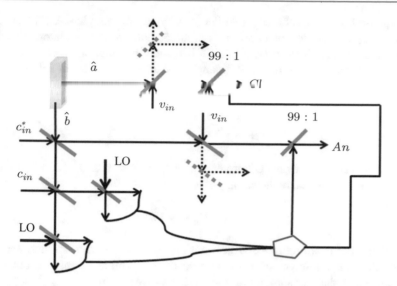

Fig. 11.10   Schematic representation of phase conjugate telecloning constructed from 50:50 and 99:1 beam splitters, where LO, v_{in}, An and Cl stand for local oscillator, vacuum input, anti-clone and clone with coherent input c_{in} and its conjugate c_{in}^* (see also [56])

let us say, $\hat{a}_{E1}$ is held by the sender and the other $\hat{a}_{E2}$ is distributed among M remote parties via $M - 1$ beam splitters with appropriately adjusted transmittance and reflectivity (see Sect. 11.4.3 and also Fig. 11.10). The induced noise modes $\hat{v}_{j,a}$ are assumed to relate to the vacuum state while the EPR entangled mode $\hat{a}_{E2}$ is mixed with $\hat{v}_{1,in}$ at the first beam splitter [56]. The outcome of the splitting, that is, the mode $\hat{a}_1$ (reflected from the first beam splitter) contains the EPR entangled mode $\hat{a}_{E2}$ up to a factor of $1/\sqrt{M}$ and $\sqrt{(M - 1)/M}$ portion of the vacuum input. Assume also that the output $\hat{c}_2$ is split at the second beam splitter and so on until it arrives at the last beam splitter.

The transformation performed by the jth beam splitter can then be expressed as

$$\hat{a}_j = \sqrt{\frac{1}{M - j + 1}}\hat{c}_{j,a} + \sqrt{\frac{M - j}{M - j + 1}}\hat{v}_{j,a}, \tag{11.190}$$

$$\hat{c}_{j+1,a} = \sqrt{\frac{M - j}{M - j + 1}}\hat{c}_{j,a} - \sqrt{\frac{j}{M - j + 1}}\hat{v}_{j,a}, \tag{11.191}$$

where $\hat{c}_{1,a} = \hat{a}_{E2}$ and $\hat{c}_{M,a} = \hat{a}_M$. Note that $\hat{c}_{1,a}$ is the radiation incident on the first beam splitter, $\hat{a}_1$ is the radiation transmitted from first beam splitter, $\hat{c}_2$ is the first reflected beam; and $\hat{a}_M$ stands for the beam transmitted through the last beam splitter, and in between the first and last beam splitter $\hat{a}_j$ and $\hat{c}_j$ stand for reflected and transmitted beams.

Suppose the input coherent state denoted by $\hat{c}_{in}$ and its phase conjugate state $\hat{c}_{in}^*$ are prepared at the sender site by amplitude and phase modulators. Once the input coherent state is properly modulated, assume the input phase conjugate state $\hat{c}_{in}^*$ to be combined with the EPR entangled beam $\hat{a}_{E1}$ on a beam splitter having transmission and reflectivity rates T and R. The transmitted field can hence be expressed as

$$\hat{c}_{1t} = \sqrt{T}\hat{c}_{in}^* - \sqrt{R}\hat{a}_{E1}, \tag{11.192}$$

which is perceived to be divided into M modes $\{\hat{b}_j; \quad j = 1, 2, 3, \ldots, M\}$ via $M - 1$ beam splitters while the reflected output would take the form

$$\hat{c}_{1r} = \sqrt{R}\hat{c}_{in}^* + \sqrt{T}\hat{a}_{E1}. \tag{11.193}$$

Combining the resulting mode with input $\hat{c}_{in}$ on a 50:50 beam splitter, and then performing the homodyne measurement on the two output beams—to obtain the amplitude and phase quadratures simultaneously—results in

$$\hat{X}_m = \frac{1}{\sqrt{2}}\left[\sqrt{R}\hat{X}_{c^*}^{in} + \sqrt{T}\hat{X}_{E1} + \hat{X}_c^{in}\right], \tag{11.194}$$

$$\hat{P}_m = \frac{1}{\sqrt{2}}\left[\sqrt{R}\hat{P}_{c^*}^{in} + \sqrt{T}\hat{P}_{E1} - \hat{P}_c^{in}\right]. \tag{11.195}$$

The sender after that distributes the measured results x_m and p_m to local modes $\hat{b}_j$ (anti-clone) and the remote modes $\hat{a}_j$ (clones). After receiving the measurement results, let each receiver displaces his mode by a means of 99:1 beam splitter with auxiliary beam the amplitude and phase of which is modulated via two independent modulators with the aid of the received signals. So upon selecting appropriate gain $(g_x = -g_p = g_1)$ for $\hat{a}_j$ mode, the displaced fields of the remote parties can be expressed as

$$\hat{a}_j' = \sqrt{\frac{1}{M - j + 1}}\hat{c}_{j,a}' + \sqrt{\frac{M - j}{M - j + 1}}\hat{v}_{j,a}, \tag{11.196}$$

$$\hat{c}_{j+1,a}' = \sqrt{\frac{M - j}{M - j + 1}}\hat{c}_{j,a}' - \sqrt{\frac{1}{M - j + 1}}\hat{v}_{j,a}, \tag{11.197}$$

where

$$\hat{c}_{1,a}' = g_1\left[\sqrt{\frac{RM}{2}}\hat{c}_{in}^{*\dagger} + \sqrt{\frac{M(1 - R)}{2}}\hat{a}_{E1}^{\dagger} + \sqrt{\frac{M}{2}}\hat{c}_{in}\right] + \hat{a}_{E2} \tag{11.198}$$

with $\hat{a}_M' = \hat{c}_{M,a}'$ as already highlighted.

In the same way, the cloning transformation of the displaced local modes $\hat{b}_j$ would follow

$$\hat{c}'_{1,b} = \left(\sqrt{1-\frac{p}{}} + g_L\sqrt{\frac{MR}{2}}\,\hat{n}^*_{in}\right)\left(\sqrt{R} + g_L\sqrt{\frac{M(R-1)}{2}}\,\hat{m}_{L1}\right) + g_L\sqrt{\frac{M}{2}}\,\hat{v}^\dagger_{in}$$

(11.199)

and $\hat{b}'_M = \hat{c}'_{M,b}$ as one might readily infer. After choosing appropriate gains,

$$g_1 = \sqrt{\frac{2M}{M(1-R)}}, \qquad g_2 = \sqrt{\frac{2R}{M(1-R)}}, \tag{11.200}$$

so that optimal clones can be obtained, it is possible to see that

$$\hat{c}'_{1,a} = \sqrt{\frac{R}{1-R}}\,\hat{c}^{*\dagger}_{in} + \sqrt{\frac{1}{1-R}}\,\hat{c}_{in} + \hat{a}_{E1} + \hat{a}_{E2}, \tag{11.201}$$

$$\hat{c}'_{1,b} = \sqrt{\frac{1}{1-R}}\,\hat{c}^*_{in} + \sqrt{\frac{R}{1-R}}\,\hat{c}^\dagger_{in}. \tag{11.202}$$

With this in mind, the fidelity can thus be expressed as

$$F^c_{1+1\to M+M} = \frac{4M^2}{4M^2(1+e^{-2r}) + (M-1)^2}, \tag{11.203}$$

$$F^{ac}_{1+1\to M+M} = \frac{4M^2}{4M^2 + (M-1)^2}. \tag{11.204}$$

One may observe that this scheme produces M anti-clones locally and M clones nonlocally due to bipartite entangled light. Note that the fidelity of the anti-clones is optimal and independent of the entanglement but the fidelity of the clones does depend on the entanglement.

The $N + N \to M + M$ quantum telecloning machine that yields M identical clones and M identical anti-clones from N copies of a coherent state and N phase conjugate copies can also be conceived in the same way. In this respect, it is possible to consider two cases of interest: telecloning that produces M clones nonlocally and M anti-clones locally or vice versa, and both clones and anti-clones are created nonlocally. Optimal cloning fidelities of such telecloning require perfect EPR entanglement, that is, an infinite two-mode squeezing as a resource. In light of this, $N + N \to M + M$ telecloning scheme would generally have a fidelity of the form

$$F_{N+N\to M+M} = \frac{4M^2N}{4M^2N(1+e^{-2r}) + (M-N)^2}. \tag{11.205}$$

Appendix

Fidelity

Fidelity can be taken as the idea of quantifying how similar a given output state is to the ideal state which might be ubiquitous in the application of the quantum features of radiation (or continuous variable quantum information processing) since it represents the numerical value of the quality of transferring information from one site to the other as in quantum gate operation (see Sect. 9.4). The fidelity of quantum state with respect to the reference state can be perceived as the probability that the state is mistaken for the reference state in measurement.

For two arbitrary states $|\Psi\rangle$ and $|\Psi_0\rangle$, the fidelity can be taken as overlap

$$F = |\langle\Psi_0|\Psi\rangle|^2, \tag{11.206}$$

where one of them is taken as the reference, and for a mixed state denoted by the density operator $\hat{\rho}$,

$$F = \langle\Psi_0|\hat{\rho}|\Psi\rangle. \tag{11.207}$$

In case there are two mixed states with density operator $\hat{\rho}_1$ and $\hat{\rho}_2$;

$$F = \left[Tr\sqrt{\sqrt{\hat{\rho}_2}\hat{\rho}_1\sqrt{\hat{\rho}_2}}\right]^2. \tag{11.208}$$

One may opt to begin with relatively simpler approach with which the fidelity is sought for in case one tries to copy an arbitrary coherent state $|\alpha\rangle$. To this effect, the density matrix of the state to be cloned is assumed to be expandable in a coherent state basis, let us say, $|\beta\rangle$ where the probability of reconstructing the state $|\alpha\rangle$ is denoted by $P(\beta)$. With this consideration, the fidelity can be expressed as

$$F = \int P(\beta)|\langle\alpha|\beta\rangle|^2 \, d\beta, \tag{11.209}$$

which can be rewritten in terms of the phase space variables[11] as

$$F = \int P(x, p) \exp\left[-\frac{(x - x_a)^2}{4} - \frac{(p - p_a)^2}{4}\right] dxdy, \tag{11.210}$$

where the variable transformation $\alpha = \frac{1}{2}(x_a + ip_a)$ and $\beta = \frac{1}{2}(x + ip)$ is inferred.

---

[11]The mathematical expression of the fidelity may look different based on how the quadrature variables are defined; and yet, leads to the same conclusion. It should also be possible to connect fidelity with quasi-statistical distributions provided in Sect. 2.3.

Besides, the probability distribution can be expressed as

$$P(x, p) = \frac{1}{2\pi \sqrt{V_Y^{out} V_P^{out}}} \exp\left[-\frac{(x - x_b)^2}{2V_X^{out}} - \frac{(p - p_b)^2}{2V_P^{out}} \right],$$ (11.211)

where the variances are linked to the input noises, and x_b and p_b are the mean coordinates for the reconstructed distribution. With the help of this description, one can see that

$$F = \frac{1}{\sqrt{(2 + V_X^{out})(2 + V_P^{out})}} \exp\left[-\frac{(x_a - x_b)^2}{2(2 + V_X^{out})} - \frac{(p_a - p_b)^2}{2(2 + V_P^{out})} \right],$$ (11.212)

which is strongly peaked when $x_a = x_b$ and $p_a = p_b$, that is, when

$$F = \frac{1}{\sqrt{(2 + V_X^{out})(2 + V_P^{out})}}.$$ (11.213)

References

1. C. Wang et al., Phys. Rev. A **71**, 044305 (2005)
2. J. Fiurasek, Phys. Rev. A **80**, 053822 (2009); U.L. Andersen, G. Leuchs, C. Silberhorn, Laser Photon Rev. **4**, 337 (2010); H.J. Kim et al., ibid. **85**, 013839 (2012)
3. M. Zukowski et al., Phys. Rev. Lett. **71**, 4287 (1993); R.E.S. Polkinghorne, T.C. Ralph, ibid. **83**, 2095 (1999); O. Glückl et al., Phys. Rev. A **68**, 012319 (2003)
4. A.M. Goebel et al., Phys. Rev. Lett. **101**, 080403 (2008)
5. S. Bose, V. Vedral, P.L. Knight, Phys. Rev. A **57**, 822 (1998); J.W. Pan et al., Phys. Rev. lett. **80**, 3891 (1998); V. Coffman, J. Kundu, W.K. Wootters, Phys. Rev. A **61**, 052306 (2000); B.M. Terhal, D.P. DiVincenzo, D.W. Leung, Phys. Rev. Lett. **86**, 5807(2001); H. de Riedmatten et al., Phys. Rev. A **71**, 050302 (2005); T.J. Osborne, F. Verstraete, Phys. Rev. lett. **96**, 220503 (2006); M. Halder et al., Nature Phys. **3**, 692 (2007)
6. S. Takeda et al., Phys. Rev. Lett. **114**, 100501 (2015)
7. C. Tian et al., Opt. Expr. **26**, 29159 (2018)
8. K. Marshall, C. Weedbrook, Entropy **17**, 3152 (2015)
9. X. Jia et al., Phys. Rev. Lett. **93**, 250503 (2004); N. Takei et al., ibid. **94**, 220502 (2005); J.H. Obermaier, P. van Loock, Phys. Rev. A **83**, 012319 (2011); M. Abdi et al., ibid. **89**, 022331 (2014)
10. A. Kuzmich, K. Molmer, E.S. Polzik, Phys. Rev. Lett. **79**, 4782 (1997); J. Hald et al., ibid. **83**, 1319 (1999); S. Tesfa, J. Mod. Opt. **55**, 1683 (2008); T. Takano et al., Phys. Rev. A **78**, 010307 (2008); Y. Liu et al., Sci. Rep. **6**, 25715 (2016); A. Nourmandipour, M.K. Tavassoly, Phys. Rev. A **94**, 022339 (20016)
11. S.L. Braunstein, C.A. Fuchs, H.J. Kimble, J. Mod. Opt. **47**, 267 (2000)
12. C.H. Bennett et al., Phys. Rev. Lett. **70**, 1895 (1993); S.L. Braunstein et al., Phys. Rev. A **64**, 022321 (2001); P. Lambropoulos, D. Petrosyan, *Fundamentals of quantum optics and quantum information* (Springer, Berlin, 2007)
13. P. Kok, B.W. Lovett, *Introduction to optical quantum information processing* (Cambridge University Press, New York, 2010)
14. S.L. Braunstein, H.J. Kimble, Phys. Rev. Lett. **80**, 869 (1998)

15. A. Furusawa et al., Science **282**, 706 (1998); J. Joo, E. Ginossar, Sci. Rep. **6**, 26338 (2016)
16. T.C. Zhang et al., Phys. Rev. A **67**, 033802 (2003)
17. P. van Loock, S.L. Braunstein, H.J. Kimble, Phys. Rev. A **62**, 022309 (2000); S.L. Braunstein, P. van Loock, Rev. Mod. Phys. **77**, 513 (2005)
18. P.L. Scorpo et al., Phys. Rev. Lett. **119**, 120503 (2017); H.S. Qureshi, S. Ullah, F. Ghafoor, J. Phys. B **53**, 135501 (2020)
19. C.H. Bennett, S.J. Wiesner, Phys. Rev. Lett. **69**, 2881 (1992); T. Schaetz et al., ibid. **93**, 040505 (2004); R. Horodecki et al., Rev. Mod. Phys **81**, 865 (2009)
20. Z. Yang et al., AIP Adv. **6**, 065008 (2016)
21. D. Bruss et al., Phys. Rev. Lett. **93**, 210501 (2004)
22. M. Ziman, V. Buzek, Phys. Rev. A **67**, 042321 (2003)
23. V. Coffman, J. Kundu, W.K. Wootters, Phys. Rev. A **61**, 052306 (2000); A. Streltsov et al., Phys. Rev. Lett. **109**, 050503 (2012); F.F. Fanchini et al., Phys. Rev. A **87**, 032317 (2013); T. Das et al., Phys. Rev. A **92**, 052330 (2015)
24. K. Mattle et al., Phys. Rev. Lett. **76**, 4656 (1996); M. Pavicic, Int. J. Quan. Inf. **9**, 1737 (2011)
25. S.L. Braunstein, H.J. Kimble, Phys. Rev. A **61**, 042302 (2000); K. Shimizu, N. Imoto, T. Mukai, ibid. **59**, 1092 (1999); J. Mizuno et al., ibid. **71**, 012304 (2005)
26. T.C. Ralph, E.H. Huntington, Phys. Rev. A **66**, 042321 (2002); X. Su et al., ibid. **74**, 062305 (2006)
27. B. Schumacher, M.D. Westmoreland, Phys. Rev. A **56**, 131 (1997); X. Li et al., Phys. Rev. Lett. **88**, 047904 (2002)
28. H.P. Yuen, M. Ozawa, Phys. Rev. Lett. **70**, 363 (1993); C.M. Caves, P.D. Drummond, Rev. Mod. Phys. **66**, 481 (1994)
29. J. Jing et al., Phys. Rev. Lett. **90**, 167903 (2003); J. Lee et al., Phys. Rev. A **90**, 022301 (2014)
30. W.K. Wootters, W.H. Zurek, Nature **299**, 802 (1982); V. Buzek, M. Hillery, Phys. Rev. A **54**, 1844 (1996); N. Gisin, Phys. Lett. A **242**, 1 (1998)
31. V. Gheorghiu, L. Yu, S.M. Cohen, Phys. Rev. A **82**, 022313 (2010)
32. B. Schumacher, Phys. Rev. A **51**, 2738 (1995)
33. H. Fan et al., Phys. Rep. **544**, 241 (2014)
34. N. Gisin, S. Popescu, Phys. Rev. Lett. **83**, 432 (1999); J. Fiurasek et al., Phys. Rev. A **65**, 040302 (2002); A.L. Linares et al., Science **296**, 712 (2002); J. Fiurasek, N.J. Cerf, Phys. Rev. A **77**, 052308 (2008)
35. D. Bruss et al., Phys. Rev. A **57**, 2368 (1998)
36. N. Gisin, S. Massar, Phys. Rev. Lett. **79**, 2153 (1997); V. Scarani et al., Rev. Mod. Phys. **77**, 1225 (2005)
37. L. Mandel, Nature **304**, 188 (1983)
38. C. Simon, G. Weihs, A. Zeilinger, Phys. Rev. Lett. **84**, 2993 (2000); J. Mod. Opt. **47**, 233 (2000)
39. S. Fasel et al., Phys. Rev. Lett. **89**, 107901 (2002)
40. N.J. Cerf, A. Ipe, X. Rottenberg, Phys. Rev. Lett. **85**, 1754 (2000)
41. S.L. Braunstein et al., Phys. Rev. Lett. **86**, 4238 (2001)
42. M. Reck et al., Phys. Rev. Lett. **73**, 58 (1994)
43. N. Imoto, H.A. Haus, Y. Yamamoto, Phys. Rev. A **32**, 2287 (1985); B.C. Sanders, ibid. **45**, 6811 (1992); B.C. Sanders, D.A. Rice, ibid. **61**, 013805 (1999); J.S.N. Nielsen et al., Phys. Rev. Lett. **97**, 083604 (2006); K. Huang, ibid. **115**, 023602 (2015)
44. F. Grosshans, P. Grangier, Phys. Rev. A **64**, 010301 (2001)
45. M.D. Reid, Phys. Rev. A **40**, 913 (1989); G.M. DAriano, F. De Martini, M.F. Sacchi, Phys. Rev. Lett. **86**, 914 (2001); W.T.M. Irvine et al., Phys. Rev. Lett. **92**, 047902 (2004)
46. J. Fiurasek, Phys. Rev. Lett. **86**, 4942 (2001)
47. N.J. Cerf, S. Iblisdir, Phys. Rev. A **62**, 040301 (2000)
48. N.J. Cerf, S. Iblisdir, Phys. Rev. A **64**, 032307 (2001)
49. U.L. Andersen, V. Josse, G. Leuchs, Phys. Rev. Lett. **94**, 240503 (2005)
50. H. Chen, J. Zhang, Phys. Rev. A **75**, 022306 (2007)
51. N.J. Cerf, S. Iblisdir, Phys. Rev. Lett. **87**, 247903 (2001)

52. M. Murao et al., Phys. Rev. A **59**, 156 (1999)
53. P. van Loock, S.L. Braunstein, Phys. Rev. Lett. **87**, 247901 (2001)
54. J. Zhang, C. Xie, K. Peng, Phys. Rev. A **73**, 042315 (2006)
55. P. van Loock, S.L. Braunstein, Phys. Rev. Lett. **84**, 3482 (2000); S. Koike et al., ibid. **96**, 060504 (2006)
56. J. Zhang et al., Phys. Rev. A 77, 022316 (2008)

Application Overview and Foresight

12

With the advent of highly efficient lasers and advance in technical capabilities, it becomes possible to routinely entangle the states of the atoms with photons of different frequencies and with themselves via nonlinear interaction (see Chap. 9). The underlying understanding and generation of light with significant quantum features in turn enable the realization of optical processes that are intractable in the regime of classical physics (see Chaps. 8 and 11). In light of this, it might be worthy at some point to examine issues related to the inherent quantum traits that can be exploited in processing quantum information. The insight that a two-state quantum system can be in the superposition of the engaged states for instance envisages quantum information processing to be different from the classical counterpart in a way that enables to perform peculiar tasks. The awareness and expectation that the correlation leading to the emergence of quantumness opens multifaceted venues for exploring the ensuing applications of nonclassical properties of light with enhanced performance can therefore be taken as the main motivation for looking into the overview and foresight of continuous variable quantum communication, computation, cryptography and error correction.

Since the required quantum dynamics needs to be unitary, it is possible to propose that the best model that captures how the quantum device processes information should rely on some unitary operation that transforms the input qubits to the output via nonlinear interaction of light with matter in the regime of cavity quantum electrodynamics (see Chap. 6). The quantum state of the processing device notably can be epitomized by a vector that can be expressed in terms of a complex linear superposition of binary states of the bits;

$$|\Psi(x,t)\rangle = \sum_x C_x |x_1, \ldots, x_m\rangle \tag{12.1}$$

with $\sum_x |C_x|^2 = 1$ and $x \in (0,1)$ in which the state vector evolves in time according to the laws of quantum mechanics (see Eq. (3.1)). One may infer that

S. Tesfa, *Quantum Optical Processes*, Lecture Notes in Physics 976,
https://doi.org/10.1007/978-3-030-62348-7_12

a desired unitary transformation on arbitrarily large number of variables can be realized by successively inciting with the pertinent Hamiltonian.

Owing to the fact that the involved transformation depends on the input and output relations but not on the intermediate state of affairs, it should be technically feasible to achieve the same unitary transformation by applying different Hamiltonian, that is, one can have various options for a given quantum information processing task. This observation and the perception that information is intrinsically physical may legitimize embedding quantum optical operations and processes into the building block of theoretical foundation of quantum communication and computation. The inspiration to ponder on this chapter arises also from already established fact that the intricacy in the foundation of quantum theory can be exploited to perform important information processing tasks that otherwise would not have been implemented reliably and efficiently (see Chaps. 10 and 11); the ambition sought for to be galvanized with quantum optics.

By taking the idea that photonic qubits or gate operations can be performed by making use of successive half and quarter wave plates, splitters and reflectors as starting point (see Chap. 9), it might be imperative outlining the way by which the potential of a continuous variable quantum features can be harnessed in enhancing the efficiency [1]. One may also need to proclaim that a quantized light can be taken as a viable scheme specially for communication with added advantage that a photonic qubit can be operational over a longer time scale due to its weak interaction with environment, and implemented with linear optical setups and corresponding photon detection mechanisms as testified in a long distance communication (see Sect. 12.2 and also [2]). With emphasis on the basics of continuous variable quantum information processing tasks, the theme of this chapter is mainly directed towards highlighting the way how the already studied nonclassical properties of light and accompanying quantum optical operations and processes can be utilized to modify the outcomes of quantum manipulations.

12.1   Application Overview

Following the existing trend of making chips, it might not be hard to conceive that the size of the microprocessor will reach the point where the engraved logic gates would be so small that they consist of barely few atoms. The underlying quantum effects at this stage can become profoundly ubiquitous and relevant. For the processing power of electronic gadgets to continue at the rate humanity demands, a new technology should replace or supplement the tendency of relying on itching the chip further and further (see Sects. 12.1.3 and 12.2). Such a relentless setback imposed on the fate of unabated desire for a more powerful and faster technology can be taken as a clear indication for the need of careful in-depth reevaluation of quantum communication and computation, and the corresponding cryptography and error correction procedures. Recent endeavors embarked on such a mission reveal that quantum aspects of light can provide a new mode of communication and computation expected to be more powerful and reliable than its classical counterpart [3].

One may also insist that this enduring effort should amount to some kind of application; the notion that becomes apparent from the way one perceives or describes the prevailing situation or measures or represents quantized light that can be invoked as a collection of noninteracting quantum harmonic oscillators, and so can be taken as a system of Gaussian continuous variable (see Sect. 2.4). For example, one can cite Gaussian operations that are effectively denoted by interaction Hamiltonian at most quadratic in the optical mode's annihilation and creation operators, and lead to linear input-output relations as in beam splitting or squeezing transformation that include homodyne detection and phase space displacement. A general operational scheme can also be construed from the architecture of Mach–Zehnder type interferometry that encompasses passive optical elements such as splitters and reflectors and nonlinear medium (resonator) in one of the arms with various arrangements of the accompanying homodyne detection in the other (see Chaps. 4 and 11). Although the quantized light can exhibit a significant number of nonclassical properties, it might be possible to emphasize that the scheme for application is mainly constructed from unique attributes related to measurement incompatibility, squeezing, nonlocality and entanglement (see Chap. 7).

The overall communication framework for instance can be portrayed as information is first encoded onto the state of the carrier; then transmitted over the noisy communication channel; and finally decoded without affecting the content of the transmitted information in a meaningful manner (or with high fidelity; see Sect. 12.1.2). Prior to directly delving into the enigmatic issues of quantum information processing, it is hence appropriate and desirable to know how the inherent error can be detected, and then corrected afterwards. In this context, the effect of the associated noise needs to be minimized to enhance the quality since quantum states are extremely fragile to maintain a large multi-qubit coherent superposition required to implement a realistic procedure for sufficiently long time (see Chap. 10). Specifically, since the state in the process of quantum manipulation can be in the superposition of the clean and logically erred states, the rational for dealing with the correction scheme is tantamount to increasing the fidelity or probability for observing the correct outcome [4]. One may therefore observe that error correcting codes that make use of entanglement to rectify unknown errors in qubits are required ingredient for a large scale quantum information processing (see Sect. 12.1.1 and also [5]).

Since resource oriented constraints are often physically motivated, a variety of physical models and resources can be considered for realizing a certain information processing task, and the success rate can be enhanced when the participating parties share a prior entanglement or correlations with nonlocal character [6]. With this in mind, it might be compelling to extend the exposition on quantum information processing to the regime of continuous variable, which is meant to widen the perspective from discrete to continuous variable, and broaden the scope from finite to infinite dimensions. The impetus to proceed with continuous variable emanates from the observation that the implementation procedures such as preparing, manipulating and measuring the quantum state can be achieved with the aid of continuous quadrature amplitudes of the quantized electromagnetic field related to entanglement that can be efficiently generated using squeezed light in unconditional fashion; the option that is hard to come by in the discrete variable qubit based implementations.[1]

12.1.1 Quantum Error Correction

One of the important facets of classical information processing is its prowess of exploiting error correcting procedures over and over again—which requires copying the data. One however can not enjoy the privilege of using copies of the quantum state (see Sect. 10.2). The pertinent quantum error correcting techniques is therefore compelled to rely on protecting the quantum states from noise and decoherence. Even then, generalizing the idea of classical error correction to the quantum realm appears to be a formidable task since a quantum state is a priori continuous; so the resulting error is also continuous; and yet, it should be admissible to distribute the content of the information to entangled state of several qubits with the aid of entanglement swapping, quantum teleportation and telecloning (see Chap. 11).

In case there are a set of error operators that the information needs to be protected against, the correction scheme would be deemed as successful for each error operator when either the information is unchanged or the pertinent error is detected. Classical error correction procedure for instance relies on redundancy: storing the information multiple times; and if the copies are later turn out to disagree in some respect, just take the majority vote.[2] To correct errors in classical scenario, it may be desired generally to qualify the code; identify the decoding procedure; determine the syndromes and the information carrying subsystems; and analyze the behavior of the error of the code and subsystem. The outlined procedure can be considered as the error correction scheme that can be surmised as the classical error correcting codes employ a syndrome measurement to diagnose which error corrupts the encoded state, whereby the emerging error can be corrected by reversing the accompanying error by applying a corrective operation based on the syndrome. In quantum error correction scheme used for communication purposes, quantum states are sent directly through a potentially noisy channel after encoding them onto a relatively larger system. In case the larger system is subjected to errors during

---

[1]Contrary to earlier claim, in continuous variable setting, when it comes to advanced quantum information procedures such as universal quantum computation or entanglement distillation tools more than a mere Gaussian operations may be required [7].

[2]Imagine that a bit of information is copied three times. Assume that the noisy error corrupts the three bit state in such a way that one of the bits equals to a logic zero but the other two to logic one when received. Assume also that the noisy errors are independent and occur with some probability. In this characterization, the error is a single bit error. In case of a double bit error, assume that the transmitted message can be equal to three zeros but this outcome is less likely. So taking the most likely out of all possible outcomes as the actual result constitutes the notion of majority vote.

transmission through quantum channel, under certain circumstances, the errors can be corrected at the receiving site, and then the input quantum state can be retrieved with unit fidelity [8].

There is also a strong evidence which shows that an infinite set of qubit errors can be corrected with the aid of quantum error correction technique that encompasses a large number of unitary operations conditioned on measurement. Quantum error correction technique mainly exploits a syndrome measurement that does not disturb the measurement of quantum information while applied in the act of retrieving the message since the state of a single qubit can be encoded onto entangled states of several qubits as in quantum telecloning. One however should note that a quantum machine is far more susceptible of making errors since a quantum system is more prone to decoherence and manipulation of quantum information can be implemented only with a certain precision but not exactly.

It is thus expected to be enthralling to highlight the underlying challenges in the process of establishing quantum error correction implementations. Particularly, in view of quantum no-cloning theorem, the way of storing the information in the redundant and equivalent manner would be one of the challenges. To correct the error one also needs to carry out measurement to get the information, which might collapse the quantum state of the system and so destroy the information encoded onto the quantum state. Hence how one extracts the information about the error without destroying the precious quantum superposition that contains the message is another challenge.

Suppose one seeks to transmit a block of qubits in some unknown quantum state over a noisy quantum channel in which each transmitted qubit has a chance to become entangled with the channel. To enhance the fidelity of transmission, one can encode the state of the qubits onto a set of qubits, and then disentangle certain number of qubits from the channel at the receiving end. The syndrome measurement can thus identify whether a given qubit has been corrupted; and if so, which one and in how many several ways that the qubit was affected. Although the error due to the noise is arbitrary, it can be expressed as the superposition of error bases that can be described by the Pauli matrices (see Sect. 9.4). This proposal insinuates that the syndrome measurement compels the qubit to pick a specific Pauli error, where the same Pauli operator is acted on the corrupted qubit to revert the effect of the error. So one of the practical routes that can be followed to overcome decoherence is to distill coherence from a large set of entangled particles, and form a subset of particles with enhanced entanglement purity [9]. In case it is allowable to protect the state of the system or part of it from decoherence, the states that is not prone to decoherence can be deployed in error correcting procedure by encoding the desired information onto this state.

The other type of quantum error can be linked to qubit leakage that would be manifested owing to the fact that most systems applied for realizing the qubits may not involve the required energy level transformation. Leakage induced errors in principle need to be corrected in the same way as the loss associated with the leakage to the vacuum state. One of physically plausible processes in coupled system scenario is dissipation of information via decoherence in which the populations of the engaged quantum states are altered during evolution due to the inevitable entanglement with the bath. In case the system is in one of its excited states, and the bath is at temperature lower than that of the system, the system gives off energy to the bath, and the higher energy eigenstates of the system decohere to the lower energy state; and the desired information may thus be destroyed in the process. The relevant idea of corresponding error correction then amounts to introducing a sufficient information so that one can able to recover the original state [10]. Such error correction procedure requires quantum operation, and so are susceptible to the same error that imposes a limitation on the scope of the correction codes.

12.1.1.1 General Consideration

In the context that error correction scheme can be perceived as a way of disentangling the system from its environment, it might be worthy noting that each qubit would be coupled independently to its own reservoir in such a way that the error can be expressed as the direct product of the errors in the individual qubits, that is, the initial state can be defined as

$$|\psi\rangle_i = \sum_{j=1}^{q} C_j |c^j\rangle \otimes |R\rangle, \tag{12.2}$$

where q stands for the encoded qubits, $|c^j\rangle$'s for the state of the codeword and $|R\rangle$ for the state of the reservoir.

The state after the error has been introduced on the other hand can be designated as superposition of errors acting on the initial state that include amplitude error $\hat{A}$, phase error $\hat{P}$ and their combination $\hat{A}\hat{P}$:

$$|\psi\rangle_f = \sum_{l,m} \hat{A}_l \hat{P}_m \sum_{j=1}^{q} C_j |c^j\rangle \otimes |R\rangle_{lm}, \tag{12.3}$$

where tracing over the reservoir variables yields

$$|\psi\rangle_n = \sum_{l,m} \langle R|R\rangle_{lm} \hat{A}_l \hat{P}_m \sum_{j=1}^{q} C_j |c^j\rangle \tag{12.4}$$

in which $|R\rangle_{lm}$ is independent of the codewords.

To detect the error, one needs to perform the relevant measurement to see if there is an overlap with one of the subspaces $\{\hat{A}_l \hat{P}_m |c^j\rangle\}$; $j = 1, \ldots, q$. Each time the overlap is measured and negative result is obtained, the pertinent space reduces in dimension due to the elimination of the subspace that comprises the error. One of these overlap measurements eventually may give a positive result, which is mathematically equivalent to projecting the state of the system onto the pertinent subspace where a successful projection effectively leads to the state generated by the superposition with the error.

Once the projection is deemed successful, one needs to correct the emerging error by applying the conjugate transpose of the original operator on the projected state. For the error correction procedure to be successful, the emerging state has to be proportional to the initial state of the codewords, that is,

$$\langle c^j | \hat{P}_m A_l A_s P_t | c^k \rangle = \delta_{mt} \delta_{ls} \delta_{jk}. \tag{12.5}$$

For example, for arbitrary qubit $|\Psi\rangle = \alpha|0\rangle + \beta|1\rangle$, the first step of the three-qubit bit flip code is to entangle this qubit with two other qubits by using two CNOT gates with input $|0\rangle$. The result of this operation looks like $|\Psi_r'\rangle = \alpha|000\rangle + \beta|111\rangle$—a tensor product of the three qubits which is different from cloning the state (see Chap. 9).

Suppose these qubits are then sent through separate channels where the first qubit is flipped in the channel and the emerging state happens to be $|\Psi'\rangle = \alpha|100\rangle + \beta|011\rangle$. To diagnose the bit flips in the three possible qubits, a syndrome diagnosis that includes four projection operators is required:

$$P_0 = |000\rangle\langle000| + |111\rangle\langle111|, \qquad P_1 = |100\rangle\langle100| + |011\rangle\langle011|;$$

$$P_2 = |010\rangle\langle010| + |101\rangle\langle101|, \qquad P_3 = |001\rangle\langle001| + |110\rangle\langle110|, \tag{12.6}$$

so one can observe that

$$\langle\Psi'|P_0|\Psi'\rangle = 0, \quad \langle\Psi'|P_1|\Psi'\rangle = 1, \quad \langle\Psi'|P_2|\Psi'\rangle = 0, \quad \langle\Psi'|P_3|\Psi'\rangle = 0, \tag{12.7}$$

where the error syndrome corresponds to P_1. This characterization may evince that a three-qubit bit flip code can help in correcting one error in case one bit flip error appears in the channel.[3]

But to include the effect of decoherence, one needs to imagine that the qubit on which the information is encoded to be coupled to the environment. Assume also that the procedure we aspire to consider works equally well in case any one of the other eight qubits is presumed to decohere. Since $|0\rangle$ and $|1\rangle$ form the bases for the original qubit, one worries about what happens to these states. The decoherence process may in general look like

$$|e\rangle|0\rangle \to |a_0\rangle|0\rangle + |a_1\rangle|1\rangle), \qquad |e\rangle|1\rangle \to |a_2\rangle|0\rangle + |a_3\rangle|1\rangle), \tag{12.8}$$

where a_i's correspond to the environment and may not be orthogonal [5].

---

[3] Sophisticated and economical quantum error correcting codes that employ five, seven and even larger codes capable of encoding more physical qubits and able to correct multi-qubit errors are also available [5, 11].

After decoherence, note that the superposition $\frac{1}{\sqrt{2}}[|000\rangle + |111\rangle]$ goes to

$$\frac{1}{\sqrt{2}}[|000\rangle + |111\rangle] \rightarrow \frac{1}{\sqrt{2}}[(|a_0\rangle|0\rangle + |a_1\rangle|1\rangle)|00\rangle + (|a_2\rangle|0\rangle + |a_3\rangle|1\rangle)|11\rangle],$$

$$(12.9)$$

which can be rewritten with the aid of the GHZ basis (see Eq. (9.8)) as

$$\frac{1}{\sqrt{2}}[|000\rangle + |111\rangle] \rightarrow \frac{1}{2\sqrt{2}}[|a_0\rangle + |a_3\rangle](|000\rangle + |111\rangle)$$

$$+ \frac{1}{2\sqrt{2}}[|a_0\rangle - |a_3\rangle](|000\rangle - |111\rangle)$$

$$+ \frac{1}{2\sqrt{2}}[|a_1\rangle + |a_2\rangle](|100\rangle + |011\rangle)$$

$$+ \frac{1}{2\sqrt{2}}[|a_1\rangle - |a_2\rangle](|100\rangle - |011\rangle). \qquad (12.10)$$

In the same way,

$$\frac{1}{\sqrt{2}}[|000\rangle - |111\rangle] \rightarrow \frac{1}{2\sqrt{2}}[|a_0\rangle + |a_3\rangle](|000\rangle - |111\rangle)$$

$$+ \frac{1}{2\sqrt{2}}[|a_0\rangle - |a_3\rangle](|000\rangle + |111\rangle)$$

$$+ \frac{1}{2\sqrt{2}}[|a_1\rangle + |a_2\rangle](|100\rangle - |011\rangle)$$

$$+ \frac{1}{2\sqrt{2}}[|a_1\rangle - |a_2\rangle](|100\rangle + |011\rangle). \qquad (12.11)$$

Once this unitary transformation is performed, one can also think of quantum error correction by including the measurement of the error via collapsing the state onto one of the four possibilities: nothing, bit flip, phase flip, or combination of bit flip and phase flip. In this way, although the error is principally continuous, it is made to be discrete in the process of quantum error correction. Assume also that the error processes which affect different qubits are independent from each other, that is, quantum error correcting code should be such that it can protect against the four possible errors.

In the process, one should be able to recognize the original state of the encoded vector by looking at the other two triples. One can restore in the meantime the original state of the encoded vector and can also keep the unitary evolution by creating a few ancillary qubits that notify which qubit has decohered. One can after that restore the original state by measuring these ancillary qubits, where the restoration procedure encompasses a unitary transformation that can be regard as being performed by the quantum processor and measurement of some of the qubits of the outcome. What actually done is comparing the three triples in the Bell basis. If these triples turn out to be the same, the output can be taken as free of decoherence, and the processor should leave the encoded qubit as it is. But if those triplets turn out to be not the same, the processor should single out which triplet is different and how it is different.

Once the corrupted state is known, the processor is supposed to restore the encoded qubit to the original state. It seems that one is getting something for nothing, that is, restoring the state of the superposition to the exact original state although some of the information can be destroyed along the way. This can happen mainly due to the expansion of one qubit to many encoded qubits, and consequent establishment of some sort of redundancy. The cost is that the mechanism that meant to implement the unitary transformation may not be exact: getting rid of the decoherence may induce extra error. Since each time we get rid of the decoherence or even check whether there is decoherence, small error would be incurred; hence choosing the rate at which the state is measured to balance the error induced by a decoherence with that induced by the restoration of decoherence would be essential.

One can also consider a more general error correction procedure, namely, the stabilizer code [12]. One can begin with the eigenvalue equation of the form $\hat{s}|\psi\rangle = |\psi\rangle$ while the error occurs on one (or multiple) qubit(s) via operator $\hat{E}$ over some qubits of the logical state;

$$\hat{s}_j\hat{E}|\psi\rangle = (-1)^m\hat{E}\hat{s}_j|\psi\rangle = (-1)^m\hat{E}|\psi\rangle, \tag{12.12}$$

where m denotes the situation in which $m = 0$ if $[\hat{E}, \hat{s}_j] = 0$ and $m = 1$ if $\{\hat{E}, \hat{s}_j\} = 0$. This entails that if the error operator commutes with the stabilizer, the state remains a 1 eigenstate of $\hat{s}_j$; and if the error operator anti-commutes with the stabilizer, the logical state flipped to a -1 eigenstate of $\hat{s}_j$. The error correction procedure is straightforward, that is, while each of the code stabilizer is sequentially measured, an error that anti-commute with the stabilizer flips the relevant eigenstate, and so measuring the parity of this stabilizer returns the required result.

12.1.1.2 Continuous Variable Quantum Error Correction

One can also seek to extend the discussion on quantum error correction to the regime of continuous variable whose realization may seem more complicated than its discrete variable counterpart as much wider class of errors are expected to occur. Even then, existing experience reveals that certain continuous variable quantum error correction protocols can be unconditionally implemented with the help of passive linear optical elements, nonlinear medium and homodyne measurement [13]. It might be advantageous thus to adapt the discrete error correcting technique of redundancy to continuous quantum variable regime.

To this end, let us take three position variables that can be described by a state $|x_1 x_2 x_3\rangle_{123}$, where the errors are associated with unitary operator $e^{-i\hat{Q}(\hat{P}_j)}$ in which

$\hat{P}_j = -i\frac{\partial}{\partial x_j}$ is the momentum operator of the jth variable, $\hat{Q}$ is a polynomial function of $\hat{P}_j$. With this designation, it may not be difficult to see that the state of the system becomes

$$|x\rangle_j \longrightarrow e^{-i\hat{Q}(\hat{P}_j)}|x\rangle_j = \frac{1}{\sqrt{2\pi}}\int_{-\infty}^{\infty} e^{-ipx-iQ(p)}|p\rangle_j dp, \qquad (12.13)$$

where

$$|p\rangle_j = \frac{1}{\sqrt{2\pi}}\int_{-\infty}^{\infty} e^{ipx}|x\rangle_j dx. \qquad (12.14)$$

If the error is acted on one of the variables such as $\hat{Q}(\hat{P}_j) = \delta x \hat{P}_j$, the state of the system would evolve according to

$$|x\rangle_j \rightarrow \frac{1}{\sqrt{2\pi}}\int_{-\infty}^{\infty} e^{-ipx-ip\delta x}|p\rangle_j dp = |x+\delta x\rangle. \qquad (12.15)$$

This process can be interpreted as the imparted error displaces the system by infinitely small amount. To attach a sensible attribute to this approach, one may assume that certain variables can be prepared at will in the state $|0\rangle_j$ by making use of some dissipative process, and the state $|x\rangle_j$ can be prepared by applying certain Hamiltonian. It is required to apply the intended procedure to the other three continuous quantum variables to introduce the majority vote.

To pave the way, take x, y and z as components of the position of a single particle in three dimensions initially prepared in the state $|xxx\rangle_{123}$ together with three ancilla variables $|x_1x_2x_3\rangle_{1'2'3'}$ initially in the state $|000\rangle_{1'2'3'}$. Once the initial state is defined, a general continuous variable error correcting procedure can follow the same steps as for discrete case. On account of this, imagine that the error occurs to one of the variables

$$|x\rangle_2 \longrightarrow e^{-i\hat{Q}(\hat{P}_2)}|x\rangle_2 = \int_{-\infty}^{\infty}\int_{-\infty}^{\infty} e^{-p(x-x')-iQ(p)}|p\rangle|x'\rangle_2 dpdx. \qquad (12.16)$$

Suppose the ancilla variables in the state $|000\rangle_{1'2'3'}$ then turns out to be

$$|000\rangle_{1'2'3'} \longrightarrow (|x\rangle_1|0\rangle_{1'})$$

$$\times \left[\int_{-\infty}^{\infty}\int_{-\infty}^{\infty} e^{-p(x-x')-iQ(p)}|p\rangle|x'\rangle_2|0\rangle_{2'}\right](|x\rangle_3|0\rangle_{3'})dpdx'. \qquad (12.17)$$

To expedite continuous variable majority voting, one can begin with the presumption that it is acceptable to perform a simple real number operations such as comparing the values of two variables to see if they are equal or not by comparing

$|x_1\rangle_1$ with $|x_2\rangle_2$ to verify if $x_1 = x_2$. It should also be possible to add the value of one variable to another, and if they are equal, perform operations such as

$$|x_1\rangle_1|x_2\rangle_2|x_3\rangle_3 \longrightarrow |x_1\rangle_1|x_2\rangle_2|x_3 + \delta x\rangle_3. \tag{12.18}$$

These operations are reversible and related to the unitary transformation on the Hilbert space, and as a result can be implemented by invoking interaction between the variables. If only one error has occurred, two of the x's should be equal. To correct the error, one thus needs to compare each $|x\rangle_i$ to the other two one by one.

The general strategy in quantum error correction procedure may therefore comprise encoding the state of the qubit onto the collective entangled state of several qubits that belong to certain dutifully chosen code with the condition that arbitrary single qubit error takes the state to orthogonal subspace uniquely associated with that particular qubit and error type. This consideration entails that every possible error in the correction code leads to a state orthogonal to the original uncorrupted state. Performing a multi-qubit measurement that can distinguish between the uncorrupted state and other states resulting from a single qubit error reveals the error syndrome. After identifying the error syndrome, the error correction protocol is implemented to restore the original qubit by applying appropriate unitary transformation of the corrupted qubit. This kind of procedure is tantamount to correcting the error by restoring the three variables to the original continuous variable codeword while leaving the ancilla in a state independent of the initial value of the variables.

Recall that a continuous variable qubit can be constructed from a coherent state by making use of Kerr nonlinearity and homodyne detection (see Sect. 9.3), where the nonlinearity can be described by interaction Hamiltonian

$$\hat{H} = \kappa \hat{a}_1^\dagger \hat{a}_1 \hat{a}_2^\dagger \hat{a}_2. \tag{12.19}$$

If two input coherent states denoted by $|\alpha\rangle_1$ and $|\beta\rangle_2$ are pumped into the setup, the interaction can lead to the output in number state

$$|\Psi\rangle_{12} = e^{-\frac{\alpha^*\alpha}{2}} \sum_{n=0}^{\infty} \frac{\alpha^2}{\sqrt{n!}} |n\rangle_1 \otimes |\beta e^{-in\kappa t}\rangle_2. \tag{12.20}$$

Since the interaction is usually very weak, one can let $e^{-in\kappa t} \approx 1 - in\kappa t$. After that, upon displacing the phase of the second mode by a constant ($\hat{a}_2 \rightarrow \hat{a}_2 + \beta$), one can see that

$$|\Psi\rangle_{12} = e^{-\frac{\alpha^*\alpha}{2}} \sum_{n=0}^{\infty} \frac{\alpha^2}{\sqrt{n!}} |n\rangle_1 \otimes |-i\beta n\kappa t\rangle_2, \tag{12.21}$$

which can be regarded as an output of the interaction Hamiltonian

$$\hat{H}' = \beta\kappa\hat{a}_1^\dagger\hat{a}_1(\hat{a}_2^\dagger + \hat{a}_2) = \sqrt{2}\beta\hat{a}_1^\dagger\hat{a}_1\hat{q}_2. \tag{12.22}$$

This process may indicate that the approximate codeword can be created in mode 2 by measuring the quadrature in mode 1 with outcome let us say q_1. In light of this suggestion, the state of the quantum system in mode 2 can be expressed as

$$|\Psi_{out}\rangle_2 = e^{-\frac{\alpha^*\alpha}{2}} \sum_{n=0}^{\infty} \frac{\alpha^2}{\sqrt{n!}} \langle q_1|n\rangle| - in\beta\kappa t\rangle_2 \equiv |\mathbf{1}\rangle, \tag{12.23}$$

which represents one of the computational bases. Once one of the code states is known, the pertinent code state $|\mathbf{0}\rangle$ can be obtained by using one of the Pauli rotations (see Sect. 9.4). In the context of quantum error correction, it is possible to generate a nine-qunat code as extension to the discrete nine-qubit code. This can be achieved by concatenation of two majority codes one for each position and momentum errors. Such an idea depends on the fact that one can distribute a single nonzero qunat over three modes according to

$$|q, 0, 0\rangle \longrightarrow \left|\frac{q}{\sqrt{3}}, \frac{q}{\sqrt{3}}, \frac{q}{\sqrt{3}}\right\rangle. \tag{12.24}$$

It might be insightful noting that this cannot be done for a discrete case since qubit values cannot be divided by a factor of arbitrary number such as $\sqrt{3}$, and vector $|\mathbf{0}\rangle$ is the position eigenstate with zero eigenvalue but not the vacuum state as usually presumed. This distribution can be realized with the help of a three-mode beam splitter having balanced three ports that can act on the three quadrature operators. Once the transmitted mode is presumed to experience small displacement at the end of one of the ports, the error correction procedure would be carried out by measuring the output with the help of passive linear optics and homodyne detection while the list of the emerging real numbers constitutes the syndrome.

12.1.2 Quantum Communication

Pertaining to the expectation that the strong correlation shared among entangled states enhances the exchange of information between the sending and receiving sites, there has been a significant effort geared towards harnessing the nonclassical features of light to facilitate perceived communication [3]. It is thus desirable to connect the idea of quantum communication with the procedure in which the ability to communicate would be enhanced by exploiting the peculiar quantum traits of the participating systems such as entanglement and nonlocality. With this background, quantum communication can be conceived as the distribution of the quantum states between two parties via the channel that preserves the quantum features of the distributed states.

In realistic situation, the nonorthogonal quantum states sent through a quantum channel would be subjected to environmentally induced noise since the channel would be noisy. Such a challenge could be circumvented with the help of quantum teleportation combined with purification of the shared entanglement or quantum error correction. Since the ingredient of quantum communication setups do not destroy the interconnectedness established between the sites, one of the most valid example of entanglement assisted processes that leads to information communication is quantum teleportation (see Sect. 11.2). Entanglement swapping can also be an essential ingredient of a long distance implementation in the sense that the two remote ends of the noisy quantum channel are endowed with a substantial entanglement after purification of the noise in different segments of the channel, and then combination of the segments via entanglement swapping (see Sect. 11.1). Besides, even though nonorthogonal quantum states cannot be reliably distinguished, the amount or content of transmitted information can be increased when the sender sends half of the shared entangled state via quantum channel as in quantum dense coding (see Sect. 11.3).

One of the main issues that needs to be dealt with in quantum communication is the security of the communication. In line with this, the notion of cryptography can be perceived as an art of rendering the message unintelligible to unauthorized party or as a technique of hiding the content of the message via sharing preestablished secret key unknown to the potential eavesdropper or as an effort of preserving the secrecy of the communication. Quantum cryptography (key distribution) can as a result be taken as a technique that allows two remote parties to realize quantum communication by sharing a secret strings of random bits. The aspect of quantum security mainly stems from the understanding that the measurement of incompatible variables inevitably affects the state of the quantum system, that is, the leakage of the information to the eavesdropper induces disturbance of the system that can be detected by authorized receiver.[4] The main presumption of secrecy rests on the fact that there are no available means to the eavesdropper to acquire the message intended for the legitimate receiver without in someway altering the involved quantum states.

In this regard, if Alice encodes the information onto the photon she sends to Bob, and he receives the photon unperturbed, he can take the photon as not measured, that is, as the eavesdropper did not acquire the information about the photon since the process of acquiring information can be taken synonymous with carrying out measurement. After exchanging or sharing the defining characteristics of the photon via public channel, Alice and Bob can verify whether someone has tampered with their communication since receiving unperturbed sets by Bob is tantamount to no

---

[4]Quantum cryptography and key distribution are often used synonymously despite the fact that quantum cryptography does also refer to other secrecy tasks such as quantum state sharing, bit commitment and random number generation [14].

perturbation; no measurement; and no eavesdropping. The presumption is that a quantum key distribution can guarantee a securer communication than the available classical schemes can offer by the way of exploiting the underlying principles of quantum theory rather than relying on the computational difficulty of solving certain mathematical functions.

A secured cryptography protocol relies in classical scenario on the availability of a common private key: the sender and receiver should meet and exchange the keys or employ the third party to provide them with the keys prior to the commencement of communication without having a clear criteria to judge whether the third party is faithful. The most difficult and costly part perhaps is to acquire a reliable key distribution. It might hence be reasonable to couple the interest that surrounds quantum key distribution with the need for generating more secured private key by utilizing quantum aspects. Note that the possibility of generating a public key via distribution of a random secured key between parties who share no secret information initially and when the parties enjoy access to quantum and ordinary classical channels is currently available [15].

The trust of generating secured key rests on the presumption that once the key is produced, the communicating parties by subsequent consultation over ordinary classical channel subjected to passive eavesdropping can tell with high probability whether the original quantum transmission has been tampered with in transit. If the transmission is not found to be disturbed, they would agree to use it as shared secret key; or else, they would discard it and possibly try again until they succeeded. The secrecy of the key distribution would thus be guaranteed by the feasibility of detecting the activity of the eavesdropper, and be able to correct it prior to the commencement of the actual communication.

Imagine that Alice sends to Bob N qubits of polarized single-photon prepared in one of the four possible states denoted by $|V\rangle$, $|H\rangle$, $|R\rangle$ and $|L\rangle$ via randomly chosen quantum channel. To do so, one can perceive that states $|V\rangle$ and $|R\rangle$ correspond to logical value 0 of Alice's random bit, and states $|H\rangle$ and $|L\rangle$ to 1. Suppose Bob upon receiving the qubits measures in a randomly chosen bases selected by Alice. Just like Alice, let he also assigns 0 to his random bit if the measurement yields $|V\rangle$ or $|R\rangle$; or else 1. After that, suppose Bob and Alice communicate via a classical channel to find out if any of the qubits is lost. Out of N qubits he received and measured, in the absence of loss, he is expected to employ correct bases in average in half of the cases. After carrying out comparison, Alice and Bob discard those bits for which they did not agree, and so would be left with $N/2$ random bits. Alice in the process is expected to prepare and send new keys to compensate for the lost bits so that sufficient keys can be generated at will. In case there is eavesdropper lurking somewhere along the quantum channel and trying to access the key, s/he needs to measure the qubits in the same way as Bob, that is, it is required to generate each detected qubit in the measured state and send to Bob. Otherwise, Alice will discard it along with the lost qubits and substitute them with new ones.

To detect eavesdropping, Bob and Alice can randomly choose $n < N$ bits out of their common strings and compare them. If about $n/4$ bits or less do not coincide, they realize that the eavesdropper was trying to get access to the key. Even then, there can be various ways by which the secrecy of the quantum key distribution could be attacked by utilizing certain cloning procedure [16]. For a cloning based attack to succeed, the best the eavesdropper can do is to imperfectly clone Alice's qubit and keep the copy while sending the original to Bob. In this way, appropriate measurement of the clone and ancilla system may enable the eavesdropper to gain information on Alice's key bit.

The information or the key can also be encoded onto the quadrature components of the light in which the key distribution encompasses; the process in which Alice modulates randomly the Gaussian beam, and then sends the result to Bob via Gaussian noisy channel. Bob after that measures either the phase or the amplitude of this beam, and informs Alice about the measurement he has made. Bob and Alice then possess two correlated sets of Gaussian variables from which they can extract common secret string. So a generic continuous variable quantum key distribution procedure between two trusted parties can be initiated by Alice who prepares the distribution of Gaussian quantum states of the light in the form of coherent or squeezed or entangled state. In the perceived protocol, Alice transfers the prepared quantum states via quantum channel to Bob who performs measurements on the quadrature components of the light field with the help of homodyne detection, where the outcome of the measurement is believed to be partially correlated with Alice's data set which she obtained in the process of preparing the distribution of quantum states [17].

It is presumed in this case as well that to estimate the secrecy of the transmission, Alice and Bob should compare the subset of their data with the aid of classical communication. In case the security thresholds for the channel loss and excess noise have not been crossed, the resulting set of raw data can be mapped onto the shared secret key. The key can be construed then as a Gaussian noise imposed on the squeezed light by displacing the intended quadrature. Since it is impossible to measure with accuracy both quadratures, Alice can encode the key onto one of the quadrature components as in squeezed state so that the eavesdropper cannot acquire information without disturbing the state. The requirement for a maximally secured cryptographic setup then would be the distribution of a quadrature measurement outcomes to be indistinguishable when either of them is used by Alice. If this condition is fulfilled, the eavesdropper cannot acquire any hint of which quadrature of the two-mode squeezed state Alice has measured [18].

The other main ingredient of quantum communication is quantum network, which can be taken as a scheme that consists of distributed processors connected via communication channels. Since quantum communication channels can be designated effectively by photons and quantum nodes, to realize quantum network, one needs to coherently transfer quantum information from the stationary matter qubits to the flying photon qubits [19]. Quantum communication channels on the other hand are usually designated by optical setups as the light can carry the content of the information to a distant location at high speed. Even with the best fibers, photon pulses however decay exponentially over a distance. This hopefully can be mitigated by making use of repeater circuits expected to periodically amplify the signal before a significant degradation happens.

A straightforward extension of this idea unfortunately does not work in the realm of quantum theory since unknown quantum signals cannot be amplified. Nonetheless, it should be possible to send a quantum state over arbitrary distance upon utilizing the mechanism of quantum repeater[5] that may require purification of the entangled state repeatedly. So once purified, the entangled states linked by a network can be employed in the implementation of quantum cryptography or in a faithful transmission of quantum state over a long distance as in quantum teleportation.

The content of the message can be stored in a collective atomic state arising from the interference of the photon emitted from different atoms since the collective atomic state can be coupled to photonic modes. This may indicate that the coupling of trapped atoms to the flying photonic qubits can provide the required interface between two parties that participate in the networking such as when optically active material qubit is excited by laser that can normally couple to infinite number of optical modes (see Sect. 9.2 and also [21]). The nonclonability of a quantum state nevertheless may limit the achievable distance between the nodes, and so poses a severe restriction on practical application.

The idea of quantum repeater in which the quantum entanglement would be distributed over small distances; stored in quantum memories at the nodes; purified; and then swapped can thus be formulated as a remedy. A quantum repeater can also be perceived as a device or a technique in which the entangled state on which the message is imprinted can be repeatedly purified over a distance connected by the quantum nodes placed close to each other and having extra quantum memory [22]. There is however a price to be paid such as creating the entanglement remotely in a heralded way, storing and swapping it many times. The basic idea of quantum repeater can thus be realized by purifying the noisy entanglement, and then transferring the entanglement over adjacent segments by means of entanglement swapping or quantum teleportation or telecloning [23].

One may also contemplate that the fragile nature of quantum correlations and inevitable photon loss in the process of communication can pose a serious challenge on the intent of outperforming the direct transmission of photons along the communication channel. It so appears advantageous introducing the idea of quantum relay that works in a similar way as quantum repeater but without entanglement purification procedure and quantum memories [14, 24]. Although this approach seems more feasible as compared to repeater, it does not allow to achieve arbitrary long distance.

The procedure of quantum teleportation related to a time-bin qubits particularly might be very appealing in some respect to illustrate the basic idea of quantum relay.

---

[5]Since the involved entanglement purification and entanglement swapping require feasible and reliable quantum logic gates, practical and efficient schemes for realizing quantum repeater by using light alone may not be straightforward [20].

Let Alice prepares a time-bin qubit as in coherent superposition of two time-bins by passing a single photon through unbalanced interferometer (see Sect. 9.3);

$$|\Psi\rangle_A = a_0|1, 0\rangle_A + a_1 e^{i\alpha}|0, 1\rangle_A, \qquad (12.25)$$

where $|1, 0\rangle_A$ represents the first time-bin (the photon that passes through the short arm of the interferometer), $|0, 1\rangle_A$ the second time-bin (the photon that passes through the long arm), α is a relative phase and $a_0^2 + a_1^2 = 1$.

Suppose Alice then sends the qubit

$$|\Phi^+\rangle_{BC} = \frac{1}{\sqrt{2}}\big[|1, 0\rangle_C|1, 0\rangle_B + |0, 1\rangle_C|0, 1\rangle_B\big] \qquad (12.26)$$

to Charlie who shares with Bob a classical communication channel and a pair of time-bin entangled qubits in the Bell state in which he projects photons A and C onto one of

$$|\Phi^\pm\rangle = \frac{1}{\sqrt{2}}\big[|1, 0\rangle|1, 0\rangle \pm |0, 1\rangle|0, 1\rangle\big],$$

$$|\Psi^\pm\rangle = \frac{1}{\sqrt{2}}\big[|1, 0\rangle|0, 1\rangle \pm |0, 1\rangle|1, 1\rangle\big]. \qquad (12.27)$$

Depending on the result of the Bell state measurement as communicated by Charlie with two classical bits, Bob can apply the appropriate transformation or error correction procedure to recover the initial state. For example, to discriminate two out of the four Bell states, for projection onto the $|\Phi^+\rangle$ singlet state, it can be shown that the detection of one photon in each output mode can realize this projection. It is possible in this case to project the state owned by Bob onto the state

$$|\Psi\rangle_B = a_0|0, 1\rangle_B - a_1 e^{i\alpha}|1, 0\rangle_B = i\hat{\sigma}_y|\Psi\rangle_A, \qquad (12.28)$$

where $\hat{\sigma}_y$ is one of the Pauli matrices. Markedly, this could be taken as the process that epitomizes quantum networking.

12.1.3 Quantum Computation

Even though the notion of computing until recently solely relies on classical understanding, one may perceive that there are tasks that a classical computer cannot perform efficiently and reliably according to known algorithm which a quantum computer is supposed to do in a polynomial time[6] such as Shor's factoring and Grover's search algorithms [25] and expected also to perform more reliably. To simulate a quantum system on a quantum computer one first needs to initialize the registry with a known initial state, and then load the input by preparing the registry in the initial state of the system to be simulated. Since the initial state can be the product state of the subsystems, one may not witness entanglement among these states; and such a step can be enforced with the help of a small number of logic gates (see Sect. 9.4). The relevant evolution or dynamics in the simulation process would be sanctioned with the aid of a sequence of logic gates whose action on the registry brings about the evolution operator perceived to be simulated efficiently when the interaction Hamiltonian is expressed as the sum. The evolved quantum registry whose state is related to the simulated system would be measured afterwards.

In case the measured state is not the eigenstate of the registry, one may be compelled to perform many repetitions of this cycle so that a reliable probability distribution for the final state of the system is obtained. Note that the manipulation of the relevant qubit by the pertinent quantum gate is the main feature of this consideration. The computational aspect of the quantum process would thus be invariably linked to the dynamical evolution of the engaged state, that is, the mechanism of computing is closely related to the physical process that can be practically realized [26].

12.1.3.1 General Framework
Quantum computation can also be perceived as a process in which the state of the system evolves according to

$$|a_j\rangle = \hat{U}|a_i\rangle \longrightarrow |f(a_i)\rangle, \tag{12.29}$$

where $\hat{U}$ is unitary operator (relate with Eq. (9.3)). The main task would thus be obtaining the Hamiltonian that induces this type of evolution, that is,

$$|\Phi_f\rangle = e^{-i\int \hat{H}dt}|\Phi_0\rangle. \tag{12.30}$$

One of the advantages for going into the quantum regime can be taken as the possibility of gaining computational saving. Imagine that the receiver's goal is to determine some information imprinted as a function of a that depends on another data b that resides with the receiver but unknown to the sender. It can also be presumed that the sender knows a but not b while the receiver knows b but not a. Suppose the aim of the receiver is to determine the value of some function $f(a, b)$

---

[6] Algorithm is said to be in a polynomial time if its running time is bounded by a polynomial with the size of the input $T(n) = O(n^k)$, and solvable in a polynomial time when the number of steps required to complete the algorithm is $O(n^k)$ for nonnegative integer k, where n is the complexity of the input. Familiar mathematical operations such as addition, subtraction, multiplication, division and computing square roots and powers, and the digits of mathematical constants such as π and e can be carried out in a polynomial time.

known to both parties. As illustration, let the sender's schedule is represented by a and that of the receiver by b. If there are n time slots, one can set the jth bit of a to 1 when the sender is available in time slot j and 0 if not, and denote similarly for b. In light of computational saving, one may aspire to know how much communication trial is required for the receiver to find the time when both parties are available, when there is j such that $a_j = b_j = 1$. The most obvious expectation would be: quantum interconnectedness between the states of the sender and receiver can enable the task to be accomplished with less communication trial than would be required classically—the conception usually dubbed as quantum saving.

It should also be possible to implement a unitary transformation over a finite number of variables to a desired degree of precision by repeated application of logic gates. Since arbitrary unitary transformation over a single continuous variable requires assigning an infinite number of parameters, such implementation cannot be approximated by a finite number of continuous quantum manipulations. Even then, it should be possible to sanction a notion of universal quantum computation over continuous variable for various subclasses of transformations.

One can also think of a linear optical quantum computing via measurement approaches that nondeterministically realize the optical nonlinearity required for two-qubit entangling gates and also deterministic version that can be realized with the help of teleportation via Bell state measurement [27, 28]. It is thus imperative to foresee that the natural way to overcome the difficulty such as realizing a perfect single-photon sources and implementing photon-photon gates might be looking beyond two-dimensional physical systems for encoding and decoding the content of quantum information such as a coherent state with opposite amplitudes in the regime of continuous variable.

Since arbitrary unitary transformation over extensive degrees of freedom requires addressing a large number of parameters, extending the discussion on a discrete quantum computation to continuous variable may not be as straightforward as expected. Even then, the notion of quantum computation with continuous variables related to a given Hamiltonian that can be described by a polynomial function of the involved operators can be conceived. A set of continuous variable quantum operations are thus regarded as universal if—after a finite number of appropriate operations—one can come up with the outcome that arbitrarily close to one of the transformations in the set. That means, a universal continuous variable quantum computation can be carried out by executing certain quantum logic gates such as displacement, squeezing, shearing and controlled-U (see Sect. 9.4). A continuous variable quantum computer is hence said to be universal in case it can simulate the action of the Hamiltonian that can be described in terms of the polynomials of $\hat{p}$ and $\hat{q}$.

While generalizing a single-mode quantum logic operations to a multi-valued domain, the fundamental memory units are no longer two-state qubits but are d-valued qudits. One may note that the qudit can have the same information as $\log_2 d$ qubits since they span the same Hilbert space, and the measurement of a qudit is presumed to yield only one value that corresponds to the eigenstate to which the d-valued quantum system collapses. There is still one major concern: the tradeoff in the processing time of executing a large number of small (2 or 4-dimension) binary gates versus a small number of large (d or d^2-dimension) multi-valued gates. Based on the perception that a large single-qubit operations are more viable than doing many small ones in sequence, the multi-valued (continuous variable) option might be more effective.

The motivation for dealing with continuous variable logic operation (computation) is therefore related to the benefit one expects as a result of the associated greater information processing capacity. Nonetheless, one of the most obvious challenge in realizing functional quantum computer is that the qubit is intrinsically open to the environment, that is, it suffers a loss of coherence. The other challenge is that the readout process collapses the states from which the qubit is constructed. Since the process of the collapse and decoherence are inherently irreversible, they are believed to impose a severe constraints on the procedure to be followed.

12.1.3.2 Quantum Optical Consideration

In the context of quantum optics, since a continuous variable quantum scheme can be denoted in terms of the quadrature amplitudes of the modes of the electromagnetic field $\hat{X}$ and its conjugate $\hat{P}$, the strategy is to look for the way of constructing the Hamiltonian out of arbitrary polynomials of $\hat{X}$ and $\hat{P}$ [29]. To this end, recall that the Hamiltonian that can be expressed in terms of $\hat{X}$ for some time t takes $\hat{X} \to \hat{X}$ and $\hat{P} \to \hat{P} - t$, and the conjugate $\hat{P}$ takes $\hat{X} \to \hat{X} + t$ and $\hat{P} \to \hat{P}$. These attributes indicate that the canonical positions and momenta render the tendency of shifting the conjugate variables by a constant. One may note that such a process accounts for a linear displacement (translation) of the quadrature amplitudes.

Consistent with such description and owing to the nature of the commutation of $\hat{X}$ and $\hat{P}$, it should be possible to construct a linear Hamiltonian of the form $a\hat{X} + b\hat{P} + c$. One may also see the possibility of performing a higher-order operations in $\hat{X}$ and $\hat{P}$ to construct more complex Hamiltonian such as a quadratic Hamiltonian

$$\hat{H} = \frac{1}{2}\left[\hat{X}^2 + \hat{P}^2\right], \tag{12.31}$$

which designates a nonlinear interaction of quantum systems. Then applying the involved commutation relation yields

$$\hat{X} \longrightarrow \cos(tX) - \sin(tP), \qquad \hat{P} \longrightarrow \cos(tP) + \sin(tX). \tag{12.32}$$

In case $\hat{X}$ and $\hat{P}$ are quadrature amplitudes, the Hamiltonian (12.31) would exemplify the phase shifter. As noted, the commutation relations between $\hat{H}$, $\hat{X}$ and $\hat{P}$ allows to construct the Hamiltonian of the form $a\hat{H} + b\hat{X} + c\hat{P} + d$.

In the same manner, one can employ the quadratic Hamiltonian of the form

$$\hat{S}_\pm = \pm\frac{1}{2}\left[\hat{X}\hat{P} + \hat{P}\hat{X}\right], \tag{12.33}$$

where

$$\frac{d\hat{X}}{dt} = i[\hat{S}_+, \hat{X}] = \hat{X}, \qquad \frac{d\hat{P}}{dt} = i[\hat{S}_+, \hat{P}] = -\hat{P}, \tag{12.34}$$

which reveal that applying $\hat{S}_+$ institutes the transformation $\hat{X} \rightarrow e^t \hat{X}$ and $\hat{P} \rightarrow e^{-t}\hat{P}$. This outcome can be interpreted as $\hat{S}_+$ stretches $\hat{X}$ and squeezes $\hat{P}$ by the same amount. In the conventional quantum optics, $\hat{S}_\pm$ stand for the squeeze operator in the linear regime.

One can also take the Hamiltonian

$$\hat{Q} = (\hat{X}^2 + \hat{P}^2)^2, \tag{12.35}$$

which designates a nonlinear optical process. Such higher-order Hamiltonian has a key property that commuting $\hat{X}$, $\hat{P}$ and $\hat{S}$ with some polynomial function in $\hat{X}$ and $\hat{P}$ results in a polynomial of lower order while commuting $\hat{Q}$ with a polynomial function in $\hat{X}$ and $\hat{P}$ increases its order. One can consider as demonstration the commutation relations that provide the third-order polynomials in $\hat{X}$ and $\hat{P}$

$$\left[\hat{Q}, \hat{X}\right] = \frac{i}{2}\left[\hat{X}^2\hat{P} + \hat{P}\hat{X}^2 + 2\hat{P}^3\right], \qquad \left[\hat{Q}, \hat{P}\right] = -\frac{i}{2}\left[\hat{P}^2\hat{X} + \hat{X}\hat{P}^2 + 2\hat{X}^3\right], \tag{12.36}$$

$$\left[\hat{X}, \left[\hat{Q}, \hat{S}\right]\right] = \hat{P}^3, \qquad \left[\hat{P}, \left[\hat{Q}, \hat{S}\right]\right] = \hat{X}^3. \tag{12.37}$$

One can also construct a Hamiltonian of arbitrary polynomials of higher-orders of $\hat{X}$ and $\hat{P}$. For a polynomial of order m with a degree of at least 3, one can see that

$$\left[\hat{P}^3, \hat{P}^m\hat{X}^n\right] = i\hat{P}^{m+2}\hat{X}^{n-1} + \text{lower} - \text{order terms}, \tag{12.38}$$

$$\left[\hat{X}^3, \hat{P}^m\hat{X}^n\right] = -i\hat{P}^{m-1}\hat{X}^{n+2} + \text{lower} - \text{order terms} \tag{12.39}$$

since a polynomial of order $m + 2$ can be constructed from monomials of order $m + 2$ (and lower) by applying some linear and single nonlinear operations a finite number of times. One can also construct polynomials of arbitrary order in $\hat{X}$ and $\hat{P}$ to any desired degree of accuracy by commutating $\hat{X}^3$ and $\hat{P}^3$ with monomials of order m. It is hence possible to highlight that commutation of a polynomial in $\hat{X}$ and $\hat{P}$ with $\hat{X}$ and $\hat{P}$ themselves can reduce the order of the polynomial by at least 1, and commutation with $\hat{H}$ and $\hat{S}$ having order 2 do not increase the order, but commutation with a polynomial of order 3 or higher increases the order by at least 1. Besides, commutation of $\hat{X}$, $\hat{P}$, $\hat{H}$ and $\hat{S}$ with applied Hamiltonian of order 3 or higher allows the construction of arbitrary Hermitian polynomials of any order in $\hat{X}$ and $\hat{P}$. This exposition may entail that simple linear operations together with nonlinear interaction between continuous variables would be sufficient to enact Hamiltonian of arbitrary polynomials.

In case of optical setup, linear operations such as translation, phase shifting, squeezing and beam splitting combined with some nonlinear mechanism such as Kerr nonlinearity could be adequate to perform arbitrary polynomial transformations. It might be insightful noting that the physical features of such transformations can be taken as the foundation of quantum computation with light, where information processing in view of polynomial transformations of continuous variables helps to perform quantum floating point manipulations. It is possible for instance to map $\hat{X}_1$ and $\hat{X}_2$ to $a\hat{X}_1 + b\hat{X}_2 + c$ using a linear operation alone, and the application of a three-variable Hamiltonian $\hat{H} = \hat{X}_1\hat{X}_2\hat{P}_3$ can take $\hat{X}_3 \rightarrow \hat{X}_3 + \hat{X}_1\hat{X}_2 t$. This operation shows that one can multiply $\hat{X}_1$ and $\hat{X}_2$, and then place the result on register $\hat{X}_3$. Note that a wide variety of ways of generating quantum floating point operations are currently admissible, which depends crucially on the strength of the squeezing and entanglement that can be generated from the pertinent nonlinearities [30].

A one-way quantum computation is as a result perceived to be attractive in the sense that the corresponding local projective measurements are easier to implement. For example, the resources required for quantum computation with linear optics can be significantly reduced when the photonic cluster states are created via nondeterministic gates. With this information, imagine a quantum system with infinite dimensional Hilbert space spanned by a continuum of orthogonal states $|s\rangle_q$; $\langle s|r\rangle_q = \delta(s - r)$ and when the two bases are related by Fourier transform

$$|s\rangle_p = \frac{1}{\sqrt{2\pi}} \int_{-\infty}^{\infty} e^{irs} |r\rangle_q \, dr, \qquad |s\rangle_q = \frac{1}{\sqrt{2\pi}} \int_{-\infty}^{\infty} e^{-irs} |r\rangle_p \, dr. \qquad (12.40)$$

Recall also that the position and momentum observable $\hat{q}$ and $\hat{p}$ such that

$$\hat{q}|s\rangle_q = s|s\rangle_q, \qquad \hat{p}|s\rangle_p = s|s\rangle_p \qquad (12.41)$$

with $[\hat{q}, \hat{p}] = i\hat{I}$ can be introduced where $\hat{p}$ is the generator of positive translation in position and $-\hat{q}$ is the generator of positive translation in momentum [31].

With this designation, one can write arbitrary position and momentum eigenstates in terms of the position translation and momentum boost operators as

$$|s\rangle_q = X(s)|0\rangle_q, \qquad |s\rangle_p = P(s)|0\rangle_p, \qquad (12.42)$$

where a single-oscillator Pauli operators

$$X(s) = e^{-is\hat{p}}, \qquad P(s) = e^{is\hat{q}} \qquad (12.43)$$

embody displacements, and

$$|0\rangle = \frac{1}{\pi^{1/4}} \int e^{-s^2/2} |s\rangle_{p,q} \, ds \qquad (12.44)$$

in the computational and conjugate bases. The action of these operators then results in

$$\hat{X}(q)|s\rangle_q = |s + q\rangle, \qquad \hat{P}(p)|s\rangle_p = e^{is\hat{p}}|s\rangle, \qquad (12.45)$$

where arbitrary pure quantum state $|\phi\rangle$ of continuous variable system can be decomposed as superposition of $|s\rangle_p$ and $|s\rangle_q$.

The Hamiltonian related to experimentally feasible interaction in quantum optics is at most quadratic in $\hat{q}$ and $\hat{p}$ and undergoes Gaussian transformation. Note that standard single-mode Gaussian transformations that can be applied in quantum computation are: rotation $(e^{i\theta(\hat{q}^2+\hat{p}^2)/2})$ is the operation that rotates the state counterclockwise in phase space by angle θ (phase shift) and can be described by $\hat{q} = \hat{q}\cos\theta - \hat{p}\sin\theta$ and $\hat{p} = \hat{q}\sin\theta + \hat{p}\cos\theta$; quadratic displacement $(e^{is\hat{q}})$ is the operation that displaces the state in phase space by s in momentum: $\hat{q} = \hat{q}$ and $\hat{p} = \hat{p} + s$ while $(e^{-is\hat{p}})$ displaces the state in phase space by s in position: $\hat{q} = \hat{q} + s$ and $\hat{p} = \hat{p}$; squeezing $(e^{-is(\hat{q}\hat{p}+\hat{p}\hat{q})/2})$ is the operation that squeezes the position quadrature by a factor of e^s while stretching the conjugate quadrature by e^{-s}: $\hat{q} = e^s\hat{q}$ and $\hat{p} = \hat{p}e^{-s}$; and shearing $(e^{is\hat{q}^2/2})$ is the operation that trims the state with respect to the $\hat{q}$ axis by gradient of s: $\hat{q} = \hat{q}$ and $\hat{p} = s\hat{q} + \hat{p}$, and can be taken as phase gate. One may need to underline that since quantum dynamics that consists solely of Gaussian operations on Gaussian states can be effectively simulated on a classical computer, some sort of a non-Gaussian element such as a single nonlinear operation at least cubic in the interaction terms is required to carry out universal quantum computation. So a set of gates such as $\exp(is\hat{q}^k/k)$ for $k = 1, 2, 3$ and the corresponding Fourier transform are thought to be sufficient for universal single-mode optical quantum computation.

As already highlighted in Sect. 9.4, a single-mode continuous variable NOT gate is related to the parity operator: $\mathrm{NOT} = (-1)^{\hat{a}^\dagger\hat{a}}$, where

$$\mathrm{NOT}|x\rangle = |-x\rangle, \qquad \mathrm{NOT}|p\rangle = |-p\rangle, \qquad \mathrm{NOT}^2 = 1 \qquad (12.46)$$

in which the corresponding Hadamard gate can be taken as the Fourier transformation

$$F|x\rangle = \frac{1}{\sqrt{2\pi}} \int_{-\infty}^{\infty} e^{ixy}|y\rangle \, dy \qquad (12.47)$$

from which follows $FF^\dagger|x\rangle = F^\dagger F|x\rangle = |x\rangle$.

The CNOT gate on the other hand can be defined by $\text{CNOT}_{\pm} = e^{\mp i \hat{x} \hat{p}}$ with property that

$$\text{CNOT}_{\pm} |x\rangle |y\rangle = |x\rangle |x \pm y\rangle, \tag{12.48}$$

which also works in the momentum space; and the X (bit-flip) gate that flips the logical value of the qubit by delaying it with respect to the local oscillator by half a cycle can be denoted by $\hat{X} = e^{i\pi \hat{a}^{\dagger} \hat{a}}$ [32]. Such assertion envisages that the optical quantum systems that include the cat and cluster states can be taken as a prominent candidate for universal quantum computation.

It might be worthy noting that a general architecture of optical quantum computer can be outlined as in the initialization part, one needs a deterministic sources of single photons to generate a single-photon pulses with precise timing and well defined polarization in which a collection of such photons constitutes the quantum register. Then as the application of a sequence of single- and two-qubit unitary transformations (quantum processor) initiates the execution of the program according to the desired quantum algorithm. After that, the result of computation would be readout by collecting the polarization states efficiently and reliably by a phase sensitive photon detectors as in heterodyne detection.

Note that quantum computation with linear optics and single photons has the advantage that the smallest unit of quantum information, the photon, is potentially free from decoherence, that is, the quantum information stored in a photon tends to stay longer. The downside is that photons do not naturally interact with each other although such an interaction is essential to apply two-qubit quantum gates. This challenge can be circumvented by using exponentially large number of optical modes since one may require an off-line resources, quantum teleportation and error correction to construct the associated protocol [27,33]. From practical point of view, the Hong-Ou-Mandel effect lies at the heart of linear optical quantum computing (see Fig. 9.3 and also [34]). Since almost any two-qubit gate is universal for quantum computing, in linear quantum optics,[7] the controlled phase and CNOT gates are often considered.

There is however a conceivable problem with teleportation when applied to a linear optics in present qubit representation since the corresponding Bell measurement is not complete and works at best half of the time. As a way out, the qubits can be chosen from the dual-rail representation when teleportation is applied to a single-rail state $\alpha|0\rangle + \beta|1\rangle$, where $|0\rangle$ and $|1\rangle$ denote the vacuum and single-photon Fock states since the CNOT gate involves only one optical mode of each qubit [35]. Note that an alternative version of a coherent state quantum computing, utilizes coherent

---

[7]Recall that a linear quantum optics can be designated by Hamiltonian $\hat{H} = \sum_{ij} c_{ij} \hat{a}_i^{\dagger} \hat{a}_j$ that commutes with the photon number operator, and has a peculiar property that a single-mode transformation of creation operators into a linear combination of other creation operators does not induce terms that are quadratic (or higher) order in creation or annihilation operators.

states as basis qubit in which one can envision to apply mesoscopic coherent states $|\alpha\rangle$ and $|-\alpha\rangle$ as computational bases [36].

The intended quantum optical computation scheme can be in general thought of as a box in which the incoming and outgoing modes of the electromagnetic fields interact, where the said box can be perceived to transform the state of the incoming modes into different states of the outgoing modes. The modes in the process may be mixed by beam splitters or they may pick a relative phase shift or undergo polarization rotation. The box may also include measurement devices whose outcomes modify the action of the optical components on the remaining modes.

12.2 Foresight

Identifying and characterizing a suitable optical candidate for practical implementation of quantum information processing tasks nowadays turns out to be the integral part of the field of modern quantum optics. Even then, before one delves into the underlying complexity of describing the unique nature of physical system that leads to quantum application, it is necessary recognizing some criteria that can be used to identify whether a certain physical system or situation can be taken as a legitimate resource for quantum information processing. In light of this, we opt to cite an outline due to DiVincenzo: scalability of the physical system from which the quantum register is constructed; the ability to prepare the state of the register in a known initial state; the ability to implement the universal set of quantum logic gates; the ability to reliably or effectively measure the states of individual qubits in the computational bases; the ability to faithfully transmit or teleport qubits between distant locations; the ability to inter-convert stationary and flying qubits [37].

With this in mind, the main platform for quantum information processing task such as quantum gate operation can be constructed from atom or ion trap, superconducting charge, nuclear magnetic resonance, spin or charge based quantum dots, nuclear spin and optical setups with corresponding advantages (see Sect. 9.4). For example, in utilizing the ensemble of atoms as qubit, it is required disrupting the cascading process so that the absorption and emission of the photon can be restricted, which can be achieved with the help of atomic dipole blockade.

Since the radiative properties of the atom in the strong coupling regime would be substantially different from the atom in a free space, the change in the atomic population dynamics can lead to a new quantum phenomenon that makes possible entangling atom(s) with photon(s) in a controlled way; although realizing the controlling procedure can be technically challenging (see Chap. 6). Such a general possibility opens the door for contriving the application of quantum optical processes presented in this book. It may also worth emphasizing that the interaction of the atom with its environment would become even more spectacular when the atom is surrounded by a conducting wall in which the emerging multiple electric images reinforce the resulting interference phenomena that may lead to generating coherently superposed atomic states, nonclassical photonic states and scalable quantum logic gates as in active Plasmonics and photonics.

It may not be hard to comprehend that a practical realization of quantum information processing protocol requires a coherent manipulation of a large number of coupled quantum systems [38]. Such coherent manipulation can be implemented by optically pumping the atomic ensemble into states with a strong atom-atom and atom-radiation interactions so that the required entanglement is induced (see Chap. 6). Combining with the resulting atomic dipole blockade such interaction is expected to considerably alleviate many stringent requirements during practical realization that can be used to generate superposition of a collective spin states in the ensemble of atoms; coherently convert these states into the states of photon wave packets of prescribed direction; and perform quantum gate operations between distant qubits (see Chap. 9).

One may also argue that trapping and cooling can be employed in building a quantum computer from a neutral atoms or ions since the atoms acquire a long radiative lifetimes at low temperature that make them an auspicious physical candidate for preserving quantum information, and so can serve as registry [39]. One may also utilize neutral atom in the ground state as the storage of information, and the Rydberg excitation as logic device. Injecting several atoms into confining region on the other hand may allow to map atom-photon entanglement onto atom-atom entanglement even when the atoms do not directly interact with each other (see Sect. 6.6). One also needs to be restricted to a modest atomic densities such that the interaction between the atoms can be safely neglected whenever they are in the sub-level of the ground state. This consideration can lead to the realization of the required long coherence lifetime [40].

Suppose the manipulation of the atoms is done by using a light with different frequencies and polarizations. In case the hyperfine sub-levels of the ground state and Rydberg states are excited in such a way that only a symmetric atomic collective states are engaged in the process [41]. On account of this, in the atom-radiation dressed state representation, if the atom that initially occupies one of the excited states, let us say, $|e\rangle$ enters an empty cavity, one may recall that the state of the system will evolve in time according to

$$|\psi_+^{out}\rangle = \cos(Gt)|e, 0\rangle + \sin(Gt)|g, 1\rangle \tag{12.49}$$

and when the atom is initially in the ground state $|g\rangle$ as

$$|\psi_-^{out}\rangle = \cos(Gt)|g, 1\rangle - \sin(Gt)|e, 0\rangle, \tag{12.50}$$

where G is the atom-radiation coupling constant (see Chap. 6). Equations (12.49) and (12.50) imply that the atom-radiation combined system would be entangled due to the confinement.

In case $Gt = \pi/4$, the atom-radiation combined system would be in either of the states

$$|\pm\rangle = \frac{1}{\sqrt{2}}[|g, 1\rangle \pm |e, 0\rangle]; \tag{12.51}$$

maximally entangled Bell state (see Sect. 9.2). It may not be difficult to infer that the main contribution of the light is to induce coherent superposition between the excited and lower energy levels of the atom so that a subsequent manipulation of this system involves projective measurement. In case the cavity contains n photons initially, one may deduce that

$$|\psi_+^{out}\rangle = \cos\left(\sqrt{n+1}Gt\right)|e, n\rangle + \sin\left(\sqrt{n+1}Gt\right)|g, n+1\rangle, \tag{12.52}$$

where such a transformation corresponds to a SWAP gate for $\sqrt{n+1}Gt = \pi$, and swaps $|e, n\rangle$ state with $|g, n+1\rangle$. By varying the phase, it should also be possible to mimic various quantum logic gates (see Sect. 9.4).

Moreover, imagine that the atom is either in the resonant ground state $|g\rangle$ or in arbitrary off-resonant excited state $|s\rangle$ that do not interact with the driving radiation. As portrayed earlier, the atom-radiation combined system can transform as

$$|0\rangle \otimes |s\rangle \longrightarrow |0\rangle \otimes |s\rangle, \qquad |0\rangle \otimes |g\rangle \longrightarrow |0\rangle \otimes |g\rangle,$$

$$|1\rangle \otimes |s\rangle \longrightarrow |1\rangle \otimes |s\rangle, \qquad |1\rangle \otimes |g\rangle \longrightarrow -|1\rangle \otimes |g\rangle, \tag{12.53}$$

where the minus sign accounts for the π phase shift, and resembles a controlled-Z gate. Besides, upon denoting $|s\rangle \equiv |0\rangle$ and $|g\rangle \equiv |1\rangle$, the rotation of the qubits can lead to a CNOT gate [42]. These examples are meant to evince the foreseeability of logic gate transformation by performing phase shift and appropriate rotation. Such understanding may also strengthen the possibility of implementing quantum information processing tasks with the aid of the cavity dynamics of selected energy levels by making use of a coherently superposed high-lying energy levels of the Rydeberg atom with the flexibility acquired pertaining to entanglement swapping among neighboring atoms of a multi-level configuration (see Sect. 6.5 and also [43]).

The proposal that photonic qubits are expected to be stable and have long coherence time on the other hand entails that they may not lose their quantum properties over long period of time (see Sect. 9.3). Photonic qubits have also advantages that include high mobility and low interaction rate with the environment that make complex insulation and cooling unnecessary. Such attributes are expected to make light an ideal carrier of the content of quantum information. Even then, there are still potential practical challenges to materialize long distance continuous variable quantum communication and construct a workable quantum computation structure. The pertinent challenge may include a source that can reliably and efficiently emit a definite number of photons has not been realized as yet, the emission of the photon is random and the circumstance that photons hardly interact with each other or their environment. Besides, the required entangling operations between photonic qubits are implemented probabilistically, which makes the expansion to a circuit comprising a large number of operations impractical. Another challenge may emerge while attempting to utilize photonic qubits as a storage due to the speed at which the light travels since many algorithms naturally require a greater storage time, which requires additional structure to slow the light.

Despite the challenges, it may not be hard to comprehend that constructing a series of horizontally and vertically polarized photons can faithfully represent a computational bit string. Upon mixing the polarized light on phase beam splitter, it should be straightforward to see that a superposed state can be induced. Consequently, one of the nonclassical features of encoding quantum information onto the states of light is based on the fact that different bases do not commute, and the measurement that generates information about the bit values of one of the bases inevitably disturbs the bit values of the other basis, which can be applied to create a secured communication channel via the technique of quantum key distribution. The ability of the qubit to span a different bit values generated from horizontal and vertical polarized states may thus allow one to think of various qubit superpositions that have a potential to be a tool for realizing various logic gates (see Chaps. 9 and 11).

To achieve the required two-qubit gate with some sort of entanglement, one requires advanced and sophisticated setups that can generate a single photons on demand or a heralded single photons. Whichever way one opts to look, an attempt to produce and delete a single-photon states efficiently and reliably may pose a formidable technological obstacle. It is not difficult to note that the idea of generating a single-photon state relies on a certain approximation scheme based on the attenuation of the source laser that can be epitomized by

$$|\psi\rangle = |0\rangle + \alpha|1\rangle + \frac{\alpha^2}{2}|2\rangle + \dots, \tag{12.54}$$

where α is taken to be very small.

In the scope of this approximation, one might be satisfied with

$$|\psi\rangle = |0\rangle + \alpha|1\rangle. \tag{12.55}$$

This consideration can be understood as there is a definite probability for generating a single-photon state. In case a photon counter is placed at the end of the setup and one cares for the click of the detector, it is possible to postselect just the single-photon part of the state, and if the source is polarized, the single-photon state can then be taken as a qubit.

One may also consider two highly attenuated coherent sources in which postselecting those events leads to having two photons. For a highly attenuated laser, the joint state of the two equal events can be expressed as

$$|\psi\rangle_{ab} = |0\rangle_a|0\rangle_a + \alpha\big[|1\rangle_a|0\rangle_b + |0\rangle_a|1\rangle_h\big]$$

$$+ \frac{u'}{2!}\big[2!|1\rangle_a|1\rangle_b + |2\rangle_a|0\rangle_b + |0\rangle_a|2\rangle_b\big] + \dots. \qquad (12.56)$$

If one goes for α^2 order terms that involve pairs of photons in one beam while the vacuum occurs in the other with the same probability, postselecting two-photon events will not remove a single-photon terms, and it is not hence possible to perform a two-qubit experiment by using a highly attenuated laser source alone.

The remedy may require application of a nonlinear medium such as parametric down converter that has a potential to generate spontaneously created photon pairs (see Chap. 8). The vacuum input thus can be transformed according to

$$|0\rangle_a|0\rangle_b \longrightarrow \frac{1}{\sqrt{G}}\big[|0\rangle_a|0\rangle_b + \chi|1\rangle_a|1\rangle_b + \chi^2|2\rangle_a|2\rangle_b + \dots, \big], \qquad (12.57)$$

where χ stands for the strength of the nonlinear interaction and G for the gain. Upon taking χ to be small, one can see that

$$|\psi\rangle_{ab} = \frac{1}{\sqrt{G}}\big[|0\rangle_a|0\rangle_b + \chi|1\rangle_a|1\rangle_b\big]. \qquad (12.58)$$

In case one postselects these events from the record in which the photons are detected simultaneously or in coincidence, there would be a record of the part of the state due to the pair of photons.

We now seek to move onto another approximation scheme that relies on a conditional photon counting strategy and designed to produce a single-photon states. To this end, one can begin with a heralded single photons that can be generated from the process of a nondegenerate down conversion via detecting one of the output modes. In this case, the conditional state in which a single photon is detected in mode a can be expressed as

$$_a\langle1|\psi\rangle_{ab} = \chi|1\rangle_b, \qquad (12.59)$$

which also stands for the case when a single-photon state is created in mode b.

The output state conditioned on photon counting procedure can be expressed in a more general form as

$$\hat{\rho}_b = Tr_a\big[|\psi\rangle_{ab}\langle\psi|_{ab}\hat{\rho}_a\big]. \qquad (12.60)$$

Such a mechanism can be demonstrated by carrying out a homodyne tomography of conditionally produced photon states [44]. A key issue in designing the conditional photon state is tantamount to perceiving the signal detector to look only at the single-photon mode; otherwise, the accompanying vacuum modes do also couple to the analyzer. This consideration envisages that the local oscillator pulse used in the homodyne detection should be accurately mode matched to the single-photon state. After that, the data collected in the homodyne detection would be employed to produce the quasi-statistical distribution of the single-photon state.

The other approach that can be utilized to produce a single-photon state is referred to as on demand. One may argue that an approximate single-photon state can be created on demand by generating light from a single isolated emitter such as a single ion, atom or quantum dot. The crux of this idea pertains to presuming that a single emitter can produce only a single photon at one given time with some dead time in between emissions since the emitter requires some time to be re-excited.

On the basis of such consideration, the output state can be written as

$$|\psi\rangle = |0\rangle + \alpha|1\rangle + \tau\left(\frac{\alpha^2}{2}|2\rangle + \ldots\right),\tag{12.61}$$

where τ—a number between 0 and 1—represents the suppression of a higher photon number terms (compare with Eq. (12.54)). In case τ is very small, α can be made large such that there is a high probability that a single photon will be emitted. Such presumption clearly leads to a phenomenon of photon anti-bunching expected of a single photon source. Due to the vast solid angle over which the emitter emits light, the most effective approach is thought to be placing the emitter in high finesse optical cavity so that the photon emission should be matched to single Gaussian mode (see Chap. 6).

It might worth emphasizing that while earnestly looking for the application of light in quantum information processing, one should make a distinction between continuous variable states and discrete variable approach based on single-photon qubit and linear optics techniques. For example, in the regime of continuous variable, the constructed circuit would be deterministic, whereas in the single-photon case the constructed circuit would be nondeterministic, and requires a very large overhead in terms of photon source and memory in order to make it near deterministic. Even though much attention is not given to it in this book, nowadays a significant progress has been made towards bridging the two approaches with the aim of realizing procedures that overcome the intrinsic limitations associated with these approaches individually.

It might also be important stressing that all required measurements are straightforward in the single-photon approach while the non-Gaussian measurement poses a serious challenge in the regime of continuous variable. It is not however obvious to assert convincingly which of these problems represent a more demanding impediment when building a large scale system. Even then, there are a number of avenues for future research such as incorporating continuous variable quantum algorithms into continuous variable model along with fault tolerance quantum computation and cryptography. Besides, the integration of discrete and continuous variable approaches as unified hybrid mechanism seems to be very favorable field of research in connection to quantum repeater, teleportation and entanglement swapping [45].

References

1. D. Gottesman, I.L. Chuang, Nature **402**, 390 (1999); E. Knill, R. Laflamme, G.J. Milburn, ibid. **409**, 46 (2001)
2. H. Takesue et al., Nat. Photonics **1**, 343 (2007); D. Stucki et al., New J. Phys. **11**, 075003 (2009); J.P. Bourgoin et al., ibid. **15**, 023006 (2013); S. Marahidharan et al., Sci. Rep. **6**, 20463 (2016); D. Huang et al., ibid., 19201 (2016); J. Zhou, D. Huang, Y. Guo, Phys. Rev. A **98**, 042303 (2018)
3. S.L. Braunstein, P. van Loock, Rev. Mod. Phys. **77**, 513 (2005); H.A. Bochor, T.C. Ralph, *A Guide to Experiments in Quantum Optics*, 2nd edn. (Wiley-VCH Verlag GmbH and Co. KGaA, Weinheim, 2004); P. Kok, B.W. Covett, *Introduction to Optical Quantum Information Processing* (Cambridge University, New York, 2010); S. Takeda, A. Furusawa, APL Photonics **4**, 060902 (2019)
4. D. Gottesman, Phys. Rev. A **57**, 127 (1998); A.P. Lund, T.C. Ralph, H.L. Haselgrove, Phys. Rev. Lett. **100**, 030503 (2008); N.C. Menicucci, ibid. **112**, 120504 (2014)
5. P.W. Shor, Phys. Rev. A **52**, 2493 (1995)
6. H. Buhrman et al., Rev. Mod. Phys. **82**, 665 (2010)
7. J.W. Pan et al., Rev. Mod. Phys. **84**, 777 (2012); C. Weedbrook et al., ibid. **84**, 621 (2012)
8. J. Chiaverini et al., Nature **432**, 602 (2004); T. Aoki et al., Nature Phys. **5**, 541 (2009); D. Nigg et al., Science **345**, 302 (2014)
9. A. Ekert, C. Macchiavello, Phys. Rev. Lett. **77**, 2585 (1996); C. Ahn, A.C. Doherty, A.J. Landahl, Phys. Rev. A **65**, 042301 (2002)
10. C.H. Bennett et al., Phys. Rev. A **54**, 3824 (1996); E. Knill, R. Laflamme, ibid. **55**, 900 (1997); V. Vedral, *Introduction to Quantum Information Science* (Oxford University, New York, 2006)
11. A.M. Steane, Phys. Rev. Lett. **77**, 793 (1996); R. Laflamme et al., ibid., 198 (1996); M. Muller et al., Phys. Rev. X **6**, 031030 (2016)
12. D. Gottesman, Phys. Rev. A **54**, 1862 (1996)
13. S.L. Braunstein, Phys. Rev. Lett. **80**, 4084 (1998); S. Lloyd, J.E. Slotine, ibid., 4088 (1998); S.L. Braunstein, Nature **394**, 47 (1998); M.M. Wilde, H. Krovi, T.A. Brun, Phys. Rev. A **76**, 052308 (2007); T.A. Walker, S.L. Braunstein, Phys. Rev. A **81**, 062305 (2010); M. Lassen, L.S. Madsen, U.L. Andersen, Nat. Photonics **4**, 700 (2010); M.H. Michael et al., Phys. Rev. X **6**, 031006 (2016); J. Dias, T.C. Ralph, Phys. Rev. A **97**, 032335 (2018)
14. N. Gisin et al., Rev. Mod. Phys. **74**, 145 (2002)
15. C.H. Bennet, G. Brassard, *Quantum Cryptography: public key distribution and coin tossing, in Proceeding of IEEE International Conference on Computers, Systems and Signal Processing, Bangalore, 1984*; A.K. Ekert, Phys. Rev. Lett. **67**, 661 (1991); C.H. Bennet, G. Brassard, N.D. Mermin, ibid. **68**, 557 (1992); C.H. Bennet, ibid. **68**, 3121 (1992); N. Lutkenhaus, Phys. Rev. A **59**, 3301 (1999)
16. R.G. Patron, N.J. Cerf, Phys. Rev. Lett. **97**, 190503 (2006); A. Leverrier, P. Grangier, Phys. Rev. A **81**, 062314 (2010)

17. T.C. Ralph, Phys. Rev. A **62**, 062306 (2000); N.J. Cerf, M. Levy, G.V. Assche, ibid. **63**, 052311 (2001); F. Grosshans, P. Grangier, Phys. Rev. Lett. **88**, 057902 (2002); F. Grosshans et al., Nature **421**, 238 (2003); G. He, J. Zhu, G. Zeng, Phys. Rev. A **73**, 012314 (2006); A. Leverrier et al., ibid. **77**, 042325 (2008); Y. Shen et al., ibid. **82**, 022317 (2010); P. Jouguet, S.K. Jacques, A. Leverrier, ibid. **84**, 062317 (2011); T. Brougham, S.M. Barnett, ibid. **85**, 032322 (2012); L.S. Madsen et al., Nature Commun. **3**, 2097 (2012); C. Weedbrook, S. Pirandola, T.C. Ralph, Phys. Rev. A **86**, 022318 (2012); P. Jouguet et al., Nat. Photonics **7**, 378 (2013); B. Heim et al., New J. Phys. **16**, 113018 (2014); E. Diamanti, A. Leverrier, Entropy **17**, 6072 (2015)
18. A.M. Lance et al., Phys. Rev. Lett. **95**, 180503 (2005); J. Lodewyck et al., Phys. Rev. A **72**, 050303 (2005); S. Lorenz et al., ibid. **74**, 042326 (2006); J. Lodewyck et al., Phys. Rev. Lett. **98**, 030503 (2007); E.M. Scott et al., Phys. Rev. A **84**, 062326 (2011)
19. J.I. Cirac et al., Phys. Rev. Lett. **78**, 3221 (1997); L.M. Duan et al., Nature **414**, 413 (2001); L.M. Duan, C. Monroe, Rev. Mod. Phys. **82**, 1209 (2010)
20. Y.B. Sheng, F.G. Deng, H.Y. Zhou, Phys. Rev. A **77**, 042308 (2008); B. Zhao et al., ibid. **81**, 052329 (2010); D. Gonta P. van Loock, Appl. Phys. B **122**, 118 (2016); F. Furrer,W.J. Munro, Phys. Rev. A **98**, 032335 (2018); K.P. Seshadreesan, H. Krovi, S. Guha, Phys. Rev. Research **2**, 013310 (2020)
21. N. Sangouard et al., Phys. Rev. A **77**, 062301 (2008); N. Gisin, S. Pironio, N. Sangouard, Phys. Rev. Lett. **105**, 070501 (2010); N. Sangouard et al., Rev. Mod. Phys. **83**, 33 (2011); J. Minar, H. de Riedmatten, N. Sangouard, Phys. Rev. A **85**, 032317 (2012)
22. H.J. Briegel et al., Phys. Rev. Lett. **81**, 5932 (1998); W. Duer et al., Phys. Rev. A **59**, 169 (1999); B. Zhao et al., Phys. Rev. Lett. **98**, 240502 (2007); Z.B. Che et al., Phys. Rev. A **76**, 022329 (2007); L. Jiang, J.M. Taylor, M.D. Lukin, ibid., 012301 (2007); H.J. Kimble, Nature **453**, 1023 (2008); M. Gao et al., Phys. Rev. A **79**, 042301 (2009); B. Zhao et al., ibid. **81**, 052329 (2010)
23. H. Yonezawa, T. Aoki, A. Furusawa, Nature **431**, 430 (2004)
24. E. Waks, A. Zeevi, Y. Yamamoto, Phys. Rev. A **65**, 052310 (2002); B.C. Jacobs, T.B. Pittman, J.D. Franson, ibid. **66**, 052307 (2002); H. de Riedmatten et al., Phys. Rev. Lett. **92**, 047904 (2004); S. Pirandola et al., Nat. Photonics **9**, 397 (2015); Y. Guo et al., Phys. Rev. A **95**, 042326 (2017)
25. P. W. Shor, *Algorithms for Quantum Computation: discrete logarithm and factoring, in Proceeding of 35th Annual Symposium on Foundations of Computer Science, Santa Fe, 1994.* SIAM Journal on Computing, vol. 26, 1484 (1997)
26. A. Barenco et al., Phys. Rev. A **52**, 3457 (1995); S. Lloyd, Phys. Rev. Lett. **75**, 346 (1995); H.F. Hofmann, Phys. Rev. Lett. **94**, 160504 (2005); P. Lambropoulos, D. Petrosyan, *Fundamentals of Quantum Optics and Quantum Information* (Springer, Berline, 2007)
27. P. Kok et al., Rev. Mod. Phys. **79**, 135 (2007)
28. Y. Wang et al., Phys. Rev. A **81**, 022311 (2010); S.A. Podoshvedov, Opt. Commun. **290**, 192 (2013)
29. S. Lloyd, S.L. Braunstein, Phys. Rev. Lett. **82**, 1784 (1999)
30. M.A. Nielsen, Phys. Rev. Lett. **93**, 040503 (2004); N.C. Menicucci et al., ibid. **97**, 110501 (2006)
31. M. Gu et al., Phys. Rev. A **79**, 062318 (2009)
32. S. Salemian, S. Mohammadnejad, Am. J. Appl. Sci. **5**, 1144 (2008)
33. N.J. Cerf, C. Adami, P.G. Kwiat, Phys. Rev. A **57**, 1477 (1998)
34. C.K. Hong, Z. Y. Ou, L. Mandel, Phys. Rev. Lett. **59**, 2044 (1987)
35. R. Raussendorf, H.J. Briegel, Phys. Rev. Lett. **86**, 5188 (2001); N. Yoran, B. Reznik, ibid. **91**, 037903 (2003); F. Verstraete, J.I. Cirac, Phys. Rev. A **70**, 060302 (2004); D.E. Browne, T. Rudolph, Phys. Rev. Lett. **95**, 010501 (2005); J. Zhang, S.L. Braunstein, Phys. Rev. A **73**, 032318 (2006)
36. P.T. Cochrane, G.J. Milburn, W.J. Munro, Phys. Rev. A **59**, 2631 (1999); H. Jeong, M.S. Kim, Phys. Rev. A **65**, 042305 (2002); T.C. Ralph et al., ibid. **68**, 042319 (2003)
37. D.P. DiVincenzo, Fortschr. Phys. **48**, 771 (2000)

38. M.D. Lukin et al., Phys. Rev. Lett. **87**, 037901 (2001); A.V. Gorshkov et al., Phys. Rev. Lett. **107**, 133602 (2011); K. Molmer, L. Isenhower, M. Saffman, J. Phys. B **44**, 184016 (2011)
39. J.I. Cirac, P. Zoller, Phys. Rev. Lett. **74**, 4091 (1995); P. Jonathan et al., Science **325**, 1227 (2009); D. Kielpinski et al., Quant. Inf. Proc. **15**, 5315 (2016)
40. E. Brion, K. Molmer, M. Saffman, Phys. Rev. Lett. **99**, 260501 (2007); L.H. Pedersen, K. Molmer, Phys. Rev. A **79**, 012320 (2009)
41. L. Isenhower et al., Phys. Rev. Lett. **101**, 010503 (2010); T. Wile et al., ibid. **104**, 010502 (2010)
42. A. Joshi, M. Xiao, Phys. Rev. A **74**, 052318 (2006)
43. D. Jaksch et al., Phys. Rev. Lett. **85**, 2208 (2000)
44. A. Ourjoumtsev, R.T. Brouri, P. Grangier, Phys. Rev. Lett. **96**, 213601 (2006); S. Grandi et al., New J. Phys. **19**, (2017)
45. L.M. Duan et al., Nature **414**, 413 (2001); J.F. Sherson et al., ibid. **443**, 557 (2006); S. Takeda et al., ibid. **500**, 315 (2013); U.L. Andersen et al., Nat. Phys. **11**, 713 (2015)

Index

Printed in the United States
By Bookmasters